CONSTANTS AND CONVERSIONS

Constants

speed of light (vacuum)	c	2.99792458×10^8 m/s (defined)
quantum of charge	e	1.602177×10^{-19} C
Boltzmann's constant	k	1.38066×10^{-23} J/K $= 8.6174 \times 10^{-5}$ eV/K
Planck's constant	h	6.62608×10^{-34} J $\cdot$ s $= 4.13567 \times 10^{-15}$ eV $\cdot$ s
$h/2\pi = \hbar$		1.054573×10^{-34} J $\cdot$ s $= 6.58212 \times 10^{-16}$ eV $\cdot$ s
	hc	1.239842×10^{-6} eV $\cdot$ m
permittivity constant	ϵ_0	$8.8541878 \times 10^{-12}$ C^2/N $\cdot$ m^2
Stefan-Boltzmann constant	σ	5.6705×10^{-8} W/m$^2 \cdot$ K^4
Avogadro's number	N_A	6.02214×10^{23} molecules/mole
electron rest mass	m	9.10939×10^{-31} kg $= 0.510999$ MeV/c$^2 = 5.48580 \times 10^{-4}$ u
proton rest mass	m_p	1.672623×10^{-27} kg $= 938.272$ MeV/c$^2 = 1.007276$ u
neutron rest mass	m_n	1.674929×10^{-27} kg $= 939.566$ MeV/c$^2 = 1.008665$ u
Bohr magneton	μ_B	9.27402×10^{-24} J/T $= 5.788383 \times 10^{-5}$ eV/T
nuclear magneton	μ_N	5.05079×10^{-27} J/T $= 3.152452 \times 10^{-8}$ eV/T
gravitational constant	G	6.673×10^{-11} N $\cdot$ m^2/kg^2

Conversions

1 eV $= 1.602177 \times 10^{-19}$ J

1 u $= 1.660540 \times 10^{-27}$ kg $= 931.494$ MeV/c^2

1 eV/c $= 5.34429 \times 10^{-28}$ kg $\cdot$ m/s

1 eV/c^2 $= 1.782663 \times 10^{-36}$ kg

K $=$ °C $+ 273.15$

From the 1986 CODATA recommended values as compiled by E. Richard Cohen, Rockwell International Science Center, and Barry N. Taylor, National Bureau of Standards.

Essentials of
MODERN PHYSICS

T. R. SANDIN
North Carolina A&T State University

ADDISON-WESLEY PUBLISHING COMPANY

Reading, Massachusetts • Menlo Park, California
New York • Don Mills, Ontario • Wokingham, England
Amsterdam • Bonn • Sydney • Singapore • Tokyo
Madrid • San Juan

To my wife, Emilie
For her invaluable help and support
And to our children
Kathy, Mike, John, Peggy, and Beth
For their faith and encouragement

Sponsoring Editor: Stuart Johnson
Production Administrator: Karen L. Garrison
Art Consultant: Loretta Bailey
Copy Editor: Jerrold A. Moore
Text Designer: Catherine L. Dorin
Illustrators: Oxford Illustrators Limited
Manufacturing Supervisor: Roy Logan
Cover Designer: Marshall Henrichs

Library of Congress Catologing-in-Publication Data

Sandin, T.R.
   Essentials of modern physics.

   Includes index.
   1. Physics.   I. title.
QC21.2.S24 1989   539   88-1268
ISBN 0-201-09256-5

Preface

Near the end of the nineteenth century, the story has it, a famous physicist stated very publicly that physics had advanced to the point where essentially all the laws of nature were understood. He claimed that the only thing that physicists could look forward to was increasing the accuracy of their measurements. The twentieth century was to be the century of the engineer, the time for taking those known laws of physics and using them to design new devices and refine old ones. After 1900, physicists would simply spend their years finding already known values to one more significant figure.

This book shows just how wrong that physicist was. The radically new modern physics of relativity and quantum mechanics first began to be understood within the first five years of the twentieth century. This new physics has since been applied to the physics of nuclei, of atoms, of collections of atoms from molecules to solids, and beyond — both to particles smaller than any nucleus and to the largest collection of atoms we're aware of: our entire universe. I've always found this to be the most exciting part of physics, partly because our knowledge and understanding of it change daily (as in the related areas of particle physics, unified theories, and cosmology), and partly because so many novel and surprising applications continue to result from it (for instance, higher temperature superconductors, positron emission tomography, and magnetic resonance imaging). I wrote this book to convey this excitement to students of engineering and science. This century has also been the century of the engineer, but new devices have not been restricted to application of the old laws of physics. Rather, modern engineering has rapidly taken advantage of the discoveries and insights of the new physics.

COVERAGE

I have included the essential standard topics of modern physics, as well as chapters on general relativity and cosmology. Some additional topics can be introduced by means of Guided Problems at the ends of the chapters. There's more than enough material here for a standard one-semester, three-credit-hour course. The extra material gives added flexibility to the course. A colleague, for example, used the pertinent chapters in a bound manuscript version of this book, along with some additional notes, to teach an entire introductory nuclear physics course.

Some material could be omitted, all or in part, without affecting continuity: Sections 3.3, 4.3, 6.2, 9.3, 14.2, 14.3, 15.1, and 19.5, as well as Chapter 21 (and then also Section 22.3). Other chemistry and physics courses may have covered what I cover in Chapter 11, Sections 12.1, 20.3, 22.2, and much of Chapter 28. In addition, the needs of the student, the judgment of the instructor, and the time available may call for deletion of details or modification of teaching methods in the solid state and nuclear chapters. For instance, Chapters 27 and 34 are very much applied physics and may be given as outside assignments. The material in Chapters 36 and 37 is not often taught in introductory modern physics courses.

LEVEL

I consider this book to be a continuation of a usual introductory study of physics. Therefore I assumed that the student has successfully completed a one-year calculus-based course in classical physics at the level of Sears, Zemansky, and Young or Halliday and Resnick. I further assumed that the student has retained much of what was taught in that course, or at least has the ability to look up the information as needed. The level of mathematics required for this book is no greater than that required for previous courses in classical physics. Just because this text goes beyond classical concepts, I saw no reason to write it with a quantum jump in complexity.

I've aimed at a balance between theory and application. For example, Chapters 2 and 13 (introducing special relativity and quantum mechanics) are more theoretical, whereas Chapters 27 and 34 (describing a number of semiconductor devices, particle detectors, and accelerators) are more applied. But the theory from Chapters 2 and 13 is soon applied and the physics behind the applications is always emphasized.

CLARITY

I've tried to write this book in a friendly and informal style, but not at the expense of scientific rigor. My aim is to get at the point of things without undue distractions. Students tell me that they learn better with short chapters (so that the material arrives in digestible chunks), so this book contains 37 chapters — two or three times the usual number. To guide the student new to modern physics, I emphasize the most important terms by bold-face type and number the more important equations. Soon after I introduce them, I explain the main concepts and make them more concrete by example questions and problems. The result is not 50–80, but about 250 worked examples throughout the text (many with multiple parts).

What is obvious to the experienced physicist can completely baffle the new student so I've avoided the "it can be shown" approach in both the examples and the text. When the units may be confusing, I explain them in the examples. Since the use of SI units is becoming more universal (and is even mandated by law in some countries), I've used them consistently except where eV-based units are simpler and more common. I designed and selected the figures in the text to aid understanding and generate interest. Counting multiple parts and using the proverbial unit conversion of 1 picture = 1000 words, I calculate that the book contains over a third of a million words worth of figures.

OBJECTIVES

Many instructors find that presenting specific objectives improves the learning experience by sharpening the focus of the instructor and making it all less of a guessing game for the student. I had the following objectives in mind during preparation of this book: The student should be able to define or discuss all the bold-face terms, should understand the meaning and units of all the symbols, should be able to work problems with all the numbered equations, and should become proficient at all the skills illustrated in the examples. (Sample skills include constructing energy-level diagrams, using integrals from tables, interpolating values, and discussing concepts.)

QUESTIONS AND PROBLEMS

"I understand the material, but I just can't work the problems" can be justifiably more than a familiar complaint if the problems demand too much too soon from the learner. This book has an unusually large number of questions and problems — almost 1100 — and a large fraction of these are the "entry-level" type of exercise that helps the student build understanding and confidence with a new concept. The book also contains many of the more challenging type of questions and problems. However, I've tried to avoid the kind of problem that can be solved with 30 seconds of physics and 30 minutes of math, in the belief that the purpose of the text is to teach physics, not to teach either frustration or methods of mathematical manipulation.

Guided Problems are problems that guide the student step-by-step to an understanding of a topic not explained in the text. These problems build on the material in the chapter they follow. For example, Problems 3.16–3.19 lead to Lorentz transformations of electromagnetic potentials and fields using matrixes, Problem 8.14 is a first introduction to gravitational redshift and black holes, Problem 19.6 concerns electron spin resonance, Problem 22.34 discusses Einstein's A and B coefficients, Problem 25.16 calculates the minimum pole strength of a magnetic monopole from flux quantization, Problems 31.31 and 31.32 are concerned with double-beta decay, and Problem 37.27 determines the radius of a neutron star. Other Guided Problems investigate topics such as radar speed measurements, Rydberg atoms, spin paramagnetism and Curie's law, adiabatic demagnetization, the Schottky barrier and diode, an SCR, and Xe-135 poisoning of a reactor.

UP-TO-DATE

One enjoyable thing about writing this book was following the physics literature (and sometimes the daily paper) to find out just what was new to make sure that this is truly a modern physics book. For instance, Supernova 1987A and the double-beta decay of selenium-82 both appear in these pages. And the section on superconductivity had to be rewritten more than once.

ANSWERS

The answers to the odd-numbered Questions and Problems are located at the end of this book. To find the method of obtaining those answers, you can turn to the *Instructor's Manual for Essentials of Modern Physics,* available *for instructors only* from the publisher. A real time-saver, this manual also contains the solution strategies and the answers for the even-numbered Questions and Problems.

TESTED

This is more like a second edition of *Essentials of Modern Physics* than a first edition. Contrary to the usual procedure, I completely wrote the book and had it locally printed, bound, and adopted before submitting it to a publisher. As a result, two colleagues have already used it in their teaching. Feedback from them and from many of the hundreds of students who have used it brought improvements (and a revised printing). Also, several reviewers have offered valuable suggestions, which have been incorporated in this edition. These reviewers include

> John Albright *Florida State University*
>
> Paul Finkler *University of Nebraska, Omaha*
>
> Donald Franceschetti *Memphis State University*
>
> Frederick Lobkowicz *University of Rochester*
>
> Bruce McCart *Augustana College*
>
> Philip Parks *Michigan Technological University*
>
> Hla Shwe *East Stroudsburg University*
>
> Albert Stwertka *U.S. Merchant Marine Academy*
>
> Francis Tam *Frostburg State College*
>
> Daniel Wilkins *University of Nebraska, Omaha*

Stuart Ahrens of North Carolina A&T State University also gave valuable assistance. Sheridan Simon of Guilford College checked all the solutions to the worked examples and the odd-numbered questions and problems.

However, I am ultimately responsible for any remaining defects. I'd appreciate your writing me if you find any mistakes, so that I can correct them. Also, please let me know if any sections give more trouble than they should, so I can try to make them clearer in future editions. You can write to me at the Physics Department, North Carolina A&T State University, Greensboro, NC 27411.

THANKS

Many hard-working, conscientious, and talented individuals at Addison-Wesley helped bring this book to publication. Some of them include Production Administrator Karen Garrison, Art Coordinator Loretta Bailey, and Editors Andy Crowley and Stuart Johnson. They, and others from Ireland to Texas, who helped produce this book have my deepest thanks.

Contents

x   Contents

PART _Two_
_____

QUANTUM PHYSICS

11 THE BOHR MODEL OF THE HYDROGEN ATOM 123

11.1	Rutherford's Nuclear Atom	123
11.2	The Hydrogen Spectrum	125
11.3	Bohr's Model: Energy	128

12 APPLICATIONS OF THE BOHR MODEL 135

| 12.1 | Bohr's Model: Spectra | 135 |
| 12.2 | Reduced Mass and Hydrogen-like Atoms | 138 |

13 THE SCHRÖDINGER WAVE EQUATION 145

| 13.1 | The Wave Function | 145 |
| 13.2 | The Time-Independent Schrödinger Wave Equation | 150 |

14 ONE-DIMENSIONAL POTENTIAL WELLS 155

14.1	The Particle in a Box	155
14.2	The Semi-Infinite Well	160
14.3	A Finite Potential Well	163

15 MORE ONE-DIMENSIONAL APPLICATIONS OF THE SWE 171

15.1	A Potential Barrier	172
15.2	Tunneling	173
15.3	The Harmonic Oscillator	177

PART *Three*

ATOMIC PHYSICS

16 THE HYDROGEN-LIKE ATOM AND THE SWE 185

16.1	Separation of Variables	185
16.2	The Radial Solutions	187
16.3	The Angular Solutions	189
16.4	Quantum Numbers	191

17 WAVE FUNCTIONS OF THE HYDROGEN-LIKE ATOM 195

17.1	Hydrogen-like Wave Functions	195
17.2	State Notations	197
17.3	Probability Density	197
17.4	The Radial Probability Density	199

PART *Five*

NUCLEAR AND PARTICLE PHYSICS

PART One

Relativity

 CHAPTER 1

Galilean Relativity

Many students ask me what the main topics in our Modern Physics course will be. I reply, "Well, we'll begin with relativity, . . ." and at that point I often see a distressed look come into their eyes. They—and you—probably have heard all sorts of terrible things about how difficult relativity is supposed to be. But actually, many of its ideas are relatively simple to understand. In this chapter, for instance, we will be reviewing (perhaps from a slightly different perspective) some of the ideas of relativity that you probably studied in your previous physics courses.

1.1
FRAMES OF REFERENCE

Suppose that we have the following situation: A railroad car is moving along straight, level tracks at a constant velocity of 3.0 m/s to the north. In the car a woman is walking up the aisle with a constant velocity of 1.0 m/s, also to the north, as shown in Fig. 1.1. Now let's ask the following question: What is the woman's velocity?

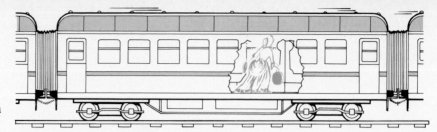

FIGURE 1.1
A woman walking in a railroad car.

The question sounds simple enough, but it has no single answer. The "correct" answer depends on the reference used. Two possible answers are:

1. With respect to the railroad car, her velocity is 1.0 m/s north.
2. With respect to the railroad tracks, her velocity is 1.0 m/s north + 3.0 m/s north, or 4.0 m/s north.

There are still other possible answers because the rails are fixed to the surface of the earth, which has velocities as it rotates both about its own axis and the sun. Also the sun and the whole solar system have a velocity with respect to our Milky Way galaxy's center, and so on. To arrive at a single correct answer, we must first decide just what the woman's velocity is to be determined with respect to. To do this we choose a place to fix a set of coordinates, for example, x, y, and z axes. This set of coordinates then forms a **frame of reference.**

In this course, we will restrict our study to inertial reference frames. An **inertial reference frame** is one to which Newton's first law of motion applies. That is, in an inertial reference frame objects at rest remain at rest and objects in motion continue in motion in a straight line at constant speed unless acted upon by a net external force. As you will see, any reference frame that has constant velocity (and no rotation) with respect to an inertial reference frame is itself an inertial reference frame.

If we can neglect the effects of the earth's rotations, a frame of reference fixed in the earth is an inertial reference frame. (See Fig. 1.2.) In

FIGURE 1.2
An approximately inertial reference frame fixed in the earth.

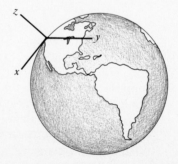

the case of the woman in the railroad car, one approximately inertial reference frame could be fixed on the railroad tracks. Another frame of reference could be moving along with the railroad car.

EXAMPLE 1.1   Use an apple remaining at rest on a flat table in your room to explain why a frame of reference fixed in the earth is not exactly an inertial reference frame.

Solution If it were an inertial reference frame, that apple remaining at rest would have a net external force on it of zero. But the net external force on the apple is *not* zero. The apple is at rest with respect to your room, but your room and the apple rotate together in a circle about the earth's axis once every 24 hours. For them to do so there must be a net centripetal force (the vector sum of the gravitational force from the earth and the normal force from the table).   ●

However, the external force in the example is small, and the forces due to the other accelerations of the earth are even smaller. So a frame of reference fixed in the earth is close to being an inertial reference frame, and we will ordinarily treat it as one.

1.2
GALILEAN COORDINATE TRANSFORMATIONS

Suppose that we've made some measurements in one frame of reference (such as the woman's velocity with respect to the railroad car) and we want to change or transform these measurements to another frame of reference (such as her velocity with respect to the railroad tracks). For instance, suppose the woman is standing on the bottom step of the railroad car exit and then steps off the moving car. Her velocity with respect to the tracks will be more important to her than her small velocity with respect to the car!

Making such transformations is relatively easy in the physics of Galileo and Newton, called **classical physics.** Let's assume that any measurements in two frames of reference are being made by observers who have previously and jointly calibrated their meter sticks and clocks. They therefore agree on the bases for measuring time and length. For simplicity, let's:

1. *Always* place the coordinate systems instantaneously together at $t = 0$.

2. Use Cartesian coordinate systems (mutually perpendicular x, y, and z axes).

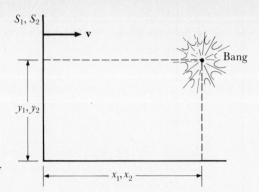

FIGURE 1.3
At $t_1 = 0 = t_2$: $x_1 = x_2$, $y_1 = y_2$, and $z_1 = z_2$ for the event.

3. Let coordinate system S_2 move in the direction of the $+x$ axis at a speed v with respect to the coordinate system S_1.

When an event occurs, perhaps a firecracker exploding, observers in both frames of reference can measure the position and time of the event in terms of their own frame of reference. Figure 1.3 shows that if the event occurs at $t_1 = 0 = t_2$ when both coordinate systems are together, then $x_1 = x_2$, and $y_1 = y_2$. Similarly, $z_1 = z_2$. Figure 1.4 shows that if the event occurs at a later time $t_1 = t_2$, then

$$
\begin{aligned}
x_2 &= x_1 - vt_1; \\
y_2 &= y_1; \\
z_2 &= z_1; \\
t_2 &= t_1.
\end{aligned}
\tag{1.1}
$$

The four relations given as Eq. (1.1) are called the **Galilean coordinate transformations.** They enable us to take positions and times measured in system S_1 and transform them to positions and times with respect to system S_2.

FIGURE 1.4
Another event at a later time.

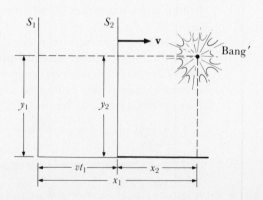

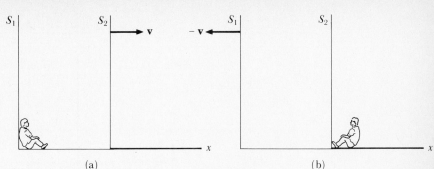

FIGURE 1.5
(a) An observer at rest in S_1 measures S_2 to be moving at v in the $+x$ direction. (b) Equivalently, an observer at rest in S_2 measures S_1 to be moving at v in the $-x$ direction.

Figure 1.4 or Eq. (1.1) and some simple algebra give us the **inverse Galilean coordinate transformations:**

$$x_1 = x_2 + vt_2;$$
$$y_1 = y_2;$$
$$z_1 = z_2;$$
$$t_1 = t_2.$$
(1.2)

Comparing Eq. (1.2) to Eq. (1.1), you can see that the subscripts have been interchanged and that the $-v$ has become a $+v$, which equals $-(-v)$. That is, v has been replaced by $-v$. This change can be made because an observer in S_1 could consider S_2 to be moving with a velocity v in the $+x$ direction, while an observer in S_2 could consider S_2 to be at rest, with S_1 moving with an opposite velocity of v in the $-x$ direction. (See Fig. 1.5.) We would see the railroad car of Section 1.1 moving north if we were standing by the tracks; if we were in the car the tracks underneath would appear to be "flowing south."

EXAMPLE 1.2   A light plane takes off in a constant 20 m/s wind to the north. It has moved 21 km north, 18 km east, and 0.6 km upward with respect to the air after 10 min. What are its coordinates with respect to the earth at this time?

Solution Since we are given values with respect to the air, we let coordinate system S_2 be moving along with the air at 20 m/s north. That means that the x direction will be north with $v = 20$ m/s, and coordinate system S_1 will be fixed in the earth. We could also let the y direction be west, as in Fig. 1.6, and the z direction be upward: the given values would be $x_2 = 21$ km, $y_2 = -18$ km, $z_2 = 0.6$ km, $t_2 = 10$ min, or 600 s, and $v = 20$ m/s.

We calculate the coordinates with respect to the earth by using the inverse Galilean coordinate transformations, Eq. (1.2): $x_1 = x_2 + vt_2 = 21$ km $+ 20$ m/s $\times 600$ s $\times 1$ km/10^3 m $= 33$ km (33 km north), $y_1 = y_2 = -18$ km (18 km east), and $z_1 = z_2 = 0.6$ km (0.6 km up). Also, observers in both coordinate systems agree on the time, $t_1 = t_2 = 10$ min, or 600 s.   ●

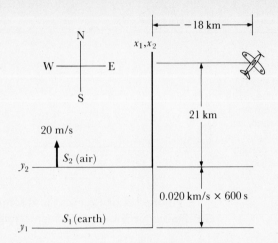

FIGURE 1.6
The x axes are first set up in the direction the air is moving. Then we choose the perpendicular y axes to be in the west direction. This makes the mutually perpendicular z axes be vertically upward in a right-handed coordinate system.

1.3
OTHER GALILEAN TRANSFORMATIONS

The question we asked in Section 1.1 was, "What is the woman's velocity?" To answer such questions we need the **Galilean velocity transformations,** that is, Eqs. (1.3) or (1.4), which follow. Using $v_{1x} = dx_1/dt_1$ and $v_{2x} = dx_2/dt_2$ (equivalent definitions for the y and z velocity components), $dt_1 = dt_2$, and recalling that v is constant, differentiation of Eq. (1.1) yields

$$v_{2x} = v_{1x} - v;$$
$$v_{2y} = v_{1y};$$
$$v_{2z} = v_{1z}.$$
(1.3)

Equation (1.3) is the component form of the vector equation

$$\mathbf{v_2} = \mathbf{v_1} - \mathbf{v}.$$
(1.4)

Interchanging the subscripts and replacing the minus sign by a plus sign gives the **inverse Galilean velocity transformation:**

$$\mathbf{v_1} = \mathbf{v_2} + \mathbf{v}.$$
(1.5)

Equation (1.5) can also be written in terms of its components, similar to Eq. (1.3). (See Fig. 1.7.)
     If we differentiate Eq. (1.3) with respect to time, remembering that v is constant, we obtain the **Galilean acceleration transformations:**

$$a_{2x} = a_{1x}, \qquad a_{2y} = a_{1y}, \quad \text{and} \quad a_{2z} = a_{1z},$$
(1.6)

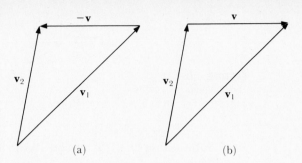

FIGURE 1.7
The Galilean velocity
transformations in vector
form: (a) $\mathbf{v}_2 = \mathbf{v}_1 - \mathbf{v}$.
(b) $\mathbf{v}_1 = \mathbf{v}_2 + \mathbf{v}$.

which is the component form of the vector equation

$$\mathbf{a}_2 = \mathbf{a}_1. \tag{1.7}$$

In classical physics, neither mass nor force depends on the frame of reference. That is, $\mathbf{F}_1 = \mathbf{F}_2$ and $m_1 = m_2$. Since $\mathbf{a}_1 = \mathbf{a}_2$, $\mathbf{F}_1 = m_1\mathbf{a}_1$ transforms to $\mathbf{F}_2 = m_2\mathbf{a}_2$. Therefore the $\mathbf{F} = m\mathbf{a}$ form of Newton's second law of motion is **invariant** (has the same mathematical form) under a Galilean transformation.

Equations (1.5) and (1.6) tell us that the acceleration is the same in any two classical physics frames of reference moving at a constant velocity with respect to one another. Suppose that S_1 is an inertial reference frame. That means that if $\mathbf{F}_1$ equals zero, then $\mathbf{a}_1$ also equals zero from Newton's first or second laws of motion. The preceding paragraph shows that $\mathbf{F}_2$ and $\mathbf{a}_2$ equal zero, too. Therefore, S_2 is also an inertial reference frame. In other words, *any frame of reference that is moving at constant velocity with respect to an inertial reference frame is also an inertial reference frame.*

EXAMPLE 1.3 Show that conservation of linear momentum is invariant under a Galilean transformation for the collision of two objects. (Linear momentum = mass × velocity.)

Solution Let the two objects have masses M and m. Give them velocities $\mathbf{V}_1$ and $\mathbf{v}_1$ before the collision and $\mathbf{V}_1'$ and $\mathbf{v}_1'$ after the collision. (See Fig. 1.8.) If the net external force is zero in an inertial reference frame, linear momentum will be conserved. That is,

$$M\mathbf{V}_1 + m\mathbf{v}_1 = M\mathbf{V}_1' + m\mathbf{v}_1'. \tag{A}$$

FIGURE 1.8
The collision of two
objects as measured in S_1.

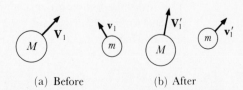

(a) Before   (b) After

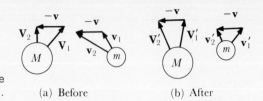

FIGURE 1.9
The transformation of the
collision of Fig. 1.8 to S_2.   (a) Before   (b) After

Substituting Eq. (1.5) into Eq. (A) (priming and capitalizing as needed), we
have

$$M(\mathbf{V_2} + \mathbf{v}) + m(\mathbf{v_2} + \mathbf{v}) = M(\mathbf{V_2'} + \mathbf{v}) + m(\mathbf{v_2'} + \mathbf{v}).$$

Subtracting the $M\mathbf{v}$ and $m\mathbf{v}$ terms from both sides gives

$$M\mathbf{V_2} + m\mathbf{v_2} = M\mathbf{V_2'} + m\mathbf{v_2'}. \tag{B}$$

Equation (B) has the same mathematical form as Eq. (A), with merely the
subscripts changed from 1 to 2, so conservation of linear momentum for
this collision is invariant under a Galilean transformation. (See Fig. 1.9.)
●

A useful application is the transformation from the "laboratory system" in which the collision is occurring to the "center-of-mass system"
where the total momentum is zero before and after the collision.

EXAMPLE 1.4   A river of width L moves uniformly with a speed v. A canoe is to be paddled
at a constant speed c with respect to the river. Calculate the time needed to
paddle directly across the river and back, and compare it to the time
needed to paddle the same distance upstream and back. Ignore the turn-
around and start-up times.

Solution Fix S_1 in the bank of the river and S_2 in the river water with the $+x$
direction being downstream. Then c is equal to v_2 because c is measured
with respect to the river water. To paddle directly across the river means
that there can be no net motion in the x direction (upstream or downstream) with respect to the bank. This is the same as saying that $v_{1x} = 0$.
Then $v_{2x} = v_{1x} - v$ gives $v_{2x} = -v$; that is, the x component of the paddling
velocity with respect to the river water must have a magnitude v and must
be directed upstream. Figure 1.10 shows that from the Pythagorean
theorem, the velocity component directly across the river will be $\sqrt{c^2 - v^2}$.
The time needed to go across will then equal $L/\sqrt{c^2 - v^2}$; the total time
needed to travel across and back perpendicular to the river's velocity, $T_\perp$,
is twice that, or

$$T_\perp = 2\frac{L}{\sqrt{c^2 - v^2}} = \frac{2L/c}{\sqrt{1 - v^2/c^2}}. \tag{C}$$

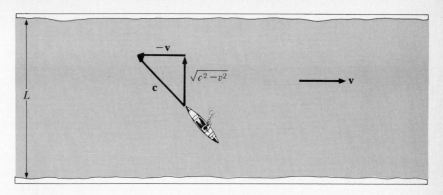

FIGURE 1.10
The canoe is to travel straight across the river. Therefore the canoe's velocity with respect to the water must have a component equal in magnitude but opposite in direction to the water's velocity.

Paddling directly upstream and downstream requires that the y and z components of all velocities equal zero. Paddling downstream, $v_{2x} = +c$ and so $v_{1x} = c + v$. Therefore the time to paddle the distance L downstream is $L/(c + v)$. Paddling upstream is in the negative x direction and so $v_{2x} = -c$, which results in $v_{1x} = -c + v$. The displacement is $-L$ and so the time to go upstream is $-L/(-c + v) = L/(c - v)$. The total time needed to travel down and up (or up and down) parallel to the river's velocity, $T_{\parallel}$, is the sum of the times; that is,

$$T_{\parallel} = \frac{L}{c + v} + \frac{L}{c - v} = \frac{2L/c}{(1 - v^2/c^2)}. \tag{D}$$

Equations (C) and (D) show that $T_{\parallel} > T_{\perp}$, unless $v = 0$ (no current in the river). Of course, the results also show that c must be greater than v, something you already know based on "common sense."

In this chapter you've seen that the Galilean transformations can be used to solve many practical problems. These problems often involve velocities or displacements with respect to a moving vehicle, the flowing water in a river, or the air when the wind is blowing. When solving these problems, we usually fix S_1 in the earth and S_2 in the vehicle, the water, or the air, with S_2 moving in the $+x$ direction. Then v is the speed of the vehicle, the river, or the wind with respect to the earth.

QUESTIONS AND PROBLEMS

1.1 Under what conditions would your car be an inertial reference frame? A noninertial frame of reference?

1.2 An object is initially at rest with respect to a frame of reference that is accelerating north. If the net external force on the object is zero, describe its motion relative to the accelerating frame of reference.

1.3 With the help of outside reading, compare the astronomical frames of reference adopted by Ptolemy and Copernicus. Which was more nearly an inertial reference frame?

1.4 Recall that $F_c = mr\omega^2$, $\omega = 2\pi/T$, and $w = mg$. Calculate the ratio of centripetal force to weight at latitude θ on the earth to show that the earth is approximately an inertial reference frame.

1.5 Suppose that S_1 is an inertial reference frame and that the Galilean coordinate transformations hold true. How do you know that S_2 is also an inertial reference frame?

1.6 If Eq. (1.1) is true, does this mean that S_1 and S_2 are necessarily inertial reference frames?

1.7 An event occurs at $x_1 = 3.0$ m, $y_1 = 1.0$ m, and $z_1 = -0.5$ m; $t_1 = 2.0$ s when $v = 4.0$ m/s. Calculate x_2, y_2, z_2, and t_2 and indicate the event on a sketch of the frames of reference.

1.8 A series of events occurs at the origin of S_2. How are their positions and times related to one another in S_1?

1.9 Repeat Example 1.2 with a 20 m/s wind to the south.

1.10 Repeat Example 1.2 with a 20 m/s wind to the east.

1.11 To measure a length in S_2, positions x_2 and x_2' are both measured at the same time, t_2 (that is, simultaneously). Then $L_2 = x_2' - x_2$. Use a Galilean coordinate transformation to find L_1, the length measured in S_1.

1.12 To measure T_2, a time interval in S_2, times t_2 and t_2' are measured at the same position, x_2. Then $T_2 = t_2' - t_2$. What then is T_1, the time interval measured in S_1?

1.13 Show that conservation of linear momentum is invariant under a Galilean transformation for a completely inelastic collision of two objects (they hit and stick together).

1.14 Show that the conservation of kinetic energy is invariant under a Galilean transformation for an elastic collision of two objects.

1.15 Does the statement that "the laws of conservation of energy and of momentum are invariant under a Galilean transformation" mean that all inertial observers will measure the same values for energy and momentum?

1.16 An airplane travels at a constant velocity 200 km southwest with respect to the ground in 1/2 hour in a constant wind of 60 km/hr to the west at a constant elevation. What are the components of the airplane's position and velocity with respect to the ground and the air? What are the accelerations?

1.17 A river is flowing south at 15 ft/s. A canoist aims northeast and paddles at 5 ft/s with respect to the water. With respect to the water and with respect to the earth, what are the components of the canoe's position and velocity after 20 s? What are the accelerations?

1.18 The *jerk* is defined as the time rate of change of acceleration. Derive the Galilean jerk transformation.

1.19 An ultralight aircraft can fly at 30 m/s in still air. The pilot wishes to fly at this airspeed in a straight line between two points, the second point being 8.0 km northeast of the first. If there is a constant 10 m/s wind blowing to the west: a) In what direction must he aim the aircraft to fly between the two points? b) How long will the trip take him?

1.20 The canoist in Example 1.4 aims the canoe directly across the river and paddles to the other side, then reverses directly back to return to the original side, then finally aims the canoe upstream and paddles back to the start. Ignoring turn-around times, what is the round-trip time?

1.21 Johnny Jones is jogging at a constant velocity of 3.0 m/s west when he sees his Aunt Susie running at a constant velocity of 7.0 m/s 30° north of east. (Both velocities are measured with respect to the earth.) What is Aunt Susie's velocity with respect to Johnny? What is Johnny's velocity with respect to Aunt Susie? (This is a relative's velocity problem, in case you hadn't noticed.)

1.22 For those who know about matrixes, suppose that position and time are represented by the column matrix

$$\begin{bmatrix} x \\ y \\ z \\ ict \end{bmatrix},$$

where $i = \sqrt{-1}$ and c is the speed of light. Show that the Galilean coordinate transformation equations become

$$\begin{bmatrix} x_2 \\ y_2 \\ z_2 \\ ict_2 \end{bmatrix} = \begin{bmatrix} 1 & 0 & 0 & i\beta \\ 0 & 1 & 0 & 0 \\ 0 & 0 & 1 & 0 \\ 0 & 0 & 0 & 1 \end{bmatrix} \begin{bmatrix} x_1 \\ y_1 \\ z_1 \\ ict_1 \end{bmatrix}$$

where $\beta = v/c$.

1.23 Following the example of Problem 1.22:

a) Find the 4×4 matrix for the inverse Galilean coordinate transformation;

b) Show that the product of the two 4×4 transformation matrixes (from Problem 1.22 and part (a) of this problem) is the identity matrix (1's on the diagonal, 0's elsewhere).

CHAPTER 2

The Special Theory of Relativity

The Galilean transformations of classical relativity are simple equations, but unfortunately too simple in certain circumstances. In this chapter we will show that they can lead to inconsistencies. To remove these inconsistencies, a young Albert Einstein came up with his special theory of relativity, basing it on two uncomplicated postulates. You will be introduced to these postulates and to the controversy over the necessity of a medium of propagation for electromagnetic waves. Finally, we will analyze the famous Michelson–Morley experiment and relate its results to that controversy and to Einstein's second postulate.

2.1
THE PRINCIPLE OF RELATIVITY

The examples and problems of Chapter 1 showed that the basic laws of classical mechanics have the same mathematical form in all inertial reference frames. As far as mechanics is concerned, there is no inertial reference frame in which its classical laws have *the* most basic form. Therefore there is no absolute frame of reference.

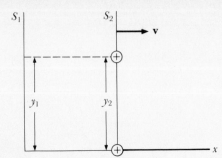

FIGURE 2.1
Two charges at rest on the
y axis of frame of
reference S_2.

Are other laws of physics also invariant under a Galilean transformation? Let's consider a simple case of two equal positive point charges, as shown in Fig. 2.1. First we will examine the system as seen by an observer in coordinate system S_2.

One charge rests at the origin of S_2. The other charge rests a distance y_2 away on the y axis of S_2. Now let's apply Maxwell's classical equations to find the electromagnetic forces the charges exert on one another. In S_2 the equal charges are at rest. Therefore they only repel one another with the electrostatic Coulomb's law force $F_2 = (1/4\pi\epsilon_0)(q^2/y_2^2)$.

Now let's consider the electromagnetic forces from the viewpoint of an observer in system S_1. This observer sees q unchanged and $y_1 = y_2$. Therefore the Coulomb's law repulsive force is unchanged.

However, in S_1 both charges are moving to the right at a speed v. (See Fig. 2.2.) Two positive charges moving to the right give two conventional currents to the right, and parallel currents attract each other. Therefore the force in S_1 and the force in S_2 differ by an amount equal to the attractive

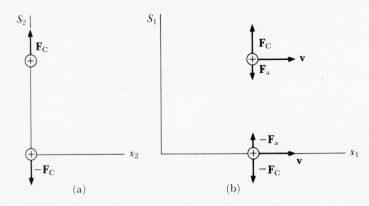

(a)   (b)

FIGURE 2.2
(a) In S_2, the charges are at rest and repel one another with a force of magnitude F_C. (b) In S_1, the charges are moving and attract one another with an additional force of magnitude F_a, giving a total force of magnitude $|F_C - F_a|$.

force of parallel currents. But according to classical physics, the forces should not differ, that is, $\mathbf{F}_1$ should equal $\mathbf{F}_2$.

This inconsistency gives us a problem. Perhaps some laws of physics depend on the choice of frames of reference (so that the choice of a particular frame is not relative). Perhaps the Galilean transformations are not the correct ones.

Albert Einstein attacked this problem in 1905. (He was 26 and had been working on the problem in his spare time while employed as a Swiss patent clerk.) Einstein proposed what is called the **special theory of relativity.** The word *special* is used because the theory applies only to inertial reference frames. The word *theory* does *not* mean that Einstein's concept was just a guess or a hypothesis. Instead it means that the concept has been validated by many different experiments, as indicated in Table 2.1.

Einstein based his theory on two fundamental postulates. The first is called the **principle of relativity,** which states that:

All laws of physics have the same mathematical form in all inertial reference frames.

As you have seen, two of the consequences of this postulate are (1) that there is no absolute frame of reference and (2) that measured values may be relative to a frame of reference. Also the Galilean transformations are not correct for all laws of physics.

EXAMPLE 2.1   We can write one of Maxwell's equations for $\mathbf{B}$ in inertial reference frame 1 as $\oint \mathbf{B}_1 \cdot d\mathbf{l}_1 = \mu_0(\epsilon_0 \partial \Phi_{E_1}/\partial t_1 + i_1)$.

a) How can we write it in inertial reference frame 2, according to Einstein's principle of relativity?   b) Therefore, does $\mathbf{B}_1 = \mathbf{B}_2$?

Solution

a) To give the equation the same mathematical form in reference frame 2, we simply replace all the 1's by 2's in the subscripts. Therefore $\oint \mathbf{B}_2 \cdot d\mathbf{l}_2 = \mu_0(\epsilon_0 \partial \Phi_{E_2}/\partial t_2 + i_2)$.

b) No, $\mathbf{B}_1$ is not necessarily equal to $\mathbf{B}_2$. The equations for $\mathbf{B}$ have the same form, but the values of $\mathbf{B}$ obtained from the equations are not necessarily the same. The example illustrated in Fig. 2.1 would have $\mathbf{B}_2 = 0$, but $\mathbf{B}_1 \neq 0$.   ●

TABLE 2.1
Agreement between Various Theories and Experiments

	Theory	Light propagation experiments							Experiments from other fields					
		Aberration	Fizeau convection coefficient	Michelson–Morley	Kennedy–Thorndike	Moving sources and mirrors	De Sitter spectroscopic binaries	Michelson–Morley, using sunlight	Variation of mass with velocity	General mass–energy equivalence	Radiation from moving charges	Meson decay at high velocity	Trouton–Noble	Unipolar induction, using permanent magnet
Ether theories	Stationary ether, no contraction	✓	✓	X	X	✓	✓	X	X		✓		X	X
	Stationary ether, Lorentz contraction	✓	✓	✓	X	✓	✓	✓	✓		✓		✓	X
	Ether attached to ponderable bodies	X	X	✓	✓	✓	✓	✓	X				✓	
Emission theories	Original source	✓	✓	✓	✓	✓	X	X			X			
	Ballistic	✓		✓	✓	X	X	X			X			
	New source	✓		✓	✓	X	X	✓			X			
Special theory of relativity		✓	✓	✓	✓	✓	✓	✓	✓	✓	✓	✓	✓	✓

Legend: ✓   The experiment agrees with the theory.
　　　　X   The experiment disagrees with the theory.
　　　　Blank   The experiment is not explained by the theory.

Source: Adapted from W. Panofsky and M. Phillips, *Classical Electricity and Magnetism,* 2nd ed. © 1962 Addison-Wesley, Reading, Mass. P. 282, Table 15.2. Reprinted with permission.

2.2
THE CONSTANCY OF THE SPEED OF LIGHT

The speed of a water wave or a sound wave relative to an observer depends on the speed of the observer relative to the medium, but not on the speed of the source. If you run out into the waves at the beach the speed of the waves relative to you is increased. If you are in a train moving away from a clanging crossing bell, the speed of the sound relative to you decreases. This change in relative speed causes the wave frequency to change and is the phenomenon called the Doppler effect.

Moving toward or away from a source of electromagnetic waves also gives a Doppler effect. However, the Doppler effect equations for electromagnetic waves, such as police radar microwaves or light waves, are different from the equations for sound waves or water waves. The reason is that sound waves or water waves are mechanical waves and require a medium to travel in, whereas electromagnetic waves are not mechanical waves and so apparently require no medium to travel through. (See Fig. 2.3.) Electromagnetic waves have oscillating electric and magnetic fields; the changing electric fields induce changing magnetic fields, which induce changing electric fields, and so on. The electromagnetic radiation we receive from the sun, stars, and other astronomical objects travels to us through the excellent vacuum of space.

Scientists at the end of the 19th century did not fully appreciate this interpretation of the results of Maxwell's equations and invented a medium for light to travel through. They called it the **luminiferous ether,** which would be present in all of space. The luminiferous ether would have to sustain a wave speed of $c = 3.00 \times 10^8$ m/s without slowing down the

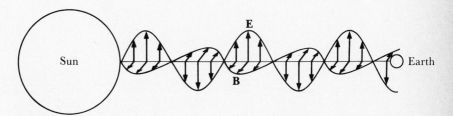

FIGURE 2.3
Light and other electromagnetic waves move through the excellent vacuum of space. (One component of **E** and its corresponding **B** represent the electromagnetic wave.) However, sound and other mechanical waves can't move through a vacuum. (We can see the sun, but can't hear it.)

motion of the planets through it. More importantly for the ideas of relativity, the speed of light and other electromagnetic waves would be measured with respect to the ether. Therefore it would be the preferred frame of reference for electromagnetism. However, various attempts to measure the speed of the earth through the ether failed. These failed attempts are the experimental bases for Einstein's second postulate of relativity, which states that:

> The speed of light in a vacuum has the same measured value in all inertial reference frames.

This postulate does away with the necessity for a luminiferous ether and the preferred frame of reference associated with it.

EXAMPLE 2.2   A spaceship moving at 2×10^8 m/s in space toward us shoots a laser beam in our direction. The crew measures the laser beam's speed to be 3×10^8 m/s with respect to their spaceship. (See Fig. 2.4.) According to Einstein, what will we measure the laser beam's speed to be?

Solution A laser beam is made of light. All inertial observers measure the speed of light in a vacuum to have the same value, according to Einstein's second postulate of relativity. We are inertial observers, so we will also measure the speed of the laser beam to be 3×10^8 m/s.   ●

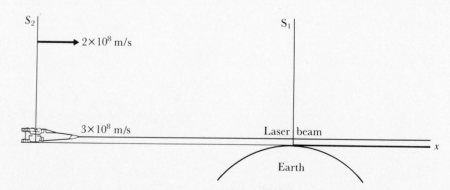

FIGURE 2.4
If we fix S_2 in the spaceship and S_1 in the Earth, $v = 2 \times 10^8$ m/s, and both v_2 and $v_1 = 3 \times 10^8$ m/s according to the second postulate of special relativity (disagreeing with the Galilean velocity transformations).

2.3
THE MICHELSON–MORLEY EXPERIMENT

One of the experiments designed to measure the speed of the earth through the ether was first accurately performed by Michelson and Morley in the United States in 1887. It has been done since with greater and greater accuracy in many different versions. There is general agreement that no speed of the earth relative to the luminiferous ether can be found.

The apparatus used in the **Michelson–Morley experiment** was the Michelson interferometer. In this device, monochromatic (one wavelength) light from a source is split into two separate beams. These beams travel two different optical paths and then come back together to interfere either constructively or destructively. If the earth were moving through the ether, the apparatus could be aligned with the source "upstream" as the ether flowed by. (See Fig. 2.5.) Ray 1' would move back and forth across the "ether river" and ray 1" would move downstream and upstream. As you can see, this is the same idea as the canoe in the river in Example 1.4. If the arms of the interferometer have equal optical lengths L, the difference in times for rays 1' and 1" will be

$$\Delta T = T_\parallel - T_\perp = \frac{2L/c}{(1 - v^2/c^2)} - \frac{2L/c}{\sqrt{1 - v^2/c^2}}. \tag{2.1}$$

This difference in times will cause a phase difference and a certain interference pattern with light and dark fringes.

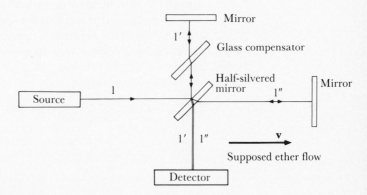

FIGURE 2.5
The Michelson interferometer. Ray 1 is split into rays 1' and 1", then recombined at the detector to form an interference pattern. Since the half-silvered mirror reflects from the side nearest the source, the compensator is introduced so that rays 1' and 1" travel through the same thickness of glass.

Michelson and Morley then carefully rotated the entire apparatus by 90°, so that ray 1′ would move downstream and upstream and ray 1″ would move back and forth across the "ether river." This rotation should have interchanged the times for rays 1′ and 1″ and the interference pattern. Although the expected change was nearly 100 times the sensitivity of their apparatus, no shift of the pattern (within experimental error) was discovered. Therefore the experimenters could not find the supposed luminiferous ether. Although this is now the most famous of the *null outcome* experiments in the search for the luminiferous ether, Einstein himself may have been unaware of it.

EXAMPLE 2.3 The speed of the earth through the ether was expected to be approximately its orbital speed around the sun, which is much less than c. Use the binomial expansion $(1 + x)^n \approx 1 + nx$ (for small x) on both terms to find an approximate relation for ΔT when $(v^2/c^2) \ll 1$.

Solution

$$\Delta T = (2L/c)[(1 - v^2/c^2)^{-1} - (1 - v^2/c^2)^{-1/2}]$$
$$\approx (2L/c)[1 + (-1)(-v^2/c^2) - 1 - (-1/2)(-v^2/c^2)]$$
$$= Lv^2/c^3.$$

$\bullet$

You have seen that the special theory of relativity is based on two simple postulates. The first states that in all inertial systems, all laws of physics have the same mathematical form. So if $\mathbf{F}_1 = d(m_1\mathbf{v}_1)/dt_1$ is the law in inertial reference frame S_1, $\mathbf{F}_2 = d(m_2\mathbf{v}_2)/dt_2$ will be the law in inertial reference frame S_2. However, this postulate does not say that the values of $\mathbf{F}$, m, $\mathbf{v}$, or even t will be the same, as measured in the two reference frames.

The second postulate states that the speed of light will be measured to be the same in all inertial reference frames. Therefore there will be no speed difference with respect to a supposed "ether river." Both $T_{\parallel}$ and $T_{\perp}$ will simply be $2L/c$, resulting in $\Delta T = 0$, the null outcome of the Michelson–Morley experiment.

QUESTIONS AND PROBLEMS

2.1 If there is no absolute frame of reference, does this mean that calculations cannot be simpler in one frame of reference than in another?

2.2 Special relativity is referred to as a theory, not a law. Does that mean it is really a doubtful sort of idea?

2.3 One of Maxwell's equations is $\oint \mathbf{E}_1 \cdot d\mathbf{l}_1 = -\partial \Phi_{B_1}/\partial t_1$, in S_1. According to the principle of relativity, what will the equation be in S_2?

2.4 What do you think the general theory of relativity applies to?

2.5 How many experiments would it take to show

with certainty that the special theory of relativity is true? To show with certainty that it is false?

2.6 Discuss the statement, "The theory of relativity has shown us that there are no absolutes and so the old ideas of morality must be discarded."

2.7 a) The speed of the wind is 3.0 m/s. You walk into the wind at 2.0 m/s (both speeds with respect to the ground). What is the wind speed relative to you?

b) The speed of light is 3.0×10^8 m/s. You travel toward a light source in a spaceship at 2.0×10^8 m/s (both speeds with respect to the earth). What is the speed of the light relative to you?

2.8 One observer moves toward a star at a speed v, another moves away from it at the same speed. Considering the speed, frequency, and wavelength of the light ($c = v\lambda$), what do the observers measure to be the same and what do they measure to be different?

2.9 Michelson and Morley repeated their experiment at different times of the year. Can you think of why? (Consider the earth's rotation around the sun.) Why do you think they also repeated it on a mountain top? (Some scientists thought the earth could drag the ether along.)

2.10 Multiple reflections gave Michelson and Morley $L = 10$ m. The earth's speed around the sun is 3×10^4 m/s. Compare $c\Delta T$ (classical) to an average wavelength of visible light.

2.11 Suppose that lengths measured parallel to the "ether river" were shortened by a factor of $\sqrt{1 - v^2/c^2}$. What then would ΔT (classical) equal? (This is called the Lorentz–Fitzgerald contraction and was used to explain the null result of the Michelson–Morley experiment.)

2.12 Discuss the statement, "Neither c nor the laws of physics are relative, so it should be the theory of absolutivity."

2.13 The speed of light in a material of index of refraction n is given by $v = c/n$. Does this fact violate either postulate of the special theory of relativity? Explain.

2.14 An observer at rest in a vacuum detects a pulse of light spread out over a spherical surface.

a) If the pulse is emitted at $t_1 = 0$, what is the radius of the sphere at a later time t_1?

b) What type of surface does a second observer detect if she is moving at a speed v with respect to the first observer, and what dimension(s) does she measure it to have at a time t_2 after emission?

2.15 Is it possible that light moving past you in flowing water might have a different speed than light moving past you in still water, or is this forbidden by a postulate of the special theory of relativity?

2.16 Does the Michelson–Morley experiment indicate that the luminiferous ether is an unnecessary concept, or does it prove that there is no such thing?

2.17 Some of the radiation from a rotating star comes out in a narrow beam. It causes an interstellar gas cloud between the earth and the star to glow as the radiation beam sweeps by (much like the beam of a rotating searchlight moving across the clouds above). This glow from the rotating star's beam can move across the cloud at a speed $v = \omega r$, which is greater than c. Does this occurrence violate Einstein's second postulate?

2.18 The speed of yellow light moving through a certain type of glass is measured to be 1.98×10^8 m/s. Does this occurrence violate Einstein's second postulate?

The Lorentz Coordinate Transformations

How can we find the frequency change that results from the relative motion of a source of electromagnetic waves and an observer? This is one of the most common applications of the special theory of relativity, so we will derive the relationship for that frequency change in this chapter. To do so, we first need to find a set of transformations that will agree with Einstein's postulates. We will then discuss the relationship between these transformations and the Galilean transformations. We will also show that the relativistic transformations can give results amazingly different from the predictions of the classical transformations and from what experience might lead us to expect.

3.1
DERIVATION OF THE TRANSFORMATIONS

Suppose that at $t_1 = t_2 = 0$ (when S_1 and S_2 are instantaneously together), a point source of light at their common origin sends out a spherical pulse of light. Since c is a constant for all observers in both S_1 and S_2 and is the same in all directions, all observers in both frames of reference must detect a

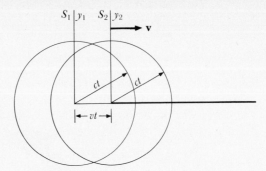

FIGURE 3.1

As this two-dimensional figure shows, Galilean transformations won't give a single spherical pulse centered at the origins, as measured in both S_1 and S_2, unless $v = 0$.

spherical wavefront expanding from their origin. Since the equation of a sphere is $x^2 + y^2 + z^2 = r^2$ and r, the radius, equals ct, we can write

$$x_1^2 + y_1^2 + z_1^2 - c^2 t_1^2 = 0; \tag{A}$$

$$x_2^2 + y_2^2 + z_2^2 - c^2 t_2^2 = 0. \tag{B}$$

It is easy to see that Galilean transformations will not satisfy both these equations. (See Fig. 3.1.) If Eq. (A) is true, we can replace y_1, z_1, and t_1 with y_2, z_2, and t_2, respectively, but replacing x_1 by $x_2 + vt_2$ does *not* give Eq. (B) for $v \neq 0$.

Hendrik Lorentz derived a transformation that *will* satisfy both these equations. We can also derive it, using a few assumptions. As usual, S_1 and S_2 are taken to be instantaneously together at $t_1 = t_2 = 0$ and S_2 is taken to be moving at a constant velocity v in the $+x$ direction with respect to S_1. Both S_1 and S_2 are inertial reference frames. Since the motion is in the $+x$ direction (perpendicular to the y and z directions) and since $y_2 = y_1$ and $z_2 = z_1$ evidently led to no problems in the spherical wavefront equations, Eqs. (A) and (B), let's attempt to keep the simple y and z transformations.

We would also like to assume that the transformations are linear, so that one set of (x_2, y_2, z_2, t_2) would correspond most simply to only one set of (x_1, y_1, z_1, t_1). For instance, we wouldn't want x_2 to equal the square root of a function of x_1. In that case x_2 might have both $(+)$ and $(-)$ values, which would mean that a person located at one position in x_1 could be found at two positions at once in S_2! Finally, we would like the relations to have some similarity to the Galilean transformations, which worked so well for classical mechanics.

A simple set of position transformations would be

$$x_2 = \gamma(x_1 - vt_1), \qquad y_2 = y_1, \quad \text{and} \quad z_2 = z_1, \tag{C}$$

and we can then rewrite Eqs. (A) and (B) as

$$x_1^2 + y_1^2 + z_1^2 - c^2 t_1^2 = 0 = \gamma^2(x_1 - vt_1)^2 + y_1^2 + z_1^2 - c^2 t_2^2, \tag{D}$$

with the substitution for x_2, y_2, and z_2. Subtracting the y_1^2 and z_1^2 terms from each side of Eq. (D) makes it clear that no solution will result unless t_2 depends both on t_1 and x_1. Let's now try $t_2 = At_1 - Bx_1$ as a simple linear possibility. Substitution yields

$$(1 - \gamma^2 + c^2B^2)x_1^2 + 2(\gamma^2 v - c^2AB)x_1 t_1 + (c^2A^2 - \gamma^2 v^2 - c^2)t_1^2 = 0. \quad \text{(E)}$$

At a given t_1, x_1 will have an infinite number of values at different points on the wavefront, not just the two values obtainable by solving Eq. (E) using the quadratic formula. But the equation can still equal zero for all values of x_1 if all the coefficients of the equation equal zero. Therefore

$$(1 - \gamma^2 + c^2B^2) = 0, \quad (\gamma^2 v - c^2AB) = 0, \quad \text{and} \quad (c^2A^2 - \gamma^2 v^2 - c^2) = 0. \quad \text{(F)}$$

We can rearrange Eqs. (F) to give

$$c^2B^2 = \gamma^2 - 1, \quad \gamma^4 = c^2A^2 c^2B^2/v^2, \quad \text{and} \quad c^2A^2 = \gamma^2 v^2 + c^2. \quad \text{(G)}$$

Substituting the first and third expressions of Eq. (G) into its second expression gives

$$\gamma^4 = (\gamma^2 v^2 + c^2)(\gamma^2 - 1)/v^2 \quad \text{or} \quad \gamma^4 = \gamma^4 + \gamma^2(c^2/v^2 - 1) - c^2/v^2,$$

which yields

$$\gamma = \frac{1}{\sqrt{1 - v^2/c^2}} = \frac{1}{\sqrt{1 - \beta^2}}, \quad (3.1)$$

where

$$\beta = \frac{v}{c}. \quad (3.2)$$

Note that γ is always greater than or equal to 1 for β less than or equal to 1. (See Fig. 3.2.)

FIGURE 3.2
For v less than or equal to c, γ is always greater than or equal to one, approaching infinity as v approaches c, and β is always less than or equal to one, linearly approaching one as v approaches c.

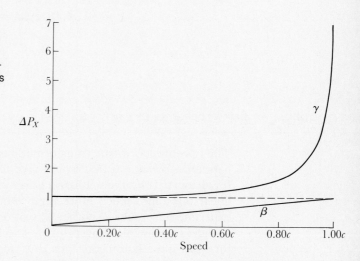

EXAMPLE 3.1 Calculate v in m/s, β, and γ for $v = 0.0010c$ and $v = 0.10c$.

Solution Since $c = 3.0 \times 10^8$ m/s, $v = 0.0010c = 3.0 \times 10^5$ m/s and $v = 0.10c = 3.0 \times 10^7$ m/s. Using $\beta = v/c$, $\beta = 0.0010$ and $\beta = 0.10$, respectively. Since $\gamma = 1/\sqrt{1 - \beta^2}$ or $1/\sqrt{1 - v^2/c^2}$, $\gamma = 1/\sqrt{1 - 10^{-6}}$ for $v = 0.0010c$. Using $(1 - x)^n \approx 1 - nx$ for $x \ll 1$, $\gamma = (1 - \beta^2)^{-1/2} \approx 1 + \beta^2/2$ for $\beta^2 \ll 1$. Therefore $\gamma \approx 1 + 10^{-6}/2 = 1.0000005$ for $v = 0.0010c$ and $\gamma \approx 1 + 10^{-2}/2 = 1.005$ for $v = 0.10c$. ●

Substituting γ back into the first and third expressions of Eq. (G) gives $B = \gamma\beta/c$ and $A = \gamma$, so the transformations become

$$x_2 = \gamma[x_1 - vt_1] = \frac{x_1 - vt_1}{\sqrt{1 - v^2/c^2}};$$

$$y_2 = y_1;$$
$$z_2 = z_1; \tag{3.3}$$

$$t_2 = \gamma[t_1 - (\beta/c)x_1] = \frac{t_1 - (v/c^2)x_1}{\sqrt{1 - v^2/c^2}}.$$

These expressions are called the **Lorentz coordinate transformations.** Since an observer at rest in S_2 sees S_1 moving at $-v$, interchanging the subscripts and replacing all v's by $-v$'s gives the **inverse Lorentz coordinate transformations.** Remembering that $(-v)^2 = v^2$ (so that γ is unchanged), we obtain

$$x_1 = \gamma[x_2 + vt_2] = \frac{x_2 + vt_2}{\sqrt{1 - v^2/c^2}};$$

$$y_1 = y_2;$$
$$z_1 = z_2; \tag{3.4}$$

$$t_1 = \gamma[t_2 + (\beta/c)x_2] = \frac{t_2 + (v/c^2)x_2}{\sqrt{1 - v^2/c^2}}.$$

3.2
THE CORRESPONDENCE PRINCIPLE AND APPLICATIONS

The Lorentz coordinate transformations are not the only possible coordinate transformations that would give the required spherical wavefronts for all inertial observers. But the Lorentz transformations are the simplest linear ones and, much more importantly, their predictions agree with experimental results. Another way of testing them is to apply the correspondence principle. This principle was first stated by Neils Bohr in 1923 for other nonclassical concepts. In a more general statement than Bohr's original one, the **correspondence principle** holds that:

Any new theory in physics must reduce to its corresponding well-established classical theory in the situations for which the classical theory is valid.

The physics of Galileo and Newton was experimentally established for objects that moved at speeds much less than the speed of light. (Even a satellite moving at 28,000 km/hr in orbit has a speed of less than $3 \times 10^{-5}c$ with respect to the earth.) From previous physics courses, you should be able to show that the attractive force between the two charges mentioned in Section 2.1 will be proportional to $\mu_0 v^2$ and the Coulomb repulsive force to $1/\epsilon_0$. The ratio of the two forces will therefore be proportional to $\mu_0 v^2/(1/\epsilon_0) = v^2(\mu_0\epsilon_0) = v^2/c^2$. This condition means that the difference between the electromagnetic forces calculated in the two frames of reference according to classical physics will be negligible for $v \ll c$. We should then find that the relativistic Lorentz transformations reduce to the classical Galilean transformations as $v/c = \beta$ approaches zero.

The Lorentz transformations do indeed agree with the correspondence principle. They reduce to the classical Galilean transformations for speeds of much less than the speed of light. As β approaches 0, γ approaches 1, Eq. (3.3) approaches Eq. (1.1), and Eq. (3.4) approaches Eq. (1.2).

EXAMPLE 3.2   Observers on an asteroid measure a spaceship to zip by at a speed of $0.60c$. (See Fig. 3.3.) All bridge and asteroid clocks are started when the bridge of the ship passes the asteroid. When the asteroid clocks show that 5.0 μs has passed, the asteroid observers measure a laser flash to occur at a position

FIGURE 3.3
A spaceship and an asteroid.

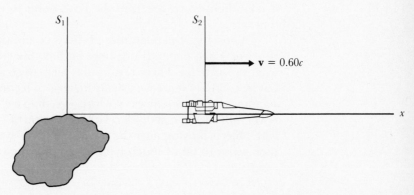

having coordinates $(3.0, 0.5, -0.2)$ km with respect to them. At what position and time does the spaceship captain measure that the laser flash occurred?

Solution Fix S_1 in the asteroid because you have one set of values with respect to the asteroid. Then fix S_2 in the bridge of the spaceship because you want to know another set of values with respect to the spaceship captain. Then $v = 0.60c$, $\beta = v/c = 0.60$, and using Eq. (3.1),

$$\gamma = 1/\sqrt{1 - (0.60)^2} = 1/\sqrt{1 - 0.36} = 1/\sqrt{0.64} = 1/0.80 = 1.25.$$

The asteroid observers are in S_1, so their measurements are $x_1 = 3.0 \times 10^3$ m, $y_1 = 0.5 \times 10^3$ m, $z_1 = -0.2 \times 10^3$ m, and $t_1 = 5.0 \times 10^{-6}$ s. All observers agree that $v = 0.60c = 0.60(3.0 \times 10^8)$ m/s $= 1.8 \times 10^8$ m/s. The spaceship captain is in S_2. We obtain his values by substituting into the Lorentz coordinate transformations, Eq. (3.3):

$$x_2 = 1.25[3.0 \times 10^3 \text{ m} - (1.8 \times 10^8 \text{ m/s})(5.0 \times 10^{-6} \text{ s})]$$
$$= 1.25[3.0 \times 10^3 \text{ m} - 0.9 \times 10^3 \text{ m}] = 1.25[2.1 \text{ km}] = 2.6 \text{ km};$$
$$y_2 = y_1 = 0.5 \text{ km};$$
$$z_2 = z_1 = -0.2 \text{ km};$$
$$t_2 = 1.25[5.0 \times 10^{-6} \text{ s} - (0.60/3.0 \times 10^8 \text{ m/s})3.0 \times 10^3 \text{ m}]$$
$$= 1.25[5.0 \times 10^{-6} \text{ s} - 6.0 \times 10^{-6} \text{ s}]$$
$$= 1.25[-1.0 \text{ } \mu\text{s}] = -1.25 \text{ } \mu\text{s}. \qquad \bullet$$

The results obtained in Example 3.2 show some of the amazing consequences of the special theory of relativity. The laser flash is detected 3.0 km in the direction of motion from the asteroid at a time when the asteroid observers say the spaceship has traveled 0.9 km. Yet the spaceship captain does not measure the flash to be $(3.0 - 0.9)$ km $= 2.1$ km ahead; he measures the flash to be 2.6 km ahead. Even more astounding, the asteroid observers say the flash occurred 5.0 μs *after* the spaceship captain passed by, but he says the flash occurred 1.25 μs *before* he arrived at the asteroid (with both sets of observers properly taking into account the time it takes light to travel from the flash point to them in measuring their times).

Therefore the quantities of distance and time are not absolute, but relative. Time depends not only on how fast you go (v), but also upon where you are (x). These results may seem to go against "common sense," but by common sense we usually mean a collection of experiences, understandings, and prejudices to which we compare new data. Since we have never moved about at $0.60c$ relative to the earth — or dealt in everyday life with objects moving near the speed of light — we cannot expect our common sense to be of much help in judging the correctness of relativistic results.

EXAMPLE 3.3   The spaceship captain sends out signals so at the instant he passes the asteroid he measures that beacons are turned on 4.0 km directly ahead and 4.0 km directly behind him (with respect to him). Where do the observers on the asteroid determine the beacons to be and when do they measure them to be turned on?

Solution Here we are given measurements from S_2 at $t_2 = 0$ and are asked for values in S_1. For the beacon behind, using Eq. (3.4) and $t_2 = 0$,

$$x_1 = 1.25(-4.0 \text{ km} + 0) = -5.0 \text{ km};$$
$$t_1 = 1.25[0 + (0.60/3.0 \times 10^8 \text{ m/s})(-4.0 \times 10^3 \text{ m})]$$
$$= -10 \times 10^{-6} \text{ s} = -10 \ \mu\text{s}.$$

For the beacon ahead,

$$x_1 = 1.25(+4.0 \text{ km} + 0) = +5.0 \text{ km};$$
$$t_1 = 1.25[0 + (0.60/3 \times 10^8 \text{ m/s})(+4.0 \times 10^3 \text{ m})]$$
$$= +10 \times 10^{-6} \text{ s} = +10 \ \mu\text{s}. \qquad \bullet$$

Again the two sets of observers disagree. Although they can agree that $x_1 = x_2 = 0$ at $t_1 = t_2 = 0$, any events away from the origin, even at $t_2 = 0$, lead to disagreement. For instance, the captain says that he turned the beacon on ahead of him at $t_2 = 0$, but the asteroid observers say it didn't happen until 10 μs later. They say that the captain turns things on ahead of him (such as clocks) late, so it is no wonder that he measured the laser flash in Example 3.2 as having occurred before $t_2 = 0$.

By the way, placing $x_1 = \pm 5.0$ km and $t_1 = \pm 10$ μs back into the Lorentz transformation for x_2 gives

$$x_2 = 1.25[\pm 5.0 \times 10^3 \text{ m} - (1.8 \times 10^8 \text{ m/s})(\pm 10 \times 10^{-6} \text{ s})] = \pm 4.0 \text{ km}.$$

Similarly, $t_2 = 0$, so we return to the original distance and time. Space and time may not be absolute, but the Lorentz transformations are consistent and there is still order in the universe.

3.3
THE DOPPLER EFFECT FOR ELECTROMAGNETIC WAVES

The **Doppler effect** is the change in the measured frequency of a source, which occurs because of motion of the source and/or observer. As we mentioned in Chapter 2, mechanical waves travel in a medium, and their relative velocities past an observer depend on the motion of the observer with respect to that medium. However, electromagnetic waves evidently require no medium (no luminiferous ether). The speed of electromagnetic waves in a vacuum is always measured to be c by any observer. Therefore the Doppler effect for electromagnetic waves can be expected to differ from that for mechanical waves, which you probably studied in a previous physics course.

Suppose that there is a source of frequency ν_0 (and corresponding period T_0) at rest on the y_2 axis in S_2. As usual, that means that the source is moving at speed v in the $+x$ direction in S_1. The coordinates of the source in S_2 are $(0, y_2, 0)$. At time t_2 the source emits a crest of a sinusoidal wave, and at time t_2' it emits the next crest of that wave. Therefore $t_2' - t_2 = T_0$, or the time between successive crests in S_2. Since the source is at rest in S_2, this time can be called the rest period. To summarize, in S_2 the first crest is emitted at $(0, y_2, 0, t_2)$ and the second crest at $(0, y_2, 0, t_2')$.

If we are at rest at the origin of S_1, we would measure different x positions and times for the two crests. Since x_2 and x_2' both equal zero, Eq. (3.4) gives us $t_1 = \gamma t_2$ and $t_1' = \gamma t_2'$. With respect to our coordinate system, the crests were emitted at t_1 and t_1'. But what is the period of the wave in S_1? The period, T, is defined as the time measured between successive crests as they reach us. This is usually not the same as $t_1' - t_1$ because the crests will usually be emitted at different distances from us.

Those different distances will give the successive crests different transit times. The first crest will take a time r_1/c to reach us and the second crest a time r_1'/c. Therefore

$$T = t_1' - t_1 + r_1'/c - r_1/c = \gamma(t_2' - t_2) + (r_1' - r_1)/c = \gamma T_0 + (r_1' - r_1)/c.$$

We now need to determine $r_1' - r_1$.

Just how far does a source travel between the emission of successive crests? In S_1, the distance is $x_1' - x_1 = v(t_1' - t_1) = v\gamma T_0$. But cT_0 is the wavelength in S_2 and $v < c$, so this distance will be reasonably close to the wavelength of an electromagnetic wave at relativistic speeds. The distance will be much less than the wavelength at lower speeds. These wavelengths, no matter how much they are altered by the Doppler effect, will ordinarily be much less than the distance from the source to the observer. (See Fig. 3.4.)

Therefore the angles θ_1 and θ_1' shown in Fig. 3.4 will be virtually the same. The triangle then gives $r_1' - r_1 = (x_1' - x_1)\cos \theta_1$. But $x_1' - x_1 = v\gamma T_0$, giving $T = \gamma T_0 + (v\gamma T_0 \cos \theta_1)/c = \gamma T_0[1 + (v/c)\cos \theta_1]$.

Taking the reciprocals of the periods to obtain the frequencies, we get

$$\nu = \frac{\nu_0}{\gamma[1 + (v/c)\cos \theta_1]} = \nu_0 \frac{\sqrt{1 - v^2/c^2}}{1 + (v/c)\cos \theta_1}. \tag{3.5}$$

In this relation, ν_0 is the frequency of the source measured in a coordinate system at rest with respect to the source (zero relative speed, thus the subscript zero). The observers measure the frequency ν, the relative speed

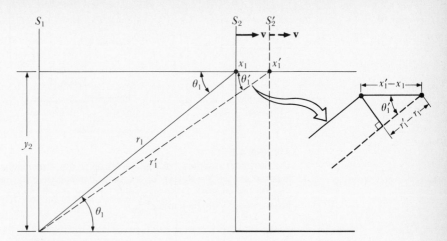

FIGURE 3.4

The observer is at rest with respect to inertial reference frame S_1 at its origin. The source is at rest with respect to inertial frame S_2 on its y axis. The crest of one sinusoidal wave is emitted at x_1 and the next crest is emitted at x_1'. The second crest travels $r_1' - r_1$ farther than the first crest to reach the observer.

v of the source with respect to them, and the angle θ_1 between the position vector to the source, $\mathbf{r}_1$, and the $+x$ axis.

EXAMPLE 3.4   Derive expressions for the frequency that you would measure if a source of light were moving

   a) directly away from you,   b) directly toward you, and
   c) perpendicular to a line from you.

Solution

a) Assume that you are at the origin of S_1 in Fig. 3.5. If the source is moving directly away from you, the source must be at the origin of S_2. The position vector $\mathbf{r}_1$ from you to the source is then parallel to the $+x$ axis, making $\theta_1 = 0$. With $\cos 0 = 1$, Eq. (3.5) becomes

$$v = v_0 \frac{\sqrt{1 - v^2/c^2}}{1 + v/c} = v_0 \frac{\sqrt{(1 + v/c)(1 - v/c)}}{1 + v/c} = v_0 \sqrt{\frac{1 - v/c}{1 + v/c}}.$$

Note that the numerator is smaller than the denominator, giving a lowered frequency when the source is moving away from you, just as for sound waves. (However, the equation is different for sound waves.)

b) If the source is moving directly toward you, it must be on your $-x$ axis, approaching you at your origin. (S_2 was assumed to move in the direction of the $+x$ axis.) The position vector is then parallel to the $-x$ axis,

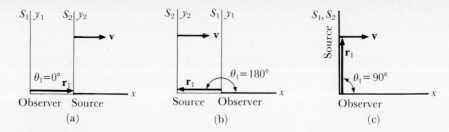

FIGURE 3.5

Our derivation assumes that you, the observer, are at rest at the origin of S_1 and the source is at rest at $x_2 = 0$ in S_2. (a) The source is at the origin of S_2, moving directly away from you. (b) The source is at the origin of S_2, moving directly toward you. (c) The source is instantaneously on the y axis of S_1.

making $\theta_1 = 180°$. With $\cos 180° = -1$, Eq. (3.5) reduces to

$$v = v_0 \frac{\sqrt{1 - v^2/c^2}}{1 - v/c} = v_0 \sqrt{\frac{1 + v/c}{1 - v/c}}.$$

For 180°, the numerator is larger than the denominator, giving the expected increase in frequency when the source is moving toward you.

c) Here $\theta_1 = 90°$, which gives the **transverse Doppler effect.** At this angle there is no relative motion toward or away from you, so the classical Doppler effect for mechanical waves would give $v = v_0$. However, the same is not true for electromagnetic waves. With $\theta_1 = 90°$, Eq. (3.5) becomes $v = v_0/\gamma = v_0 \sqrt{1 - v^2/c^2}$. The frequency decreases for this relativistic transverse Doppler effect.   ●

Quite often Eq. (3.5) will be applied when the relative speed of source and observer, v, is considerably less than the speed of light. Examples include radar-speed measurements, satellite-speed measurements, and measurement of the motion of many astronomical objects. For $v \ll c$, the binomial expansion $(1 + x)^n \approx 1 + nx + n(n - 1)x^2/2$ can be applied to Eq. (3.5); it can also be applied to the answers of Example 3.4.

You have seen in this chapter that the Lorentz and Galilean transformations can give quite different results (except when v/c approaches zero). Based on two simple postulates, time and space are no longer absolute quantities; both are relative and interrelated. Doppler shifts in the frequencies of electromagnetic waves occur not only for relative motion toward or away from an observer, but also for strictly transverse motion. How can we be confident that the Lorentz transformations are the correct ones? Because their often surprising predictions agree with experimental results.

QUESTIONS AND PROBLEMS

3.1 Why don't we need to consider whether the point source of the spherical pulse is fixed in S_1, in S_2, or in neither?

3.2 Apply the correspondence principle to show that the Galilean transformation will satisfy the two spherical wavefront equations in the limit.

3.3 What difficulty do you see with a transformation of the form $x_2 = A \sin B(x_1 - vt_1)$?

3.4 Although there are an infinite number of values of x_1 on the spherical wavefront at a given t_1, x_1 is still limited to a certain range of values. What is that range?

3.5 Solve for v in terms of γ and c.

3.6 At what speed will $\gamma = 5$?

3.7 What does γ approach as $\beta \to 1$? What does v approach as $\beta \to 1$?

3.8 What would happen to distances and times if S_2 could travel faster than c?

3.9 One observer is at rest and measures an event occurring at $(1.0, -3.0, 0)$ km and $t = 4.0\ \mu s$. What are the position and time coordinates measured by an observer moving by at $0.98c$ in the $+x$ direction? Compare to the Galilean results.

3.10 An observer moving past another observer at $0.80c$ measures an event to occur at $(-5.0, 3.0, 2.0)$ km and time of $1.0\ \mu s$. What are the position and time coordinates as measured by the fixed observer? Compare to the Galilean results.

3.11 We assume $t_1 = t_2$ at $x_1 = x_2 = 0$. At what other values of x_1 and x_2 will $t_1 = t_2$?

3.12 An observer is trying to measure the length of an object moving by her with $\gamma = 2.0$. She does this by simultaneously measuring the positions of the left and right ends of her meter stick. Suppose the left end is at the origin and the measurement is made at the observer's $t = 0$. What are the positions and times of the measurement as measured from the moving object?

3.13 At $x_1 = x_2 = 0$ and $t_1 = t_2 = 0$, a clock ticks aboard an extremely fast spaceship ($\gamma = 100$, so $v \approx c$). The captain hears it tick again 1.0 s later. Where and when do we measure the second tick to occur?

3.14 A spaceship zips by two satellites at a relative speed of $0.80c$. All clocks begin when the bridge of the spaceship passes the first satellite (as measured by the satellite observers). The first satellite receives reflected light pulses from the second satellite and finds that they take $6.0\ \mu s$ to travel out and return by first-satellite clocks. Where on the $+x$ axis does the spaceship captain measure the second satellite to be as he passes the first, and when does he measure that the second satellite's clock was turned on?

3.15 Algebraically derive the inverse Lorentz coordinate transformations from the Lorentz coordinate transformations.

3.16 For those who have worked with Problems 1.22 and 1.23, show that the Lorentz coordinate transformation matrix is

$$\begin{bmatrix} \gamma & 0 & 0 & i\beta\gamma \\ 0 & 1 & 0 & 0 \\ 0 & 0 & 1 & 0 \\ -i\beta\gamma & 0 & 0 & \gamma \end{bmatrix}.$$

3.17 Referring to Problems 1.22, 1.23, and 3.16,
a) Find the 4×4 matrix for the inverse Lorentz coordinate transformation.
b) Show that the product of the two 4×4 transformation matrixes (from Problem 3.16 and part (a) of this problem) is the identity matrix (1's on the diagonal and 0's elsewhere).

3.18 The electromagnetic potential has four components, A_x, A_y, A_z, and iV/c, where **A** is the magnetic vector potential and V is the electric potential. (The magnetic field equals **curl A** and the electric field equals $-\textbf{gradient}\ V - \partial \textbf{A}/\partial t$.) The components transform by the 4×4 matrix of Problem 3.16. Write the four equations to transform these four components from S_1 to S_2.

3.19 The electromagnetic field is represented by the following 4×4 matrix:

$$\begin{bmatrix} 0 & B_z & -B_y & -iE_x/c \\ -B_z & 0 & B_x & -iE_y/c \\ B_y & -B_x & 0 & -iE_z/c \\ iE_x/c & iE_y/c & iE_z/c & 0 \end{bmatrix}.$$

To transform this matrix from S_1 to S_2, you must multiply it by the inverse Lorentz coordinate transformation matrix and then multiply the Lorentz coordinate transformation matrix given in Problem 3.16 by that result. (The inverse Lorentz coordinate transformation matrix is simply the 4×4 matrix of Problem 3.16 with each i replaced by a $-i$.) Show that the x components of **E** and **B** are unchanged but that $B_{2z} = \gamma(B_{1z} - vE_{1y}/c^2)$ and $E_{2z} = \gamma(E_{1z} + vB_{1y})$.

3.20 A spaceship captain is brought up before a Galactic Empire court on the charge of running an interstellar red traffic light. He pleads innocent, claiming the light looked green to him. If "red" light is approximately 4.5×10^{14} Hz and "green" light 5.8×10^{14} Hz, how fast must he have been going?

3.21 The light from quasar Q0000-26 appears to have its frequencies decreased by a factor of 5.11. Assume that the quasar is moving directly away from us and that the frequency shift is due completely to the special relativistic Doppler effect (actually, it probably isn't). How fast is the quasar moving away from us?

3.22 Using mirrors, we can compare light emitted by moving atoms for the cases when the atoms are moving directly toward and directly away from us at the same speed. Find the average of these two Doppler-shifted frequencies. (This method was the basis of what is now called the Ives–Stilwell experiment.)

3.23 A spaceship has a speed that we measure to be $0.20c$. At one particular position, we measure no Doppler shift in the frequency of its radio transmissions. Specify this position.

3.24 In a Mössbauer-effect experiment, the change in frequency of a 3.5×10^{18} Hz electromagnetic wave can be detected if the source and absorber have a relative speed of 0.03 mm/s with respect to each other. Calculate the frequency shift $\nu - \nu_0$ for motion both toward and away from each other at this speed.

3.25 Prove that $\Delta\nu/\nu \approx -(v/c)\cos\theta_1 + (v^2/c^2)(\cos^2\theta_1 - 1/2)$, ignoring powers greater than v^2/c^2.

3.26 Relative speeds of objects moving directly toward one another are often measured with radar as follows: 1) Waves of a certain frequency, ν_0, are sent out from the source; 2) ν waves per time strike the object moving toward the source at a relative speed v; 3) the object reflects each wave as it hits, acting as a second source with $\nu_0' = \nu$; 4) the original source receives the reflected waves at frequency ν'; and 5) the beat frequency, $\nu' - \nu_0$, is measured and the relative speed obtained from it.

In terms of ν_0, v, and c, for $v \ll c$ (ignoring v^2/c^2 and higher power terms in the binomial expansion):   a) Give the expression for ν. b) Then find the expression for ν'.   c) Finally, find the expression for the beat frequency.   d) If the source frequency is 10.0 GHz and the beat frequency is 2.2 kHz, calculate the relative speed.   e) How do the answers change for motion directly away from one another?

3.27 a) Express $\cos\theta_1$ in terms of v, t_1, and y_1. b) For a source passing at a speed of $0.60c$, plot ν/ν_0 from $t_1 = -5y_1/c$ to $+5y_1/c$ to see how the frequency changes with time.   c) Find the limits as $t_1 \to -\infty$ and $+\infty$ and explain their meaning. d) Find ν as a function of ν_0 and t_1 for a satellite that passes 200 km overhead, moving at 7.8 km/s.

3.28 Will all sources having a velocity component toward an observer give a Doppler-shifted increase in the frequency?

3.29 Considering only the Doppler effect, if a distant source exhibits no Doppler shift in frequency, does that mean that its relative velocity must be zero?

CHAPTER

Relativistic Lengths, Times, and Velocities

In this chapter we will continue to apply the Lorentz coordinate transformations. We will derive three sets of equations that again give results quite different from the predictions of classical physics. We will find that lengths on an object that we measure in the direction of motion become increasingly smaller and that time intervals on the object become increasingly larger as the object moves at higher speeds with respect to us. Finally, we will derive and use the Lorentz velocity transformations to show, among other things, that "$c + c = c$" in the physics of special relativity.

4.1
LENGTH CONTRACTION

Suppose that we are at rest in S_1 and want to measure the length of an object moving past us, as shown in Fig. 4.1. We would first fix S_2 in the object. Then we could find the distances to the two ends of the object, x_1' and x_1. To do this correctly, we *must* measure x_1' and x_1 at the same time, or $t_1' = t_1$. (If a spaceship were moving past us and we measured the nose to be

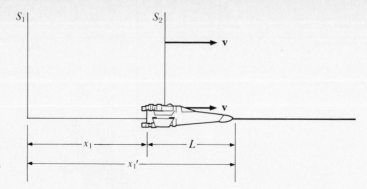

FIGURE 4.1

Measuring the length of a moving object.

at x_1' at one time, then waited to measure the position of the tail at x_1, the difference $x_1' - x_1$ would not be the measured length. $x_1' - x_1$ could, in fact, be zero or even negative!)

Applying the Lorentz coordinate transformations to our two distances, we obtain

$$x_2' = \gamma(x_1' - vt_1') \quad \text{and} \quad x_2 = \gamma(x_1 - vt_1).$$

Subtracting, we obtain $(x_2' - x_2) = \gamma(x_1' - x_1)$, remembering that $t_1' = t_1$ for a correct measurement. Note that $x_2' - x_2$ is the length as measured in S_2. Since the object is at rest with respect to S_2, let's call this length L_0, where the subscript "zero" refers to zero speed. Then we can represent $(x_1' - x_1)$ as the length L of the moving object. This gives us $L_0 = \gamma L$, or

$$L = \frac{L_0}{\gamma} = L_0 \sqrt{1 - v^2/c^2}. \tag{4.1}$$

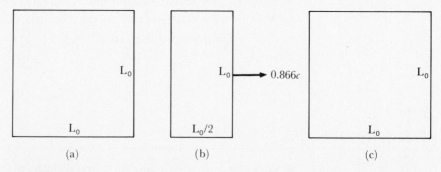

FIGURE 4.2

(a) A square at rest with respect to us. (b) A similar square as we measure it when it moves past us at 0.866c. (c) That similar square in (b) as measured by a person moving at the same velocity as the square. Incidentally, that person would measure the square at rest with respect to us to have the dimensions of (b).

For motion with $v < c$, $\gamma > 1$ or $\sqrt{1 - v^2/c^2} < 1$. Equation (4.1) tells us that L is always less than L_0. This decrease in the measured lengths of moving objects is called **length contraction.**

Recall that $y_2 = y_1$ and $z_2 = z_1$. Therefore *any lengths measured perpendicular to the direction of the motion will not be changed by the motion.* Length contraction occurs only in the direction of the motion of the object. (See Fig. 4.2.)

EXAMPLE 4.1   A 0.125-m³ cubical box is placed in the cargo hold of a spaceship, which then flies past us at $0.80c$. If some edges of the box are parallel to the motion of the spaceship, determine the box's dimensions as we'd measure them.

Solution A 0.125-m³ cube has sides of $(0.125 \text{ m}^3)^{1/3} = 0.50$ m. Also, $v = 0.80c$ gives $\sqrt{1 - v^2/c^2} = 0.60 = 1/\gamma$. The lengths of those edges perpendicular to the direction of motion would be unchanged. The length of those edges parallel to the motion would be $L = (0.50 \text{ m})(0.60) = 0.30$ m, (from Eq. (4.1)). (With the y and z dimensions unchanged and the x dimension contracted, the volume would be $(0.50 \text{ m})^2(0.30 \text{ m}) = 0.075 \text{ m}^3$.)   ●

What would a person in the cargo hold of the spaceship measure? Since the box is at rest with respect to that person, he or she would measure the box to remain cubical, 0.50 m on a side, 0.125 m³ in volume.

EXAMPLE 4.2   A spaceship captain sees us attempting to measure lengths in the direction of her motion with a measuring tape she measures to be 40 m long. Her spaceship is moving past us at $0.60c$. How long do we say the measuring tape is?

Solution First, $v = 0.60c$ gives $\sqrt{1 - v^2/c^2} = 0.80 = 1/\gamma$. The spaceship captain sees herself at rest, with us and the measuring tape moving past her at $0.60c$. Therefore $L = 40$ m. We want to find L_0 (since the tape is at rest with respect to us). From Eq. (4.1), $L_0 = 40 \text{ m}/0.80 = 50$ m.   ●

Note that Example 4.1 shows that we measure spaceship lengths contracted in the direction of motion, and Example 4.2 shows that the spaceship captain also measures our lengths contracted in the direction of motion. This must be so, because neither coordinate system is preferred. Also we see each other measuring with devices (meter sticks, measuring tapes) having decreased lengths in the direction of motion. Finally, one of us will see the other measuring x' and x at different times, not simultaneously, because one t depends on the other x.

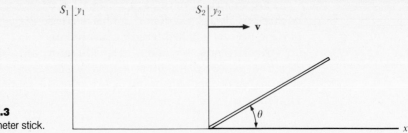

FIGURE 4.3
A leaning meter stick.

EXAMPLE 4.3   A crew member in a moving spaceship places a meter stick so that it makes an angle of 20.0° with the direction of motion. We measure the stick to make a 35.0° angle. (See Fig. 4.3.) How fast is the spaceship moving relative to us?

Solution The meter stick has length components in the x direction (parallel to the motion) and in the y direction (perpendicular to the motion), L_x and L_y. The perpendicular component will be unchanged. Therefore, the y component in the spaceship's frame of reference, $L_{0y} = (1.00 \text{ m})\sin 20.0° = 0.342$ m, also equals L_y in our frame of reference.

In the spaceship's frame of reference, $L_{0x} = (1.00 \text{ m})\cos 20.0° = 0.940$ m. In our frame of reference $L_x = L_y/\tan \theta_1 = 0.342$ m/$\tan 35.0° = 0.488$ m (using some trigonometry). Therefore Eq. (4.1) becomes $0.488 \text{ m} = (0.940 \text{ m})\sqrt{1 - v^2/c^2}$. This solves algebraically to $v = \sqrt{1 - (0.488/0.940)^2}\,c = 0.85c = 2.6 \times 10^8$ m/s.   ●

Length contraction is not an optical illusion, but is an actual, indirectly verified occurrence, necessary for the consistency of relativity. On the other hand, the visual appearance of a rapidly moving object would not just be that of an object shortened in the direction of motion. If our eyes could detect something moving past us at a speed close to the speed of light, the object would appear to be rotated. As the two-dimensional diagrams of Fig. 4.4 show, we can't see point A when the cube is at rest, but we can see point B. However, if the cube moves fast enough, it can move out of the way of the light from point A and at the same time run into the light from point B, so that we could see A but not B. Also, light from the farther points on the object will take a longer time to reach our eyes. Therefore visible farther points will appear less advanced in the direction of motion than visible nearer points. All this, plus the actual measured shortening of edges parallel to the velocity, would make the cube appear to have its trailing face rotated toward us. To summarize, the **visual appearance of a rapidly moving object** would be that of the object rotated, with off-axis distortions.

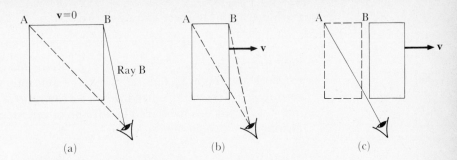

FIGURE 4.4
(a) A two-dimensional view of a cube at rest. Light from B can reach the eye but light from A cannot. (b) A two-dimensional view of the cube moving at a relativistic speed. Light starts from A and B along the dashed lines. (c) The motion of the cube has moved it out of the way of the light traveling from A to the observer. The same motion has blocked out the light from B.

4.2
TIME DILATION

The time at which an event occurs depends not only on the frame of reference chosen, but also on the position of the event. Therefore it seems reasonable that time intervals would be measured differently by observers moving at different speeds. These time intervals may be the period of a repetitive atomic process, the length of a biological process, the time between the ticks of a mechanical clock, or any other difference in time between two events.

Suppose that we are at rest in S_1 and a clock is moving by in S_2, as shown in Fig. 4.5. The clock is at rest in S_2, but has a speed v with respect to us. The clock ticks once at time t_2 and again at time t'_2. Both ticks occur at the same position in S_2: $x_2 = x'_2$. Applying the inverse Lorentz coordinate

FIGURE 4.5
(a) The clock is at rest with respect to inertial reference frame S_2 and ticks at t_2. (b) The clock, at the same position in S_2, ticks again at t'_2.

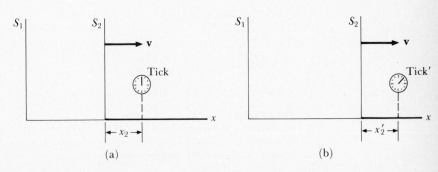

transformations to these two times, we obtain

$$t_1' = \gamma[t_2' + (\beta/c)x_2'] \quad \text{and} \quad t_1 = \gamma[t_2 + (\beta/c)x_2].$$

Subtracting gives us $t_1' - t_1 = \gamma(t_2' - t_2)$. (Remember that $x_2' = x_2$.) Since the clock is at rest in S_2, let's call $t_2' - t_2$ the time interval T_0. The subscript zero again refers to zero speed. Then we can represent $t_1' - t_1$ as the time interval T, which is measured in a frame of reference where the clock is moving at a speed v. This gives us

$$T = \gamma T_0 = \frac{T_0}{\sqrt{1 - v^2/c^2}}. \tag{4.2}$$

As you can see, T is always greater than T_0 for v greater than zero but less than c. This measured increase in time intervals of a moving object is called **time dilation.**

In the following examples and problems, we use the term *mean life-time* to mean *average lifetime*. We must use the term because any particular unstable particle in a specific frame of reference has a range of lifetimes. The *mean muon* then is a muon that happens to live exactly one mean muon lifetime.

EXAMPLE 4.4   When a charged pion is at rest, its mean lifetime is 2.6×10^{-8} s, or 26 ns. What is the mean lifetime of charged pions when they are moving at $0.98c$? (See Fig. 4.6.)

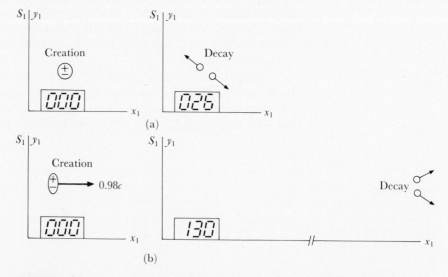

FIGURE 4.6
(a) The average charged pion at rest decays 26 ns after its creation. (b) The average charged pion moving at 0.98c with respect to a frame of reference decays 130 ns after its creation, as measured in that frame of reference.

Solution First, at $v = 0.98c$, $\gamma = 1/\sqrt{1 - v^2/c^2} = 5.0$. The rest time interval, T_0, is 26 ns and, using Eq. (4.2), this interval dilates to $(5.0)(26 \text{ ns}) = 130$ ns.

As the charged pions move at $0.98c$ past an observer, that observer would measure their mean lifetime to be five times as long as when they are at rest. However, any observer moving along with the charged pions would have no relative velocity with respect to them and would still measure their mean lifetime to be 26 ns. To this moving observer, the particles are still at rest.

EXAMPLE 4.5 Muons are created in a process near the top of a mountain 4630 m above sea level. The mean muon traveling at $v = 0.99c$ downward with $\gamma = 7.1$ will decay at sea level. What is the mean muon lifetime when it is at rest?

Solution In our frame of reference, the mean muon travels 4630 m at $0.99c = 2.97 \times 10^8$ m/s in a time of about $4630 \text{ m}/(2.97 \times 10^8 \text{ m/s}) = 1.56 \times 10^{-5}$ s $= 15.6$ μs. Since this muon is moving in our frame of reference, 15.6 μs $= T$. Then with $\gamma = 7.1$ in Eq. (4.2), we obtain $T_0 = T/\gamma = 15.6 \, \mu\text{s}/7.1 = 2.2 \, \mu\text{s}$. That is, $T_0 = 2.2 \, \mu$s is the mean lifetime when the muon is at rest. ●

The following question might now occur to you: "How can a muon, which measures its own lifetime to be 2.2 μs, travel 4630 m at 2.97×10^8 m/s?" After all, $(2.97 \times 10^8 \text{ m/s})(2.2 \times 10^{-6} \text{ s}) = 650$ m, not 4630 m. But relativity must be consistent. You must remember that not only does time dilate, but also lengths contract. (See Fig. 4.7.) In the frame of reference of the muon, the mountain is moving upward at $v = 0.99c$, giving $\gamma = 7.1$. From Eq. (4.1), the muon perceives the mountain's height as $L = L_0/\gamma = 4630 \text{ m}/7.1 = 650$ m above sea level. This 650 m perceived by the muon is the same 650 m calculated by the muon using speed $\times$ time, so you can see the necessary internal consistency of the concepts of length contraction and time dilation.

Lacking starships traveling at $0.80c$, experimenters have used subatomic particles such as muons to directly validate the concept of time dilation and indirectly validate the concept of length contraction for near light-speed velocities. However, very precise experiments have shown time dilation even for speeds much less than c. For example, as the temperature is raised in a solid, the atoms vibrate about their equilibrium positions with a higher average v^2/c^2. This results in a time dilation that changes the period and frequency of the electromagnetic radiation emitted from or absorbed by the atoms. This frequency change is measured for gamma rays through the Mössbauer effect. Time dilation has even been measured by atomic clocks flown around the world in commercial jetliners!

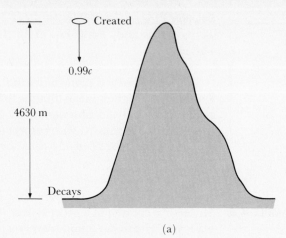

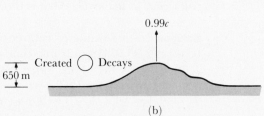

FIGURE 4.7
(a) The mean muon travels
4630 m down at 0.99c in
15.6 μs in our frame of
reference. (b) In the
muon's frame of reference,
the length-contracted
mountain moves 650 m up
at 0.99c in 2.2 μs.

EXAMPLE 4.6   Jack and Jill are 25-year-old twins. Jack must stay on earth with a head
injury, but astronaut Jill travels at 0.98c to a star 24.5 light years away and
returns immediately. (See Fig. 4.8.) Ignoring the end-point acceleration
times, find the twins' ages when she returns. (One light year = 1 c · yr, the
distance light travels in one year.)

Solution From our earth-bound frame of reference, Jill travels a total of
49 light years (out and back) at 0.98c in a time interval of 50 years =
49 c · yrs/0.98c. Therefore 50 years of earth time have passed, so Jack is
(25 + 50) years = 75 years old (if he has lived that long). This 50 years is
dilated time, however. Jill's time interval is T_0. Since $\gamma = 5.0$ for $v = 0.98c$,
$T_0 = T/\gamma = 50$ years/5.0 = 10 years. Jill therefore has only aged 10 years,
so she is (25 + 10) years = 35 years old. She is 40 years younger than her
twin!   ●

Questions that might come to you in reading about the twins is,
"Since the choice of frame of reference is relative, why don't we place Jill
in S_1? She then sees the earth move away and return, and therefore it is Jack
who has traveled out and back at 0.98c. He should be the one who is 40

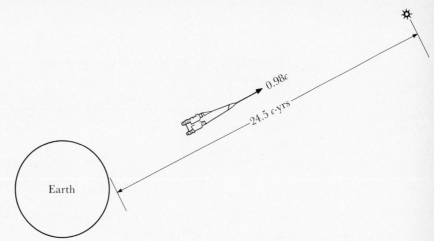

FIGURE 4.8
Jill travels a round-trip
distance of 49 $c \cdot$ yrs at
0.98c. Jack ages 50 years
during Jill's trip, while Jill
only ages 10 years.

years younger. Since they both can't be 40 years younger, does this prove
relativity wrong?''

 This apparent contradiction is called **the twin paradox.** Recall, however, that we are dealing with the special theory of relativity, which refers
to inertial reference frames. In the twin paradox, the earth is an approximately inertial reference frame, but Jill's spaceship isn't. It must be accelerated to start the trip, be decelerated and accelerated to stop and start
again when turning around at the star, and then be decelerated when it
returns to earth. The choice of frames of reference is relative in the special
theory of relativity *only* if the frames of reference are all inertial. Therefore an attempt to use the special theory in a noninertial frame of reference
invites incorrect results. Jack does age more rapidly than Jill. Experiments
(such as the clocks in jetliners) confirm this prediction.

4.3
THE LORENTZ VELOCITY TRANSFORMATIONS

It wasn't difficult to obtain the Galilean velocity transformations from the
coordinate transformations. We just differentiated both sides of the equations with respect to time. In the relativistic case, however, $dt_2 \neq dt_1$, so the
derivation takes a bit more work. Differentials of Eq. (3.1), the Lorentz
coordinate transformations, are

$$dx_2 = \gamma(dx_1 - v\, dt_1), \qquad dy_2 = dy_1, \qquad dz_2 = dz_1,$$

and

$$dt_2 = \gamma[dt_1 - (\beta/c)dx_1].$$

Dividing dx_2, dy_2, and dz_2 by dt_2 to obtain the velocity components v_{2x}, v_{2y}, and v_{2z} gives

$$v_{2x} = \frac{dx_1 - v\, dt_1}{dt_1 - (\beta/c)dx_1} = \frac{dx_1/dt_1 - v}{1 - (\beta/c)dx_1/dt_1},$$

and

$$v_{2y} = \frac{dy_1}{\gamma[dt_1 - (\beta/c)dx_1]} = \frac{dy_1/dt_1\,\sqrt{1 - v^2/c^2}}{1 - (\beta/c)dx_1/dt_1},$$

with similar results for v_{2z}. Realizing that $dx_1/dt_1 = v_{1x}$, $dy_1/dt_1 = v_{1y}$, and $dz_1/dt_1 = v_{1z}$, we finally obtain

$$v_{2x} = \frac{v_{1x} - v}{1 - (\beta/c)v_{1x}} = \frac{v_{1x} - v}{1 - v \cdot v_{1x}/c^2};$$

$$v_{2y} = \frac{v_{1y}}{\gamma[1 - (\beta/c)v_{1x}]} = \frac{v_{1y}\,\sqrt{1 - v^2/c^2}}{1 - v \cdot v_{1x}/c^2}; \qquad (4.3)$$

$$v_{2z} = \frac{v_{1z}}{\gamma[1 - (\beta/c)v_{1x}]} = \frac{v_{1z}\,\sqrt{1 - v^2/c^2}}{1 - v \cdot v_{1x}/c^2}.$$

The expressions in Eq. (4.3) are the **Lorentz velocity transformations.**

As usual, the easiest way to obtain the **inverse Lorentz velocity transformations** is to interchange the numbers in the subscripts and replace all v by $-v$. This gives

$$v_{1x} = \frac{v_{2x} + v}{1 + (\beta/c)v_{2x}} = \frac{v_{2x} + v}{1 + v \cdot v_{2x}/c^2};$$

$$v_{1y} = \frac{v_{2y}}{\gamma[1 + (\beta/c)v_{2x}]} = \frac{v_{2y}\,\sqrt{1 - v^2/c^2}}{1 + v \cdot v_{2x}/c^2}; \qquad (4.4)$$

$$v_{1z} = \frac{v_{2z}}{\gamma[1 + (\beta/c)v_{2x}]} = \frac{v_{2z}\,\sqrt{1 - v^2/c^2}}{1 + v \cdot v_{2x}/c^2}.$$

Remember that $y_2 = y_1$ and $z_2 = z_1$. Although the y and z positions don't depend on the frame of reference, the y and z velocity components do. Velocity is the time rate of change of the displacement, and time depends both on the frame of reference and the x position within that frame. Therefore any change in the x position will affect the time. That is why the y and z velocity components in one frame of reference depend on the x component of the velocity in the other frame of reference.

EXAMPLE 4.7 From a spaceship traveling past us at speed v, a laser beam shoots out at an angle θ_2 to the direction of motion of the ship. (The laser beam, being a beam of light, has speed c with respect to the ship.) (See Fig. 4.9.)

a) What speed do we measure for the laser beam?   b) What angle do we measure?

Solution

a) This answer is obvious from the second postulate of relativity. All observers measure the speed of the laser beam as c, the speed of light. But let's show that Eq. (4.4) is consistent with the second postulate of relativity by working through the problem. Both c and θ_2 are measured in S_2; we're in S_1. Let's choose the y and z axes so that $v_{2x} = c \cos \theta_2$, $v_{2y} = c \sin \theta_2$, and $v_{2z} = 0$. Then, from Eq. (4.4),

$$v_{1x} = \frac{c \cos \theta_2 + v}{1 + vc \cos \theta_2/c^2} = c\, \frac{\cos \theta_2 + v/c}{1 + v \cos \theta_2/c};$$

$$v_{1y} = \frac{c \sin \theta_2 \sqrt{1 - v^2/c^2}}{1 + vc \cos \theta_2/c^2} = c\, \frac{\sin \theta_2 \sqrt{1 - v^2/c^2}}{1 + v \cos \theta_2/c};$$

and

$$v_{1z} = 0.$$

Then we use $v_1 = \sqrt{v_{1x}^2 + v_{1y}^2}$ (since $v_{1z} = 0$), along with $\sin^2 \theta_2 + \cos^2 \theta_2 = 1$ and some algebra to obtain $v_1 = c$.

b) $\tan \theta_1 = v_{1y}/v_{1x}$. We use the velocity components derived in part (a) to find

$$\theta_1 = \tan^{-1} \left(\frac{\sin \theta_2 \sqrt{1 - v^2/c^2}}{\cos \theta_2 + v/c} \right).$$

●

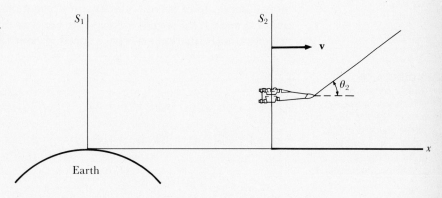

FIGURE 4.9
A spaceship and laser beam.

EXAMPLE 4.8   Two subatomic particles moving at $0.99c$ and $0.98c$ in a laboratory collide head-on. What was their relative velocity?

Solution The $0.99c$ and $0.98c$ speeds were measured with respect to the laboratory, so let's fix S_1 in the laboratory. Then fix S_2 in one particle so that S_2 is moving in the $+x$ direction (with that particle) at $v = 0.99c$. (See Fig. 4.10.) We want v_{2x}, the x component of the velocity of the other particle as measured in S_2. In S_1 the other particle will be moving at $-0.98c$, so $v_{1x} = -0.98c$. Then Eq. (4.3) gives

$$v_{2x} = \frac{-0.98c - 0.99c}{1 - (0.99c)(-0.98c)/c^2} = \frac{-1.97c}{1.9702} = -0.9999c. \qquad \bullet$$

The four 9's in the solution to Example 4.8 help show us that we can't use the relativistic addition of velocities to obtain a velocity having a magnitude greater than the speed of light. Even if the head-on collision had been between light beams, the relative speed of one light beam with respect to the other would have been the absolute value of

$$\frac{(-c - c)}{[1 - (c)(-c)/c^2]},$$

or $2c/(1 + 1) = c$. In classical physics, $c + c = 2c$, but in the physics of special relativity, "$c + c = c$"!

You have seen in this chapter that when an object moves at great speed with respect to us, we can expect to measure things that Galileo would never have expected to happen. On the object, length components in the direction of the motion shrink, but length components perpendicular to that direction don't change. If we could actually see an object moving near the speed of light, it would mainly appear to be rotated. We would measure time intervals on the object to increase, so all its processes would appear to slow down. As a result, rapidly moving unstable particles live longer than if they were at rest in our frame of reference. And finally, velocities do not add in the simpler manner of classical physics. Instead, the more complicated Lorentz velocity transformations must be used for relativistic speeds. If the speeds involved are less than or equal to c, no Lorentz velocity transformation can give a speed greater than c.

FIGURE 4.10
(a) The collision as measured in the laboratory system. (b) S_1 is fixed in the laboratory and S_2 is fixed in particle A. Then v_{2x} is the velocity of particle B with respect to particle A.

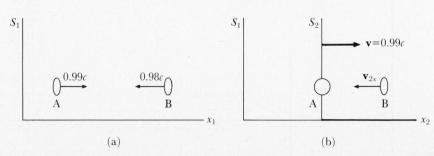

QUESTIONS AND PROBLEMS

4.1 If V is the volume of a container, prove that $V = V_0/\gamma = V_0\sqrt{1 - v^2/c^2}$.

4.2 If A is an area, under what conditions will
a) $A = A_0$?
b) $A = A_0/\gamma = A_0\sqrt{1 - v^2/c^2}$?

4.3 A meter stick is aligned along the direction of motion. How long do we measure it to be if it moves past us at $0.98c$?

4.4 A spaceship crew member is smoking a cigar, which we measure to be 4.0 cm long (in the direction of motion) as he moves past us at $0.80c$. How long does he measure his cigar to be?

4.5 How fast must a meter stick move past us to become a yardstick by our measurement?

4.6 Consider the meter stick of Example 4.3. Find general relations among θ_2, θ_1, and γ. Also find the measured length in terms of θ_2 and γ.

4.7 We measure x_1' and x_1 at $t_1' = t_1$ (simultaneously). Show that a moving observer won't agree that we are measuring both positions simultaneously.

4.8 A spaceship moves past us at $0.80c$, and we measure it to be 300 m long. How long would we measure it to be if we were moving along with it?

4.9 In the frame of reference of a certain subatomic particle moving rapidly past us, its lifetime is 4.0 μs, during which time it would measure us to move 1200 m with $\gamma = 5$. How far do we measure it to move?

4.10 We measure the distance between the start of an atomic process and its finish to be 2.0 cm when the atom moves past us at $v = 0.80c$. What is the distance as measured in the atom's frame of reference?

4.11 A subatomic particle is created at the top of a 2000-m mountain, moves at $0.98c$, and decays at the bottom (sea level). a) How tall is the mountain in the particle's frame of reference? b) What is the particle's lifetime as measured in both frames of reference?

4.12 Refer to Fig. 4.11. a) How fast must the cube be moving for us to see the light from point A? b) Will we then see the light from point B?

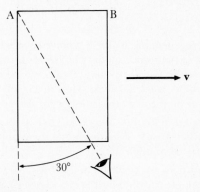

FIGURE 4.11

4.13 Describe the visual appearance of a rapidly moving flying saucer, seen edge-on.

4.14 By what percent does T differ from T_0 at the speed of a fast subsonic plane (3×10^2 m/s)?

4.15 A spaceship crew member takes 30 min to smoke a cigar (our measurement) as his spaceship moves past us at $0.80c$. How much time does he measure?

4.16 A pendulum on a spaceship moving past us at $0.60c$ takes 2.0 s (ship time) to make one complete oscillation. What do we measure its period to be?

4.17 The mean lifetime of charged antikaons at rest is 12 ns. How fast are they moving if we measure their mean lifetime to be 0.12 μs?

4.18 The mean lifetime of an Ω^- particle as it moves past us at a speed such that $\gamma = 9.0$ is 7.4×10^{-10} s as we measure it.
a) How far does a mean Ω^- particle travel in our frame of reference?
b) What is the mean lifetime of an Ω^- particle at rest?
c) How far does one of these particles moving past us measure itself to move in one mean lifetime? Calculate this answer two ways (using length contraction and time dilation).

4.19 A free neutron at rest has a mean lifetime of 900 s.
a) How fast would free neutrons have to move past us for us to measure their mean lifetimes to equal 1800 s?

b) How far would we measure them to travel in the 1800 s?

c) What is this distance measured in the neutron's frame of reference?

4.20 A crew member makes 2.0 universal credits per click of time while on duty on her spaceship. The spaceship moves past us at $0.60c$. We see the crew member work 5.0 clicks. How much will the paymaster on the spaceship pay her for that work?

4.21 The mean lifetime of a radioactive isotope is 2.7 ms when at rest. What is it when the isotope is given a speed of $0.50c$?

4.22 How does astronaut Jill in the "twin paradox" example explain how she can travel 49 light years at $0.98c$ in only 10 years?

4.23 Neutral pions are traveling at $0.98c$ with respect to us. They have a mean lifetime of 8.3×10^{-17} s when at rest. What do we measure their mean lifetime to be?

4.24 Discuss length contraction and time dilation for a) $v = c$ and b) $v > c$.

4.25 You are approaching a source of light. Use the Lorentz velocity transformation to show that the speed of light relative to you is independent of your speed of approach.

4.26 Spaceship A moves past us at $0.60c$; then spaceship B follows in the same direction at $0.80c$. a) What do we measure as their relative speed of approach? b) What do they measure as their relative speed of approach?

4.27 Spaceship A moves past us at $0.60c$, then fires a missile at a velocity it measures to be $0.80c$ at a $45°$ angle to the rear. What velocity components do we measure?

4.28 Spaceship A approaches us at $0.80c$ from the east and spaceship B approaches us at $0.60c$ from the west. What do they measure as their relative speed of approach? What do we measure as their relative speed of approach?

4.29 Particle 1 moving at $0.82c$ with respect to the lab collides head-on with particle 2 at a relative speed of $0.99c$ with respect to particle 2. What is the velocity of particle 2 with respect to the lab?

4.30 Two earth satellites, each moving at 9.00×10^3 m/s approach one another head-on. What do they measure as their relative speed of approach?

4.31 As a spaceship moves past us at $0.80c$, the ship fires a laser beam, which we measure to be perpendicular to the ship's path. For the laser beam: a) What velocity components does the ship measure? b) What angle does the ship measure?

4.32 Spaceship A has a clock whose time between ticks is exactly 1 s. Spaceship A approaches us at $0.80c$ from the north while spaceship B approaches us at $0.60c$ from the west. What is the time between ticks, as measured by the observers in spaceship B?

4.33 Spaceship A measures its length to be 200 m. Spaceship A approaches us at $0.90c$ from the west, while spaceship B approaches us at $0.95c$ from the east. What is the length of spaceship A as measured a) by us? b) by the crew of spaceship B?

4.34 Spaceship A approaches us at $0.90c$ from the west, while spaceship B moves away from us at $0.60c$ to the east. Spaceship B's crew measures a coffee break on spaceship A to take 15 minutes. How long do we measure the coffee break to last?

4.35 Spaceship A approaches us at $0.80c$ from the east, while spaceship B leaves us at $0.90c$ to the west. Spaceship A broadcasts signals at exactly 10 MHz. a) At what frequency do we receive spaceship A's signals? b) At what frequency does spaceship B receive spaceship A's signals?

4.36 a) Relate the transverse Doppler effect to time dilation. b) If $T = \gamma T_0$ and $v = 1/T$, why isn't $v = v_0/\gamma$ true at all values of θ_1?

4.37 "There are no Lorentz acceleration transformations because Lorentz transformations are for inertial reference frames, which have no acceleration." Why is this statement false?

4.38 Find the Lorentz acceleration transformations.

CHAPTER **5**

Relativistic Dynamics

Have you ever heard that E equals mc^2? Who hasn't? Well, in this chapter, we show the origin of this famous relation. First, we apply the Lorentz velocity transformations to show that the mass we measure for a moving object increases in the same way that time intervals on the object increase. This increase in mass then leads to modification of the classical $\mathbf{F} = m\mathbf{a}$ equation. Relating force through a distance to work and energy brings us to $E = mc^2$ and the realization that c is the upper limit on speed for objects which have a mass when they are at rest. Finally, we derive relativistic relations between energies and momentum and are introduced to some useful non–SI units.

5.1
RELATIVISTIC MASS AND MOMENTUM

In classical mechanics, the **linear momentum** of a particle of mass m moving at a velocity $\mathbf{v}$ is defined to be

$$\mathbf{p} = m\mathbf{v}. \tag{5.1}$$

FIGURE 5.1
(a) The collision as observed in inertial reference frame S_1 (not to scale). (b) The collision as observed in inertial reference frame S_2 (not to scale).

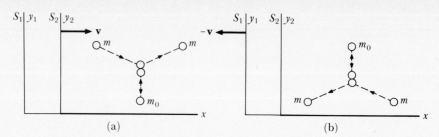

(a)   (b)

The total linear momentum of a system is conserved if the net external force on the system is zero. These concepts are also valid in relativistic mechanics, but we will find that the mass depends on the relative speed of the particle and observer.

Suppose that we have the following situation: Observers in S_1 and S_2 are given identical balls that will make a perfectly elastic collision with each other. When the observers move past one another, each will throw a ball with $v_x = 0$ and v_y's that each measures to have the same magnitude (say, 30 m/s) but opposite directions. After a completely symmetric collision, the balls will rebound with the opposite velocity, $-v_y$. Both S_1 and S_2 will have a relativistic relative speed (v close to c) but v_y will be classical ($v_y \ll c$). In Fig. 5.1(a), the observer in S_1 throws a ball that is practically at rest in S_1, so we will call its mass m_0. The ball thrown by the observer in S_2 is measured to be moving at a relativistic speed in S_1, so we will call its mass m. The observer in S_1 throws the ball upward at speed v_y. The y component of the other ball's velocity as measured in S_2 is $-v_y$. This velocity must be transformed to S_1 using the inverse Lorentz transformation, Eq. (4.4), and becomes $\sqrt{1 - v^2/c^2}(-v_y)$ (recall that $v_{2x} = 0$). In the symmetric collision, the momentums are equal and opposite, so $m_0 v_y = -m\sqrt{1 - v^2/c^2}(-v_y)$. This gives

$$m = \frac{m_0}{\sqrt{1 - v^2/c^2}} = \gamma m_0. \tag{5.2}$$

Equation (5.2) is the **relativistic mass transformation.** It tells us that our measurement of the mass of an object gives us a mass that increases as our relative speed v increases. We see that masses increase in the same way that time intervals increase. (See Fig. 5.2.)

Since $\mathbf{p} = m\mathbf{v}$ and m increases with v, linear momentum is no longer directly proportional to velocity. Increasing the velocity from $0.98c$ to $0.99c$ only increases v by 1 percent, but $1/\sqrt{1 - v^2/c^2} = \gamma$ increases from 5.0 to 7.1. Therefore the momentum increases by more than 40 percent, due mostly to the relativistic increase in mass.

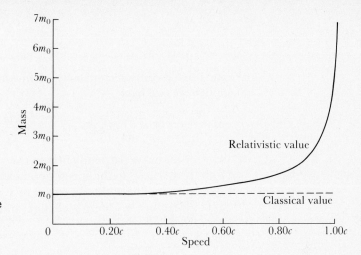

FIGURE 5.2
The variation in the measured mass with relative speed. The variation in measured time interval would have the same curve, with all *m* replaced by *T*.

EXAMPLE 5.1   If a spaceship were to move past the earth at $0.60c$, what mass would the crew measure for the standard kilogram (which is at rest on earth)?

Solution When at rest, the standard kilogram has a mass of exactly 1 kg by definition. The relative speed of the standard kilogram with respect to the spaceship is $0.60c$, which gives $1/\sqrt{1 - v^2/c^2} = \gamma = 1.25$. Therefore $m = \gamma m_0 = (1.25)(1 \text{ kg}) = 1.25$ kg.   ●

The mass m_0 is called the **rest mass.** It is the smallest mass an object has because as the object speeds up, the mass increases.

EXAMPLE 5.2   Calculate the rest mass of a negatively charged particle having a measured momentum of 1.92×10^{-21} kg · m/s at a speed of $0.99c$.

Solution First, $p = mv$ gives

$$m = \frac{p}{v} = \frac{1.92 \times 10^{-21} \text{ kg} \cdot \text{m/s}}{0.99(3.00 \times 10^8 \text{ m/s})} = 6.46 \times 10^{-30} \text{ kg.}$$

Then to obtain m_0, the rest mass, $m = \gamma m_0$ gives

$$m_0 = \frac{m}{\gamma} = m \sqrt{1 - v^2/c^2}$$
$$= (6.46 \times 10^{-30} \text{ kg}) \sqrt{1 - (0.99)^2} = 9.1 \times 10^{-31} \text{ kg.}$$

With this rest mass and a negative charge, evidently the particle is an electron.   ●

5.2
FORCE

We can use Newton's second law to define force by the relation $\mathbf{F} = d\mathbf{p}/dt$. Since $\mathbf{p} = m\mathbf{v}$, we have

$$\mathbf{F} = \frac{d}{dt}(m\mathbf{v}) = m\frac{d\mathbf{v}}{dt} + \mathbf{v}\frac{dm}{dt} = m\mathbf{a} + \mathbf{v}\frac{dm}{dt}.$$

If the force is always perpendicular to the velocity, the force can't do any work on the particle, so the speed won't change. This happens in uniform circular motion. There the direction of $\mathbf{v}$ changes, but the magnitude v doesn't. Therefore m doesn't change and $dm/dt = 0$. Substituting for $m = \gamma m_0$, we have

$$\mathbf{F} = \gamma m_0 \mathbf{a} = \frac{m_0}{\sqrt{1 - v^2/c^2}}\mathbf{a} \qquad (\text{if } \mathbf{F} \perp \mathbf{v}). \tag{5.3}$$

EXAMPLE 5.3   Find the radius of the orbit of a charged particle moving perpendicular to a uniform magnetic field. (From previous physics courses, you may recall that the magnitude of the force is qvB and when a particle moves in a circle the magnitude of the centripetal acceleration is $a = v^2/r$.)

Solution Substituting the magnitudes of the force and centripetal acceleration into $\mathbf{F} = \gamma m_0 \mathbf{a}$, we obtain $qvB = \gamma m_0 v^2/r$, or $r = \gamma m_0 v/qB = mv/qB = p/qB$. (See Fig. 5.3.)   ●

The solution to Example 5.3, mv/qB, is the same as in classical physics, except that as the speed increases, the mass increases. This relation was verified experimentally by 1909.

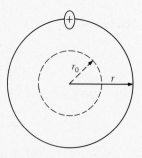

FIGURE 5.3
A charged particle moving rapidly and perpendicular to a magnetic field moves in a radius γ times larger than the classical m_0v/qB radius, explaining the huge size of many of today's high-energy particle accelerators.

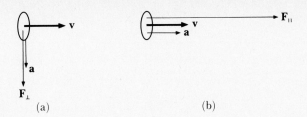

FIGURE 5.4
For a particle of rest mass m_0 moving at a velocity $\mathbf{v}$, a force of smaller magnitude is needed to give it an acceleration (a) perpendicular to $\mathbf{v}$ than to give it an acceleration of the same magnitude (b) parallel to $\mathbf{v}$.

But if the force is parallel to the velocity, the particle speed and mass won't be constant. Then

$$\frac{dm}{dt} = \frac{d}{dt}\frac{m_0}{(1 - v^2/c^2)^{1/2}} = \frac{m_0}{(1 - v^2/c^2)^{3/2}}\frac{v}{c^2}\frac{dv}{dt}.$$

Therefore

$$\mathbf{F} = \frac{m_0}{(1 - v^2/c^2)^{1/2}}\mathbf{a} + \frac{m_0}{(1 - v^2/c^2)^{3/2}}\frac{v^2}{c^2}\mathbf{a},$$

which reduces to

$$\mathbf{F} = \frac{m_0}{(1 - v^2/c^2)^{3/2}}\mathbf{a} = \gamma^3 m_0\mathbf{a} \qquad \text{(if } \mathbf{F} \parallel \mathbf{v}). \tag{5.4}$$

(See Fig. 5.4.)

EXAMPLE 5.4 A proton has a rest mass of 1.67×10^{-27} kg. Find the magnitude of the force necessary to give it an acceleration of 1.0×10^{15} m/s^2 in the direction of motion

    a) when $v = 0.90c$ and   b) when $v = 0.99c$.

Solution

    a) For $v = 0.90c$, $1/\sqrt{1 - v^2/c^2} = \gamma = 2.3$, so $F = \gamma^3 m_0 a$ becomes

$$F = (2.3)^3(1.67 \times 10^{-27} \text{ kg})(1.0 \times 10^{15} \text{ m/s}^2) = 2.0 \times 10^{-11} \text{ N}.$$

    b) For $v = 0.99c$, $1/\sqrt{1 - v^2/c^2} = \gamma = 7.1$, so

$$F = (7.1)^3(1.67 \times 10^{-27} \text{ kg})(1.0 \times 10^{15} \text{ m/s}^2) = 6.0 \times 10^{-10} \text{ N},$$

or 30 times the $0.90c$ force!

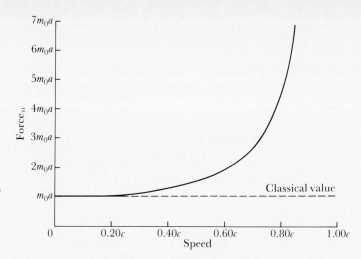

FIGURE 5.5
The force needed to accelerate a particle in its direction of motion approaches infinity as its speed approaches the speed of light.

As you can see, from the solution to Example 5.4 and from Fig. 5.5, F increases rapidly as v gets close to c. In fact, to approach c would require a force approaching infinity accelerating a mass approaching infinity (for particles with a nonzero rest mass).

5.3
ENERGY

Work may be done on a body to increase its kinetic energy, KE. In classical physics, $KE = \frac{1}{2}mv^2$. In relativistic physics, that expression is not generally correct, even if m is the relativistic mass. To obtain the correct expression let's start an object from rest with a net external force F in the $+x$ direction. Then the work done by F will be stored in the form of **kinetic energy.** That is,

$$KE = \int F\, dx = \int m_0(1 - v^2/c^2)^{-3/2}\, \frac{dv}{dt}\, dx,$$

with F substituted from Eq. (5.4) and $a = dv/dt$. Since $(dv/dt)dx = (dx/dt)dv = v\, dv$ for this motion in the $+x$ direction, we obtain

$$KE = m_0 \int_0^v (1 - v^2/c^2)^{-3/2}\, v\, dv,$$

which integrates to

$$KE = m_0 c^2 (1 - v^2/c^2)^{-1/2} - m_0 c^2 = \gamma m_0 c^2 - m_0 c^2.$$

With $\gamma m_0 = m$, we can write the expression for KE most simply as

$$KE = (m - m_0)c^2, \tag{5.5}$$

or even as $KE = \Delta mc^2$, where $\Delta m = m - m_0$ is the relativistic mass increase.

EXAMPLE 5.5 In a color TV picture tube, the electrons have a kinetic energy of 25 keV = 4.0×10^{-15} J. Calculate their

a) mass increase and   b) speed.

Solution

a) From $KE = (m - m_0)c^2$, the mass increase is

$$(m - m_0) = \frac{KE}{c^2} = \frac{4.0 \times 10^{-15} \text{ J}}{(3.00 \times 10^8 \text{ m/s})^2} = 4.4 \times 10^{-32} \text{ kg}.$$

This is a 4.8% increase in the mass and must be considered in calculating the trajectories of the electrons within the tube.

b) Since $m - m_0 = \gamma m_0 - m_0 = (\gamma - 1)m_0$, we have

$$\gamma = \frac{m - m_0}{m_0} + 1 = \frac{4.4 \times 10^{-32} \text{ kg}}{9.1 \times 10^{-31} \text{ kg}} + 1 = 1.048 = 1/\sqrt{1 - v^2/c^2},$$

which then gives

$$v = c\sqrt{1 - (1/1.048)^2} = 0.30c = 9.0 \times 10^7 \text{ m/s}. \qquad \bullet$$

The correspondence principle says that

$$KE = m_0 c^2 (1 - v^2/c^2)^{-1/2} - m_0 c^2$$

should reduce to $\frac{1}{2}m_0 v^2$ for $v \ll c$. If $v \ll c$, $v/c \ll 1$. Therefore we can use just the first two terms in the binomial expansion $(1 + x)^n \approx 1 + nx$ to evaluate $(1 - v^2/c^2)^{-1/2}$ as

$$1 + (-1/2)(-v^2/c^2) = 1 + \frac{1}{2}(v^2/c^2).$$

Then the relativistic expression for the kinetic energy becomes

$$KE = m_0 c^2 [1 + \frac{1}{2}(v^2/c^2)] - m_0 c^2 = \frac{1}{2}m_0 v^2,$$

which is the classical expression when we neglect the very small remaining terms of the binomial expansion.

When we write $KE = \Delta mc^2$, we see that an increase in the kinetic energy of a body from 0 to KE results in an increase in the mass of the body. The work could have been used instead to increase the potential energy — for example, to compress a spring. In this case also the mass would have increased by an amount equal to the work done divided by c^2, or the change in potential energy divided by c^2.

When an object is at rest it can be assigned a **rest energy** E_0. When an object is in motion, it will have a **total energy** E, so that

$$KE = E - E_0. \tag{5.6}$$

Combining Eqs. (5.5) and (5.6), we obtain $E - E_0 = mc^2 - m_0 c^2$, so we can say that

$$E = mc^2 \tag{5.7}$$

and

$$E_0 = m_0 c^2. \tag{5.8}$$

EXAMPLE 5.6   In an accelerator, protons are given a kinetic energy of 500 MeV. What is their

a) total energy?   b) mass?

Solution

a) The total energy of the protons is E; $KE = E - E_0$ solves to $E = E_0 + KE$ (that is, their total energy is their rest energy plus their kinetic energy). The expression $E_0 = m_0 c^2$ tells us that

$$
\begin{aligned}
E_0 &= (1.67 \times 10^{-27}\ \text{kg})(3.00 \times 10^8\ \text{m/s})^2 \\
&= (1.50 \times 10^{-10}\ \text{J})(1\ \text{eV}/1.60 \times 10^{-19}\ \text{J}) \\
&= (9.38 \times 10^8\ \text{eV})(1\ \text{MeV}/10^6\ \text{eV}) \\
&= 938\ \text{MeV}.
\end{aligned}
$$

(the numbers used give 939 MeV due to round-off errors). Then $E = E_0 + KE = 938\ \text{MeV} + 500\ \text{MeV} = 1438\ \text{MeV}$.

b) The expression $E = mc^2$ gives

$$m = \frac{E}{c^2} = \frac{(1438 \times 10^6\ \text{eV})(1.60 \times 10^{-19}\ \text{J/eV})}{(3.00 \times 10^8\ \text{m/s})^2} = 2.56 \times 10^{-27}\ \text{kg}. \quad \bullet$$

Many articles and books claim that the famous equation $E = mc^2$ means that energy can be converted to mass and mass to energy. This claim is incorrect. Rather, $E = mc^2$ is the mathematical statement of **the equivalence of mass and energy.** Therefore $E = mc^2$ simply says that anything that has a mass m has an energy E $(=mc^2)$, and anything that has an energy E has a mass m $(=E/c^2)$. That is, energy and mass are just two equivalent ways of describing the same thing. In $E = mc^2$, the term c^2 is then merely a conversion factor between mass units and energy units. (In SI units, $c^2 = 9.0 \times 10^{16}$ J/kg.) The two classical laws of conservation of mass and con-

186,000 miles per second is not only a good idea

PD

IT'S THE LAW

FIGURE 5.6
According to Einstein's special theory of relativity, c (which equals 3.00×10^8 m/s, or 186,000 mi/s) is nature's speed limit.

Source: T-shirt and poster design courtesy of Softwear Unlimited International, P.O. Box 265, Twisp, WA 98856.

servation of energy then merge into the one relativistic law of **conservation of mass–energy.**

Multiplying both sides of $m = m_0 / \sqrt{1 - v^2/c^2}$ by c^2 and using $E = mc^2$ and $E_0 = m_0 c^2$, we obtain

$$E = \frac{E_0}{\sqrt{1 - v^2/c^2}} = \gamma E_0. \tag{5.9}$$

Equation (5.9) and other expressions from relativity theory that involve $\gamma = 1 / \sqrt{1 - v^2/c^2}$ show us that for objects with a nonzero rest mass, c is the upper limiting speed. (See Fig. 5.6.) If v were equal to c, these objects would have an infinite energy and an infinite mass (as well as zero volume and infinitely dilated time intervals), as measured by a stationary observer. Since we don't find these properties in nature, we say that these objects can only approach c but not reach it. This doesn't forbid the existence of particles that have zero rest mass and which only move at $v = c$. We will discuss these particles later. Some physicists also have suggested that it is possible to have particles called tachyons, which always have a speed greater than c. There is, however, no experimental evidence for tachyons.

Again starting with $m = m_0 / \sqrt{1 - v^2/c^2}$, if we square both sides and rearrange terms we arrive at $m^2 = m^2 v^2/c^2 + m_0^2$. Recognizing mv as the magnitude of the linear momentum p, we get $m^2 = p^2/c^2 + m_0^2$. Multiplying by c^4 and using $E^2 = m^2 c^4$ and $E_0^2 = m_0^2 c^4$, we obtain

$$E^2 = p^2 c^2 + E_0^2. \tag{5.10}$$

(See Fig. 5.7.)

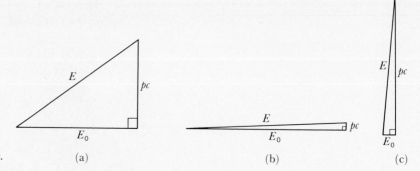

FIGURE 5.7
(a) A right triangle illustrating the relation $E^2 = p^2c^2 + E_0^2$. (b) The triangle for $v \ll c$. (c) The triangle for $v \to c$.

EXAMPLE 5.7   A proton moves by at $0.70c$. Calculate its

a) total energy,   b) kinetic energy, and   c) momentum.

Solution

a) Example 5.6 tells us that $E_0 = 938$ MeV. Also, $1/\sqrt{1 - v^2/c^2} = \gamma = 1.40$ for $v/c = 0.70$. Therefore

$$E = \gamma E_0 = (1.40)(938 \text{ MeV}) = 1313 \text{ MeV} = 1.31 \text{ GeV}.$$

b) $KE = E - E_0 = 1313$ MeV $- 938$ MeV $= 380$ MeV.

c) Solving $E^2 = p^2c^2 + E_0^2$ gives

$$p = \frac{\sqrt{E^2 - E_0^2}}{c} = \frac{[\sqrt{(1313)^2 - (938)^2} \text{ MeV}](1.60 \times 10^{-13} \text{ J/MeV})}{3.00 \times 10^8 \text{ m/s}}$$

$$= 4.9 \times 10^{-19} \text{ kg} \cdot \text{m/s}. \qquad \bullet$$

Of course, the solution to part (c) of Example 5.7 is only the magnitude of the linear momentum of the proton. Its direction will be the direction of motion. Unless otherwise indicated, we will use the word *momentum* to refer to the magnitude of the linear momentum, p.

Suppose that a particle is described as a 1-MeV proton. The 1-MeV energy given is (by common usage) the kinetic energy of the proton. The proton in Example 5.7 (which has a rest energy of 938 MeV, a kinetic energy of 380 MeV, and a total energy of 1.31 GeV) would therefore be described as a 380-MeV proton.

Since it's so easy to determine the work done and thus the kinetic energy in electron volts for charged atomic and subatomic particles, eV-based units are most often used in atomic, nuclear, and solid-state physics. For instance, a proton is given a kinetic energy of 1 MeV by moving it (with its charge of $+1e$) from rest through a potential drop of 1 megavolt. An electron is given a kinetic energy of 1 MeV by moving it (with its charge of $-1e$) from rest through a potential rise of 1 megavolt. An alpha particle has a charge of $+2e$ and so only has to move from rest through a potential drop of $\frac{1}{2}$ megavolt to gain a kinetic energy of 1 MeV.

We can often omit the cumbersome unit conversions of previous examples if we use $E/c^2 = m$ to specify mass units of eV/c^2, keV/c^2, MeV/c^2, GeV/c^2, TeV/c^2, and so on. Also $p = \sqrt{E^2 - E_0^2}/c$ tells us that momentum can then have units of eV/c, keV/c, MeV/c, GeV/c, TeV/c, and so on. (Remember: $k = 10^3$, $M = 10^6$, $G = 10^9$, and $T = 10^{12}$.)

EXAMPLE 5.8   The rest mass of a proton is 938 MeV/c^2. For a 200-MeV proton, calculate

a) the total energy,   b) the total mass,   c) the momentum, and   d) the speed.

Solution We are given $m_0 = 938$ MeV/c^2 and $KE = 200$ MeV.

a) Equation (5.6) solves to $E = E_0 + KE$. From $E_0 = m_0 c^2$, we have $E_0 = (938\ MeV/c^2)c^2 = 938$ MeV. (Note that in these units, the energy and the mass have the same numerical value.) Then $E = E_0 + KE = 938$ MeV $+ 200$ MeV $= 1138$ MeV.

b) From $E = mc^2$, $m = E/c^2 = 1138$ MeV/c^2. (The 1138 follows because m and E have the same numerical value in these units.)

c) From $E^2 = p^2 c^2 + E_0^2$,

$$p = \sqrt{E^2 - E_0^2}/c = \sqrt{(1138)^2 - (938)^2}\ MeV/c = 644\ MeV/c.$$

d) Solving $E = E_0 / \sqrt{1 - v^2/c^2}$ gives

$$v = c\sqrt{1 - (E_0/E)^2} = c\sqrt{1 - (938/1138)^2} = 0.57c,$$

or more simply,

$$v = \frac{p}{m} = \frac{(644\ MeV/c)}{(1138\ MeV/c^2)} = 0.57c. \qquad \bullet$$

If we want the answers in SI units, the conversions aren't difficult. For instance,

$$1\ MeV = (10^6\ eV)(1.6018 \times 10^{-19}\ J/eV)$$
$$= 1.6018 \times 10^{-13}\ J,$$

$$1\ MeV/c = 1.6018 \times 10^{-13}\ J/2.9979 \times 10^8\ m/s$$
$$= 5.3443 \times 10^{-22}\ kg \cdot m/s, \text{ or } N \cdot s,$$

and

$$1 \text{ MeV}/c^2 = 1.6018 \times 10^{-13} \text{ J}/(2.9979 \times 10^8 \text{ m/s})^2$$
$$= 1.783 \times 10^{-30} \text{ kg}.$$

We have found that linear momentum can still be defined as mass times velocity but that the mass now increases with speed. This property means that the time rate of change of momentum may involve terms both for acceleration and for change in mass. The relativistic expression for kinetic energy differs more and more from the classical one as speed increases and equals the mass increase times c^2. The reason is that mass and energy are equivalent with a unit conversion of c^2 in the relation $E = mc^2$.

Most objects have a rest mass and corresponding rest energy. All existing objects have some mass and energy, but no object can have infinite mass and energy. Therefore the speed of objects with a nonzero rest mass can only approach, but not equal, the speed of light, c. The expression $E^2 = p^2 c^2 + E_0^2$ relates total energy, linear momentum, and rest energy. For small particles, eV, eV/c, and eV/c^2 units (and their metric multiples) are often more convenient to work with than SI units.

QUESTIONS AND PROBLEMS

5.1 Calculate the mass of an electron moving at   a) 2.4×10^7 m/s and   b) 2.4×10^8 m/s.

5.2 A positive particle moving at $0.98c$ has a momentum of 2.47×10^{-18} kg · m/s. What is the rest mass of the particle?

5.3 At what speed will an electron have a momentum of 1.00×10^{-21} kg · m/s?

5.4 How fast would you have to move to double your mass? By whose measure?

5.5 A spaceship moving at $0.80c$ has a mass of 3.6×10^7 kg, as measured by a stationary observer. What is its rest mass?

5.6 What is the mass of a proton moving at a) $0.10c$? b) $0.90c$?

5.7 At what speed has the mass of a particle increased by 1% over its rest mass?

5.8 What force is necessary to give a proton moving in a circle a centripetal acceleration of 1.0×10^{15} m/s² when   a) $v = 0.90c$?   b) $v = 0.99c$?

5.9 What average force will speed up an electron from $0.895c$ to $0.905c$ in 2.00 ns?

5.10 Explain in words why
a) more force is needed to accelerate a particle if **F** is parallel to **v** than if **F** is perpendicular to **v**. (Assume that the magnitude of the acceleration is the same.)
b) this condition proves that the force is not in general parallel to the acceleration at relativistic speeds. When is the force parallel to the acceleration?

5.11 Calculate the ratio of the accelerating force needed for an acceleration parallel to **v** to the force needed for an equal magnitude acceleration perpendicular to **v** for   a) $v = 0.01c$,   b) $v = 0.10c$, and   c) $v = 0.99c$.

5.12 At what speed will the force needed for an acceleration be 1% greater than the classical value for   a) **F** ⊥ **v**?   b) **F** ∥ **v**?

5.13 What is the mass increase of an 80-kg sprinter running at 10 m/s?

5.14 Early scientists tried to measure the increase in mass of water as it was heated (thinking the increase was due to the addition of "caloric"). Since water has a specific heat of 1 kcal/kg · K = 4186 J/kg · K, what was the fractional increase of the water's mass per kelvin?

5.15 A spring has a force constant of 100 N/m. What is its mass increase when it is stretched 0.20 m?

5.16 How much kinetic energy would an electron have if its mass were 10 times its rest mass?

5.17 What is your rest energy?

5.18 Criticize the statement, "In the processes of fission and fusion, the masses of the end products are less than those of the initial atoms, thereby proving that mass has been converted to energy."

5.19 What is the conversion factor between mass units in kg and energy units in MeV?

5.20 a) Prove that the rest energy of an electron is 0.511 MeV. b) Calculate the total energy and mass of a 1.743-MeV electron. c) Calculate the speed of a 1.743-MeV electron. d) Calculate the momentum of a 1.743-MeV electron.

5.21 a) When can you say that $p = KE/c$ exactly? b) When can you say that $p = KE/c$ approximately? c) When can you say that $p = \sqrt{2E_0 KE}/c$ approximately?

5.22 At the Stanford Linear Accelerator, 50-GeV electrons are produced. Calculate the total energy, γ, mass, speed, and momentum of these electrons.

5.23 At Fermilab, 1-TeV protons are produced. Calculate the total energy, γ, mass, speed, and momentum of these protons.

5.24 The J or ψ particle has a rest energy of 3.1 GeV. Calculate its rest mass in kg.

5.25 A pion (rest mass of 140 MeV/c^2) has a momentum of 100 MeV/c. Calculate its total energy, mass, kinetic energy, γ, and speed.

5.26 A positron (rest mass of 0.511 MeV/c^2) has a speed of 0.80c. Calculate its total mass and energy, kinetic energy, and momentum.

5.27 A muon has a rest mass of 106 MeV/c^2. Calculate the momentum by two methods, one including Eq. (5.1), of a 54-MeV muon.

5.28 Prove that $v/c = pc/E$.

5.29 Spaceship A moves past the earth at 0.80c to the west. Spaceship B approaches A moving to the east. Both spaceships measure their relative speed of approach to be 0.98c. What mass would the crews of both spaceships measure for the standard kilogram, kept at rest on the earth a) according to classical physics, and b) according to the special theory of relativity?

5.30 A spaceship has a length of 100 m and a mass of 4.0×10^9 kg as measured by the crew. When it passes us, we measure the spaceship to be 75 m long. What do we measure its momentum to be?

5.31 Spaceship A moves past us at 0.98c, catching up with spaceship B moving in the same direction at constant speed. When spaceship B passed us, we measured its mass to be 3.6×10^9 kg, a value 20% greater than that measured by its own crew. What does the crew of spaceship A measure for the mass of spaceship B?

PART *Two*

Quantum Physics

CHAPTER 6

Planck's Constant

The special theory of relativity was the first revolutionary idea of 20th-century physics that we have discussed. In this chapter we begin to introduce you to a second revolutionary idea: quantum theory. We will discuss the experimental results for the frequency and wavelength distributions of a certain type of electromagnetic radiation and then show how classical physics failed in an attempt to explain these distributions. We will then show how Max Planck's quantization assumption did produce an equation that agreed with the experimental results. This assumption involved a fundamental constant of quantum theory, now called Planck's constant.

6.1
BLACKBODY RADIATION

You know from observing hot solids (such as the filament in a light bulb or an electric heating element) that if the solid gets hot enough it will begin to emit appreciable amounts of visible light. As the temperature rises, the object first becomes red hot, then more orange, and so on, until it may even appear to be blue–white. This phenomenon tells us that the predominant frequencies of electromagnetic radiation increase (predominant wavelengths decrease) as the temperature increases. Also, much more radiant energy is emitted if the object is at a higher temperature.

Experiments show that some objects emit radiant energy better than others at a given temperature and that good emitters are also good absorbers. The best absorber of radiation is a rough black surface. A body that absorbed all the radiation it received would be called an **ideal blackbody.** An ideal blackbody would also be the best possible emitter of electromagnetic radiation.

An excellent approximation to an ideal blackbody would be a small hole leading into a rough cavity in a solid object. (See Fig. 6.1.) Unless the rear surface of the cavity were specially prepared, virtually no radiation would reflect back through the hole. The electromagnetic waves entering the small hole would bounce around inside, with some energy absorbed at each reflection, until the energy became completely absorbed.

In the 1800s, experimenters determined for the ideal blackbody that the intensity emitted per unit wavelength interval peaked out at a wavelength, λ_p, that was inversely proportional to the absolute temperature of the cavity. That is,

$$\lambda_p T = 2.898 \text{ mm} \cdot \text{K}. \tag{6.1}$$

Equation (6.1) is called the **Wien displacement law** because it shows that as the temperature increases, the peak wavelength is displaced downward to smaller values. As you might expect, the intensity per unit wavelength emitted from the small hole has the same type of distribution as the energy density per unit wavelength, u_λ, in the cavity. In fact, the intensity is just $c/4$ times the energy density.

Alternatively, we may graph the energy density per unit frequency, u_ν, as a function of frequency. Such curves also have a single peak for each

FIGURE 6.1
A cross-section of a cavity in a solid object. The small hole acts approximately as an ideal blackbody.

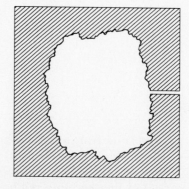

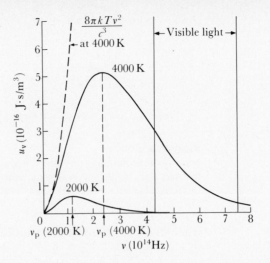

FIGURE 6.2

The frequency variation of the energy density per unit frequency at two different temperatures in a cavity. The values are also directly proportional to the intensity per unit frequency emitted by an ideal blackbody. The dashed line is the classical physics prediction.

temperature, as shown in Fig. 6.2. However, since $v = c/\lambda$, each curve peaks out at a frequency that is directly, not inversely, proportional to the absolute temperature. It may surprise you that this frequency, v_p, is not equal to c/λ_p. The reason is that v and λ are not directly proportional, and so a unit frequency interval is not a constant multiple of a unit wavelength interval.

The frequency equivalent of the Wien displacement law is

$$v_p = (58.79 \text{ GHz/K})T. \tag{6.2}$$

(Recall that G is the 10^9 metric prefix.) To summarize, Eq. (6.2) gives the position of the peak of the curve of intensity per unit frequency or u_v (energy density per unit frequency) versus v (the frequency of the electromagnetic waves emitted by and within the cavity) at an absolute temperature T.

The total energy density is the area under a u_v versus v curve (or alternatively, under a u_λ versus λ curve). (See Fig. 6.3.) This energy density will be proportional to the total energy emitted from the cavity in a given time. According to the experimental **Stefan–Boltzmann law,** this emitted energy is proportional to T^4. Therefore, if $u(T)$ is the total energy density at an absolute temperature T, we can write

$$u(T) = \int_0^\infty u_v \, dv = \int_0^\infty u_\lambda \, d\lambda = BT^4,$$

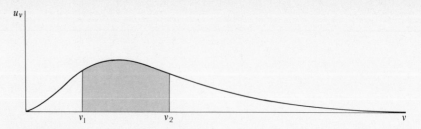

FIGURE 6.3

The area under the curve between v_1 and v_2 gives the energy per volume in the cavity for that frequency range. That energy per volume multiplied by $c/4$ gives the intensity emitted by an ideal blackbody for that frequency range.

where B is a constant. A more useful version of the Stefan–Boltzmann law is

$$I(T) = (56.71 \text{ nW/m}^2 \cdot \text{K}^4)T^4, \tag{6.3}$$

where n is the 10^{-9} metric prefix, and $I(T)$ is the intensity (power per unit surface area) emitted by an ideal blackbody at an absolute temperature T.

EXAMPLE 6.1 An ideal blackbody radiates 548 J at room temperature (20 °C) in one day. How much energy would it emit in a day at 1000 °C?

Solution The Stefan–Boltzmann law tells us that the energy is proportional to T^4. Since T is the absolute temperature, the given T's are $(20 + 273)$ K = 293 K, and $(1000 + 273)$ K = 1273 K. Then

$$E(293 \text{ K}) = 548 \text{ J}, \quad \text{and} \quad \frac{E(1273 \text{ K})}{E(293 \text{ K})} = \frac{(1273 \text{ K})^4}{(293 \text{ K})^4}.$$

So

$$E(1273 \text{ K}) = (548 \text{ J})\left(\frac{1273}{293}\right)^4 = 1.950 \times 10^5 \text{ J}. \qquad \bullet$$

Example 6.1 shows that increasing the absolute temperature by a factor of something over 4 increases the energy emitted by a factor of over 350.

EXAMPLE 6.2 At what frequency and wavelength is the greatest amount of energy per unit frequency and wavelength emitted at both room temperature and 1000 °C for the ideal blackbody in Example 6.1?

Solution We use Eqs. (6.2) and (6.1). First, $v_p = (58.79 \text{ GHz/K})T$, and for

$T = 293$ K,

$$\nu_p = (58.79 \times 10^9 \text{ Hz/K})293 \text{ K} = 1.72 \times 10^{13} \text{ Hz} = 17.2 \text{ THz}.$$

Then

$$\lambda_p = \frac{2.898 \text{ mm} \cdot \text{K}}{T} = \frac{2.898 \times 10^{-3} \text{ m} \cdot \text{K}}{293 \text{ K}}$$
$$= 9.89 \times 10^{-6} \text{ m} = 9.89 \ \mu\text{m},$$

which shows that at room temperature blackbody radiation peaks far into the infrared at 9.89 μm. For $T = 1273$ K,

$$\nu_p = (58.79 \times 10^9 \text{ Hz/K})1273 \text{ K} = 74.8 \text{ THz},$$

and

$$\lambda_p = \frac{2.898 \times 10^{-3} \text{ m} \cdot \text{K}}{1273 \text{ K}} = 2.26 \ \mu\text{m}.$$ ●

A 1000-°C blackbody has its peak wavelength and frequency in the infrared. Therefore we would detect more energy per unit frequency or wavelength at the low-frequency (red) end of the visible spectrum than at the high-frequency (violet) end. (The visible spectrum extends from about 43×10^{13} Hz (0.70 μm, red) to about 75×10^{13} Hz (0.40 μm, violet).) A 1000-°C blackbody appears to have an orange color.

EXAMPLE 6.3 What is the surface area of the blackbody in Examples 6.1 and 6.2?

Solution $I(T)$, the intensity at absolute temperature T, is the power per surface area, and power is energy per time. Therefore

$$I(T) = 548 \text{ J}/(1 \text{ day} \times 24 \text{ hr/day} \times 60 \text{ min/hr} \times 60 \text{ s/min} \times A)$$
$$= (56.71 \times 10^{-9} \text{ W/m}^2 \cdot \text{K}^4)(293 \text{ K})^4,$$

which gives $A = 1.52 \times 10^{-5} \text{ m}^2 = 15.2 \text{ mm}^2.$ ●

6.2
PLANCK'S HYPOTHESIS

In the 1800s much of what Fig. 6.2 shows was measured experimentally. However, a theoretical derivation of the curves could not be obtained using classical physics. Lord Rayleigh first obtained an expression, reasoning something like this: Suppose that the electromagnetic waves are in equilibrium in a cubic cavity of side L. Equilibrium means no net energy transfer. No net energy transfer occurs for standing waves. Standing waves with nodes at the walls mean that $n_x \lambda/2 = L$ for waves in the x direction (where $n_x = 1, 2, 3, \ldots$). Then $\nu = c/\lambda$ gives $\nu = n_x c/2L$ as possible frequencies for waves in the x direction. Similar relations hold for the y and z directions.

Let's now draw a coordinate system with n_x, n_y, and n_z plotted on the axes instead of x, y, and z. These n's can only be positive integers and correspond to discrete frequencies. However, for the frequencies of interest, the integers are so large as to practically form a continuum. This condition means that we can determine the number of kinds of oscillations occurring in the range between ν and $\nu + d\nu$ by using a spherical shell of radius n (where $n = \sqrt{n_x^2 + n_y^2 + n_z^2}$) and thickness dn; n corresponds to ν by the relation $n = (2L/c)\nu$ and dn to $d\nu$ by $dn = (2L/c)d\nu$. (See Fig. 6.4.) There can be only one set of n's in a cube one unit on a side in this type of space. For example, a cube from $n_x = 2000.5$ to 2001.5, n_y from 1641.5 to 1642.5, and n_z from 1904.5 to 1905.5 would be one unit on each side and would contain only the single set; $n_x = 2001$, $n_y = 1642$, $n_z = 1905$.

For each set of n's there can be one frequency and two mutually perpendicular transverse polarizations, which means that there are two kinds of oscillations per unit n-volume. The volume of the shell of radius n and thickness dn is $(4\pi n^2/8)dn$. (The 8 is there because only the positive values of n_x, n_y, and n_z can be taken, so only the totally positive eighth of the space is used.) The number of oscillations in the shell is then 2 per volume times $(4\pi n^2/8)dn$, which equals 2 times $4\pi[(2L/c)\nu]^2/8$ times $(2L/c)d\nu = (8\pi L^3/c^3)\nu^2\, d\nu$.

Since $L^3 = V$, the volume in normal space, the number of possible oscillations between ν and $\nu + d\nu$ per volume equals $(8\pi\nu^2/c^3)d\nu$. Classically, the energy for each kind of oscillation equals kT, where k is Boltzmann's constant, that is, $k = 1.3807 \times 10^{-23}$ J/K $= 8.617 \times 10^{-5}$ eV/K. Therefore the energy per volume between ν and $\nu + d\nu$, $u_\nu d\nu$, equals $(8\pi kT\nu^2/c^3)d\nu$.

FIGURE 6.4

The positive octant of n space, containing allowed integral values of n_x, n_y, and n_z.

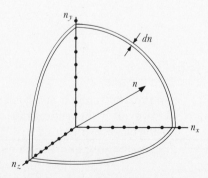

The fraction of this energy escaping the cavity through the very small hole has the same frequency dependence as this relation. The relation fits well the very low frequency end of Fig. 6.2. However, the simple parabola it gives doesn't bend over to give a peak at v_p. Instead, it continues to increase with v^2 without limit as the frequency increases. This breakdown of classical physics for high-frequency electromagnetic radiation was given the colorful name of "the ultraviolet catastrophe."

Wien then tried to improve the result by saying that the energy should be distributed with a term $C_1 e^{(-C_2 v/T)}$. He chose constants C_1 and C_2 to make his curve fit the experimental data behind Fig. 6.2 as well as possible. His arguments for this term sounded reasonable, but were not rigorous. Then Max Planck found that he could improve the agreement by putting in a (-1) at the correct place. The (-1) came from a startling assumption.

In classical physics, a harmonic oscillator (like a mass on the end of a Hooke's law spring) can have a continuous range of energies. In 1900, after six years of struggle with blackbody theory, Planck assumed that the energies of electromagnetic oscillators in equilibrium were quantized and could have only allowed energies of

$$E_n = nhv \tag{6.4}$$

where $n = 0, 1, 2, 3, \ldots$ and v is the oscillator frequency. Now called **Planck's constant**, h was determined by fitting Planck's equation for the blackbody radiation curve to the experimental data. The current experimental value is $h = 6.626 \times 10^{-34} \text{ J} \cdot \text{s} = 4.136 \times 10^{-15} \text{ eV} \cdot \text{s}$. (Modern results have modified Eq. (6.4): The lowest energy is not zero; even at absolute zero temperature some oscillation continues.)

The Boltzmann distribution was used to give the number N_n of oscillators with energy nhv, $N_n = N_0 e^{(-nhv/kT)}$. This expression gives the total number of oscillators as $\sum_{n=0}^{\infty} N_0 e^{(-nhv/kT)}$. Since each oscillator has an energy nhv, the total energy equals $\sum_{n=0}^{\infty} nhv N_0 e^{(-nhv/kT)}$. This total energy divided by the total number of oscillators yields the average energy of an oscillator:

$$\overline{E} = \frac{\sum\limits_{n=0}^{\infty} nhv N_0 \exp(-nhv/kT)}{\sum\limits_{n=0}^{\infty} N_0 \exp(-nhv/kT)}$$

$$= \frac{0 + hv \exp(-hv/kT) + 2hv \exp(-2hv/kT) + \cdots}{1 + \exp(-hv/kT) + \exp(-2hv/kT) + \cdots},$$

where $\exp(x)$ means e^x. If we let $y = \exp(-hv/kT)$, we then have

$$\overline{E} = hvy \left(\frac{1 + 2y + 3y^2 + \cdots}{1 + y + y^2 + \cdots} \right) = hvy \left[\frac{(1-y)^{-2}}{(1-y)^{-1}} \right] = \frac{hvy}{1-y} = \frac{hv}{(1/y - 1)},$$

so that $\overline{E} = hv/[\exp(hv/kT) - 1]$ for the average energy of each oscillation of frequency v. Multiplying this last expression by $(8\pi v^2/c^3)dv$, the number of oscillations per volume between v and $v + dv$, we finally get the energy per volume between v and $v + dv$:

$$u_v\, dv = \frac{(8\pi hv^3/c^3)}{\exp(hv/kT) - 1}\, dv. \tag{6.5}$$

As $v \rightarrow \infty$, $u_v dv \rightarrow 0$, and there is no ultraviolet catastrophe. As $v \rightarrow 0$, $\exp(hv/kT) - 1 \approx 1 + hv/kT - 1 = hv/kT$ (using the expansion $\exp(x) \approx 1 + x$ for very small x). Substituting into Eq. (6.5), we obtain $u_v\, dv \approx 8\pi kT(v^2/c^3)dv$ for very low frequencies, which is Rayleigh's classical result.

EXAMPLE 6.4   Derive Eq. (6.2) from Eq. (6.5).

Solution The radiated energy has the same frequency distribution, so to find the maximum of u_v versus v we must set the first derivative equal to zero:

$$0 = du_v/dv = d\{(8\pi hv^3/c^3)[\exp(hv/kT) - 1]^{-1}\}/dv \qquad \text{at } v = v_p,$$

or

$$0 = 3v_p^2[\exp(hv_p/kT) - 1]^{-1}$$
$$- v_p^3[\exp(hv_p/kT) - 1]^{-2}[\exp(hv_p/kT)](h/kT),$$

so

$$3[1 - \exp(-hv_p/kT)] = hv_p/kT.$$

Solving the equivalent equation, $3(1 - e^{-x}) = x$ by trial and error, iteration, graph, or some other method yields $x = 0$ and $x = 2.82144 = hv_p/kT$. Substituting $k = 1.38066 \times 10^{-23}$ J/K and $h = 6.6261 \times 10^{-34}$ J $\cdot$ s, we get $v_p = (58.79 \text{ GHz/K})T$.   ●

EXAMPLE 6.5   Find the amount of energy per volume in the cavity at 2000 K for light within the frequency range of from 5.00×10^{14} Hz to 5.04×10^{14} Hz.

Solution Strictly speaking, the solution requires integrating Eq. (6.5) from 5.00×10^{14} Hz to 5.04×10^{14} Hz. The integral isn't exact, so the integration can be done only by numerical methods, such as Simpson's rule. However, since the frequency range is small compared to the actual frequency $(0.04 \ll 5.00)$, we can replace dv by $\Delta v = 0.04 \times 10^{14}$ Hz and v by $\overline{v} = 5.02 \times 10^{14}$ Hz and still arrive at a very good approximation. When we do so, $u_{\overline{v}}\Delta v = (8\pi h\overline{v}^3/c^3)\Delta v/[\exp(h\overline{v}/kT) - 1]$, with $h = 6.626 \times 10^{-34}$ J $\cdot$ s, $c = 3.00 \times 10^8$ m/s, $k = 1.381 \times 10^{-23}$ J/K, and $T = 2000$ K, evaluates to 1.84×10^{-6} J/m^3 = $1.84\ \mu$J/m^3, or J/cm^3.   ●

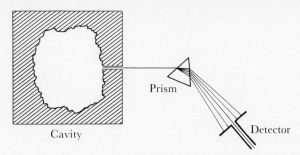

FIGURE 6.5
Measuring the amount of
energy emitted by an ideal
blackbody in a certain
wavelength and frequency
range.

Using $\lambda = c/v$ twice, we find that this frequency range corresponds to the wavelength range of from 595 nm to 600 nm (in the visible portion of the spectrum). Figure 6.5 indicates how the intensity in a given range is measured.

EXAMPLE 6.6   Calculate the intensity of the light emitted by a 2000-K ideal blackbody betwen 595 nm and 600 nm.

Solution The intensity emitted by an ideal blackbody is $c/4$ times the energy density in the cavity, so all we need to do is multiply the answer to Example 6.5 by $c/4 = 7.495 \times 10^7$ m/s. That is, this intensity equals $(1.84 \times 10^{-6}$ J/m$^3)(7.495 \times 10^7$ m/s$) = 138$ W/m^2.   ●

In the 19th century, physicists using classical physics were unable to find a theoretical explanation for the experimental blackbody curves. Max Planck began this century by assuming that the energy levels of an oscillator were quantized, thus initiating quantum theory. This assumption gave a final equation that fit the experimental results then available and correctly predicted the results of future experiments. However, Planck and others were not satisfied with what he called "an act of desperation," because there was no basic reason for the assumption. In effect, Planck's assumption was an elegant "fudge factor" added onto classical theory to yield the "correct answer."

QUESTIONS AND PROBLEMS

6.1 A chef has a number of copper-bottomed pans he keeps shined to perfection. Discuss the efficiency of radiative heat transfer from the cooking range to the pans.

6.2 Two objects are identical, except that one is smooth and white and the other is rough and black.

a) Which one will heat up or cool down faster if initially at a temperature different from its surroundings?

b) The difference in heating or cooling rates will be more pronounced in a vacuum. Why?

6.3 The wavelength of the middle of the visible

spectrum is 550 nm. What is the temperature of an ideal blackbody with an electromagnetic radiation output that would peak at this wavelength?

6.4 Radiation from space that is characteristic of an ideal blackbody at 2.7 K has been detected. At what frequency and wavelength are we receiving the most energy per unit frequency and wavelength? (This radiation is usually interpreted to be the result of the "Big Bang" beginning of the universe.)

6.5 The radiation in Problem 6.4 seems to have an asymmetry that can be interpreted as caused by the motion of our solar system with respect to the original source of the radiation. Does that mean a preferred inertial system has been established?

6.6 When the peak frequency from an ideal blackbody is tripled, what happens to the power output?

6.7 A physics text gives the filament area and power for a light bulb and asks for the filament temperature. What is wrong with the problem?

6.8 Can you determine the energy you radiate every 24 hours?

6.9 The area of a hole leading to a large cavity is 2.0 mm². How long would it take to radiate 500 J of energy if the cavity is at 100 °C?

6.10 A blackbody emits 400 J in an hour at 500 K. What is the peak frequency when it emits 25 J in an hour?

6.11 What must you do to the temperature to get a blackbody to emit twice the energy in half the time?

6.12 An ideal blackbody emits 8.72×10^3 W/m². At what wavelength is the greatest power per unit wavelength radiated?

6.13 a) Use $c = \nu\lambda$ to convert $u_\nu d\nu$ to $u_\lambda d\lambda$ (the energy per unit volume between λ and $\lambda + d\lambda$). b) Derive Eq. (6.1) from the result in part (a).

6.14 Find the energy density in the cavity at 5000 K for the light within the wavelength range of from 400 nm to 404 nm.

6.15 At what frequency would we measure the maximum intensity per unit frequency to occur for an ideal blackbody having a temperature of 4000 K and a velocity of $0.10c$ toward us?

6.16 The output intensity of a star can be well approximated, near its peak, by that of a 38,400-K blackbody. At what wavelength would we measure its intensity per unit wavelength to have a maximum if it were moving away from us at $0.0050c$? (Recall that $c = \nu\lambda$.)

6.17 The output of our sun, especially in the visible region, is often approximated by a 5800-K ideal blackbody. a) At what frequency does its intensity per unit frequency have its maximum? b) What wavelength does this previous answer correspond to? ($c = \nu\lambda$.) c) At what wavelength does its intensity per unit wavelength have its maximum? d) What frequency does the answer to (c) correspond to? e) Answers (a) and (d) do not agree, nor do answers (b) and (c). Explain why. f) Calculate the radius of this solar blackbody if sunlight reaches the earth (orbit radius = 1.50×10^{11} m) with an intensity of 1.39 kW/m².

6.18 The entire universe acts like a cavity that is expanding. Use the first paragraph of Section 6.2 to explain why expanding a cavity will expand in direct proportion the wavelengths within the cavity.

6.19 Calculate the total energy density of a cavity if the peak wavelength emitted from the cavity is 182 μm.

6.20 If the energy density within a cavity is 2.19×10^{-5} J/m³, calculate the frequency at which its intensity per unit frequency is a maximum.

Photons

In this chapter we'll again encounter experimental results that involve electromagnetic radiation and could not be explained by classical physics. In Chapter 6 we considered the emission of radiation by an ideal blackbody. In this chapter we will look at the absorption of radiation in the photoelectric effect. We will see how Albert Einstein extended Max Planck's quantization idea to bring forth an elegantly simple explanation of the experimental results. This explanation involved quanta of electromagnetic energy called photons.

7.1
THE PHOTOELECTRIC EFFECT — EXPERIMENTAL

In 1864 Maxwell predicted that oscillating electric currents would produce electromagnetic waves that would move through a vacuum at the speed of light. In 1887 Heinrich Hertz was attempting to confirm this prediction when he noticed that a spark could more easily be induced to jump a gap when the gap was illuminated. Continuing this research, other

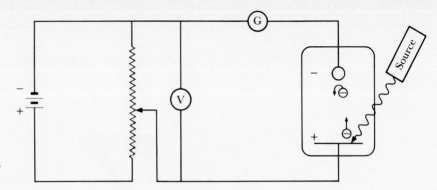

FIGURE 7.1
Experimental arrangement
for studying the photoelectric effect.

workers showed that when light is incident on a metal surface, electrons
are emitted from the surface. This phenomenon is called the **photoelectric
effect.**

Figure 7.1 shows a method of studying the photoelectric effect.
Monochromatic light (light of one frequency) of variable intensity from
source S shines on a clean metal surface in a vacuum tube. Electrons, called
photoelectrons, are emitted from that electrode with some kinetic energy
and travel to the top electrode. This action produces a small current,
which can be measured by a very sensitive galvanometer, G. The battery is
connected across a voltage divider circuit so that the potential difference
between the electrodes can be varied. This potential difference is measured by the voltmeter, V.

When the voltage is zero, the photoelectrons can easily travel to the
collecting electrode. However, giving the collecting electrode a negative
charge will set up an electric field to oppose the motion of the photoelectrons. In moving against this field, the photoelectrons do work. The total
work done is equal to the charge, $-e$, times the potential difference. This
work comes from the kinetic energy of the photoelectrons.

As the retarding potential difference (negative bias) is increased,
fewer and fewer electrons will have sufficient kinetic energy to reach the
top electrode. Therefore the current decreases as the retarding potential
difference increases. The current will drop to zero when the work done,
$(-e)(-V_0)$, equals the maximum kinetic energy of the photoelectrons,
KE_{max}. Therefore

$$KE_{max} = eV_0, \tag{7.1}$$

where V_0 is the absolute value of the potential difference measured by the
voltmeter when the current drops to zero (if both electrodes are made of
the same material) and is called the **stopping potential.** Photoelectrons are
usually not relativistic ($v \ll c$).

EXAMPLE 7.1   In the photoelectric effect, the current first drops to zero for a retarding potential difference of 1.25 V for a given light frequency. What is the maximum

a) kinetic energy and   b) speed of the emitted photoelectrons?

Solution

a) $KE_{max} = eV_0$ gives us $KE_{max} = (e)(1.25 \text{ V}) = 1.25 \text{ eV}$ as the simplest answer. If we use SI units,

$$KE_{max} = (1.60 \times 10^{-19} \text{ C})(1.25 \text{ J/C}) = 2.00 \times 10^{-19} \text{ J}.$$

b) The rest energy of an electron is 0.511 MeV. Since $1.25 \text{ eV} \ll 0.511 \text{ MeV}$, the classical expression can be used for KE: $KE = \frac{1}{2}m_0 v^2$ gives

$$v = \sqrt{2\,KE/m_0}$$
$$= \sqrt{2(2.00 \times 10^{-19}\text{J})/9.11 \times 10^{-31} \text{ kg}}$$
$$= 6.63 \times 10^5 \text{ m/s}.$$

Can you solve for v using 1.25 eV and $m_0 = 0.511 \text{ MeV}/c^2$?   ●

Classical physics runs into many difficulties when trying to explain various results of photoelectric experiments, such as:

1. For a constant light frequency, v, V_0 is independent of the light intensity, I. (See Fig. 7.2.) Classical physics would suggest that light of greater intensity (energy per area per time) should give electrons greater kinetic energy, so that KE_{max} and therefore V_0 should increase with I.

2. Even at extremely low intensity, photoemission occurs with essentially no delay as soon as the light source is turned on. Classical

FIGURE 7.2
For a constant light frequency, the stopping potential is independent of the intensity. Also, the photocurrent at a given potential is directly proportional to the intensity.

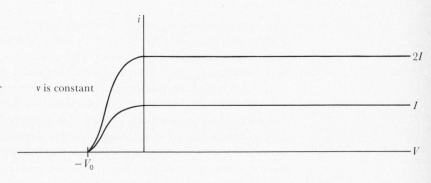

FIGURE 7.3
The stopping potential and
therefore the maximum
kinetic energy of the
photoelectrons increase
linearly with the light
frequency.

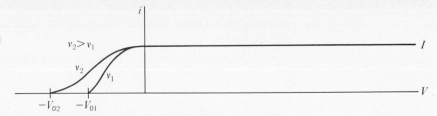

physics would say that the energy is spread over the entire wavefront. Thus a particular electron would have to gather energy in
from a certain rather small surrounding area until the electron
finally had enough energy to break free from the surface. For
extremely weak light, this occurrence could take days in the
classical model.

3. For a given frequency, v, and retarding potential, V, the photocurrent measured by G is directly proportional to the light intensity
(this property is utilized in photoelectric devices). Since the power
output is proportional to the photocurrent and the power input is
proportional to the intensity, classical physics will agree with this
result.

4. The maximum kinetic energy of photoelectrons is linearly related
to the frequency of the light. Below a certain **threshold frequency,** no photoelectrons will be emitted. Classical physics has no
explanation for this behavior. (See Fig. 7.3.)

7.2
THE PHOTOELECTRIC EFFECT — THEORETICAL

In summary, classical physics can't explain three of the four main results of
the photoelectric-effect experiments. What was needed was some revolutionary thinking. For that scientific revolutionary, Albert Einstein, 1905
was an amazing year. He proposed his special theory of relativity and
explained both Brownian motion and the photoelectric effect.

Planck's derivation of the blackbody radiation curve equation involved the assumption that the electromagnetic oscillators had quantized
energy levels. Einstein then suggested that downward transitions between
these levels should result in the emission of bundles, or **quanta,** of electromagnetic energy. (See Fig. 7.4.) These discrete quanta of electromagnetic
energy are called **photons** and have an energy E that equals Planck's

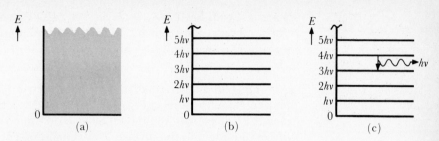

FIGURE 7.4
For an electromagnetic harmonic oscillator, (a) classical physics said that E can have a continuous range of possible values; (b) Planck said that $E = nh\nu$, so that E can have only certain allowed values; and (c) Einstein said that photons with $E = h\nu$ are emitted in transitions between adjacent allowed energy levels.

constant times the frequency of the electromagnetic wave,

$$E = h\nu. \tag{7.2}$$

Often it is the wavelength, λ, of the electromagnetic wave that is given or desired. In that case,

$$E = \frac{hc}{\lambda}. \tag{7.3}$$

The product $hc = 1.240$ eV $\cdot \mu$m in some useful non–SI units.

EXAMPLE 7.2 An educational FM station broadcasts at 90.1 MHz with a radiated power of 10 kW. How many photons does it emit each second?

Solution The station sends out 10×10^3 J/s. The energy of each photon emitted, by Eq. (7.2), is

$$E = h\nu = (6.626 \times 10^{-34}\,\text{J} \cdot \text{s})(90.1 \times 10^6\,\text{Hz}) = 5.97 \times 10^{-26}\,\text{J}.$$

Therefore

$$\frac{(10 \times 10^3\,\text{J/s})}{5.97 \times 10^{-26}\,\text{J/photon}} = 1.7 \times 10^{29}\,\text{photons/s}. \qquad \bullet$$

With this huge number of photons leaving the station in Example 7.2 each second, in most cases the discreteness of the tiny individual bundles of energy won't be noticed. That is, classical physics will explain most of the details of this station's signal broadcast and detection.

EXAMPLE 7.3 Silicon (Si) films become better conductors if illuminated by photons with energies greater than 1.14 eV (this behavior is called photoconductivity). What wavelength does this correspond to?

Solution From $E = hc/\lambda$, $\lambda = hc/E = 1.240 \text{ eV} \cdot \mu\text{m}/1.14 \text{ eV} = 1.09 \ \mu\text{m}$ (infrared radiation). (Since 1.14 eV is the minimum energy, $1.09 \ \mu\text{m}$ is the maximum wavelength for photoconduction for Si.)   ●

Einstein suggested that the photoelectric effect occurred when a photon gave up all its energy, $h\nu$, to an electron near the surface of the metal. The electron would then have to do some work in overcoming the forces binding it to the surface. Any remaining energy would be the electron's kinetic energy when free of the surface. The maximum kinetic energy of a photoelectron thus becomes

$$KE_{\text{max}} = h\nu - \phi. \tag{7.4}$$

The term ϕ is called the **work function.** It is the *minimum* amount of work that must be done to free an electron from the surface. This quantum idea then explains the results of the photoelectric effect experiments:

1. At a constant ν, KE_{max} and therefore V_0 don't depend on the intensity of the light. That is, they don't depend on how many photons per time per area are arriving.

2. As soon as a photon reaches the surface, it can be absorbed and an electron emitted.

3. The intensity is directly proportional to the number of photons per time, as is the number of electrons emitted per time. Therefore the photocurrent is directly proportional to the intensity.

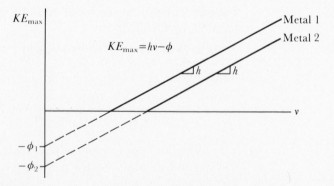

FIGURE 7.5
The maximum kinetic energy of the photoelectrons as a function of the frequency of the incident light for two different metals. The slopes of the two straight lines are the same (the slopes equal Planck's constant, h), but the intercepts are different (being proportional to the work function, ϕ).

4. The expression $KE_{max} = hv - \phi$ and Fig. 7.5 show KE_{max} to be linearly related to v for the reasons Einstein gave. The slope of the graph is h. Different metals will have the same slope, but different intercepts will result from different work functions, as shown in Fig. 7.5.

EXAMPLE 7.4   Zinc has a work function of 4.3 eV.

a) What is the threshold (minimum) frequency for photoemission? The maximum wavelength?

b) What frequency will result in electrons with kinetic energies as large as 1.25 eV?

Solution We start with 4.3 eV $= \phi$.

a) At the threshold, $KE_{max} = 0$. Therefore $KE_{max} = hv - \phi$ gives

$$v_{min} = \frac{\phi}{h} = \frac{4.3 \text{ eV}}{4.136 \times 10^{-15} \text{ eV} \cdot \text{s}} = 1.04 \times 10^{15} \text{ Hz}.$$

Then

$$\lambda_{max} = \frac{c}{v_{min}} = \frac{3.00 \times 10^8 \text{ m/s}}{1.04 \times 10^{15} \text{ Hz}} = 0.29 \times 10^{-6} \text{ m} = 0.29 \text{ } \mu\text{m},$$

which is in the ultraviolet.

b) Here $KE_{max} = 1.25$ eV. Solving $KE_{max} = hv - \phi$ gives

$$v = \frac{(KE_{max} + \phi)}{h} = \frac{1.25 \text{ eV} + 4.3 \text{ eV}}{4.136 \times 10^{-15} \text{ eV} \cdot \text{s}} = 1.3 \times 10^{15} \text{ Hz}.$$ ●

Example 7.4 explains why a negatively charged zinc plate will maintain its charge inside a normally lit room but loses it quickly in direct sunlight. Light from the sun has lots of energy at frequencies above 1.04×10^{15} Hz in the ultraviolet; ordinary artificial lighting and sunlight that has passed through window glass have very little ultraviolet energy. (Therefore the sunlight gives sunburn and skin cancer, but the ordinary artificial light doesn't.)

Sometimes photons with energies of less than ϕ can cause photoemission. This occurs only if very many photons per time per area arrive on a surface. Then there is a probability that two, or even more, photons can be absorbed at the same time by the same electron. Such absorption can happen with very high intensity laser light, for example. If the total energy of the two (or more) photons exceeds the work function, a photoelectron can be emitted with some kinetic energy. (See Fig. 7.6.)

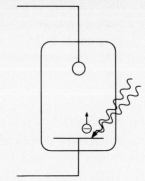

FIGURE 7.6
The two-photon photoelec-
tric effect.

The photoelectric effect (absorbing a photon) is not the only way to get electrons out of a surface. Other methods used include:

1. **Secondary emission,** which involves banging some particles other than photons into the surface to knock electrons free.
2. **Field-aided emission,** that is, applying an electric field to the surface to help pull the electrons out.
3. **Thermionic emission,** by heating the surface until some electrons have enough energy to escape.

There are also methods that involve tunneling, a quantum mechanical process that we will discuss later.

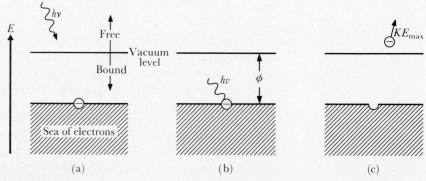

FIGURE 7.7
The (single photon) photoelectric effect: (a) A photon of energy $h\nu$ approaches a surface, and the electrons have an energy of at least ϕ less than that required to be completely free of the surface; (b) The photon and electron interact, and the photon disappears, with its energy being completely absorbed by the electron; and (c) The electron uses at least ϕ of the $h\nu$ energy to escape from the surface, leaving the electron with as much as $h\nu - \phi$ of kinetic energy.

The photoelectric effect is a good example of conservation of energy. The main difference from the classical physics explanation is Einstein's suggestion that the energy in the incoming electromagnetic radiation is quantized in small bundles of energy hv. Let's consider the single photon photoelectric effect, as shown in Fig. 7.7. When each bundle of energy hv is absorbed, at least ϕ of that is the work done to free a photoelectron. As much as $hv - \phi$ is left for the kinetic energy of the freed photoelectron.

QUESTIONS AND PROBLEMS

7.1 The maximum speed of the photoelectrons in an experiment as illustrated by Fig. 7.1 is 2.00×10^5 m/s. a) What is their kinetic energy? b) What negative bias voltage will stop them? c) Why can this problem be solved nonrelativistically?

7.2 Extremely weak light of frequency 8.0×10^{14} Hz has an intensity of 1.0×10^{-10} W/m². How many photons cross each perpendicular square meter each second?

7.3 How many photons are broadcast during a typical song by your favorite radio station?

7.4 A 1.0 mW laser emits 632.8 nm light. At what rate is it emitting photons?

7.5 The visible light range is from 400 nm to 700 nm. What is the corresponding photon energy range?

7.6 What is the frequency and wavelength of a 1.00-eV photon?

7.7 The work function of sodium is 2.3 eV. Calculate the maximum kinetic energy and speed of the photoelectrons if a clean sodium surface is illuminated by 7.5×10^{14} Hz light.

7.8 What would happen if sodium ($\phi = 2.3$ eV) were illuminated by high intensity light with a frequency of 5.4×10^{14} Hz?

7.9 Explain why most photoelectrons have kinetic energies of less than $hv - \phi$.

7.10 Why does the work function depend on surface cleanliness?

7.11 Carefully explain the physical significance of each term in Eq. (7.4).

7.12 Some physics texts state that the maximum kinetic energy of photoelectrons is directly proportional to light frequency. Why is this statement incorrect?

7.13 Laser 1 has a higher intensity than laser 2 but operates at a lower frequency. How could this condition be possible?

7.14 What is the stopping potential when the frequency equals the work function divided by Planck's constant?

7.15 Data taken from a particular metal surface indicate that the stopping potential equals 0.95 V for 313 nm light and 0.35 V for 369 nm light. Calculate Planck's constant and the work function.

7.16 Metal 1 has a work function of 2.8 eV, and metal 2 has a work function of 2.9 eV. Describe a graph of KE_{max} versus v for the two metals and determine the horizontal and vertical intercepts.

7.17 State the physical significance of the slope and intercepts of Fig. 7.5.

7.18 If the work function for potassium is 2.24 eV, what is the maximum speed of the photoelectrons when a clean potassium surface is illuminated with 546 nm light?

7.19 The work function of Mg is 3.68 eV. What happens when a clean Mg surface is illuminated with visible (400 nm to 700 nm) light?

7.20 When 546 nm light illuminates a surface, the

stopping potential is 0.42 volts. What will it be for 492 nm light?

7.21 The work function of Cs is 1.81 eV. What is the maximum speed of photoelectrons if a clean Cs surface is illuminated by a 1.0 mW, 632.8 nm He–Ne laser?

7.22 Again, the work function of Cs is 1.81 eV. A certain monochromatic light gives photoelectrons from a clean Cs surface with a maximum kinetic energy of 0.74 eV. What are the intensities emitted by ideal blackbodies having a peak frequency and wavelength that are the same as the frequency and wavelength of this monochromatic light?

7.23 The position vector to a source of photons makes an angle θ_1 with our x axis. The source moves at a velocity **v** parallel to our $+x$ axis. If the photons have an energy E_0 with respect to the source, what do we measure their energy to be?

7.24 Write a paragraph, from outside reading, that relates a partial explanation of Olber's paradox to photons and the Doppler effect.

7.25 The work function of Mg is 3.68 eV. Monochromatic light from a laser ejects photoelectrons in a two-photon process. The stopping potential for these photoelectrons is 1.47 V. a) Calculate the laser's frequency. b) What maximum wavelength could be used to emit photoelectrons from Mg in a two-photon process?

CHAPTER

Other Photon Processes

The interaction of electromagnetic radiation and electrons in the photoelectric effect was explained by the quantization of electromagnetic wave energy. In this chapter we will examine two more interactions involving electrons and those quantized bundles of electromagnetic energy called photons. The first is the production of x-ray photons in an electron tube. The second is what happens when a photon bumps into an electron. Both interactions are going on before you every time you watch TV or a CRT.

8.1
X-RAYS BY BREMSSTRAHLUNG

In 1895, Wilhelm Roentgen was performing experiments involving high voltage electrical discharges in low-pressure gas tubes. He found a new type of penetrating radiation, which he called x-rays. We now know that **x-rays** are electromagnetic waves with a frequency higher than the ultraviolet, or roughly 10^{16} to 10^{20} Hz.

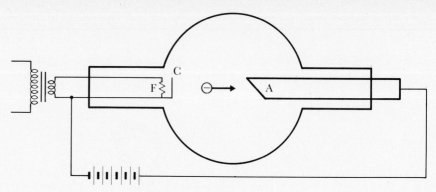

FIGURE 8.1
An x-ray tube and simplified related circuitry.

X-rays are produced in a typical x-ray tube, as shown in Fig. 8.1. The filament, F, heats the cathode, C, to emit electrons by thermionic emission. The electrons are accelerated by a high voltage potential difference (typically tens of kV). The rapidly moving electrons then hit the anode target, A, where they are stopped quickly.

According to classical physics, a charged particle that is decelerated (like the electron hitting the target) should radiate electromagnetic energy. According to quantum physics, this electromagnetic energy should occur in quanta of energy $E = h\nu$. Therefore x-ray quanta are produced in the target by decelerated electrons. This is called the **bremsstrahlung process.** *Bremsstrahlung* is the German word for *braking radiation.*

When an electron is slowed down and a photon is created (as shown in Fig. 8.2), mass–energy is conserved. If KE_i is the initial electron kinetic energy and KE_f is the final electron kinetic energy,

$$KE_i - KE_f = h\nu. \tag{8.1}$$

The most energetic photon will be produced when the electron is completely stopped and all its kinetic energy goes to produce one photon.

In the x-ray tube, the kinetic energy gained by the electron equals eV where e is the quantum of charge and V is the high voltage applied to the x-ray tube. Therefore

$$eV = h\nu_{max} = \frac{hc}{\lambda_{min}}. \tag{8.2}$$

FIGURE 8.2
An electron interacts with a nucleus, losing kinetic energy and emitting a photon.

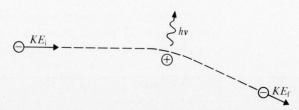

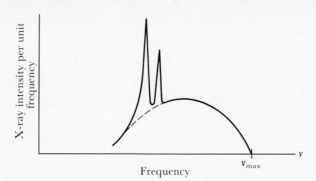

FIGURE 8.3

A typical x-ray spectrum from an x-ray tube. The peaks are characteristic of the target material in the tube.

For solving most practical problems of this type, we use $h = 4.136 \times 10^{-15}$ eV $\cdot$ s and $hc = 1240$ keV $\cdot$ pm. Quite often, x-ray wavelengths are given in angstroms (1 Å $= 10^{-10}$ m), but the Å is not an SI unit, so we'll use nm (10^{-9} m) and pm (10^{-12} m).

The graph of x-ray intensity versus frequency often looks like Fig. 8.3. The two peaks superimposed on the smooth bremsstrahlung curve result from the x-rays emitted by the inner atomic electrons of the target metal. We will explain this process in Section 20.4.

EXAMPLE 8.1 An electron is slowed down quickly from 1.00×10^7 m/s to 0.50×10^7 m/s, and a single photon is created in the process. What is its frequency and wavelength?

Solution We obtain the photon frequency by solving $KE_i - KE_f = h\nu$ to arrive at $\nu = (KE_i - KE_f)/h$. Since $v^2 \ll c^2$, we don't need relativistic equations. So $KE = \frac{1}{2}m_0 v^2$ gives

$$\nu = \frac{m_0}{2h} (v_i^2 - v_f^2) = \frac{9.11 \times 10^{-31} \text{ kg}}{2(6.626 \times 10^{-34} \text{ J} \cdot \text{s})} (1.00 - 0.25) \times 10^{14} \text{ m}^2/\text{s}^2$$

$$= 5.2 \times 10^{16} \text{ Hz}.$$

Then

$$\lambda = \frac{c}{\nu} = \frac{3.00 \times 10^8 \text{ m/s}}{5.2 \times 10^{16} \text{ Hz}} = 5.8 \times 10^{-9} \text{ m} = 5.8 \text{ nm}.$$

This is a very low-frequency (and energy), long-wavelength x-ray. ●

EXAMPLE 8.2 A potential difference of 35 kV is applied to an x-ray tube. Calculate the limiting value of the frequency and wavelength of the x-ray spectrum.

Solution The simplest way to solve this problem is to use electron volt units in $eV = h\nu_{max}$. Then

$$\nu_{max} = \frac{eV}{h} = \frac{e(35 \times 10^3 \text{ V})}{h} = \frac{35 \times 10^3 \text{ eV}}{4.136 \times 10^{-15} \text{ eV} \cdot \text{s}} = 8.5 \times 10^{18} \text{ Hz}.$$

We can now obtain λ_{min} by using $\lambda_{min} = c/\nu_{max}$, but let's use $eV = hc/\lambda_{min}$ for the practice:

$$\lambda_{min} = \frac{hc}{eV} = \frac{1240 \text{ keV} \cdot \text{pm}}{e(35 \text{ kV})} = 35 \text{ pm}.$$

●

Note that ν_{max} represents the maximum frequency of the x-rays emitted in the bremsstrahlung process. It is not the frequency at which the intensity curve peaks. You can find the position of the peaks on the intensity curve in tables if you know the target metal.

8.2
THE COMPTON EFFECT

In classical physics, when an electromagnetic wave of frequency ν is incident on a charged particle, the oscillating electric field of the electromagnetic wave exerts a force at the frequency ν on the particle. The charged particle then oscillates at that same frequency and reradiates energy in all directions except along its direction of oscillation. This phenomenon explains, for example, why you see your reflection in a mirror with the colors unchanged.

However, if an x-ray beam is directed at a material that has many free electrons in it, we find that the scattered beam has two wavelengths and two frequencies at a given angle. The two wavelengths are λ_o, the original wavelength, and λ_s, a somewhat longer wavelength. The scattered wavelength λ_s depends on the scattering angle θ, but not on the material used. A. H. Compton explained this effect in 1922. We can do so by using our knowledge of conservation of mass–energy, conservation of momentum, special relativity, and photon energy.

A photon is a quantum of light energy and moves at speed c. Then $m_0 = m\sqrt{1 - v^2/c^2}$, with $v = c$, tells us that $m_0 = 0$, and $E_0 = m_0c^2$ tells us that $E_0 = 0$. The photon has zero rest mass and zero rest energy. There is no such thing as a photon at rest, it is either moving at $v = c$ or it doesn't exist. Since $E_0 = 0$, $E^2 = E_0^2 + p^2c^2 = 0 + p^2c^2$, or $E = pc$. The photon therefore has a momentum $p = E/c$. But $E = h\nu$ for photons, so the momentum of a photon is $h\nu/c$. Also, $\nu/c = 1/\lambda$, so we can also write the momentum as h/λ. In summary,

$$p = \frac{E}{c} = \frac{h\nu}{c} = \frac{h}{\lambda} \qquad \text{(photon)}. \qquad (8.3)$$

EXAMPLE 8.3   An x-ray photon has a wavelength of 100 pm. Calculate its momentum in SI and eV/c units using Eq. (8.3).

Solution Evaluating Eq. (8.3), we get

$$p = \frac{h}{\lambda} = \frac{6.626 \times 10^{-34}\,\text{J} \cdot \text{s}}{100 \times 10^{-12}\,\text{m}} = 6.6 \times 10^{-24}\,\text{N} \cdot \text{s} \quad \text{or} \quad \text{kg} \cdot \text{m/s}.$$

We can then convert the result directly into eV/c units. Another way to solve the problem using Eq. (8.3) would be to first find the frequency:

$$\nu = \frac{c}{\lambda} = \frac{3.00 \times 10^8\,\text{m/s}}{1.00 \times 10^{-10}\,\text{m}} = 3.00 \times 10^{18}\,\text{Hz}.$$

Then

$$p = h\nu/c = (4.136 \times 10^{-15}\,\text{eV} \cdot \text{s})(3.00 \times 10^{18}\,\text{Hz})/c = 1.24 \times 10^4\,\text{eV}/c.$$

                                                               ●

EXAMPLE 8.4   Find the rest mass, rest energy, mass, and energy of a 100-pm photon.

Solution The rest mass and rest energy of a photon are zero. From Example 8.3, $\nu = 3.00 \times 10^{18}$ Hz. Therefore the energy of the photon is

$$E = h\nu = (6.626 \times 10^{-34}\,\text{J} \cdot \text{s})(3.00 \times 10^{18}\,\text{Hz}) = 1.99 \times 10^{-15}\,\text{J}.$$

Its mass is

$$m = \frac{E}{c^2} = \frac{1.99 \times 10^{-15}\,\text{J}}{(3.00 \times 10^8\,\text{m/s})^2} = 2.21 \times 10^{-32}\,\text{kg}.$$

                                                               ●

Since 2.21×10^{-32} kg is more than 1/50 the mass of an electron, a 100-pm photon moving at the speed of light can give an electron a good bump in a collision.

The photon is a quantum of electromagnetic wave energy, but it does have some particle-like properties: It is a bundle of energy that has momentum and mass, and in the **Compton effect** it collides with an electron, as shown in Fig. 8.4.

In classical physics a perfectly elastic collision involved conservation of kinetic energy and conservation of linear momentum. In modern physics such a collision involves conservation of mass–energy and conservation of linear momentum. Before the collision the system's energy is the energy

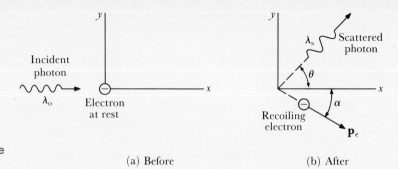

FIGURE 8.4
The Compton effect: A photon collides with a free electron.

(a) Before

(b) After

of the incident photon plus the rest energy of the electron. After the collision the system's energy is the energy of the scattered photon plus the total energy of the electron as it recoils.

From $E = hc/\lambda$, the photon energies are hc/λ_o and hc/λ_s. The scattered photon has a wavelength and frequency that are different from those of the incident photon: It is a different photon. In the Compton effect, the incident photon is annihilated and the scattered photon is created in the scattering process.

Conservation of mass–energy gives us

$$\overset{\text{photon}}{\frac{hc}{\lambda_o}} + \overset{\text{electron}}{m_0 c^2} = \overset{\text{photon}}{\frac{hc}{\lambda_s}} + \overset{\text{electron}}{(m_0 c^2 + KE)}. \tag{A}$$

Then

$$KE = \frac{hc}{\lambda_o} - \frac{hc}{\lambda_s}, \tag{B}$$

which simply states that the electron's recoil kinetic energy is the photon energy before minus the photon energy after.

We can represent conservation of linear momentum by the vector triangle shown in Fig. 8.5. Applying the law of cosines to the triangle gives

$$p_e^2 = \left(\frac{h}{\lambda_o}\right)^2 + \left(\frac{h}{\lambda_s}\right)^2 - 2\left(\frac{h}{\lambda_o}\right)\left(\frac{h}{\lambda_s}\right)\cos\theta. \tag{C}$$

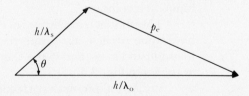

FIGURE 8.5
Conservation of linear momentum in the Compton effect: The vector sum of the momentums after the collision equals the momentum before.

For the electron, $E^2 = p_e^2 c^2 + (m_0 c^2)^2$ and $E = KE + m_0 c^2$ yield

$$p_e^2 = \left(\frac{KE}{c}\right)^2 + 2KEm_0.$$ (D)

Equating Eqs. (C) and (D), then substituting Eq. (B) to eliminate KE, gives

$$\left(\frac{h}{\lambda_o}\right)^2 + \left(\frac{h}{\lambda_s}\right)^2 - 2\left(\frac{h}{\lambda_o}\right)\left(\frac{h}{\lambda_s}\right)\cos\theta = \left(\frac{h}{\lambda_o} - \frac{h}{\lambda_s}\right)^2 + 2m_0 c\left(\frac{h}{\lambda_o} - \frac{h}{\lambda_s}\right).$$ (E)

A bit of algebra helps us to finally reduce Eq. (E) to

$$\Delta\lambda = \lambda_s - \lambda_o = \frac{h}{m_0 c}(1 - \cos\theta).$$ (8.4)

For a free electron, $h/m_0 c = 2.426 \times 10^{-12}$ m $= 2.426$ pm, which is called the Compton wavelength of the electron.

EXAMPLE 8.5   100.0-pm x-rays are scattered from a metal. What are the changes in the wavelength and scattered wavelengths for scattering angles of 10°, 90°, and 180°?

Solution $\Delta\lambda$ is the change in the wavelength. From Eq. (8.4), at 10°: $\Delta\lambda =$ (2.426 pm)$(1 - \cos 10°) = 0.037$ pm. At 90°: $\Delta\lambda = $ (2.426 pm)$(1 - \cos 90°) = 2.426$ pm. And at 180°: $\Delta\lambda = $ (2.426 pm)$(1 - \cos 180°) = 4.852$ pm. The scattered wavelengths are $\lambda_s = \lambda_o + \Delta\lambda = 100.0, 102.4,$ and 104.9 pm.   ●

Complete backscattering ($\theta = 180°$) gives the greatest momentum change and therefore the greatest wavelength gain. The maximum recoil kinetic energy that can be given to an electron also occurs at 180° in the Compton effect. For a 100.0-pm photon,

$$KE = hc\left(\frac{1}{\lambda_o} - \frac{1}{\lambda_s}\right) = \frac{hc(\lambda_s - \lambda_o)}{\lambda_o\lambda_s} = \frac{hc\Delta\lambda}{\lambda_o\lambda_s}$$

$$= \frac{(1240 \text{ keV} \cdot \text{pm})(4.852 \text{ pm})}{100.0 \text{ pm} \times 104.9 \text{ pm}} = 0.574 \text{ keV}.$$

When Compton-effect experiments are performed (as illustrated in Fig. 8.6), two peaks in the intensity versus wavelength curve are found, as shown in Fig. 8.7. One scattered wavelength is at λ_s—in agreement with Eq. (8.4). The other scattered wavelength gives a peak at the original wavelength λ_o—within experimental error. This virtually unshifted peak is caused by scattering from electrons bound to the atoms in the target. For

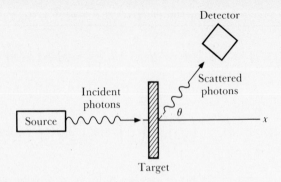

FIGURE 8.6
A Compton-effect experiment.

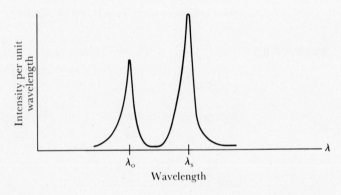

FIGURE 8.7
Intensity as a function of wavelength for photons scattered in the Compton effect through a large angle.

these bound electrons, the electron rest mass m_0 in $h/m_0 c$ must be replaced by the mass of the whole atom. Since the mass of the whole atom is usually 10^4 or more larger than m_0, $\Delta\lambda$ is smaller by 10^4 or more, making $\lambda_s \approx \lambda_o$.

You've seen that the energy lost by an electron when it slows down quickly can be emitted in the form of a photon. If the voltage is high enough (as for example in a color TV set), it can be an x-ray photon. After production, the x-ray photons can have their energy reduced and they can be scattered from their original paths by collisions with free electrons. This explains in part how you are shielded from the x-rays emitted from your TV or CRT. Shielding involves photon absorption, which we discuss in Chapter 9.

QUESTIONS AND PROBLEMS

8.1 Why must engineers and scientists shield against x-ray production in high-voltage equipment?

8.2 A color TV set operates with a picture tube voltage of 26,000 V. What is the minimum wavelength x-ray produced?

8.3 The maximum frequency bremsstrahlung process is sometimes referred to as "the inverse of the photoelectric effect." Why?

8.4 An electron moving at 1.50×10^7 m/s is decelerated with the creation of a single x-ray photon having a wavelength of 4.0 nm. Calculate the electron's new speed.

8.5 An electron is decelerated to a speed of 3.0×10^6 m/s with the creation of a single 2.0×10^{17}-Hz photon. What was the electron's initial speed?

8.6 In emitting a single photon in a bremsstrahlung event, an electron's mass decreases from three times its rest mass to twice its rest mass. Calculate the wavelength of the photon.

8.7 An electron is stopped with the emission of a single 10^{20}-Hz photon. How fast was the electron moving?

8.8 A 0.20-MeV electron emits a photon in decelerating to a kinetic energy of 0.15 MeV. Calculate the a) energy, b) frequency, and c) wavelength of the x-ray photon.

8.9 How many kilovolts must be applied to an x-ray tube to get x-rays with frequencies as high as 2.0×10^{19} Hz?

8.10 What potential difference applied to an x-ray tube will result in x-rays as short as 62 pm?

8.11 Discuss the x-ray spectrum produced by stopping 26-keV electrons in a color TV set.

8.12 An electron is stopped in a series of decelerations from a speed of 2.90×10^8 m/s. How many 10^{20}-Hz x-ray photons could possibly be created in the process?

8.13 A theorist suggests that photons have a small rest mass. What problems do you see with this idea?

8.14 Photons must have mass since they have energy and since $m = E/c^2$. Objects with mass are attracted by the force of gravity. Therefore a photon moving away from a massive body must do work against the gravitational field, losing kinetic energy.

a) Find the expression for the mass of a photon in terms of its frequency; in terms of its wavelength.

b) Show that v decreases and λ increases when a photon is moving away from a massive body (these changes in v and λ are called the *gravitational redshift*).

c) Show that, if $\Delta v \ll v$, then $\Delta v/v \approx -gy/c^2$ for the gravitational redshift at constant g.

d) Show that if $(M/R) > (c^2/G)$, then not even photons can escape the massive body. (This result is wrong by a factor of 2 according to general relativity.) If not even light can escape the body because its mass has been compressed into such a small radius, it is called a *black hole*.

e) Calculate the maximum radius of a black hole with a mass of 2.0×10^{30} kg (one solar mass) from $(M/R) > (c^2/G)$.

8.15 Calculate the gravitational force of attraction on a 14.4-keV photon at the earth's surface. (This attractive force can do work on the photon. The resulting energy and frequency change of 14.4-keV photons was detected via the Mössbauer effect in the Pound–Rebka experiment.)

8.16 Calculate the momentum of a yellow-light photon (let $\lambda = 550$ nm).

8.17 Calculate the maximum momentum x-ray created in an x-ray tube operated at 32 kV.

8.18 A photon has a momentum of 1.44×10^4 eV/c. a) Calculate its energy, frequency, and wavelength. b) Where in the electromagnetic spectrum does this place the photon?

8.19 Show that the mass of a photon can be written as $m = p/c$.

8.20 Calculate the momentum of a photon having a mass of 3.0×10^{-30} kg.

8.21 Equation (8.4) is written in terms of wavelength. Use Eq. (8.4) to find an equation for v_s in terms of v_o, θ, and constants.

8.22 A proton has a rest mass of 1836 electron rest masses. Calculate the maximum wavelength shift for Compton scattering from free protons.

8.23 We know that 50.0-pm x-rays are incident on a sample. At what angle should we look for
a) 53.1-pm x-rays?   b) 50.0-pm x-rays?

8.24 By what percent is the wavelength changed for Compton scattering of yellow (550 nm) light for perpendicular reflection? Discuss the practical consequences of this percent change in the wavelength.

8.25 If 1.00-MeV photons are Compton scattered at a 5° angle, what is their new energy? Explain the missing energy.

8.26 Some 2.00-MeV photons are incident on a target. At what angle will 0.50-MeV, Compton-scattered photons be discovered?

8.27 An x-ray photon has a momentum of 6.626×10^{-23} N · s. After a Compton scattering through 60°, what is the new photon's momentum? Explain the change.

8.28 Will two Compton scatterings at $\theta/2$ together produce the same wavelength change as one scattering at θ?

8.29 What is the magnitude of the maximum possible change in the momentum of photons that are Compton scattered after having been emitted from an x-ray tube operating at 50 kV?

8.30 Each photon in a beam has the same momentum as a proton moving at 1.00 m/s. When the beam strikes a metal surface, the stopping potential is 0.86 V. Calculate the work function of the metal.

8.31 An ideal blackbody emits 9.072×10^5 W/m². Calculate the momentum of the photons that have the frequency at which the blackbody has its greatest energy per unit frequency.

8.32 A star, emitting like an ideal blackbody, has an intensity at its surface of 64 MW/m². It will emit a few photons with a wavelength 10^5 times smaller than the wavelength at which it emits its greatest intensity per unit wavelength. What scattered wavelengths will be detected when those photons are Compton scattered through 90° by a metal?

CHAPTER 9

Some Gamma-ray Processes

Photons completely disappear with the creation of matter and antimatter. Matter is totally annihilated by antimatter with complete conversion of all initial rest energy into kinetic energy. Can such events occur, or are they just science fiction? We will give and discuss the answer in this chapter. Also, we will introduce photon absorption, a subject with many radiation-shielding and medical and technological x-ray and gamma-ray practical applications.

9.1
PAIR PRODUCTION

X-ray photons are created by processes that usually involve electrons; x-ray frequencies are approximately in the 10^{16} Hz to 10^{20} Hz range. **Gamma rays** (γ rays) are also high-frequency electromagnetic waves; γ-ray photons are created by processes that usually involve subatomic particles.

Figure 9.1 shows that γ rays roughly fall into the 10^{18} Hz and up frequency range. Any 10^{19}-Hz electromagnetic wave is composed of pho-

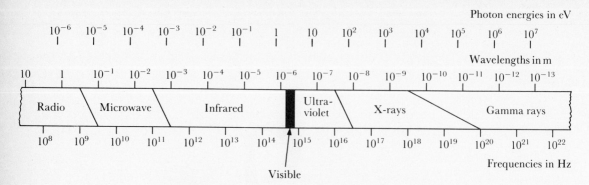

FIGURE 9.1

The electromagnetic spectrum: The frequencies, wavelengths, and energies found in nature extend over such a wide range that we have to use the logarithmic scale to show all important bands. The boundaries between bands are rather arbitrary and therefore should not be taken as exact.

tons. Whether they are x-ray or γ-ray photons can only be determined by finding the source of the photons. The photons themselves are identical.

It's possible for a γ ray to disappear and for two particles to appear in its place under certain circumstances, such as near the nucleus of an atom. In this process, called **pair production,** several conservation laws must be satisfied including conservation of mass–energy, linear momentum, and charge. For instance, two negative electrons can't be produced. A photon has zero charge, so the total charge of the pair produced must also add to zero.

Experiments have confirmed that antiparticles exist for particles in nature. **Antiparticles** have the same rest mass as their particle but opposite charge. The antiparticle of the electron is the **positron.** It has a rest mass of 9.11×10^{-31} kg, or 0.511 MeV/c^2, the same as that of the electron. However, the positron has a charge of $+1.60 \times 10^{-19}$ C, or $+e$, the opposite of the $-e$ charge of the negative electron. In 1928 P. A. M. Dirac theoretically predicted these properties of the positron. Later, Iréne Curie and Frédérick Joliot saw the tracks of positrons emerging from a radioactive source but interpreted them as negative electrons being attracted to that source. Thus Carl Anderson gained the credit for the first correct identification of a positive electron on August 2, 1932, although he was unaware of Dirac's prediction.

The minimum energy of a photon needed to create a particle and its antiparticle is $m_0 c^2 + m_0 c^2$, by conservation of mass–energy. That is,

$$h\nu_{\min} = 2m_0 c^2. \tag{9.1}$$

In this case the particle and the antiparticle would have negligible kinetic energy after the creation process.

EXAMPLE 9.1 Calculate the energy, frequency, and wavelength of the photon that will create an electron–positron pair with about zero kinetic energy.

Solution The photon doesn't need to supply any kinetic energy in this case, so the minimum energy and frequency and maximum wavelength are

$$E = h\nu_{min} = 2m_0c^2 = 2(0.511 \text{ MeV}/c^2)c^2 = 1.022 \text{ MeV};$$

$$\nu_{min} = \frac{1.022 \text{ MeV}}{h} = \frac{1.022 \times 10^6 \text{ eV}}{4.136 \times 10^{-15} \text{ eV} \cdot \text{s}} = 2.47 \times 10^{20} \text{ Hz};$$

$$\lambda_{max} = \frac{c}{\nu_{min}} = \frac{3.00 \times 10^8 \text{ m/s}}{2.47 \times 10^{20} \text{ Hz}} = 1.21 \times 10^{-12} \text{ m}.$$

Pair production by one photon can't occur in empty space without violation of conservation of linear momentum. To understand the reason, consider the minimum energy case. In it the photon carries a momentum $p = h/\lambda_{max}$ into the process, but the electron and positron have virtually zero momentum after the process is completed. Some momentum before and zero momentum after the event violates conservation of momentum. Even if the photon has an energy greater than $h\nu_{min}$ and the electron and positron move rapidly apart, we could make a Lorentz transformation into the center-of-mass frame of reference of the electron and positron. In this frame of reference, the vector sum of the momentum of the two particles is zero after their creation (because the velocity of the center of mass is zero in the center-of-mass frame of reference) but the initial photon momentum is not zero. Again, conservation of momentum is violated.

However, linear momentum can be conserved if there is a heavy nucleus nearby, which can move to conserve momentum. (See Fig. 9.2.) The nucleus will move with $v \ll c$, so that its kinetic energy is $\frac{1}{2}M_N v_N^2 = p_N^2/2M_N$. Because of its large mass its kinetic energy will be negligible, and Eq. (9.1) will hold. For example, a lead (Pb) nucleus has a rest mass of about $2 \times 10^5 \text{ MeV}/c^2$. Therefore, in carrying away 1.022 MeV/c of momen-

(a) Before   (b) After

FIGURE 9.2
Minimum-energy, single-photon pair production: A photon of energy $h\nu_{min}$ approaches a heavy nucleus. The photon disappears, a particle–antiparticle pair is produced, and the nucleus recoils. The particle, antiparticle, and nucleus have negligible kinetic energy.

tum, its kinetic energy will be only

$$\frac{p_N^2}{2M_N} = \frac{(1.022 \text{ MeV}/c)^2}{2(2 \times 10^5 \text{ MeV}/c^2)},$$

or about 3×10^{-6} MeV $= 3$ eV, and 3 eV is negligible compared to 1.022 MeV.

EXAMPLE 9.2 Electron–positron pair production occurs with a 1.234 MeV γ-ray photon. Account for the energy.

Solution First, 1.022 MeV goes into the creation of the electron and positron, leaving 1.234 MeV $-$ 1.022 MeV $= 0.212$ MeV. A negligible amount of this remainder goes to the kinetic energy of the heavy nucleus, so almost all the 0.212 MeV is shared between the kinetic energy of the electron and the kinetic energy of the positron.   ●

9.2

PAIR ANNIHILATION

Pair production is going on constantly, especially in the upper atmosphere as a result of cosmic-ray activity. But here on earth we find very few positrons in nature. The reason is that in nature there are so many electrons around, and electrons annihilate positrons. In some ways, this process is the opposite of pair production. It is called **pair annihilation** and also occurs for other particle–antiparticle pairs.

In the positron–electron case, the two oppositely charged particles briefly attract one another to form a kind of "atom" called **positronium.** Within 10^{-10} s, the two particles spiral into one another and annihilate. (See Fig. 9.3.) In their place, two (or occasionally three) γ rays are found. In

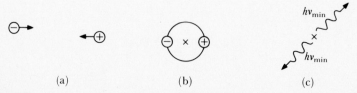

(a)   (b)   (c)

FIGURE 9.3
Minimum-energy, two-photon pair annihilation: (a) A particle and an antiparticle approach each other with negligible kinetic energy; (b) for a short time they may orbit about their common center of mass, radiating a negligible amount of energy; and (c) then they annihilate one another, with all their mass–energy going to two equal-energy, opposite-momentum photons.

the positronium frame of reference, the two γ rays move away from each other in opposite directions with equal energies, frequencies, and wavelengths. Therefore the vector sum of their linear momentum is zero, which is necessary to conserve linear momentum. (In the positronium frame of reference, the initial linear momentum of the system is zero.) Just one γ ray could not have zero momentum, so two or three must be created. For the much more common two γ-ray case, when the initial kinetic energies of the particle and its antiparticle are negligible, conservation of mass–energy yields $m_0 c^2 + m_0 c^2 = 2h\nu_{min}$, or

$$m_0 c^2 = h\nu_{min}. \qquad (9.2)$$

Particle–antiparticle annihilation is an everyday occurrence in many labs. For example, if one or both particles have appreciable kinetic energy before the collision, the total kinetic energy will be shared by the γ rays. Then measurements of γ-ray annihilation angles and energies can be used to obtain information about free-electron velocities in metals by bombarding the metals with positrons of known energy. Also, **positron emission tomography** (PET) is a medical diagnostic tool that utilizes the emitted γ rays.

EXAMPLE 9.3 Show that the minimum γ-ray energy in two γ-ray pair annihilation equals the rest energy of one of the particles. Give examples.

Solution Equation (9.2) gives $h\nu_{min} = m_0 c^2$, which immediately proves the statement; $h\nu_{min}$ is the minimum gamma-ray energy, and $m_0 c^2$ is the particle rest energy. Therefore, as examples, for electron and positron the minimum γ-ray energy is 0.511 MeV, and for proton and antiproton it is 938.3 MeV. ●

9.3
PHOTON ABSORPTION

You have now studied three processes that can result in the annihilation of a photon: the photoelectric effect, the Compton effect, and pair production. In all three cases, an electron is moving after the event, so the moving electron can be used to detect the photon. At low energies the photoelectric effect is very strong, but its probability drops off rapidly at higher energies. As photons increase in energy and momentum, the Compton effect becomes less important in scattering them out of a beam. Pair production doesn't begin until 1.022 MeV, but it becomes more and more effective after that.

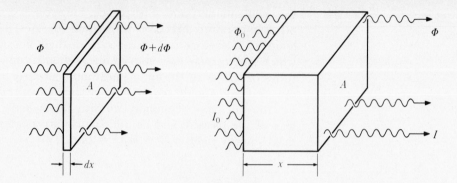

FIGURE 9.4
Absorption of photons:
$d\Phi$ is negative.

If we let Φ equal the **photon flux** (the number of photons per perpendicular area per time), the intensity (the energy per perpendicular area per time) equals the photon flux times the energy per photon. That is,

$$I = \Phi h\nu. \tag{9.3}$$

When Φ photons per area per time move through a thickness of material dx (measured in the direction of motion), $d\Phi$ will be removed. This $d\Phi$ is a decrease in Φ, and will be proportional to Φ and to the infinitesimal thickness dx. The proportionality constant, μ, is called the **absorption coefficient** and has units of reciprocal length, such as m^{-1} or cm^{-1}. A large μ means a large photon absorption. Therefore, $d\Phi = -\mu\Phi\,dx$. (See Fig. 9.4.) If we let the material extend from $x = 0$ to $x = x$, the incident photon flux be Φ_0, and the transmitted photon flux be Φ, we can divide both sides of $d\Phi = -\mu\Phi\,dx$ by Φ and integrate. That is,

$$\int_{\Phi_0}^{\Phi} \frac{d\Phi}{\Phi} = -\mu \int_0^x dx$$

gives

$$\ln\frac{\Phi}{\Phi_0} = -\mu x \quad \text{or} \quad \frac{\Phi}{\Phi_0} = e^{-\mu x},$$

or

$$\Phi = \Phi_0 e^{-\mu x}. \tag{9.4}$$

Thus Φ drops off exponentially from Φ_0, approaching zero only in the limit of an infinitely thick absorber. Since $I = \Phi h\nu$, multiplying both sides of Eq. (9.4) by $h\nu$ gives

$$I = I_0 e^{-\mu x} \tag{9.5}$$

where I_0 is the incident intensity, and I is the transmitted intensity. (See Fig. 9.5.)

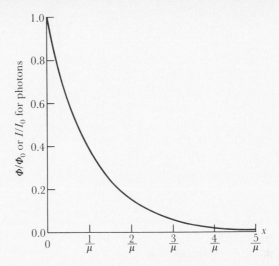

FIGURE 9.5
The fraction of the photon flux or of the intensity that will pass through an absorber of thickness x in terms of the reciprocal of the absorption coefficient. For example, one-half will pass through if the thickness is $0.69/\mu$.

EXAMPLE 9.4 At what absorber thickness will the photon intensity and photon flux drop to $1/e = 36.8\%$ of their incident values?

Solution Equations (9.4) and (9.5) give $\Phi/\Phi_0 = I/I_0 = e^{-\mu x}$. Here we want $\Phi/\Phi_0 = I/I_0 = 1/e = e^{-1}$. Therefore, $e^{-1} = e^{-\mu x}$, or $1 = \mu x$, so $x = 1/\mu$. ●

When the thickness of a material equals the reciprocal of the absorption coefficient, 36.8 percent of the incident beam will remain. After another equal thickness is added, 36.8 percent of that 36.8 percent, or 13.5 percent, will remain. When $x = 4.6/\mu$, only 1 percent remains.

The absorption coefficient is not a universal constant. It depends mainly on the material and on the incident photons' energy (that is, frequency and wavelength). If we ignore certain types of resonant processes, which give sharp peaks in the absorption curve, the total μ will equal $\mu_{\text{photoelectric}} + \mu_{\text{Compton}} + \mu_{\text{pair production}}$, as shown in Fig. 9.6.

EXAMPLE 9.5 The absorption coefficient for 2.0-MeV photons in iron (Fe) is 33 m^{-1}. If there are 5.0×10^6 of these 2.0-MeV photons incident normally on each square meter of an iron plate each second, what iron thickness will decrease the intensity to 8.0×10^5 MeV/m$^2 \cdot$ s?

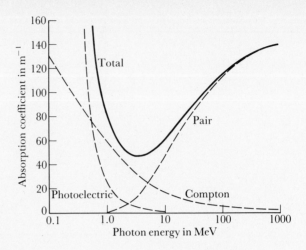

FIGURE 9.6

Dependence of absorption coefficients on photon energy for lead.

Source: Adapted from I. Kaplan, NUCLEAR PHYSICS, 2/e, © 1962, Addison-Wesley Publishing Co., Inc., Reading, Mass. P. 416, Fig. 15.7.

Solution We want to find the x that will make $I = 8.0 \times 10^5$ MeV/m² · s. We are given $h\nu = 2.0$ MeV/photon, $\mu = 33$ m⁻¹, and $\Phi_0 = 5.0 \times 10^6$ photons/m² · s. We can either use $I = \Phi h\nu$ to find Φ and then use $\Phi = \Phi_0 e^{-\mu x}$ to find x or use $I_0 = \Phi_0 h\nu$ to find I_0 and then use $I = I_0 e^{-\mu x}$ to find x. Let's use the second method:

$$I_0 = \Phi_0 h\nu = (5.0 \times 10^6 \text{ photon/m}^2 \cdot \text{s})(2.0 \text{ MeV/photon})$$
$$= 1.0 \times 10^7 \text{ MeV/m}^2 \cdot \text{s}.$$

Then $I_0/I = e^{+\mu x} = 1.0 \times 10^7/8.0 \times 10^5 = 12.5$. Taking the natural log of both sides, solving for x, and substituting for μ gives

$$x = (\ln 12.5)/33 \text{ m}^{-1} = 0.077 \text{ m} = 77 \text{ mm}. \qquad \bullet$$

Figure 9.7 shows that below 1.022 MeV (the onset of pair production), μ increases rapidly with decreasing photon energy. Since the x-ray range is below 1.022 MeV, "hard" (high-energy) x-rays will have a lower absorption coefficient than "soft" (low-energy) x-rays (ignoring certain resonant processes). Therefore a greater thickness of absorber will be needed to decrease hard x-rays by the same fraction as soft x-rays in a particular material. The basis for many medical and technological x-ray applications is the fact that different materials usually have different absorption coefficients.

You now know that it's possible to create equal amounts of matter and antimatter in a pair production process. But if we start with a single photon, conservation laws require that the total charge remain zero, that

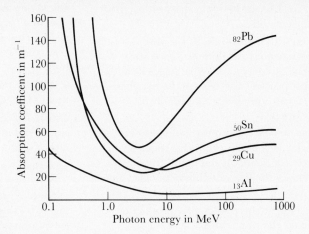

FIGURE 9.7
Total absorption coefficients for several metals.

Source: Adapted from I. Kaplan, NUCLEAR PHYSICS, 2/e, © 1962, Addison-Wesley Publishing Co., Inc., Reading, Mass. P. 417, Fig. 15.8.

the photon energy is greater than $2m_0c^2$, and that there be another particle about to conserve linear momentum. However, antimatter doesn't last long in our universe (unless somehow isolated), because when a particle meets its antiparticle they annihilate one another.

It would be wrong to say that these two processes involve the "creation of mass from energy" or the "total conversion of mass into pure energy." Since $E = mc^2$ and mass–energy is conserved in these processes, there is as much mass or energy before the processes as there is after. The mass and energy have just been redistributed.

You have also seen that the photon flux and intensity decrease exponentially with absorber thickness. Therefore, if a single thickness of absorber cuts the photon flux in half, a double thickness will not drop it to zero, but will just cut it in half again (to $\Phi_0/4$). The photon flux and intensity only approach zero as the absorber's thickness approaches infinity. The rate at which photons are absorbed is governed by the absorption coefficient, a quantity that depends mainly on the type of material and the photon energy.

QUESTIONS AND PROBLEMS

9.1 Calculate the energy, frequency, and wavelength of the photon that may create a proton–antiproton pair.

9.2 Electrons moving in a circle in a particle accelerator emit 10^{19}-Hz electromagnetic waves. Would these be considered x-rays or γ rays?

9.3 Can a γ ray near a heavy nucleus produce two protons? Explain.

9.4 A photon with momentum 211.4 MeV/c produces a muon and an antimuon, each with negligible kinetic energy. a) What happens to the photon in the process? b) What is the rest mass of a muon? Of an antimuon? c) Are any other particles involved?

9.5 Two equal-energy photons collide head-on and annihilate each other, producing an electron–positron pair. Calculate the minimum frequency of either photon for this two-photon pair production process.

9.6 It's possible for two-photon pair production to occur if the density of photons is very high. Consider such a process in the center-of-mass frame of reference and explain why no heavy nucleus is needed.

9.7 Occasionally a pair annihilation process near a heavy nucleus will produce a single photon.
a) Why is the heavy nucleus necessary? b) What minimum energy will the photon have for positronium decay?

9.8 Occasionally a pair annihilation process will produce three photons. In terms of the positronium frame of reference, a) sketch a diagram of the three γ rays leaving, and b) calculate their minimum energy if they have equal wavelengths.

9.9 In the positronium or center-of-mass system, the electron and positron each initially has 27 keV of energy. Calculate the energy, momentum, wavelength, and frequency of the two resulting γ rays.

9.10 A proton and antiproton collide head-on with equal kinetic energies. Two γ rays with wavelengths of 1.00 fm are produced. Calculate the kinetic energy of the proton.

9.11 What is the maximum wavelength produced by a) an electron–positron annihilation? b) by a proton–antiproton annihilation?

9.12 Give the charge and rest mass of the antiproton.

9.13 Use the momentum of a photon to show that the radiation pressure equals I/c for complete

absorption and $2I/c$ for complete reflection from a surface perpendicular to a beam of electromagnetic waves. *Hint:* Pressure = force/perpendicular area, and force = time rate of change of momentum.

9.14 Show that at $x = 4.6/\mu$, 99% of the photon flux has been absorbed or scattered out.

9.15 If 1.0 cm of a certain material is to absorb or scatter out 90% of the initial photon flux, what must the absorption coefficient be?

9.16 Derive an expression involving I_0, μ, x, A, ρ (the density), and c (the specific heat) for the maximum rate of temperature rise of an absorber due to its absorption of photons. What will make this a high estimate?

9.17 a) Calculate the thickness of an absorber needed to drop the intensity to one-half the original value. b) What will I/I_0 equal after two of these thicknesses?

9.18 Some 0.50-MeV photons are incident normally on a surface at a flux of 2.0×10^6 photons/m² · s. If the intensity is 7.0×10^5 MeV/m² · s on the other side of the 5.0-mm thick absorber, calculate the absorption coefficient.

9.19 Sketch I versus x or Φ versus x for a) soft x-rays and b) hard x-rays.

9.20 From a graph like that in Fig. 9.7, how can you tell which photons will require the thickest shielding for a given percentage of absorption?

9.21 Two photon beams have the same incident flux, but beam 2 has twice the absorption coefficient of beam 1. What thickness absorber will give the two beams the same transmitted power per area?

9.22 Given that 6.0×10^4 of 40-pm x-ray photons/m²/s are incident normally on a 0.30-mm thick absorber that has an absorption coefficient of 5.2×10^4 m⁻¹ for these photons, calculate the transmitted intensity.

9.23 Using Fig. 9.7; determine the thickness of $_{13}$Al that will absorb 99% of the photons in a beam of 1.0-MeV/c γ rays.

9.24 Refer to Fig. 9.6: 1.2×10^6 MeV/m² · s of 2-Mev photons are incident normally on 5-mm thick lead shielding. How many photons per square meter are annihilated every minute a) in Compton scattering? b) in the photoelectric effect? c) in pair production?

9.25 Let's assume that 5.00 photons/m² · s are incident normally on a 2.0-cm thick Pb absorber. Calculate the transmitted intensity if these photons are created in the pair annihilation of pions and antipions of negligible kinetic energy and a rest mass of 140 MeV/c^2.

9.26 A beam of photons with just barely enough individual photon energy to create electron–positron pairs undergoes Compton scattering from free electrons before pair production occurs.
a) Can the Compton-scattered photons create electron–positron pairs? b) Calculate the momentum of the photons found at a 20° angle from the original beam.

9.27 Every second, 300 photons are incident normally on each square meter of a 3.0-cm thick absorber. Each of these photons has only 75% of the energy needed to create an electron–positron pair. The absorption coefficient for these photons in this absorber is 20 m^{-1}. Calculate the transmitted photon intensity.

The Wave Nature of Particles

In this chapter we will present more deviations from classical physics and from our everyday experience. It seems that small things in nature are ambivalent, sometimes acting as if they are particles and sometimes acting as if they are waves. Luckily, they don't act in both ways at the same time. We'll show how waves can be added to give particle-like properties. And we'll demonstrate that there is a fundamental limit to how precisely we can measure certain physical quantities.

10.1
DE BROGLIE WAVES

You've already seen how electromagnetic waves sometimes behave with particle-like properties. Small bundles, or quanta, of energy called photons have localized collisions with electrons, transferring energy and momentum somewhat like balls do on a pool table. The next step is to ask whether one of nature's beautiful symmetries allows particles to possess wavelike properties. Prince Louis de Broglie of France suggested that this might be

FIGURE 10.1
The upper half of the photo shows the diffraction pattern for 71-pm x-rays passing through aluminum foil. The lower half, with a different scale, shows the diffraction pattern for 600-eV electrons from aluminum.

Source: Courtesy of Education Development Center, Inc., Newton, Mass.

possible in his doctoral dissertation in 1924. Reportedly, his professors doubted the idea, but then Einstein heard about it. Einstein told the professors that there might be something to the idea, so de Broglie received his doctorate. In 1927 Davisson and Germer in the United States and G. P. Thomson in England provided experimental verification. Figure 10.1 gives an illustration of particles exhibiting wavelike properties.

You saw that $p = h/\lambda$ for a photon. This equation has a wavelike property (λ) on one side of the equation and a particle-like property (p, the momentum of a single quantum) on the other side. The two properties are connected by Planck's constant, h. We can also describe the equation $E = h\nu$ in a similar manner. In fact, de Broglie suggested that $p = h/\lambda$ can be solved to give the **wavelength of a particle:**

$$\lambda = \frac{h}{p} = \frac{h}{mv}.$$

$$(10.1)$$

EXAMPLE 10.1   Calculate the wavelength of

a) a 50.0-kg woman leisurely jogging at 2.0 m/s,   b) a 2.0-MeV electron, and   c) a 200-eV electron.

Solution

a) Eq. (10.1) gives

$$\lambda = \frac{h}{mv} = \frac{6.626 \times 10^{-34} \, \text{J} \cdot \text{s}}{50 \, \text{kg} \times 2.0 \, \text{m/s}} = 6.6 \times 10^{-36} \, \text{m}.$$

b) First use

$$E = KE + E_0 = 2.0 \, \text{MeV} + 0.511 \, \text{MeV} = 2.5 \, \text{MeV}.$$

Then solve $E^2 = p^2c^2 + E_0^2$ for p:

$$p = \sqrt{E^2 - E_0^2}/c = \sqrt{(2.5)^2 - (0.511)^2} \, \text{MeV}/c = 2.45 \, \text{MeV}/c.$$

Finally, Eq. (10.1) gives

$$\lambda = \frac{h}{p} = \frac{4.136 \times 10^{-15} \, \text{eV} \cdot \text{s}}{(2.45 \times 10^6 \, \text{eV})/(3.00 \times 10^8 \, \text{m/s})}$$

$$= 5.1 \times 10^{-13} \, \text{m} = 0.51 \, \text{pm}.$$

c) Since $200 \, \text{eV} \ll E_0$, we can use classical results. That is, $KE = p^2/2m_0$, so $p = \sqrt{2m_0 KE}$, and

$$\lambda = \frac{h}{p} = \frac{h}{\sqrt{2m_0 KE}}$$

$$= \frac{6.626 \times 10^{-34} \, \text{J} \cdot \text{s}}{[2(9.11 \times 10^{-31} \, \text{kg})(200 \, \text{eV})(1.60 \times 10^{-19} \, \text{J/eV})]^{1/2}}$$

$$= 8.7 \times 10^{-11} \, \text{m} = 87 \, \text{pm}. \qquad \bullet$$

Example 10.1 shows that wavelengths of large objects are very, very small, and thus their wave effects will be undetectable. For instance, to get an appreciable amount of wave diffraction, the magnitude of λ should be fairly close to that of the aperture width. Even if this width is as large as $10^3 \lambda$, there is no way the jogging woman will get through a 6.6×10^{-33} m opening! There is, therefore, no experimental evidence that $\lambda = h/p$ actually applies to macroscopic objects. However, 5.1×10^{-10} m could easily be the space between atoms in a crystal, and 8.7×10^{-8} m apertures can be fabricated. Small objects like electrons can and do experience wave effects.

Because of diffraction effects, the resolving power of a visible light ($\lambda = 400$ nm to 700 nm) microscope is limited. Electron microscopes can have wavelengths 10,000 times smaller than visible light wavelengths with only a few hundred volts potential difference, as shown in part (c) of Example 10.1. Diffraction effects also limit the narrowness of light beams used in integrated circuit construction, but particle beams have smaller wavelengths and therefore less diffraction and narrower or more precisely defined beams. X-ray diffraction is used to study periodic structures such as crystals or DNA, but particle diffraction also occurs and both electron and neutron diffraction studies are common.

10.2
THE PRINCIPLE OF COMPLEMENTARITY

It may seem that modern physics confuses, rather than enlightens, with its **wave–particle duality** (with waves having particle-like properties and particles having wavelike properties). For instance, if you are throwing baseballs at a piece of plywood with two holes cut in it, classical physics simply states that all of the baseballs that pass completely through one hole or the other will continue in a normal trajectory beyond.

As Fig. 10.2(a) shows, the baseballs that pass cleanly through the holes will reach either region A or region B. But suppose that you send waves from a single source toward two apertures. When the waves pass through the apertures, classical physics not-so-simply states that the waves will spread out (diffract), as shown in Fig. 10.2(b). Then the diffracted waves will constructively and destructively interfere with one another to form the maxima and minima of an interference pattern.

The two descriptions can't hold at the same time for the same process. But relations such as $E = h\nu$ or $\lambda = h/mv$ show that both descriptions

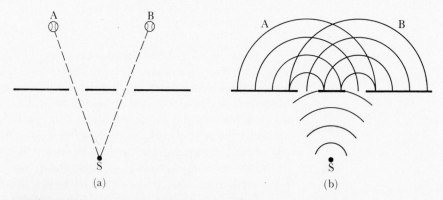

FIGURE 10.2
(a) The baseballs reach either region A or region B. (b) The waves diffract to form an interference pattern spanning both regions beyond the holes.

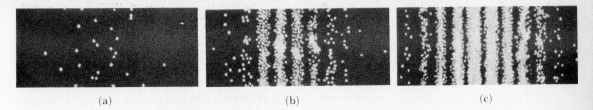

(a)   (b)   (c)

FIGURE 10.3

Formation of an interference pattern for particles diffracted through two slits (a) after 28 particles, (b) after 1000 particles, and (c) after 10,000 particles.

Source: Elisha Huggins, *Physics I.* © 1968, Benjamin/Cummings Publishing Company. P. 510.

are necessary. The **principle of complementarity** was first stated by Niels Bohr in 1928. We can now state that the wave description and the particle description are complementary. We need both descriptions to complete our model of nature, and the two complete, not compete. We will never need to use the wave and particle aspects at the same time to describe any single part of an occurrence.

Suppose that we shine light from a single monochromatic source through two slits and that this light forms an interference pattern on a metal surface. The explanation of this pattern is based on the wave model of light. Now suppose that we begin to detect photoelectrons leaving the metal surface. The explanation of this occurrence requires the particle-like model of light (the photon in the photoelectric effect). A complete description requires use of both the particle and wave models. The waves diffract and interfere; the particles are then absorbed to eject electrons. No electrons will be ejected at the zeros of the interference pattern, and the most electrons per unit time will be ejected at the maxima. In other experiments it is even possible to see electrons ejected from a surface (particle-like behavior) form interference patterns (wavelike behavior), which result from the regularity of the surface atoms that act as a sort of diffraction grating.

EXAMPLE 10.2   Suppose that an electron beam is spread to pass through two apertures. Will the result resemble Fig. 10.2(a) or 10.2(b)?

Solution The result depends on the wavelength of the electrons and the aperture size. If λ is far less than the aperture size, almost no diffraction will occur, and the electrons will act like the baseballs in Fig. 10.2(a). However, as λ approaches the aperture size, more and more diffraction will occur. If the diffraction patterns begin to overlap appreciably, interference will increase until a situation like that in Fig. 10.2(b) occurs. If you want to collect individual electrons, you'll find none at the zeros and most at the maxima. (See Fig. 10.3.)   ●

10.3
WAVE PACKETS

A wave is an entity that extends throughout space, but a particle is located at one position in classical physics. To give a photon its particle-like properties, we consider it to be represented by a **wave packet,** that is, by the sum of a number of waves of infinite extent and precise wavelength and frequency. We must then choose the amplitudes of these waves so that the waves will all cancel out, except in the region occupied by the wave packet. Fourier analysis is the branch of mathematics that deals with calculating these amplitudes.

Suppose that we consider two traveling waves represented by

$$\psi_1 = A \sin[kx - \omega t] \quad \text{and} \quad \psi_2 = A \sin[(k + dk)x - (\omega + d\omega)t],$$

where A is the amplitude, $k = 2\pi/\lambda$ and is called the wave number (**k**, with a direction, is called the wave vector), and ω is the angular frequency ($\omega = 2\pi\nu$). The waves interfere to give a resultant wave represented by $\psi = \psi_1 + \psi_2$. Using

$$\sin \theta + \sin \phi = 2 \cos \left(\frac{\phi - \theta}{2} \right) \sin \left(\frac{\phi + \theta}{2} \right),$$

we have

$$\psi = 2A \cos[(dk)x - (d\omega)t]\sin[kx - \omega t]. \tag{A}$$

(We ignore the $dk/2$ and $d\omega/2$ added to k and ω in the sine term because they are infinitesimal compared to k and ω, respectively.) If we graph Eq. (A) we find that we have an "envelope" equal to $2A \cos[(dk)x - (d\omega)t]$, modulating the sine wave given by $\sin[kx - \omega t]$. (See Fig. 10.4.)

To see how fast the sine wave moves, we take a particular point on the wave (which means letting $kx - \omega t$ be constant). Then we calculate dx/dt,

FIGURE 10.4
A sketch (not to scale) of a wave packet formed by a single beat of two traveling waves. The envelope has the long wavelength $2\pi/dk$, while the amplitude-modulated sine wave inside has the much shorter wavelength $2\pi/k$.

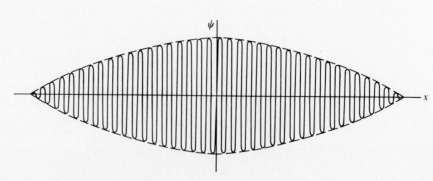

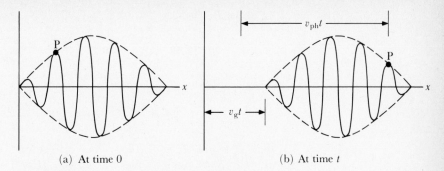

(a) At time 0   (b) At time t

FIGURE 10.5
A wave packet with $v_{ph} > v_g$. (a) At $t = 0$, the left end of the packet is at $x = 0$, and we label a crest of a wave in the packet as P. (b) At $t = t$, the wave packet as a whole has moved a distance $v_g t$. That particular crest has moved the greater distance $v_{ph} t$.

which gives

$$v_{ph} = \frac{\omega}{k},$$ (10.2)

where v_{ph} is the magnitude of the **phase velocity.** The wave packet itself doesn't move at v_{ph}. To see how fast the envelope moves, we let $(dk)x - (d\omega)t$ be constant and calculate dx/dt, so

$$v_g = \frac{d\omega}{dk}$$ (10.3)

where v_g is the magnitude of the **group velocity.**

The phase velocity has no limit, even approaching infinity under some circumstances. But information, energy, and particles move with the group velocity, and the maximum magnitude of v_g is c.

EXAMPLE 10.3 Dispersion occurs when ω and k aren't directly proportional. Prove that if there is no dispersion, the magnitude of the group velocity equals that of the phase velocity.

Solution With no dispersion, ω is directly proportional to k, or $\omega = Ck$, where C is a constant. Then $v_{ph} = \omega/k = Ck/k = C$. Also, $v_g = d\omega/dk = d(Ck)/dk = C$. Therefore $v_{ph} = C = v_g$.   ●

A wave packet with no dispersion moves with $v_{ph} = v_g$ and all the waves shown in Fig. 10.4 move unchanged. If $v_{ph} > v_g$, the waves inside would appear at the left of the wave packet, move through and disappear at the right of the wave packet, while the wave packet itself moves more slowly to the right as illustrated in Fig. 10.5.

For a particle,

$$\omega = 2\pi v = \frac{2\pi E}{h} \quad \text{and} \quad k = \frac{2\pi}{\lambda} = \frac{2\pi}{h/p} = \frac{2\pi p}{h}.$$

Thus

$$d\omega = \frac{2\pi \, dE}{h} \quad \text{and} \quad dk = \frac{2\pi \, dp}{h},$$

so that $v_g = d\omega/dk = dE/dp$. The differential of $E^2 = p^2 c^2 + E_0^2$ is $2E \, dE = 2p \, dp \, c^2$, which gives $dE/dp = p/(E/c^2) = p/m = v$. Therefore the group velocity equals the particle velocity. Note that the constant term E_0^2 dropped out when we took the differential, so $v_g = v$ even for zero rest mass particles.

10.4
THE HEISENBERG UNCERTAINTY PRINCIPLE

We can't have just one frequency if we want to have a wave packet. In Section 10.3, to get a series of wave packets we needed an ω and an $\omega + d\omega$. Suppose instead that we use waves of frequencies $v - \Delta v/2$ and $v + \Delta v/2$. The $\sin(\phi + \theta)/2$ term in their sum will tell us that the sine wave has a frequency v, and we see that the sum is of two waves with a spread Δv. The two waves beat together with a beat frequency equal to their frequency difference, Δv. The "beat" is the wave packet, and the time it takes the packet to go past is the period of the beat, $1/\Delta v$. So if we call the beat period Δt, the frequency spread is $\Delta v = 1/\Delta t$. Adding more waves could result in a larger frequency spread, so let's write it as $\Delta v \, \Delta t \geq 1$.

If the wave packet is to represent a quantum, then $E = hv$ or $\Delta E = h\Delta v$. Multiplying both sides of $\Delta v \, \Delta t \geq 1$ by h, we obtain

$$\Delta E \, \Delta t \geq h. \tag{10.4}$$

Equation (10.4) is one version of the **Heisenberg uncertainty principle**. In words, it states that if an object is known to exist in an energy state for a time Δt, the spread (or uncertainty) of the energy in that state will be at least $h/\Delta t$.

EXAMPLE 10.4   The lifetime of a typical atomic state is 10^{-8} s. What is the uncertainty in the energy of such a state?

Solution   Solving Eq. (10.4), we get

$$\Delta E \geq \frac{h}{\Delta t} = \frac{4.136 \times 10^{-15} \, \text{eV} \cdot \text{s}}{10^{-8} \, \text{s}} = 4 \times 10^{-7} \, \text{eV}.$$

Therefore the spread in energy of the state is at least 4×10^{-7} eV.   ●

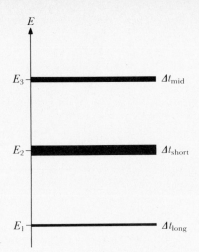

FIGURE 10.6
The longer the lifetime of an energy state, the smaller is its spread in energy.

We can't measure the energy exactly but know that its spread is inversely proportional to its lifetime, as shown in Fig. 10.6. *Exactly* would mean *no* uncertainty; that is, $\Delta E = 0$. But $\Delta E = 0$ would require $\Delta t \geq h/0$, or an infinite lifetime. There is a fundamental uncertainty in the measurement of the energy of an unstable state that cannot be eliminated by even the best of measuring instruments.

If the wave packet is moving in the x direction, in time Δt it will travel a distance Δx, where $\Delta x = v_x \, \Delta t = (p_x/m)\Delta t$. If there is an uncertainty in the energy during Δt, there will also be an uncertainty in the momentum because $E^2 = p_x^2 \, c^2 + E_0^2$. Again, the differential is $2E \, dE = 2p_x \, dp_x \, c^2$. Letting dE and dp_x become ΔE and Δp_x, we have $\Delta p_x = E \, \Delta E/(c^2 p_x)$. Replacing E/c^2 by m and multiplying by Δx yields

$$\Delta x \, \Delta p_x = [(p_x/m)\Delta t](m \, \Delta E/p_x) = \Delta E \, \Delta t.$$

Since $\Delta E \, \Delta t \geq h$, we now have the second version of the Heisenberg uncertainty principle:

$$\Delta x \, \Delta p_x \geq h. \tag{10.5}$$

In words, Eq. (10.5) states that the product of the uncertainty in the position and the uncertainty in the momentum in the same direction is at least Planck's constant. (We have been using the Δ's to mean the width of the spread. If we define the Δ's more precisely in terms of the root-mean-square deviation, we may replace h by $h/4\pi$.) Both the position and the momentum are being measured over the time interval Δt; therefore this version of the Heisenberg uncertainty principle is for simultaneous measurements of position and momentum.

There is nothing sacred about the x axis in Eq. (10.5), so

$$\Delta y \, \Delta p_y \geq h, \qquad \Delta z \, \Delta p_z \geq h, \quad \text{and} \quad \Delta r \, \Delta p_r \geq h. \tag{10.5a}$$

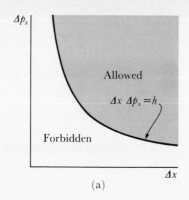
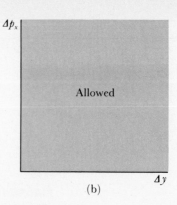

FIGURE 10.7
(a) The product $\Delta p_x \Delta x$ must be greater than or equal to h. (b) The product $\Delta p_x \Delta y$ can have any value.

But we derived these equations for motion in one direction; they say nothing about uncertainties in mutually perpendicular directions. For example, there is no specific minimum value for $\Delta x \, \Delta p_y$. However, according to Eqs. (10.5) and (10.5a), if we want to make Δx and Δp_y equal to zero, Δp_x and Δy would have to be infinite. That is, we would have no idea of what the motion of the particle is in the x direction nor any idea of its position in the y direction if $\Delta x = 0 = \Delta p_y$. Figure 10.7 illustrates the possible values of some uncertainties.

EXAMPLE 10.5   Calculate the minimum uncertainty in the momentum of a ^{4}He atom confined to a 0.40-nm region.

Solution We know only that the ^{4}He atom is somewhere in the 0.40-nm region; therefore $\Delta x = 0.40$ nm. Equation (10.5) gives us $\Delta p_x \geq h/\Delta x$. Using the equal sign to obtain the minimum, we have

$$(\Delta p_x)_{\min} = \frac{h}{\Delta x} = \frac{6.626 \times 10^{-34} \text{ J} \cdot \text{s}}{0.40 \times 10^{-9} \text{ m}} = 1.66 \times 10^{-24} \text{ kg} \cdot \text{m/s}. \quad \bullet$$

Example 10.5 gives a reasonable picture of what happens to ^{4}He atoms at low temperatures if we try to make them stay in one region by solidifying the He. Even at temperatures approaching absolute zero, the ^{4}He atoms have considerable momentum. Since ^{4}He has a mass of 6.7×10^{-27} kg, a momentum spread of 1.66×10^{-24} kg $\cdot$ m/s means that at some time the ^{4}He atom probably has a momentum of at least that much, or a speed of at least

$$v = \frac{p}{m} = \frac{1.66 \times 10^{-24} \text{ kg} \cdot \text{m/s}}{6.7 \times 10^{-27} \text{ kg}} = 250 \text{ m/s},$$

which is over 500 mi/hr! So even as $T \rightarrow 0$ K, this large **zero-point motion** persists because of the Heisenberg uncertainty principle. The associated kinetic energy is so large that ^{4}He will not solidify even as $T \to 0$ K, unless more than 20 atm of external pressure are applied. This pressure pushes the atoms close enough together that their attractive binding forces will be large enough to hold the solid crystal together.

Both versions of the uncertainty principle have Planck's constant on the right-hand side of the equation. But Planck's constant is very small compared to macroscopic quantities. If a 0.25-kg baseball thrown at 40.0 m/s has its speed measured to within 0.25 percent, or 0.10 m/s, then $\Delta p_x = m \, \Delta v_x = (0.25 \text{ kg})(0.10 \text{ m/s}) = 2.5 \times 10^{-2}$ kg $\cdot$ m/s. Equation (10.5) tells us that we can't simultaneously know the position of the baseball to better than 6.63×10^{-34} J $\cdot$ s$/2.5 \times 10^{-2}$ kg $\cdot$ m/s $= 2.7 \times 10^{-32}$ m. A fantastically small distance, 2.7×10^{-32} m is about 10^{17} times smaller than the smallest nucleus and 10^{22} times smaller than the smallest atom in the baseball. Obviously, this distance doesn't limit at all our ability to know the position of the baseball. The correspondence principle for quantum mechanics can be stated that as we move from the microscopic to the macroscopic, we move from quantum phenomena to classical plenomena. We can quite satisfactorily describe the motion of a baseball in classical-physics terms. However, we have to describe the motions of the atoms and the motions within the atoms of the baseball in quantum-physics terms.

In classical physics we could, in principle, obtain all the positions, velocities, masses, forces, and so on, for all particles at one time and then calculate exactly where they'd be and what they'd be doing at any later time. This clockwork universe seemed extremely predictable and orderly and was a great comfort to many people. However, in this chapter you have seen that small particles also have wavelike behavior, and that this wavelike behavior must be used to completely describe those particles. The particle is represented by a wave packet, and as such must have uncertainties in its energy, position, and momentum. The exact predictability of classical physics has been done away with by the Heisenberg uncertainty principle. There are fundamental limitations on our ability to simultaneously measure position and momentum, no matter how much we refine our instruments. For small particles, this means that we can only predict approximately where and what will occur at any later time.

QUESTIONS AND PROBLEMS

10.1 Use $E^2 = p^2c^2 + E_0^2$ and show that the smaller the rest mass, the shorter the wavelength for a given total energy.

10.2 Calculate the wavelength of a) a $\frac{1}{40}$-eV electron, b) a 2.0-GeV proton.

10.3 Calculate the wavelength of a 2000-kg car

moving at 30 m/s and discuss the result.

10.4 What potential difference can you move an electron through from rest to give it a 100-pm wavelength?

10.5 What is the speed of a neutron that has a 100-pm wavelength?

10.6 Express Eq. (10.1) for λ in terms of the kinetic energy for a) photons, b) far relativistic particles ($E \gg E_0$), c) relativistic particles, and d) classical particles.

10.7 If your wavelength were 1.0 m, you would undergo considerable diffraction in moving through a doorway. At what speed would you have to move to have this wavelength? How many years would it take you to move 0.8 m (one step) at this speed? Should you worry about self-diffraction as you walk through doorways? Why or why not?

10.8 Electrons move down a channel in a crystal that is 0.20 nm wide. Give a speed at which the electrons will act like particles and another at which the electrons will act like waves.

10.9 What do two waves do when they arrive at the same point? What do two particles do? What do electrons do?

10.10 What are the advantages of electron microscopes over visible-light microscopes? Can you think of any disadvantages? (Consider, for example, focussing and detection.)

10.11 As we make λ smaller and smaller so as to increase the resolving power of a microscope and thereby minimize the position uncertainty Δx, what happens to the momentum associated with λ? This momentum is partially transferred to whatever we are looking at as we bounce photons or particles off it to see it, which makes the object's momentum uncertain by Δp_x. So what happens to Δp_x as Δx is decreased? Why is it said that in measuring a system you change it?

10.12 A stream of electrons with a wavelength λ moves through a slit of width $a = \lambda$. Those electrons diffracted to the side must acquire a sideways momentum as they pass through the slit. Why doesn't the slit, then, acquire a net opposite momentum from all the electrons? Are you using

the wave or particle properties of the electron?

10.13 Calculate the phase velocity of a relativistic free particle and show that $v_{ph} > c$, if $m_0 > 0$.

10.14 Show the $v_{ph} \rightarrow \infty$ for a free particle as $v_g \rightarrow 0$, if $m_0 > 0$.

10.15 For an electron in a solid, $\omega = (\omega_{max}/2) \cdot (1 - \cos ak)$. a) Calculate the group velocity. b) Calculate the phase velocity. c) Show that $v_g \approx 2v_{ph}$ for small k (*Hint:* $\sin \theta \approx \theta$, $\cos \theta \approx 1 - \theta^2/2$ for small θ).

10.16 Prove that $v_g = v_{ph} + k\, dv_{ph}/dk$.

10.17 For water waves, it is possible to obtain $\omega = (\text{constant}) \sqrt{k}$. Show that the magnitude of the group velocity is one-half that of the phase velocity.

10.18 A metastable (long-lived) atomic state has an energy uncertainty of 8.3×10^{-11} eV. What is the lifetime of this state?

10.19 A nuclear energy state has a lifetime of 98 ns. What is its energy spread?

10.20 In an electronic tuned circuit, Q may be defined as the ratio of the frequency to the bandwidth, or $v/\Delta v$. Similarly, for an energy state E we may define $Q = E/\Delta E$. What is the maximum Q for a state that has an energy of 14.4 keV and a lifetime of 450 ns?

10.21 The position of an atom is certain to within 0.50 nm in the x direction. What can you say about its a) x component of momentum? b) y component of momentum?

10.22 The momentum of an electron is determined with an uncertainty of 2.2×10^{-31} kg · m/s in the z direction. What can you say about the a) x position? b) z position?

10.23 The uncertainty in the position of a certain particle is equal to its de Broglie wavelength. What is the minimum percentage of uncertainty in its momentum in the same direction?

10.24 A scientist claims to have developed a new method of isolating individual particles, which will enable him to detect their radial position to within 0.20 nm and their radial momentum to within 10 eV/c, simultaneously. Discuss the possibility of his claim.

10.25 Assume that the uncertainty in the momentum equals the minimum possible momentum and is due completely to the uncertainty in the speed. Then calculate the minimum energy an electron would have if it were confined in a nucleus to a 3.3×10^{-15}-m region.

10.26 Compare the Heisenberg uncertainty principle and the classical clockwork universe to free will and determinism, using outside readings as necessary.

CHAPTER 11

The Bohr Model
of the Hydrogen Atom

Physicists construct models of atoms because individual atoms are too small to see in any detail. "Good" models are those having properties that agree with experimental results. In this chapter you will first read about experimental alpha particle scattering from atoms. You will then meet Rutherford's nuclear model of the atom, which explained that scattering data. Following that we will introduce you to experimentally derived equations for the wavelengths and frequencies emitted and absorbed by gases of hydrogen atoms. Finally, we will begin to describe the Bohr model of the hydrogen atom, which explained that data.

11.1
RUTHERFORD'S NUCLEAR ATOM

Before 1911 the most commonly accepted model of the atom was J. J. Thomson's **"plum pudding" atom.** In that model the neutral atom was thought to be composed of a massive positive sphere of charge, with electrons oscillating about fixed centers within the positive sphere. Application of Gauss's law to this model of the atom shows that the net charge

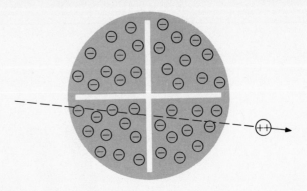

FIGURE 11.1
The plum-pudding model of the atom: An alpha particle moves almost undeflected through the atom because $F = qE$ is never large.

enclosed within any chosen volume will be close to zero and that the resultant electric fields within the "plum pudding" atom should be rather small.

Rutherford (along with Geiger and Marsden) was studying how **alpha particles** (⁴He nuclei composed of 2 protons and 2 neutrons) from radium are scattered by some materials. Using samples of thin gold foils, they expected the small electric fields of the "plum pudding" gold atoms to give very small deflections to the energetic alpha particles. (See Fig. 11.1.) However, they found that some alpha particles were deflected through large angles. A few even bounced straight back through 180°.

This led Rutherford to his **nuclear model of the atom** in 1911. In this model, all the positive charge of the atom is concentrated in a small volume at the atom's center, called the **nucleus.** Rutherford's analysis of the data indicated that the radius of the nucleus is less than 10^{-14} m. The electrons in this model ordinarily orbit about the nucleus with radii of less than 10^{-9} m. Since the nuclear particles have masses of about 1800 times the electron mass, and since volume is proportional to the radius cubed, we see that the vast majority of the mass of the atom is located in only about

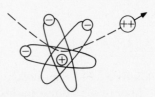

FIGURE 11.2
Rutherford scattering: Near the tiny nucleus, $F = qE$ is very large, and the alpha particle can be scattered through a large angle.

$(10^{-14} \text{ m}/10^{-9} \text{ m})^3 = 10^{-15}$ of the total volume of the atom. Thus the density of nuclear matter is roughly 10^{15} times the density of ordinary solids.

The electric field outside a spherical charge distribution is inversely proportional to the square of the distance from its center. Therefore the huge repulsive fields needed to explain the large-angle alpha particle scattering (see Fig. 11.2) are found close to the massive nucleus. (In moving through the low-mass electron cloud toward the nucleus, the alpha particle is hardly more affected than is a bullet moving through a swarm of mosquitoes.)

EXAMPLE 11.1   The radius of a gold (Au) nucleus is approximately 7×10^{-15} m. Gold contains 79 protons. Recall that the electric field $E = (1/4\pi\epsilon_0)(q/r^2)$ outside a spherical charge distribution. Use $(1/4\pi\epsilon_0) = 9 \times 10^9 \text{ N} \cdot \text{m}^2/\text{C}^2$ to calculate the electric field at the surface of the gold nucleus.

Solution   Since $q = 79e = 79(1.60 \times 10^{-19} \text{ C})$, we get

$$E = 9 \times 10^9 \frac{\text{N} \cdot \text{m}^2}{\text{C}^2} \frac{79(1.60 \times 10^{-19}\text{C})}{(7 \times 10^{-15} \text{ m})^2} = 2 \times 10^{21} \text{ N/C}. \qquad \bullet$$

The result in Example 11.1 is a huge electric field and easily explains how the alpha particle could be strongly deflected. The question it raises is, "How can the nucleus be held together when such huge fields are pushing its positive charges apart?" We will delay answering this question until you study the attractive strong nuclear force in Section 28.2.

Rutherford was not the first to suggest a planetary model of the atom, with electrons orbiting the nucleus like planets about the sun. Hantaro Nagaoka had made a similar suggestion in 1903. However, Rutherford's scattering experiment verified numerically such a model—verification that was previously lacking.

11.2
THE HYDROGEN SPECTRUM

In Rutherford's model of the atom, negative electrons orbit a positive nucleus. The centripetal force to hold the electrons in their orbits is provided by the Coulomb's law attraction of opposite charges. However, centripetal force means centripetal acceleration, and accelerated charges should radiate energy (as in bremsstrahlung). As we know from decay of earth satellite orbits, orbiting objects that lose energy should spiral inwardly toward the center. Therefore from the moment of their creation,

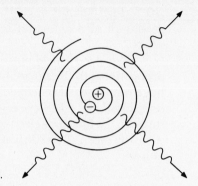

FIGURE 11.3
Classical physics predicts
that the electron should
continuously radiate
electromagnetic waves
and spiral into the nucleus.

atoms should radiate electromagnetic energy of a continuously increasing frequency, while their electrons spiral into ever smaller orbits, finally falling into the nucleus. (See Fig. 11.3.) Calculations based on classical physics show that this decay should happen in less than 10^{-8} s.

But this decay simply doesn't happen. Atoms don't even continuously radiate electromagnetic energy, much less have a lifetime of less than 10^{-8} s. So classical physics again fails to explain a microscopic phenomenon adequately.

The simplest atom is the hydrogen atom. We study it for that reason and because about 90 percent of the atoms in our galaxy are hydrogen atoms. The hydrogen atom has one proton in its nucleus and one electron orbiting about the nucleus. If we "excite" hydrogen gas by passing an electrical discharge through it, the H_2 molecules disassociate into individual H atoms, and light is emitted. Passing the light through a prism or diffraction grating shows that it doesn't form a continuous spectrum, but occurs only at certain frequencies or wavelengths. A number of experiments have shown that these frequencies are given by

$$\nu = (3.29 \times 10^{15} \text{ Hz}) \left| \frac{1}{n_f^2} - \frac{1}{n_i^2} \right|, \qquad (11.1)$$

where n_f and n_i are positive integers. For emitted light, $n_i > n_f$. We may also write Eq. (11.1) as

$$\frac{1}{\lambda} = (1.097 \times 10^7 \text{ m}^{-1}) \left| \frac{1}{n_f^2} - \frac{1}{n_i^2} \right|. \qquad (11.2)$$

In 1885 a Swiss high school teacher named Balmer found a version of Eq. (11.2) for $n_f = 2$ from the "spectral lines" of visible hydrogen light by just playing around with the numbers.

EXAMPLE 11.2 Visible light has wavelengths in the 400-nm to 700-nm range. Calculate the visible wavelengths of hydrogen light.

Solution For visible light, $n_f = 2$, so let's use $n_i = 3, 4, 5, 6, 7, \ldots$ Evaluating Eq. (11.2) for $n_i = 3$ gives $1/\lambda = (1.097 \times 10^7 \text{ m}^{-1})(1/2^2 - 1/3^2)$, or $\lambda = (36/5)/(1.097 \times 10^7 \text{ m}^{-1}) = 656$ nm. (This wavelength is called the H_α line and is red light.) Similarly, for $n_i = 4$, $\lambda = (16/3)/(1.097 \times 10^7 \text{ m}^{-1}) = 486$ nm (the H_β line, a sort of blue-green). Continuing, $n_i = 5$ gives $\lambda = 434$ nm (the H_γ line, a blue-violet color); $n_i = 6$ gives $\lambda = 410$ nm (the H_δ line, a difficult-to-see violet). For $n_i = 7$, $\lambda = 397$ nm. Therefore $n_i \geq 7$ gives wavelengths in the ultraviolet.   ●

The quantity $1.097 \times 10^7 \text{ m}^{-1}$ is called the **Rydberg constant,** and the series of spectral lines obtained with $n_f = 2$ is called the **Balmer series.** Only the Balmer series has lines in the visible part of the spectrum. The longest wavelength in the $n_f = 1$ **Lyman series** is $\lambda = (4/3)/(1.097 \times 10^7 \text{ m}^{-1}) = 122$ nm, which is in the ultraviolet. The shortest wavelengths in the $n_f = 3$ **Paschen series,** $n_f = 4$ **Brackett series,** and $n_f = 5$ **Pfund series** occur with $n_i \to \infty$ and are $\lambda = 820$ nm, 1459 nm, and 2279 nm, respectively, all of which are in the infrared. The value for $n_i \to \infty$ is the shortest wavelength (highest frequency) line for each series and is called the **series limit.**

The H_α, H_β, H_γ, and H_δ lines are present in the spectrum of the sun, but not as bright lines against a dark background (as they appear in the light from a hydrogen discharge tube). Rather they are dark lines in the continuous, bright solar spectrum. These equivalent lines mean that hydrogen atoms absorb electromagnetic energy only at certain frequencies and wavelengths, which are the same as those given by Eqs. (11.1) and (11.2). We illustrate these lines schematically in Fig. 11.4.

For absorption, $n_f > n_i$. Then, n_i labels the series, and n_f approaches infinity to give the series limit. However, hydrogen gas must be extremely

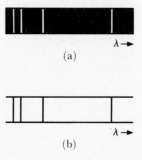

(a)

(b)

FIGURE 11.4
(a) The visible emission spectrum of hydrogen. The lines are images of a hydrogen source, which is much taller than it is wide. Different wavelengths (therefore different colors) are refracted or diffracted to different positions by a prism or diffraction grating. (b) The visible absorption spectrum of a gas of hydrogen atoms. The absorption lines are at the same wavelength positions as the emission lines.

TABLE 11.1

Type \ Series	Lyman	Balmer	Paschen	Brackett	Pfund
Emission					
n_f	1	2	3	4	5
n_i	≥ 2	≥ 3	≥ 4	≥ 5	≥ 6
Absorption					
n_f	≥ 2	≥ 3	≥ 4	≥ 5	≥ 6
n_i	1	2	3	4	5

hot to absorb at wavelengths other than those of the Lyman ($n_i = 1$) series. Table 11.1 summarizes the values of n_i and n_f for the first five series.

EXAMPLE 11.3   What is the energy of the ultraviolet photon absorbed by the hydrogen atom for which the values of n_i and n_f are 1 and 5, respectively?

Solution Equation (11.1) gives

$$v = (3.29 \times 10^{15} \text{ Hz})|1/5^2 - 1/1^2|$$
$$= (3.29 \times 10^{15} \text{ Hz})(24/25) = 3.16 \times 10^{15} \text{ Hz,}$$

and

$$E = hv = (4.136 \times 10^{-15} \text{ eV} \cdot \text{s})(3.16 \times 10^{15} \text{ Hz}) = 13.06 \text{ eV.} \qquad \bullet$$

11.3
BOHR'S MODEL: ENERGY

Equations (11.1) and (11.2), along with similar equations for some other atoms, often gave excellent values for the positions of the spectral lines of these atoms. But they weren't equations that had been derived from more fundamental concepts. These fundamental concepts were first provided by Bohr's theory of the hydrogen atom in 1913. Bohr developed this theory while working in Rutherford's laboratory. His theory started with Rutherford's nuclear model of the atom. He allowed the electron in the hydrogen atom to move in a circular orbit, as shown in Fig. 11.5. If the electron revolves with the nucleus at the center of the orbit, we can apply $F = ma$, where $F = (1/4\pi\epsilon_0)Ze^2/r^2$ (the Coulomb force of attraction be-

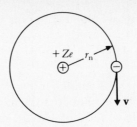

FIGURE 11.5
The nuclear atom with one electron: $Z = 1$ for hydrogen.

tween the Z positive protons and the one negative electron) and $a = v^2/r$ (the centripetal acceleration). Therefore

$$\frac{1}{4\pi\epsilon_0}\frac{Ze^2}{r^2} = m\frac{v^2}{r}, \tag{A}$$

where $Z = 1$ for hydrogen.

The trouble with Rutherford's nuclear atom was that, classically, the v^2/r acceleration should result in a constant radiation of electromagnetic energy and decaying orbits. **Bohr's first postulate** stated that only those orbits are possible that have an angular momentum, $|\mathbf{r} \times \mathbf{p}| = mvr$, times the angle of one revolution, 2π radians, equal to a positive integer times Planck's constant. That is, $mvr2\pi = nh$. We can now see that this postulate is equivalent to stating that a whole number of de Broglie electron wavelengths must be fitted around any orbit. (See Fig. 11.6.) This condition is like setting up standing waves, and pure standing waves transmit no net energy. Therefore

$$n\frac{h}{mv} = 2\pi r. \tag{B}$$

We can now solve for the orbital speed, v. From Eq. (B), $v = nh/(m2\pi r)$. Substituting back into Eq. (A) gives

$$\frac{1}{4\pi\epsilon_0}\frac{Ze^2}{r^2} = m\frac{[nh/(m2\pi r)]^2}{r} = n^2\frac{h^2}{m4\pi^2 r^3}. \tag{C}$$

FIGURE 11.6
"Snapshots" taken at five different times of an electron standing wave fitted around a Bohr orbit. This wave gives $n = 4$, the third excited state.

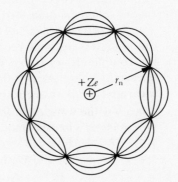

If we solve Eq. (C) for the orbit radii, r_n, we obtain

$$r_n = \frac{n^2}{Z}\left(\frac{\epsilon_0 h^2}{m\pi e^2}\right) = \frac{n^2}{Z}(52.9 \text{ pm}). \tag{11.3}$$

Remember that n is a positive integer; that is, n may only equal 1, 2, 3, 4, . . . For hydrogen, $Z = 1$, so the first three allowed hydrogen orbits have radii of $1^2(52.9 \text{ pm})/1 = 52.9$ pm, $2^2(52.9 \text{ pm})/1 = 212$ pm, and $3^2(52.9 \text{ pm})/1 = 476$ pm. These are the only orbits possible in the Bohr model in this range of radii. There is no $r = 100$ pm radius orbit, for example.

EXAMPLE 11.4   Calculate the attractive force and the electron's speed, $nh/(m2\pi r)$, in the smallest orbit of the Bohr hydrogen atom.

Solution Equation (11.3) shows that $n = 1$ for the smallest orbit. Since $Z = 1$ for hydrogen, $r = 52.9$ pm. Then

$$F = \frac{1}{4\pi\epsilon_0}\frac{Ze^2}{r^2} = 9.0 \times 10^9 \frac{\text{N} \cdot \text{m}^2}{\text{C}^2}\frac{1(1.60 \times 10^{-19} \text{ C})^2}{(52.9 \times 10^{-12} \text{ m})^2} = 8.2 \times 10^{-8} \text{ N}.$$

This force is 9.2×10^{21} times the earth weight of an electron, so the centripetal acceleration of the electron is 9.2×10^{21} g's! The speed is

$$v = \frac{nh}{m2\pi r_1} = \frac{1(6.626 \times 10^{-34} \text{ J} \cdot \text{s})}{(9.11 \times 10^{-31} \text{ kg})2\pi(52.9 \times 10^{-12} \text{ m})} = 2.19 \times 10^6 \text{ m/s},$$

or about $(1/137)c$.   ●

The speed in Example 11.4 is small enough that we can neglect the change in mass with speed. The value $(1/137. . . .)$ is called α, the **fine structure constant.** It equals $h/(m2\pi r_1 c)$, which in turn equals $e^2/(2\epsilon_0 hc)$. The term α contains fundamental constants of electromagnetism (ϵ_0 and e), quantum theory (h), and relativity (c) and is often found in the advanced theory called *quantum electrodynamics* (QED).

We can now calculate the sum of the kinetic and potential energies of the atom. It is customary to measure the energy of the atom with respect to the completely separated nucleus and electron at rest, and to call this energy E. Therefore $E = 0$ when the nucleus and electron are completely separated and both are at rest. In orbit the kinetic energy of the electron is $\frac{1}{2}mv^2$, but since $mv^2/r = Ze^2/(4\pi\epsilon_0 r^2)$, the kinetic energy becomes $KE = Ze^2/(8\pi\epsilon_0 r)$. The electrical potential energy of two charges is equal to $q_1 q_2/(4\pi\epsilon_0 r)$. Here $q_1 = +Ze$ and $q_2 = -e$, giving us an electric potential energy $PE = -Ze^2/(4\pi\epsilon_0 r)$. Other energies, such as the gravitational potential energy, are negligible, so

$$E = KE + PE = \frac{Ze^2}{8\pi\epsilon_0 r} - \frac{Ze^2}{4\pi\epsilon_0 r} = \frac{-Ze^2}{8\pi\epsilon_0 r}. \qquad \text{(D)}$$

Finally, substituting Eq. (11.3), $r_n = n^2(\epsilon_0 h^2/m\pi e^2)/Z$, into Eq. (D) gives

$$E_n = -\frac{Z^2}{n^2}\frac{me^4}{8\epsilon_0^2 h^2} = -\frac{Z^2}{n^2}(13.60 \text{ eV}). \qquad (11.4)$$

The integer n is called the **principal quantum number** because Eq. (11.4) shows that the total energy (not counting the rest energy) of the hydrogen atom is principally determined by this integer and because the energy is quantized. The total energy is negative because the electron and nucleus form a bound system. That is, energy must be added to the system to completely separate the electron and nucleus. The minimum energy that must be added to separate them completely is called the **binding energy.** Therefore, the binding energy equals $-E_n$. (See Fig. 11.7.)

The lowest value the energy can have is in the $n = 1$ state. The lowest energy state is called the **ground state.** The energy of the ground state for a hydrogen atom is $-1^2(13.60 \text{ eV})/1^2 = -13.60 \text{ eV}$, from Eq. (11.4). The next lowest value is for $n = 2$. The $n = 2$ state is called the first **excited state** and has an energy for hydrogen of $E_2 = -1^2(13.60 \text{ eV})/2^2 = -3.40$ eV. The second excited state has $n = 3$, which gives $E_3 = -13.60$ eV$/9 = -1.51$ eV; the third excited state has $n = 4$, which gives $E_4 = -13.60 \text{ eV}/16 = -0.85$ eV, and so on, for hydrogen. The energy differ-

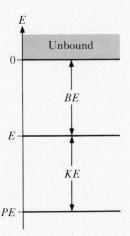

FIGURE 11.7

In the Bohr atom, $KE = -PE/2$. The total energy is $E = KE + PE = PE/2$, or $-KE$. The binding energy is $BE = -E = +KE = -PE/2$. Both PE and E are negative; KE and BE are positive.

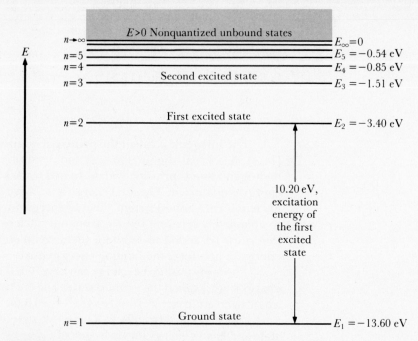

FIGURE 11.8
The energy-level diagram for the Bohr hydrogen atom. Energy is plotted vertically, but is not linear so that some of the larger n states can be included. States for $5 < n < \infty$ are not shown.

ence between an excited state and a ground state is called the **excitation energy.** For example, the excitation energy equals -3.40 eV $- (-13.60$ eV$) = 10.20$ eV for the first excited state in hydrogen as seen in Fig. 11.8.

EXAMPLE 11.5   Find the principal quantum number, the total energy, the binding energy, and the excitation energy of the fifth excited state of hydrogen.

Solution The fifth excited state is the fifth state above $n = 1$, so its principal quantum number is $1 + 5 = 6$. For hydrogen, $Z = 1$, so Eq. (11.4) gives the total energy to be $E_6 = -1^2(13.60$ eV$)/6^2 = -0.38$ eV. Then the binding energy equals $-E_6 = -(-0.38$ eV$) = +0.38$ eV. That is, you must add a minimum of 0.38 eV to separate completely the electron and proton. Since the ground state energy of hydrogen is -13.60 eV, the excitation energy is the difference between the excited state energy and the ground state energy: -0.38 eV $- (-13.60$ eV$) = 13.22$ eV. In other words, you must add 13.22 eV to a hydrogen atom in its ground state to excite it to its fifth excited state.   ●

In this chapter you saw how Rutherford arrived at the essentials of the atomic model accepted to this day: A very small positive nucleus surrounded by electrons. We then discussed the simplest atom, hydrogen, and showed that hydrogen atoms in a gas emit and absorb electromagnetic radiation only at certain allowed values of frequency and wavelength. That is, v and λ are quantized. Finally, you saw how Bohr took Rutherford's model, made an assumption (which we now see to be equivalent to fitting standing electron waves around circular orbits), and arrived at quantized energy levels. In the next chapter, we will continue to examine and apply Bohr's model.

QUESTIONS AND PROBLEMS

11.1 Use $F = ma$ and $F = qE$ to find the maximum acceleration experienced by an alpha particle (mass $= 6.7 \times 10^{-27}$ kg) at the surface of a gold nucleus.

11.2 What kinetic energy must an alpha particle have if its center is to approach to within 9×10^{-15} m of the center of a gold nucleus? (*Hint:* Use classical conservation of energy.)

11.3 Calculate the two longest wavelengths and the shortest wavelength (series limit) in the Lyman series.

11.4 Calculate the longest and shortest wavelengths (the shortest is the series limit) in the Paschen series.

11.5 The frequency of an electron orbit in the Bohr model is $v_n = v_n/(2\pi r_n)$. Show that v_n approaches $(E_{n+1} - E_n)/h$ for large orbits. That is, quantum results approach classical results for large quantum numbers (a version of the correspondence principle).

11.6 Show that $E_n = -(KE)_n$ in the Bohr model, so that $(KE)_n \ll m_0 c^2$ for all orbits in hydrogen.

11.7 Calculate the separation between the electron and the proton in an $n = 120$ orbit of hydrogen (found under very special conditions).

11.8 Is an orbit of radius 22.2 nm allowed in the Bohr model of hydrogen? Explain.

11.9 Find the principal quantum number, the

binding energy, and the excitation energy of the third excited state of hydrogen.

11.10 Which excited state of hydrogen has an excitation energy of 12.09 eV?

11.11 Use the constants on the end papers of this book to check the 52.9 pm of Eq. (11.3).

11.12 Use the constants on the end papers of this book to check the 13.60 eV of Eq. (11.4). (The discrepancy will be explained in Section 12.2.)

11.13 Start with Bohr's first postulate to show that $v/c = (137n)^{-1}$ for hydrogen atoms, then find the relativistic γ as a function of n to the order n^{-2}, and finally evaluate γ for the ground state and first two excited states of H.

11.14 What frequency would you measure for the transition from the second excited state to the first excited state from a beam of hydrogen atoms moving toward you at a speed of 2.0×10^6 m/s?

11.15 The maximum intensity per unit wavelength of an ideal blackbody occurs at the same wavelength as the electromagnetic radiation absorbed in a transition from the fourth excited state to the sixth excited state in a hydrogen atom. Calculate the intensity emitted by the blackbody.

11.16 Suppose that, on the average, an electron in a hydrogen atom will stay in any excited state 10^{-8} s.

a) Calculate the spread in energy of that

excited state and compare it to the energy of the state.

b) Calculate how many revolutions (as a function of the principal quantum number) the electron will make about the nucleus on the average in the Bohr model.

11.17 Use Eq. (11.1) to find the allowed energies of photons emitted or absorbed for a hydrogen atom. Relate your answer (to three significant figures) to Eq. (11.4).

11.18 Is the following statement true for all n?

"The principal quantum number of the $(n - 1)$th excited state is n."

11.19 Calculate the photon flux if the intensity of light resulting from transitions from the first excited state to the ground state in hydrogen is 1.3 nW/m².

11.20 Find the principal quantum number, which excited state it's in, and the binding energy for that state that can be excited by the absorption of a 13.39-eV photon.

CHAPTER

Applications of the Bohr Model

Bohr was a careful revolutionary. He did not "throw out the baby with the bath water" as many impetuous revolutionaries do. He quantized the angular momentum, but he did not do away with Coulomb's law, centripetal acceleration, or Newton's laws. We will show in this chapter how he also maintained conservation of energy, but used the Einstein quantization of electromagnetic energy. He then refined his model by considering the motion of the nucleus. You will also learn how to extend Bohr's model to systems that are similar to hydrogen.

12.1
BOHR'S MODEL: SPECTRA

Bohr's second postulate stated that the electron doesn't radiate electromagnetic energy as long as it remains in one of the allowed states; and if it changes to another state, a photon of energy $|\Delta E| = h\nu$ will be emitted or absorbed. Let n_i be the principal quantum number of the initial state and n_f be the principal quantum number of the final state. For photon emission or

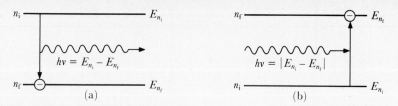

FIGURE 12.1
(a) The electron drops from the n_i state to the n_f state with the emission of a single photon of energy $E_{n_i} - E_{n_f}$. (b) The electron is excited from the n_i state to the n_f state with the absorption of a single photon of energy $|E_{n_i} - E_{n_f}|$.

absorption, Bohr's second postulate then is

$$hv = |E_{n_i} - E_{n_f}|. \tag{12.1}$$

Since only certain energies are possible in the hydrogen atom, the photons can have only certain allowed frequencies (and wavelengths). Therefore the emission and absorption spectra will be line spectra, not continuous spectra. Also, the lines will be at the same v and λ for both emission and absorption. (See Fig. 12.1.)

EXAMPLE 12.1   Find the energy, frequency, and wavelength of the photon emitted when the electron falls from the second excited state in hydrogen to the first excited state.

Solution From Fig. 11.8, or using $Z = 1$, $n_i = 3$, and $n_f = 2$ in $E_n = -1^2(13.60 \text{ eV})/n^2$, we find that the energies are $E_{n_i} = -1.51$ eV and $E_{n_f} = -3.40$ eV. Therefore $hv = |E_{n_i} - E_{n_f}| = -1.51$ eV $- (-3.40$ eV$) = 1.89$ eV. The energy of the emitted photon is $hv = 1.89$ eV; its frequency is

$$v = 1.89 \text{ eV}/h = \frac{1.89 \text{ eV}}{4.136 \times 10^{-15} \text{ eV} \cdot \text{s}} = 4.57 \times 10^{14} \text{ Hz}.$$

Then the wavelength is

$$\lambda = \frac{c}{v} = \frac{3.00 \times 10^8 \text{ m/s}}{4.57 \times 10^{14} \text{ Hz}} = 0.656 \times 10^{-6} \text{ m} = 656 \text{ nm}.$$

These quantities are also those of the energy, frequency, and wavelength of the photon that would be absorbed in exciting hydrogen from its first excited state to its second excited state.   ●

Note that the wavelength calculated in Example 12.1 is that of the H_α line calculated in Example 11.2 (the longest wavelength in the Balmer series). Its wavelength has now been determined from an equation that has

some basic physics behind it, not from one determined by merely playing
with numbers.

We can now use Bohr's results to derive Eqs. (11.1) and (11.2).
Substituting $E_n = -Z^2(13.60 \text{ eV})/n^2$ into $h\nu = |E_{n_i} - E_{n_f}|$ gives us $h\nu = Z^2(13.60 \text{ eV})|1/n_f^2 - 1/n_i^2|$. Dividing by $h = 4.136 \times 10^{-15} \text{ eV} \cdot \text{s}$, we
have $\nu = Z^2(3.29 \times 10^{15} \text{ Hz})|1/n_f^2 - 1/n_i^2|$. Since $\nu = c/\lambda$, dividing by
$c = 2.998 \times 10^8 \text{ m/s}$ gives us $1/\lambda = Z^2(1.097 \times 10^7 \text{ m}^{-1})|1/n_f^2 - 1/n_i^2|$.
Finally, substituting $Z = 1$ for hydrogen yields Eqs. (11.1) and (11.2). (See
Fig. 12.2.)

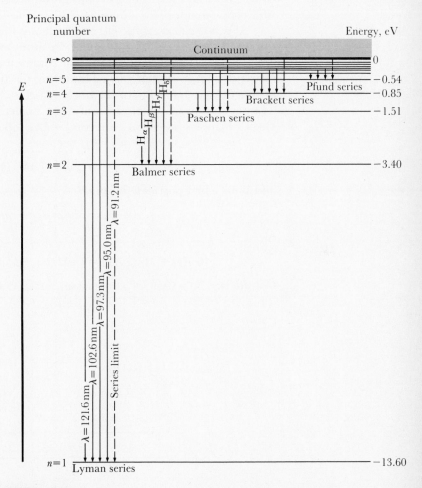

FIGURE 12.2

The causes of the Lyman, Balmer, Paschen, Brackett, and Pfund emission series in the
hydrogen atom. All transitions that have an $n_f = 1$ final state are in the Lyman series. These
transitions therefore start from $n_i = 2$ to ∞ (the series limit) states. The other four series end
on an $n_f = 2$, $n_f = 3$, $n_f = 4$, or $n_f = 5$ state, respectively, and start on all $n_i > n_f$.

12.2
REDUCED MASS AND HYDROGEN-LIKE ATOMS

If we actually substitute the best available values of the constants in Eq. (11.4), we come up with something closer to 13.61 eV than to 13.60 eV. We can easily check this disagreement with accurately measured hydrogen light. The disagreement occurs because in deriving Eq. (11.4) we assumed that the nucleus remained completely fixed in some inertial reference frame as the electron orbited about it. Bohr realized that this condition, of course, does not exist. The two objects don't rotate about one or the other; they rotate about their common center of mass.

If m is the mass of the electron and M is the mass of the nucleus, the distance from the center of mass of the atom to the electron will be $r_e = Mr/(M + m)$. The distance from that common center of mass to the nucleus will be $r_N = mr/(M + m)$. Therefore, r_e is the radius of the electron's orbit and r_N is the radius of the nucleus' orbit about their center of mass. The separation of the electron and the nucleus is $r = r_e + r_N$. Another check of these statements is that $mr_e = Mr_N$. (See Fig. 12.3.)

The expression $F = ma$ then becomes

$$(1/4\pi\epsilon_0)Ze^2/r^2 = mr_e\omega^2 = [mM/(m + M)]r\omega^2,$$

and Bohr's first postulate becomes

$$(mv_e r_e + Mv_N r_N)2\pi = nh$$

or

$$(mr_e^2\omega + Mr_N^2\omega)2\pi = [mM/(m + M)]r^2\omega 2\pi = nh.$$

Considering a system of two masses, m_1 and m_2, let's now define the

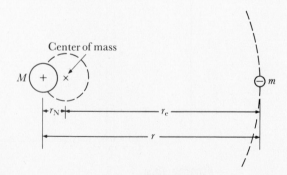

FIGURE 12.3
The nucleus and the electron rotate about their common center of mass. The radius of the nucleus's orbit is r_N, the radius of the electron's orbit is r_e, and the separation of the electron and nucleus is r. This sketch is not to scale.

reduced mass, m_r, by the equation

$$m_r = \frac{m_1 m_2}{m_1 + m_2}.$$ (12.2)

If we let $m_1 = M$ and $m_2 = m$, $F = ma$ becomes

$$\frac{1}{4\pi\epsilon_0} \frac{Ze^2}{r^2} = m_r r\omega^2$$ (A)

and Bohr's first postulate becomes

$$m_r r^2 \omega 2\pi = nh.$$ (B)

Eliminating the angular speed, ω, from Eqs. (A) and (B) gives us the separation

$$r = \frac{n^2}{Z} \frac{\epsilon_0 h^2}{m_r \pi e^2}.$$ (C)

This separation, however, is *not* the radius of the electron's orbit, r_e. Substituting the expression for r from Eq. (C) along with $m_r = Mm/(M + m)$ into $r_e = Mr/(M + m)$ gives

$$r_e = \frac{n^2}{Z} \frac{\epsilon_0 h^2}{m \pi e^2}.$$ (D)

Comparing Eq. (D) with Eq. (11.3), you can see that $r_e = r_n$. As long as we have an electron orbiting the nucleus, its Bohr orbital radii *will* equal $n^2 (52.9 \text{ pm})/Z$. The 52.9 pm value will change *only* if the mass of the orbiting particle is different from the electron mass, $0.511 \text{ MeV}/c^2$. If that mass is different, Eq. (11.3) shows us that r_n is inversely proportional to the mass of the orbiting particle but independent of the reduced mass.

Energies, wavelengths, and frequencies, however, *do* depend on the reduced mass. The total kinetic energy of the atom is

$$KE = \tfrac{1}{2}m r_e^2 \omega^2 + \tfrac{1}{2}M r_N^2 \omega^2.$$ (E)

Substituting the expressions for r_e and r_N as a function of M, m, and r into Eq. (E) converts it to

$$KE = \tfrac{1}{2}m_r r^2 \omega^2.$$ (F)

Comparing Eq. (F) to Eq. (A), you can see that the kinetic energy again equals $(1/8\pi\epsilon_0)Ze^2/r$. Since the electric potential energy remains $-(1/4\pi\epsilon_0)Ze^2/r$, the energy (kinetic plus potential) is $-(1/8\pi\epsilon_0)Ze^2/r$. Substituting Eq. (C) into the preceding expression finally gives us $E_n = -Z^2(m_r e^4/8\epsilon_0^2 h^2)/n^2$, which is Eq. (11.4) with m replaced by m_r; it shows us that the energies are directly proportional to the reduced mass. Also, $h\nu = |E_{n_i} - E_{n_f}|$ shows us that the frequencies are directly proportional to the energy difference.

Therefore all energies and frequencies in the Bohr model are proportional to the reduced mass. That is,

$$\frac{E_A}{E_B} = \frac{\nu_A}{\nu_B} = \frac{(m_r)_A}{(m_r)_B} \qquad (12.3)$$

for otherwise equivalent energy levels and transitions. But wavelengths are inversely proportional to frequencies and therefore are also inversely proportional to the reduced mass in the Bohr model. That is,

$$\frac{\lambda_A}{\lambda_B} = \frac{(m_r)_B}{(m_r)_A}. \qquad (12.4)$$

For ^{1}H, or ordinary hydrogen with a nucleus of one proton, $M/m = (938.3 \text{ MeV}/c^2)/(0.5110 \text{ MeV}/c^2) = 1836.2$. Therefore the reduced mass for ^{1}H is $m_r = (1836.2m)(m)/(1836.2m + m) = 1836.2m^2/1837.2m = 0.99946m$. Replacing m in Eq. (11.4) by $0.99946m$ gives $13.61 \text{ eV} \times 0.99946m/m = 13.60 \text{ eV}$.

Deuterium, ^{2}H, or "heavy hydrogen," has a nucleus composed of one proton bonded to one neutron and a nuclear rest mass of $1875.6 \text{ MeV}/c^2$. For deuterium,

$$M/m = (1875.6 \text{ MeV}/c^2)/(0.5110 \text{ MeV}/c^2) = 3670.4.$$

Thus $m_r = (3670.4m)(m)/(3670.4m + m) = 0.99973m$. Therefore, if we let atom A be ^{2}H and atom B be ^{1}H, the reduced mass ratio in Eq. (12.3) becomes $(0.99973m)/(0.99946m) = 1.00027$; we then multiply all the energies and frequencies of ^{1}H by 1.00027 to obtain the corresponding values for ^{2}H.

By contrast, if we solve for λ_A in Eq. (12.4), we find that all wavelengths of ^{1}H are to be divided (not multiplied) by 1.00027 to obtain the corresponding values for ^{2}H. For example, the ^{1}H$_\alpha$ spectral line is found at 656.28 nm in air, while the ^{2}H$_\alpha$ spectral line is at $656.28 \text{ nm}/1.00027 = 656.10 \text{ nm}$. This is how Harold Urey first discovered ^{2}H in 1932. (Incidentally, the wavelengths calculated previously in this book have been wavelengths in a vacuum. The wavelengths in air equal the vacuum wavelengths divided by the index of refraction of the air, which is about 1.00028 under standard conditions.)

EXAMPLE 12.2   A positronium atom is composed of a positron and electron as shown in Fig. 12.4.

a) Calculate the reduced mass for the positronium atom.

b) Determine its ground state electron orbit radius.

c) Find its ground state energy.

d) Calculate the equivalent of the ^{1}H$_\alpha$ wavelength (in air).

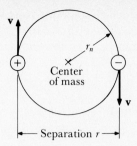

FIGURE 12.4

In the Bohr model of positronium, the electron and positron rotate about their common center of mass, which is located midway between them because they have equal mass. Therefore their separation is twice either orbit radius.

Solution

a) For positronium, $M = m$. Therefore

$$m_r = \frac{m_1 m_2}{m_1 + m_2} = \frac{mm}{m + m} = \frac{m^2}{2m} = \frac{m}{2}.$$

b) As long as the orbiting particle has the same mass as an electron, the orbit radii are unchanged from $n^2(52.9\text{ nm})/Z$. The "nucleus" of positronium still has a charge, like hydrogen, of $+1e$, so $Z = 1$. The value $n = 1$ for the ground state gives $r_1 = 52.9$ pm, the same as for H.

c) Letting atom A be positronium and atom B be ^{1}H in Eq. (12.3), the reduced mass ratio is $(m/2)/(0.99946m) = 0.50027$. The ground state energy of hydrogen is -13.60 eV, so the ground state energy for positronium is $E_A = -13.60\text{ eV} \times 0.50027 = -6.80$ eV.

d) The paragraph preceding this example tells you that the ^{1}H$_\alpha$ wavelength in air is 656.28 nm. Then $\lambda_A = 656.28\text{ nm}/0.50027 = 1311.85$ nm is given by Equation (12.4). (This positronium equivalent of the ^{1}H$_\alpha$ line is in the infrared. The detection of radiation at wavelengths $1/0.50027$ times those of ^{1}H proved the existence of the positronium atom.) ●

A positive core orbited by a single electron makes up a **hydrogen-like atom.** One example, as you have seen, is positronium. In positronium, energies and frequencies are about halved, and wavelengths are about doubled compared to ^{1}H because $m_r = m/2$. Ionized atoms that have lost all but one of their electrons are also hydrogen-like atoms. Examples include He$^+$, Li^{2+}, and Be^{3+} ions. Their nuclear charge is $+Ze$, where $Z = 2, 3$, and 4, respectively (one more than the number of superscripted $+$'s). We can now finally see why Z was left in Eqs. (11.3), (11.4), and other equations. These relations can then be used (often ignoring the relatively small change due to m_r) to determine the energies and spectra of hydrogen-like atoms. However, if Z becomes too large, the electron speed will approach c too closely to use the nonrelativistic Bohr model.

EXAMPLE 12.3 For B^{4+}, calculate:

a) the binding energy of the ground state of the final electron in the ion;

b) the energy, frequency, and wavelength of the photon needed to excite that electron from the first excited state to the second excited state;

c) the excitation energy of the fourth excited state. The nucleus is so massive that 13.61 eV should be used rather than 13.60 eV.

Solution

a) For B, $Z = 5$. Therefore $E_1 = -5^2(13.61\ \text{eV})/1^2 = -340.25\ \text{eV}$, making the binding energy $+340$ eV.

b) The first excited state has $n_i = 2$, and the second excited state has $n_f = 3$, so

$$hv = |E_{n_i} - E_{n_f}| = 5^2(13.61\ \text{eV})|1/2^2 - 1/3^2|$$
$$= (340\ \text{eV})(5/36) = 47.3\ \text{eV}.$$

Therefore

$$v = \frac{47.3\ \text{eV}}{h} = \frac{47.3\ \text{eV}}{4.136 \times 10^{-15}\ \text{eV} \cdot \text{s}} = 1.143 \times 10^{16}\ \text{Hz}$$

(just in the x-ray range), and

$$\lambda = \frac{c}{v} \quad \text{or} \quad \frac{hc}{47.3\ \text{eV}} = 26.2\ \text{nm}.$$

c) The fourth excited state has $n = 5$; so $E_5 = -5^2(13.61\ \text{eV})/5^2 = -13.61\ \text{eV}$. Since the ground state energy is -340 eV, the excitation energy wanted is $-13.61\ \text{eV} - (-340.25\ \text{eV}) = 327\ \text{eV}$.   ●

Note that E_5 for B^{4+} is approximately equal to E_1 for H. Because other near equivalencies occur, spectra for hydrogen-like atoms include some wavelengths that are nearly the same as those for hydrogen. (See Fig. 12.5.)

There are other situations that are less exactly hydrogen-like. For example, the alkali metals have one electron outside a number of inner electrons. This one electron behaves somewhat like the electron in an excited state in hydrogen. Also, trapped electrons and holes, as well as electron–hole pairs in semiconductors act somewhat like hydrogen. Other examples of hydrogen-like behavior are also found in nature.

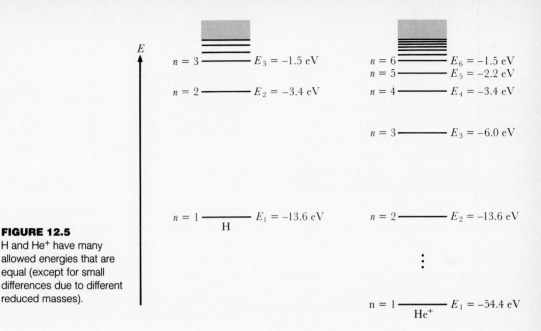

FIGURE 12.5
H and He$^+$ have many allowed energies that are equal (except for small differences due to different reduced masses).

In this chapter you have seen how Bohr's second postulate enabled him to derive the experimental equations for frequency and wavelength. You also saw that energies and frequencies are directly proportional and wavelengths are inversely proportional to the reduced mass. However, we showed that the orbiting particle's radius is independent of the reduced mass and is inversely proportional to the mass of that particle. Finally, we demonstrated that for low-mass ions in which all the electrons but one have been removed we can use the hydrogen-like model by setting Z equal to the number of protons in their nuclei.

QUESTIONS AND PROBLEMS

12.1 Two ways of exciting an atom are 1) absorption of a photon and 2) an inelastic collision with a particle. In both cases the atom recoils, conserving linear momentum.

a) For a 12.09-eV photon striking a 939-MeV/c^2 hydrogen atom initially at rest, show that the recoil energy $p^2/2M$ is negligible.

b) For a particle of kinetic energy KE_{th} and mass m which "hits and sticks" to a hydrogen atom

of mass M (giving the hydrogen atom an excitation energy E_{ex}), show that $E_{ex} = KE_{th} M/(M + m)$. Use nonrelativistic conservation of linear momentum and energy.

c) If $E_{ex} = 12.09$ eV, find KE_{th} for   i) an electron and   ii) a proton.

12.2 For single-photon excitation of the first excited state of hydrogen, only 10.2-eV photons will work. However, any single electron with

$KE > 10.2$ eV may work. Explain.

12.3 Calculate the energy, frequency, and wavelength of the photon absorbed in exciting a hydrogen atom from its ground state to its third excited state.

12.4 Calculate the energy, frequency, and wavelength of the photon emitted when an electron makes a transition in H from the fourth excited state to the first excited state.

12.5 A negative muon ($m_0 = 105.7$ MeV/c^2) is captured by a proton to form a muonium atom. Calculate the reduced mass, the ground state energy, the ground state muon orbit radius, and the wavelength of the photon emitted in the transition from the first excited state to the ground state.

12.6 a) Calculate the reduced mass of tritium, ^{3}H. The nucleus has a mass of 2808.9 MeV/c^2.
b) Calculate the wavelength of the $^3H_\alpha$ line in air.

12.7 For Li^{2+}, calculate a) the binding energy of the first excited state, b) the excitation energy of the first excited state, and   c) the energy, frequency, and wavelength of the photon emitted when the first excited state is de-excited.

12.8 The visible part of the Balmer emission series of hydrogen is a result of the $n_i = 6, 5, 4$, or 3 to $n_f = 2$ transitions. a) Show that approximately the same wavelengths are found in the spectrum of He$^+$. b) Explain why these wavelengths differ by a factor of 1.0004.

12.9 If two charges in a dielectric are separated by r, $F = (1/4\pi\kappa\epsilon_0)q_1q_2/r^2$, where κ is the dielectric constant. How would the dielectric constant enter into the expressions for the radius and energy for hydrogen-like atoms if the nucleus and electron were separated by a dielectric?

12.10 A Be^{3+} ion is in its second excited state.
   a) Calculate all the possible wavelengths the ion could emit in de-exciting to its ground state (ignore the reduced mass correction for Be^{3+}, using "13.60 eV").
   b) The Be^{3+} nucleus has a rest mass of 8393 MeV/c^2. Recognizing that the "13.60 eV"

term was calculated using $m_r = 0.99946m$, you should multiply your results for part (a) by what factor to include the m_r for Be^{3+}?

12.11 Spectra for hydrogen-like atoms include some wavelengths that are nearly the same as that of hydrogen. Why aren't they exactly the same?

12.12 Physicists have succeeded in creating U^{91+} ($Z = 92$). Why will the -13.6-eV value from the Bohr model as presented be very inaccurate for U^{91+}?

12.13 a) The Pickering series was given by $v = (3.29 \times 10^{15}$ Hz$)[1/4 - 1/(n/2)^2]$ and was attributed to hydrogen by some before Bohr. Bohr realized that this wasn't a hydrogen series, but belonged to another element, ionized so that it had but one electron. What element was it and what was the final state for emission?
   b) Carried out to more significant figures, the 3.29 . . . in the Pickering series didn't agree with the 3.29 . . . for hydrogen, anyway. Why not?

12.14 At what voltage must an x-ray tube be operated if its shortest wavelength x-ray is to have the same wavelength as the photon emitted in the transition from the fifth excited state to the ground state in B^{4+}?

12.15 At what speed and in which direction (toward or away) must deuterium atoms (^{2}H) be moving for their spectrum to exactly match the ^{1}H spectrum?

12.16 Calculate the momentum of the photon emitted when Li^{2+} decays from its second excited state to its ground state and give the direction of the momentum relative to the direction of the recoiling atom.

12.17 The nonrelativistic Bohr model can't be used accurately if v gets too close to c. Express the ground state v as a function of Z and c. Then find the relativistic γ as a function of Z to order Z^2.

12.18 Show that, when the motion of the nucleus is considered, Bohr's first postulate is no longer equivalent to fitting a whole number of electron wavelengths around the electron's orbit.

CHAPTER

The Schrödinger Wave Equation

You found that Newton's laws of motion were essential to your understanding of mechanics and that Maxwell's equations were essential to your understanding of electromagnetism. In this chapter we'll write an equation that is the basis of all nonrelativistic quantum phenomena: the Schrödinger wave equation. It is a formidable differential equation, involving partial derivatives with respect to three coordinate variables and to time. We begin by interpreting the meaning of the equation's solutions. Then we simplify the equation, first by removing its time dependence, then by reducing it to a one-dimensional equation.

13.1
THE WAVE FUNCTION

The Bohr model of the hydrogen atom and experimental results agreed closely as far as energies, frequencies, and wavelengths were concerned. However, this model was only a first step. Bohr's postulate that the angular momentum of the hydrogen atom times 2π should equal nh for its circular orbits predicts magnetic effects that don't agree with experimental results.

145

And trying to extend the Bohr model to other systems, such as the rest of the atoms in the periodic table, most often results in failure. Also, experiments have shown that (even in the hydrogen atom) some transitions between energy states are likely to occur, but other transitions are unlikely to occur. The Bohr model can't explain this behavior. Something more basic is needed.

You've seen that particles often have wavelike properties. In fact, Bohr's angular momentum postulate for a stationary nucleus is the equivalent of fitting a whole number of electron standing waves around a circular orbit. However, the radial position of the electron is determined precisely by $r_n = n^2(52.9 \text{ pm})$. At the same time, the radial component of its momentum is determined precisely ($p_r = 0$ because it moves in a circle, perpendicular to the radius). Thus $\Delta r \, \Delta p_r = 0$, and the Bohr model of the hydrogen atom violates the Heisenberg uncertainty principle. We need a more basic equation that expresses the wavelike properties of particles correctly.

The wave equation we need is called the Schrödinger wave equation (SWE). It is not a derived equation but has been verified by the correct results it gives. It was developed in 1926 by Erwin Schrödinger and is equivalent to a matrix formulation published a year earlier by Heisenberg. The **three-dimensional, time-dependent, nonrelativistic Schrödinger wave equation** is

$$-\frac{\hbar^2}{2m} \nabla^2 \Psi + V \Psi = i\hbar \frac{\partial \Psi}{\partial t}, \tag{A}$$

where $\hbar$ (called "h-bar") is the symbol for $h/2\pi$, or Planck's constant divided by 2π; m is the mass of the particle; ∇^2 is a mathematical operator, equal to $(\partial^2/\partial x^2 + \partial^2/\partial y^2 + \partial^2/\partial z^2)$ in Cartesian coordinates; and V now stands for the potential energy function of the system. (In previous chapters I let V stand for the electric potential, but virtually all physicists use V in the SWE to stand for potential energy. I grudgingly bow to convention and follow suit.) Finally, i is equal to $\sqrt{-1}$; therefore $i^2 = -1$.

The term Ψ (capital psi) is a function of x, y, z, and t in Cartesian coordinates and is called the **wave function**. In general, Ψ is a complex function like $x + iy$. That is, Ψ may have both real and imaginary parts. Since it's rather useless to try to detect something imaginary, the wave function determined by a SWE is not, by itself, an observable quantity. The SWE is a wave equation, and the amplitude of Ψ is the amplitude of the wave.

If we see light form an interference pattern, we see the intensity, not the amplitude, of the light. At the maxima of the pattern the intensity is greatest, and at the zeros the intensity is zero, as in Fig. 13.1. The intensity of a wave is proportional to the amplitude squared. At the maxima the probability of finding a photon is greatest, and at the zeros the probability

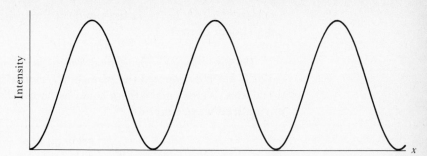

FIGURE 13.1
The intensity as a function of position in an interference pattern. If the pattern were for visible light, we would see the brightest light at the maxima and no light at the zeros.

is zero. (See Fig. 13.2.) The probability of finding a photon at a certain position is proportional to the intensity and therefore proportional to the amplitude squared.

Similarly, for the SWE, we can interpret the amplitude of Ψ as the amplitude of a probability wave, with the square of Ψ being proportional to the probability per unit volume of finding the particle. We write the square of the complex function Ψ as $\Psi^*\Psi$ (called "sigh-star-sigh"), where Ψ^* is the complex conjugate of Ψ. To obtain Ψ^*, merely replace all the i's in Ψ with $-i$'s. Then $\Psi^*\Psi\,dv$ is interpreted as being proportional to the probability of finding the particle in the volume dv.

EXAMPLE 13.1   If $\Psi = \psi e^{-i\omega t}$, where ψ(small psi) is a function of x, y, and z, show that the probability of finding the particle in an infinitesimal volume dv is independent of time.

Solution The probability desired is proportional to

$$
\begin{aligned}
\Psi^*\Psi\,dv &= (\psi e^{-i\omega t})^*(\psi e^{-i\omega t})dv \\
&= (\psi^* e^{-(-i)\omega t})(\psi e^{-i\omega t})dv \\
&= \psi^*\psi e^{+i\omega t - i\omega t}\,dv \\
&= \psi^*\psi e^0\,dv = \psi^*\psi\,dv.
\end{aligned}
$$

FIGURE 13.2
Part of an interference pattern for visible light from two rectangular slits.

Source: F. W. Sears, M. W. Zemansky, and H. D. Young, *University Physics,* 7th ed. © 1987, Addison-Wesley Publishing Co., Inc., Reading, Mass. P. 918. Reprinted with permission.

Since $\psi*$, ψ, and dv are independent of time, the probability desired is independent of time.   ●

Finding solutions for many different problem types would be much easier if $\Psi*\Psi\, dv$ equaled the probability, rather than just being proportional to it. We can make them equal by "normalizing" Ψ. That is, Ψ is a **normalized wave function** if

$$\int_{\text{all space}} \Psi*\Psi\, dv = 1. \tag{13.1}$$

If Ψ is normalized, the probability, P, of finding the particle in a volume v is given by

$$P = \int_v \Psi*\Psi\, dv \qquad \text{for } \Psi \text{ normalized.} \tag{13.2}$$

Remember: Small v = volume, and capital V = potential energy.

EXAMPLE 13.2   Suppose that ψ in Example 13.1 in spherical coordinates is $\psi = Ae^{-\alpha r^3 - i\delta}$, where A, α, and δ are real, A has units of (length)$^{-3/2}$, and $\alpha = 1$ in units of (length)$^{-3}$. For spherical coordinates, $dv = 4\pi r^2\, dr$ when $\Psi*\Psi$ depends only on r (not on θ or ϕ). Normalize ψ and find the probability that the particle is between $r = 1$ and $r = 3$ units of length.

Solution   Example 13.1 showed that $\Psi*\Psi = \psi*\psi$. Here,

$$\psi*\psi = (Ae^{-r^3+i\delta})(Ae^{-r^3-i\delta}) = A^2e^{-2r^3}.$$

Substituting this expression into Eq. 13.1 gives $\int_{r=0}^{r=\infty} A^2e^{-2r^3}\, 4\pi r^2\, dr$, where r goes from 0 to ∞ to include all space. Integrating, with the substitution $u = 2r^3$, we have

$$A^2\left(\frac{4\pi}{6}\right)\int_0^\infty e^{-2r^3}6r^2\, dr = A^2\left(\frac{2\pi}{3}\right)\int_0^\infty e^{-u}\, du$$

$$= A^2\left(\frac{2\pi}{3}\right)(-e^{-u})\Big|_0^\infty$$

$$= A^2\left(\frac{2\pi}{3}\right)[-e^{-\infty}-(-e^0)]$$

$$= A^2\left(\frac{2\pi}{3}\right)(0+1)$$

$$= A^2\left(\frac{2\pi}{3}\right) = 1.$$

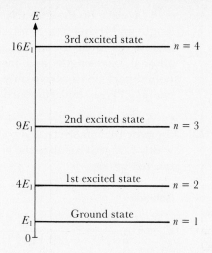

FIGURE 14.5
The four lowest allowed energy levels for a particle in a one-dimensional box, where
$E_1 = h^2/8mL^2$.

is, n is the number of the particle's half wavelengths being fitted between the walls. In the limit of large quantum numbers, the number of humps within any finite Δx becomes so large that $\psi^*\psi\,\Delta x$ approaches the classical value, $(1/L)\,\Delta x$.

Now that we have ψ_n, we can substitute back into the SWE to find the allowed energies. We use

$$\frac{d^2}{dx^2}\sqrt{\frac{2}{L}}\sin\frac{n\pi}{L}x = \sqrt{\frac{2}{L}}\left(\frac{n\pi}{L}\right)^2\left[-\sin\frac{n\pi}{L}x\right] = \frac{-n^2\pi^2}{L^2}\,\psi_n,$$

and

$$\frac{-\hbar^2}{2m}\frac{d^2\psi_n}{dx^2} + 0\psi_n = E_n\psi_n$$

becomes

$$\frac{-\hbar^2}{2m}\left(\frac{-n^2\pi^2}{L^2}\right)\psi_n = E_n\psi_n \quad\text{or}\quad E_n = n^2\,\frac{\pi^2\hbar^2}{2mL^2}.$$

With $\hbar = h/2\pi$, the allowed energies are

$$E_n = n^2\,\frac{h^2}{8mL^2}. \tag{14.2}$$

Like the energies of the Bohr atom, these energies are quantized. The energies can have only those values given by Eq. (14.2) with $n = 1$ (the ground state), $n = 2$ (the first excited state), $n = 3$ (the second excited state), and so on. (See Fig. 14.5.)

A wave on an infinitely long string can propagate with any wave-length and frequency. Similarly, a free particle can have any momentum and energy. But if the string has boundaries, standing waves of only certain allowed wavelengths and frequencies can be maintained. Similarly, a bound particle can have only certain allowed values of momentum and energy. These similarities are due to the wavelike properties of particles.

EXAMPLE 14.2   Suppose that an electron is confined in an infinitely deep potential well of width 1.0 nm. Calculate the energies of the ground state and the first two excited states.

Solution

$$E_n = n^2 \frac{h^2}{8mL^2} = n^2 \frac{(6.626 \times 10^{-34} \, \text{J} \cdot \text{s})^2}{8(9.11 \times 10^{-31} \, \text{kg})(1.0 \times 10^{-9} \, \text{m})^2} \left(\frac{1 \, \text{eV}}{1.602 \times 10^{-19} \, \text{J}} \right)$$

$$= n^2(0.376 \, \text{eV}).$$

The ground state quantum number is $n = 1$, which gives $E_1 = 1^2(0.376 \, \text{eV}) = 0.38 \, \text{eV}$. For the first excited state, $n = 2$, so $E_2 = 2^2(0.376 \, \text{eV}) = 1.5 \, \text{eV}$; for the second excited state, $n = 3$, which gives $E_3 = 3^2(0.376 \, \text{eV}) = 3.4 \, \text{eV}$.   ●

Note that as the box becomes larger (that is, as L increases), both the allowed kinetic energies of the particle and the energy differences between states become smaller. Or, as the box becomes smaller, these energy terms become larger. This concept is important to an understanding of the behavior of several different quantum systems.

It seems unlikely that infinitely deep potential wells exist in nature, but they may (in the binding of quarks, for example). And absolutely abrupt changes in V are even less likely. Since $F_x = -\partial V/\partial x$, an abrupt change in V would correspond to an infinite force. However, if V is much larger than E and if L is much larger than the distance over which V changes from 0 to V, the particle in a box solutions will be a good approximation of an actual quantum mechanical system.

14.2
THE SEMI-INFINITE WELL

The **semi-infinite well** has an infinite potential energy wall at $x = 0$ and a finite wall of height V at $x = L$, as illustrated in Fig. 14.6. Let's consider the possibility of bound states, that is, states with energy $E < V$, states for which the particle can't escape the well. In the well, we've seen that $\psi_n =$

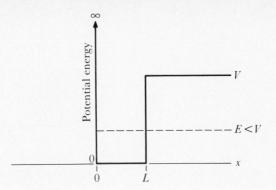

FIGURE 14.6
A bound-state energy for a
semi-infinite well.

$B \sin k_n x$ is a wave function that satisfies both the SWE and the boundary condition $\psi = 0$ at $x = 0$. Substituting $B \sin k_n x$ and $V = 0$ into the SWE gives $k_n = \sqrt{2mE_n}/\hbar$. The E_n is *not* given by Eq. (14.2), because we no longer have infinite walls on both sides.

For $x \geq L$, outside the wall, we can write the SWE as

$$\frac{\hbar^2}{2m} \frac{d^2\psi_n}{dx^2} = (V - E_n)\psi_n.$$

Substitution back into this SWE shows that solutions of the form $Ce^{K_n x}$ and $De^{-K_n x}$ are possible, if $K_n = \sqrt{2m(V - E_n)}/\hbar$. We can't have the probability approaching infinity as x approaches infinity, so another boundary condition is that the wave function must be finite at infinity. This rules out the $Ce^{K_n x}$ solutions and leaves us with $\psi_n = De^{-K_n x}$ for $x \geq L$.

Our other boundary conditions are that ψ_n must be continuous at $x = L$ (so that $d\psi_n/dx$ will be finite) and that $d\psi_n/dx$ also must be continuous at $x = L$ (so that the $d^2\psi_n/dx^2$ term in the SWE will not become infinite).

EXAMPLE 14.3 Find two equations relating the coefficients B and D for bound states in a semi-infinite well.

Solution Since the wave function must be continuous at the boundary, $x = L$, we have

$$B \sin k_n L = De^{-K_n L}. \tag{A}$$

Since the first derivatives of the wave function must also be continuous at $x = L$; we get

$$k_n B \cos k_n L = -K_n De^{-K_n L}. \tag{B}$$

●

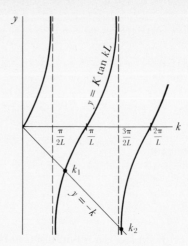

FIGURE 14.7
If bound states exist in the semi-infinite well, we can determine their wave numbers and energies from the intersections.

If we drop the subscripts and divide Eq. (A) by Eq. (B), we'll be able to find the energies of the bound states (if there are any). The division gives

$$K \tan kL = -k, \tag{14.3}$$

where

$$k = \sqrt{2mE}/\hbar \quad \text{and} \quad K = \sqrt{2m(V-E)}/\hbar.$$

In Eq. (14.3), kL is in radians. We can solve this transcendental equation by a number of methods, one of which involves plotting $y = K \tan kL$ and $y = -k$ on the same graph, as shown in Fig. 14.7. If there are solutions, they'll be wherever the lines intersect at $k > 0$. If the bound states exist, the first intersection gives k_1 and E_1, the second intersection yields k_2 and E_2, and so on.

EXAMPLE 14.4   If the first intersection in Fig. 14.7 occurs at $k_1 = 2/L$, what is the ground state energy and how does it compare to that of the particle in a box?

Solution We start with $k_n = \sqrt{2mE_n}/\hbar$, which gives

$$E_1 = \frac{h^2 k_1^2}{8\pi^2 m} = \frac{h^2 (2/L)^2}{8\pi^2 m} = \frac{(2/\pi)^2 h^2}{8mL^2} = \frac{0.4h^2}{8mL^2}.$$

The ground state energy of a particle in a box is, by Eq. (14.2) with $n = 1$, equal to $h^2/8mL^2$. The particle in this semi-infinite well has a ground state energy of about 40 percent of that of a particle in a box of the same size.   ●

The energies for a semi-infinite well are all smaller than those of the equivalent infinite well or box because the wave function doesn't go to zero for $x \geq L$ for the particle in a semi-infinite well. This behavior means that the particle has some probability of spending time out of the well and that the effective width of the well becomes larger than L. The expression $E_n = n^2h^2/8mL^2$ shows that larger L's mean smaller energies.

It's easy to convert the particle in a semi-infinite well to the particle in a box: Just let V approach infinity. As $V \rightarrow \infty$, $K = \sqrt{2m(V - E)}/\hbar \rightarrow \infty$, and the $y = K \tan kL$ lines approach straight vertical lines through $k = \pi/L$, $2\pi/L$, $3\pi/L$, . . . , $n\pi/L$. Then $n\pi/L = k_n = \sqrt{2mE_n}/\hbar$ gives $E_n = n^2h^2/8mL^2$, which is Eq. (14.2) for the particle in a box. Also, $K \rightarrow \infty$ gives $\psi_n \rightarrow 0$ for $x \geq L$.

One thing that's not obvious from Fig. 14.7 is that there is a limit to the number of bound solutions to the semi-infinite well problem. Since $E_n = \hbar^2k_n^2/2m$, E_n increases as k_n increases. If $E > V$, then K becomes imaginary, and we can no longer plot $y = K \tan kL$. What this situation means, of course, is that if $E > V$, the particle is no longer trapped in the well.

14.3
A FINITE POTENTIAL WELL

Suppose that V equals some constant finite value outside the well for $x \leq 0$ and $x \geq L$. Also, $V = 0$ in the resulting **finite potential well,** as shown in Fig. 14.8. Let's first consider the $E_n < V$ (bound particle) case. Then $\psi_n = Ce^{K_n x}$ for the negative regions of x and $De^{-K_n x}$ for the positive regions outside the well. It's important to note that these wave functions for ψ_n aren't equal to zero at $x = 0$ and $x = L$. Therefore ψ_n can look like Fig. 14.8. As you can see, there is now some probability of finding the particle outside the well on both sides.

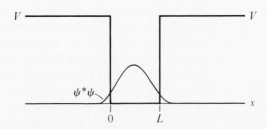

FIGURE 14.8
A potential well with finite walls. The probability per unit distance is sketched for the ground state.

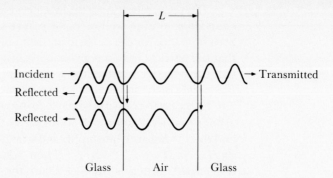

FIGURE 14.9
Light waves incident normally on a thin film of air between glass. The two reflected waves
have been moved downward for clarity. The reflected waves are 180° out of phase, giving
a reflection minimum and a corresponding transmission maximum.

The simple solutions for the particle in a box and the not-so-simple
solutions for the particle in a semi-infinite well become even more complicated for a finite well because the wave function doesn't become zero at
either $x = 0$ or $x = L$. As a result, the wave function is neither a simple sine
function nor a simple cosine function within the well. We can say that
the particle in a finite well has a probability of spending some time out
of the well on both sides, making the effective width of the well even larger
than it became for the semi-infinite well. Therefore the energy values
and the level separations are smaller for a finite well than for a semi-
infinite well (which were, in turn, smaller than those for a box) of the
same L.

You may have previously studied what happens when a light wave
strikes a thin insulating film perpendicular to its boundary. (See Fig. 14.9.)
If there is a change of index of refraction (a change of wave speed) at the
boundaries of the film, some reflection and some transmission will occur.
In addition, certain thicknesses of the film will give maxima and minima in
the reflection and transmission due to interference effects. For instance,
$2L = n\lambda$ (where n is an integer, L is the thickness of the film, and λ is the
wavelength in the film) expresses the condition for maximum transmission
(minimum reflection) for a thin film of air with glass on both sides.

A similar situation occurs if a particle arrives at a region where there
is a potential well. The particle is initially moving outside the well, with
$E > V$. The total energy, E, equals $KE + V$, and is conserved. So when V
drops to zero in the well, the kinetic energy increases from $E - V$ to E and
the particle abruptly speeds up. Then when the particle passes the other
edge of the well, the potential energy returns to V and the KE to $E - V$,
with the particle abruptly slowing back down to its original speed.

Let's consider particles incident from the left in region I of Fig. 14.10. In region I, the potential energy is V, and the solution to the pertinent SWE is

$$\psi_I = Ae^{ik_1x} + Be^{-ik_1x}.$$

The term Ae^{ik_1x} describes a particle moving in the $+x$ direction (an incident particle), and the term Be^{-ik_1x} describes a particle moving in the $-x$ direction (a reflected particle). The wave number for region I, k_I, equals $\sqrt{2m(E-V)}/\hbar$. In region II, the potential well, $V = 0$, so

$$\psi_{II} = Ce^{ik_{II}x} + De^{-ik_{II}x},$$

where $k_{II} = \sqrt{2mE}/\hbar$. In region III, there are no particles incident from the left, so there is no $e^{-ik_{III}x}$ term. Also, $k_{III} = \sqrt{2m(E-V)}/\hbar = k_I$. Therefore

$$\psi_{III} = Fe^{ik_1x}.$$

Both ψ and $d\psi/dx$ must be continuous at $x = 0$ and $x = L$. At $x = 0$,

$$A + B = C + D \quad \text{and} \quad k_I A - k_I B = k_{II} C - k_{II} D.$$

At $x = L$,

$$Ce^{ik_{II}L} + De^{-ik_{II}L} = Fe^{ik_1L} \quad \text{and} \quad k_{II}Ce^{ik_{II}L} - k_{II}De^{-ik_{II}L} = k_I Fe^{ik_1L}.$$

We have just obtained four equations with five unknowns. However, dividing all equations by A makes it possible to solve for F/A. This is a tedious bit of math and is probably best done with determinants.

The term F/A is the ratio of the amplitude of the transmitted wave function to that of the incident wave function. Probabilities are proportional to $\psi^*\psi$, and the probability of transmission past this potential well is $T = (F/A)^*(F/A)$. We call T the **transmission coefficient**; it is dimension-

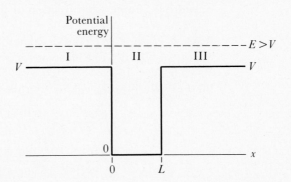

FIGURE 14.10
Particles are incident from region I and move across the well (region II) to be transmitted to region III.

less and gives the probability of a particular particle getting past the potential well or, alternatively, the fraction of those particles with energy E that won't be reflected by the potential well.

If we do the tedious math referred to, we eventually arrive at

$$T = \frac{1}{\cos^2 k_{II}L + (1/4)(k_{II}/k_I + k_I/k_{II})^2 \sin^2 k_{II}L}$$

or

$$T = \left[1 + \frac{(V \sin k_{II}L)^2}{4E(E - V)} \right]^{-1}, \qquad (14.4)$$

where

$$k_{II} = \frac{\sqrt{2mE}}{\hbar}$$

and $k_{II}L$ is in radians. Recalling that $\sin n\pi = 0$, we first note in Eq. (14.4) that $T = 1$ whenever $k_{II}L = n\pi$, which we can write as $(2\pi/\lambda_{II})L = n\pi$, or $2L = n\lambda_{II}$, as for the thin insulating film and light rays.

The difference in path lengths between particles reflected at the left side of the well and particles reflected at the right side equals $2L$. When $2L$ equals a whole number of the wavelengths of the particles in the well, their reflected wave functions will be 180° out of phase, will completely cancel one another, and all particles will be transmitted. This occurrence is a simple explanation of the Ramsauer–Townsend effect, in which electrons of a certain wavelength pass through inert gas atoms almost as if they weren't there.

EXAMPLE 14.5   If a beam of 5.0-eV electrons are incident on a 15.0-eV deep potential well of width 0.20 nm, what fraction of the electrons will be transmitted?

Solution Here, $V = 15.0$ eV. Remember, 5.0-eV electrons have a kinetic energy of 5.0 eV. The electrons are incident in region I where $E = KE + V = 5.0$ eV $+ 15.0$ eV $= 20.0$ eV. Since energy is conserved, $E = 20.0$ eV in all three regions. With $m = 0.511 \times 10^6$ eV$/c^2$,

$$k_{II} = \sqrt{\frac{2mE}{\hbar}} = \sqrt{\frac{2(0.511 \times 10^6 \text{ eV}/c^2)(20.0 \text{ eV})}{6.582 \times 10^{-16} \text{eV} \cdot \text{s}}}$$

$$= \frac{6.869 \times 10^{18} \text{ rad/s}}{2.9979 \times 10^8 \text{ m/s}} = 2.291 \times 10^{10} \text{ rad/m}.$$

Then

$$\sin(k_{II}L) = \sin[(2.291 \times 10^{10} \text{ rad/m})(0.20 \times 10^{-9} \text{ m})]$$
$$= \sin(4.582 \text{ rad}) = -0.9916.$$

Since $E - V = KE = 5.0$ eV, we have

$$T = \left\{ 1 + \frac{[(15.0 \text{ eV})(-0.9916)]^2}{4(20.0 \text{ eV})(5.0 \text{ eV})} \right\}^{-1} = \frac{1}{1.553} = 0.64, \text{ or } 64\%. \quad \bullet$$

In Example 14.5, if 64 percent of the electrons are transmitted, 36 percent are reflected. Since $\sin^2 k_{II} L$ is almost 1, this data gives almost a minimum in the transmission.

For the classical particle in a box, any energy is allowed, and the probability per unit length is constant. But quantum mechanics says that only certain energies are possible and that the probability per unit length is a $\sin^2$ function. For a semi-infinite or finite potential well (with $E < V$), the classical probability of finding the particle outside the well is zero because $E < V$ corresponds to a negative kinetic energy and an imaginary speed. Quantum mechanics again disagrees with classical mechanics and predicts a finite possibility of finding the particle outside the well. Also, small particles incident on a potential well act much like waves incident on a thin film, not like classical particles. With all this disagreement, which is right, quantum mechanics or classical mechanics? Microscopic systems that can be approximated by a potential well give experimental results that agree with the quantum solution, not the classical one, in the areas of disagreement.

QUESTIONS AND PROBLEMS

14.1 Sketch a wave function that is discontinuous at a boundary. Then sketch $d\psi/dx$ and $d^2\psi/dx^2$ for this wave function.

14.2 Rearranging a SWE gives $d^2\psi/dx^2 = (2m/\hbar^2)(V - E)\psi$. If $d\psi/dx$ were discontinuous at a boundary, what would $d^2\psi/dx^2$ be there? For what V is this result possible?

14.3 For an infinitely deep well, the wave function is continuous at the boundaries. Is the first derivative of the wave function continuous there?

14.4 A particle is in a box with infinitely rigid walls. Suppose we move the walls to $x = -L/2$ and $x = +L/2$ just for this problem.

a) Is $\psi_n = A \cos k_n x$ a possible solution? What must k_n equal?

b) Is $\psi_n = B \sin k_n x$ a possible solution? What must k_n equal?

c) What are the allowed energies for $\psi_n = A \cos k_n x$? For $\psi_n = B \sin k_n x$?

14.5 What is the probability of finding the particle in the right half of a box?

14.6 What is the probability of finding the particle in a box between $x = L/4$ and $3L/4$ a) in the ground state? b) in the first excited state?

14.7 What is the approximate probability of finding the particle in a box between $x = 0.50 L$ and $x = 0.51 L$ a) in the ground state? b) in the third excited state?

14.8 Sketch $\psi^*\psi$ for $n = 6$ for the particle in a box.

14.9 Let the particle be a neutron confined to a box of width 3.5×10^{-15} m. Find the wave functions and energies of the ground state and the first excited state.

14.10 Find the excitation energy of the third excited state for an electron confined to a box of width 200 pm.

14.11 What are the similarities and differences of the allowed energies for the particle in a box and the Bohr hydrogen atom?

14.12 It takes 4.0 eV to excite an electron from the ground state to the first excited state in a box. What does this tell you about the box?

14.13 A long molecule used in a laser has a transition between the first excited state and the ground state. The electron in the molecule that makes the transition acts like a particle in a box 3.78 nm long.

a) Calculate the wavelength of the photon emitted in this transition.

b) By a chemical reaction, a radical is added to each molecule, thereby increasing the length. How will this change your answer to (a)?

14.14 In a hydrogen atom the electron is restricted to a diameter of about 10^{-10} m in the ground state. In an H_2 molecule, the distance increases. Consider the particle-in-a-box model. What will this increase do to the energy?

14.15 Let ΔE equal the energy difference between adjacent states and E equal the energy of the lower of the two adjacent states. a) What is the largest value of $\Delta E/E$ for the particle in a box? b) What is the limit of $\Delta E/E$ for large quantum numbers? c) Which answer, (a) or (b), is similar to the classical answer?

14.16 Let $KE_n = E_n = p_n^2/2m$. a) Show that the allowed values of p_n for the particle in a box are $nh/2L$. b) What then are the allowed wavelengths for the particle in a box?

14.17 In measuring p_x for the particle in a box, we would disturb the system by a minimum of one state either way, so that $\Delta p_x = p_{n+1} - p_{n-1}$, at least. We know the particle is somewhere in the box, so the minimum $\Delta x = L$. Use the answer to Problem 14.16(a) to show that $\Delta p_x \Delta x \geq h$.

14.18 To make sure that a particle in a box is in a particular state, we let it move across the box at least one time. Therefore $\Delta t = L/v_n$, at least, where $v_n = p_n/m$ and p_n is given in Problem

14.16(a). In measuring the particle we will change its energy, so let it change one state either way, at least. That is, the minimum $\Delta E = E_{n+1} - E_{n-1}$. Show that $\Delta E \Delta t \geq h$.

14.19 The $E = 0$ solution for the SWE for a particle in a box is $\psi = Ax + B$. a) Show that this statement is true by substitution into the SWE. b) Apply the boundary conditions to show why the probability of the $E = 0$ state is zero.

14.20 Why isn't $k = 0$ an acceptable solution of Eq. (14.3)?

14.21 Refer to Example 14.4. Calculate V.

14.22 As $E_1 \to V$, the only bound solution for the particle in a semi-infinite well will occur for $k \to \pi/2L$. a) Explain why, using Fig. 14.7. b) Show that $E_1 \to 25\%$ of the ground state energy of a particle in a box.

14.23 If the deuteron is modeled by a proton in a semi-infinite box of width 4×10^{-15} m, with a ground state binding energy of $V - E_1 = 2.2$ MeV, prove that $E_1 = 5.8$ MeV and that there are no excited states.

14.24 As L gets larger, the lowest energy levels of the semi-infinite well approach those of an equal-width box. Explain why, using Fig. 14.7.

14.25 Show that neither $\psi_n = A \cos k_n x$ nor $\psi_n = B \sin k_n x$ satisfies the boundary conditions for a bound particle in a finite potential well with walls at $x = 0$ and $x = L$.

14.26 Compare a particle in a finite well to one in an infinite well. a) Where is the SWE the same? b) Where is the SWE different?

14.27 a) In Example 14.5, what are the wavelengths of the particles in the three regions? b) Compare the wavelength in region II with $2L$. Explain the meaning of your answer.

14.28 If the width of a well is $n\lambda_{II}/2$, there is a transmission maximum. What widths give a reflection maximum?

14.29 Show that as the depth of a well approaches zero (no well!), the probability of transmission across the well approaches 100%.

14.30 If a beam of 3.7-MeV protons are incident on a 2.2-MeV-deep potential well, what widths of the well will give a) no reflection? b) 90% transmission?

14.31 If a potential well extends from 0 to ∞, there is no region III; there also are no reflected particles in region II. a) Show that $A + B = C$ and that $(k_I/k_{II})(A - B) = C$. b) Show that the reflection coefficient, $R = (B/A)*(B/A)$, equals $[(k_{II} - k_I)/(k_{II} + k_I)]^2$ for this "step potential."

CHAPTER

More One-Dimensional Applications of the SWE

Like relativity, quantum mechanics is probably far removed from your everyday life. As a result, its predictions also may well violate your common sense. In this chapter, we will describe how one student tried to prove a quantum mechanics prediction wrong, but instead wound up with the world's highest scientific honor for proving it right. The phenomenon he couldn't initially believe was called tunneling. Also, you know from your previous physics courses that repetitive motion (simple harmonic motion is its least complicated form) is very common in nature. We will explain the differences between classical and quantum harmonic oscillators in this chapter. And you will see how accurate Planck's guess was when he quantized the harmonic oscillator when deriving the blackbody radiation curves.

15.1
A POTENTIAL BARRIER

If we have a potential energy that is zero for $0 \leq x \leq L$ and is V elsewhere, we have a potential well. But suppose we "turn the well upside down." Then the potential energy is V for $0 \leq x \leq L$ and is zero elsewhere, and we have a type of **potential barrier**. In Fig. 15.1, we see such a barrier, with regions I, II, and III, similar to Fig. 14.10.

Let's first consider the situation where incident particles have enough energy to get over the barrier, that is, $E > V$. Just as in Section 14.3, we have

$$\psi_I = Ae^{ik_1x} + Be^{-ik_1x} \qquad \psi_{II} = Ce^{ik_{II}x} + De^{-ik_{II}x} \quad \text{and} \quad \psi_{III} = Fe^{ik_1x}.$$

The difference here is that by switching the values of the potential energy in the regions, we've also switched the values of k_I and k_{II}. For this potential barrier with $E > V$, $k_I = \sqrt{2mE}/\hbar$ and $k_{II} = \sqrt{2m(E - V)}/\hbar$. Again requiring that the wave functions and their first derivatives be continuous at both $x = 0$ and $x = L$, we have the same equations as previously, which eventually yield the same expression for T, but with the different k_{II}. That is,

$$T = \frac{1}{\cos^2 k_{II}L + (1/4)(k_{II}/k_I + k_I/k_{II})^2 \sin^2 k_{II}L}$$

or

$$T = \left[1 + \frac{(V \sin k_{II}L)^2}{4E(E - V)} \right]^{-1}, \qquad (15.1)$$

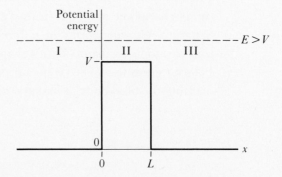

FIGURE 15.1
A one-dimensional potential energy barrier, with $E > V$.

where

$$k_{\rm II} = \frac{\sqrt{2m(E - V)}}{\hbar}$$

and $k_{\rm II}L$ is in radians. Again, $T = 1$ when $k_{\rm II}L = n\pi$, or $2L = n\lambda_{\rm II}$.

Example 15.1   If a beam of 15.0-eV electrons are incident on a 5.0-eV potential barrier of width 0.20 nm, what fraction of the electrons will be transmitted?

Solution Here V = 5.0 eV, and 15.0 eV is the kinetic energy of the electrons in region I. Since $V = 0$ in region I, $E = KE + V = 15.0$ eV $+ 0 =$ 15.0 eV. Because energy is conserved, $E = 15.0$ eV in all three regions. In region II, $E - V = 15.0$ eV $- 5.0$ eV $= 10.0$ eV. With $m = 0.511 \times 10^6$ eV/c^2,

$$k_{\rm II} = \frac{\sqrt{2m(E - V)}}{\hbar} = \frac{\sqrt{2(0.511 \times 10^6 \text{ eV}/c^2)(10.0 \text{ eV})}}{6.582 \times 10^{-16} \text{ eV} \cdot \text{s}}$$

$$= \frac{4.857 \times 10^{18} \text{ rad/s}}{2.9979 \times 10^8 \text{ m/s}} = 1.620 \times 10^{10} \text{ rad/m},$$

or just $1/\sqrt{2}$ times the $k_{\rm II}$ in Example 14.5. Then

$$\sin(k_{\rm II}L) = \sin[(1.620 \times 10^{10} \text{ rad/m})(0.20 \times 10^{-9} \text{ m})]$$
$$= \sin(3.240 \text{ rad}) = -0.0985.$$

We finally arrive at the fraction of electrons transmitted:

$$T = \left\{ 1 + \frac{[(5.0 \text{ eV})(-0.0985)]^2}{4(15.0 \text{ eV})(10.0 \text{ eV})} \right\}^{-1} = \frac{1}{1.0004} = 0.9996, \text{ or } 99.96\%. \quad \bullet$$

In Example 15.1, if 99.96 percent of the electrons are transmitted only 0.04 percent are reflected. Since $\sin^2 k_{\rm II}L$ is almost 0 and T is almost 1, there is almost a maximum in the transmission.

15.2
TUNNELING

Let's now consider the same potential energy barrier as in Section 15.1, except for energies $E < V$. (See Fig. 15.2.) We have a free particle of energy E (all of it kinetic) incident on a potential energy barrier of height V. In classical physics the particle can't move over the barrier if E is less than V. For example, a 0.01-kg mass moving at 10 m/s has a kinetic energy

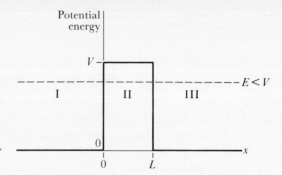

FIGURE 15.2
A one-dimensional
potential energy barrier
with $E < V$.

of 0.5 J. If it came to a 1.0-m higher hill, the potential energy increase would have to be $mgy = 0.98$ J for the object to make it to the top and over the hill. However, with a total energy of only 0.5 J, the object could never make it. A classical particle of energy E can't get by a potential energy barrier of height V if $E \leq V$.

Surprisingly, this isn't true in quantum mechanics. As in Section 15.1, $\psi_I = Ae^{ik_1x} + Be^{-ik_1x}$ to the left of the barrier and $\psi_{III} = Fe^{ik_1x}$ to the right of the barrier. Within the barrier (see Section 14.2), $\psi_{II} = Ce^{Kx} + De^{-Kx}$, where $K = \sqrt{2m(V-E)}/\hbar$. Again requiring that the wave functions and their first derivatives be continuous at both $x = 0$ and $x = L$, we obtain equations similar to those in Sections 14.3 and 15.1, which eventually yield a similar expression for T, but with K replacing ik_{II}.

Because ψ_{II} isn't zero at either $x = 0$ or $x = L$, Fe^{ik_1x} isn't zero. That is, when a quantum particle is incident on a potential energy barrier, there is some probability that the particle will appear on the other side of the barrier. This effect is called **tunneling.** (See Fig. 15.3.) We say that the quantum particle tunnels through the barrier. In doing so, it does no work (E is constant). The particle doesn't actually bore a hole through the barrier; it just appears on the other side.

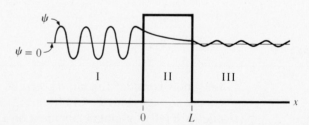

FIGURE 15.3
Tunneling: There is some probability that a particle incident from the left side will appear on the right side of the barrier, even though E is less than V.

Ivar Giaver said that he found this prediction of quantum physics so preposterous in terms of classical physics that he devised an experiment to prove that tunneling couldn't occur. Well, it did occur, and his experiment won him a share of the 1973 Nobel prize in physics. The other two winners that year were Leo Esaki for the tunnel diode and Brian Josephson for tunneling effects in superconducting junctions. All three Nobel laureates did their prize-winning research while students!

To arrive at the expression for the tunneling transmission coefficient T, you need to realize that the equations for the potential well and $E > V$ potential barrier include

$$e^{ik_{II}L} + e^{-ik_{II}L} = 2 \cos k_{II}L \quad \text{and} \quad e^{ik_{II}L} - e^{-ik_{II}L} = 2i \sin k_{II}L$$

terms. For $E < V$ tunneling, K replaces ik_{II}. Then

$$e^{KL} + e^{-KL} = 2 \cosh KL \quad \text{and} \quad e^{KL} - e^{-KL} = 2 \sinh KL,$$

or the hyperbolic cosine and sine functions of KL in place of the cosine and sine functions of $k_{II}L$. That is,

$$T = \frac{1}{\cosh^2 KL + (1/4)(K/k_I - k_I/K)^2 \sinh^2 KL}$$

or

$$T = \left[1 + \frac{(V \sinh KL)^2}{4E(V - E)} \right]^{-1}, \tag{15.2}$$

where

$$K = \frac{\sqrt{2m(V - E)}}{\hbar}$$

and KL is in radians. The function $\sinh KL$ is zero only at $KL = 0$ and increases rapidly as KL increases. Therefore $T < 1$ for $KL > 0$.

Example 15.2   If a beam of 5.0-eV electrons are incident on a 15.0-eV potential barrier of width 0.20 nm, what is the probability that any particular electron will tunnel through the barrier

    a) according to classical physics, and

    b) according to quantum mechanics?

Solution

    a) Zero, because tunneling is forbidden in classical physics.

    b) We start with $KE = 5.0$ eV $= E$, $V = 15.0$ eV, and $V - E = 10.0$ eV. These quantities give $K = \sqrt{2m(V - E)}/\hbar = 1.620 \times 10^{10}$ rad/m, as in

Example 15.1, and sinh KL = sinh(3.240 rad) = 12.75. Then we use Eq. (15.2) to obtain the probability of tunneling:

$$T = \left[1 + \frac{(15.0 \text{ eV} \times 12.75)^2}{4(5.0 \text{ eV})(10.0 \text{ eV})}\right]^{-1} = \frac{1}{183.8} = 5.4 \times 10^{-3} = 0.54\%.$$ •

In Example 15.2, a 5.0-eV electron will be reflected from a 15.0-eV, 0.20-nm potential barrier 99.46 percent of the time. Actually, a 0.20-nm barrier is extraordinarily thin. Thicker and/or taller barriers would cause the wave function beyond the barrier to drop much closer to zero. That is, tunneling becomes rapidly less probable as the height and/or thickness of the barrier increase. The tunneling particle must be small: Electrons tunnel through thin insulating layers, but people don't tunnel through walls.

Tunneling has applications in many areas of solid state physics, as, for example, that shown in Fig. 15.4. Tunneling also shows how alpha particles can spontaneously leave some nuclei, and it explains some molecular oscillations. Many other applications have been found for tunneling, even being used in some models of the creation of the universe and the eventual fate of black holes.

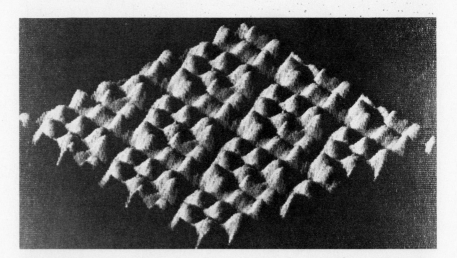

FIGURE 15.4

A scanning tunneling microscopy image of a silicon surface. An extraordinarily sharp metal tip is positioned about 1 nm above the surface. As the tip is moved horizontally across the surface, it is also moved vertically to keep the tunneling current constant. Many such traversals yield the representation shown: Each bump is a silicon atom.

Source: Photo courtesy of Dr. H. Rohrer, International Business Machines Corporation.

FIGURE 15.5
Can the electron trapped in potential well A move to potential well B with no addition of energy?

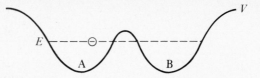

Example 15.3 An electron is trapped in a crystal in potential energy well A, as shown in Fig. 15.5. The electron is given no added energy. How is it possible for the electron to move to the nearby empty potential energy well B? When will this event be most probable?

Solution Even though $E < V$, the electron can tunnel from A to B. The probability that it will do so is greatest if A and B are close to each other and if E is near V. Electrons tunnel between potential wells in many systems. ●

15.3
THE HARMONIC OSCILLATOR

One simple **harmonic oscillator** that you may remember from a previous physics course is a spring–mass system. (See Fig. 15.6.) Hooke's law, $F = -k'x$, gives the force exerted on the mass by the spring. The term k' is the force constant of the spring. The mass freely oscillates at a classical angular frequency $\omega_c = \sqrt{k'/m}$. The potential energy stored in the spring is $V = \frac{1}{2}k'x^2 = \frac{1}{2}m\omega_c^2x^2$.

In quantum mechanics, the SWE for a one-dimensional harmonic oscillator is

$$-\frac{\hbar^2}{2m}\frac{d^2\psi_n}{dx^2} + \frac{1}{2}m\omega_c^2x^2\psi_n = E_n\psi_n. \qquad (15.3)$$

FIGURE 15.6
A classical harmonic oscillator consisting of a mass m connected to a spring of negligible mass and a force constant k'.

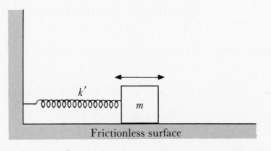

Frictionless surface

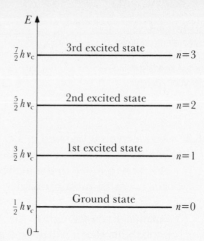

FIGURE 15.7
The four lowest allowed energy states for a one-dimensional harmonic oscillator.

Equation (15.3) isn't at all easy to solve. Complete analysis of it yields the result:

$$d^2\psi_n/dx^2 = [(m^2\omega_c^2/\hbar^2)x^2 - (2n+1)m\omega_c/\hbar]\psi_n.$$

Substituting back into the SWE, we obtain

$$E_n = (n + \tfrac{1}{2})\hbar\omega_c = (n + \tfrac{1}{2})h\nu_c. \tag{15.4}$$

Just as Planck assumed (in deriving the blackbody radiation curve), the energy levels of a harmonic oscillator are quantized, and each level is $\Delta E = h\nu_c$ away from its adjacent level(s). (See Fig. 15.7.) However, Eq. (15.4) differs from Planck's assumption in one important way: The minimum energy of a harmonic oscillator is *not* zero. The minimum energy in one dimension is $\tfrac{1}{2}h\nu_c$. (The minimum energy is $\tfrac{3}{2}h\nu_c$ for a three-dimensional isotropic harmonic oscillator.) For example, molecules have mass and interact with one another with springlike forces. Therefore the minimum kinetic energy of each molecule (at absolute zero temperature) is not zero. This **zero-point energy** is most important at low temperatures. As we pointed out in Section 10.4, this energy is large enough to keep liquid helium from ever freezing at normal pressures.

Example 15.4   The wave function of the first excited state of a one-dimensional harmonic oscillator is

$$\psi_1 = (\alpha/4\pi)^{1/4}2\sqrt{\alpha}xe^{-\alpha x^2/2}.$$

Find the constant α and show that the value of the energy agrees with that in Eq. (15.4).

Solution We start with

$$\frac{d\psi_1}{dx} = \left(\frac{\alpha}{4\pi}\right)^{1/4} 2\sqrt{\alpha}(e^{-\alpha x^2/2} - \alpha x^2 e^{-\alpha x^2/2}).$$

Then

$$\frac{d^2\psi_1}{dx^2} = \left(\frac{\alpha}{4\pi}\right)^{1/4} 2\sqrt{\alpha}(-\alpha x - 2\alpha x + \alpha^2 x^3)e^{-\alpha x^2/2} = \psi_1(-3\alpha + \alpha^2 x^2).$$

Substituting into the SWE, Eq. (15.3), we have

$$-\left(\frac{\hbar^2}{2m}\right)\psi_1(-3\alpha + \alpha^2 x^2) + \frac{1}{2}m\omega_c^2 x^2\psi_1 = E_1\psi_1$$

or

$$\left(\frac{1}{2}m\omega_c^2 - \frac{\hbar^2\alpha^2}{2m}\right)x^2 = E_1 - \frac{\hbar^2 3\alpha}{2m}. \tag{A}$$

Equation (A) must be true for all x. It has the form $ax^2 = b$. However, $ax^2 = b$ is true only for $x = -\sqrt{b/a}$ and $x = +\sqrt{b/a}$ (not *all* x) unless both a and b equal zero. Therefore

$$\frac{1}{2}m\omega_c^2 - \frac{\hbar^2\alpha^2}{2m} = 0 \quad \text{or} \quad \alpha = \frac{\omega_c m}{\hbar}.$$

Also

$$E_1 - \frac{\hbar^2 3\alpha}{2m} = 0 \quad \text{or} \quad E_1 = \frac{\hbar^2}{2m}3\alpha = \frac{\hbar^2}{2m}3\left(\frac{\omega_c m}{\hbar}\right) = \frac{3}{2}\hbar\omega_c.$$

This value of the energy is in agreement with that of Eq. (15.4), which gives (for $n = 1$) $E_1 = (1 + \frac{1}{2})\hbar\omega_c = \frac{3}{2}\hbar\omega_c$. ●

Note that the quantum number, n, for the ground state is zero. For the first excited state, $n = 1$; for the second excited state, $n = 2$; and so on. These values are different from the values of n for the ground state or various excited states of both the hydrogen atom and the particle in a box.

Many oscillating systems don't have Hooke's law forces. For small oscillations, however, we can often closely approximate these systems by a simple harmonic oscillator. And just as Fourier analysis enables us to express complicated repetitive waveforms in terms of sines and cosines, so we can describe complicated oscillating systems by sums of harmonic oscillator wave functions. Therefore we can use the simple harmonic oscillator as the starting point for solving more complicated problems.

We began this chapter by considering microscopic particles incident on a rectangular potential energy barrier. When the particles have a kinetic energy greater than the barrier height, the probability of transmission depends on the energy and oscillates between 1 and a minimum value. When the particles have a kinetic energy less than the barrier height, tunneling is possible. We showed that the probability of tunneling decreases rapidly as the barrier height and/or thickness increases. We also presented the expressions for the SWE and the allowed energies of the one-dimensional harmonic oscillator. We now know that its ground state quantum number equals zero and that its ground state energy equals $\frac{1}{2}h\nu_c$; Planck failed to guess the $\frac{1}{2}h\nu_c$.

QUESTIONS AND PROBLEMS

15.1 Suppose that L approaches infinity, doing away with region III and changing the potential barrier to a potential step up. For $E > V$, explain why $D = 0$.

15.2 a) Continue Problem 15.1 to show that the reflection coefficient, $R = (B/A)*(B/A)$ equals $[(k_I - k_{II})/(k_I + k_{II})]^2$. b) Compare your answer to part (a) to that of Problem 14.31(b).

15.3 In terms of interference of reflected particle waves, why does $T = 1$ when $2L = n\lambda_{II}$ for a potential barrier with $E > V$?

15.4 The reflection coefficient $R = 1 - T$. Prove that $R \approx (V \sin k_1 L/2E)^2$ for $E \gg V$ for a potential barrier.

15.5 For 15.0-eV electrons incident on a 5.0-eV potential barrier, a) what barrier thicknesses will give no reflection? b) what barrier thicknesses will give a maximum in the reflection?

15.6 For a potential barrier, $E > V$.

a) Prove that T approaches $1/[1 + (k_1L/2)^2]$ as E approaches V.

b) Evaluate this T for 10-eV electrons incident on a 0.20-nm thick barrier.

c) Graph T as a function of k_1L.

d) What is the classical value of T for $E = V$?

15.7 What are the differences between highway construction workers tunneling through a mountain and an electron tunneling through a potential energy barrier?

15.8 The potential energy of an atom in a

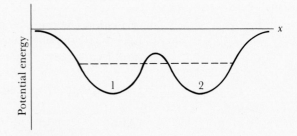

FIGURE 15.8
The dashed line is the sum of the kinetic and potential energies.

molecule (for example, the nitrogen atom in NH_3) is sketched in Fig. 15.8. How is it possible for the atom to oscillate back and forth between positions 1 and 2?

15.9 Suppose that L approaches infinity, doing away with region III and changing the potential barrier to a potential step up. For $E < V$, explain why $C = 0$.

15.10 Continue Problem 15.9, showing that $R = (B/A)*(B/A) = 1$. Explain how this result is possible despite the fact that $D \neq 0$.

15.11 Redo Example 15.2 for barriers of thickness a) 0.40 nm and b) 2.0 nm.

15.12 For $E > V$, $T = 1$ when the barrier thickness equals $n\pi/k_{II}$. Is there a similar result for $E < V$?

15.13 a) Show that T approaches $1/[1 + (k_1L/2)^2]$

as E approaches V for tunneling through a rectangular barrier. b) Compare this result with that of Problem 15.6(a).

15.14 Figure 15.9 is a graph of the potential energy of electrons at the surface of a metal when a strong electric field is applied normal to the surface. As the field is increased, the slope at the right becomes steeper and the height of the barrier is lowered. How do these changes affect the rate at which electrons tunnel out of the surface?

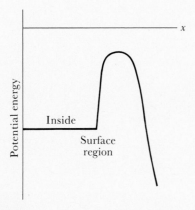

FIGURE 15.9
The potential energy for strong-field emission.

15.15 Superconductor A has a thin layer of insulator on it. Then superconductor B is deposited over the insulator. a) How can electrons be transferred between the two superconductors with no loss in energy? b) Why must the insulating layer be thin?

15.16 A baseball is headed toward a plate-glass picture window. Will it tunnel through?

15.17 Tunneling can occur in junction diodes if the junction region, which ordinarily blocks electron motion in certain voltage ranges, is very thin. Why does it have to be very thin?

15.18 The wave function of the first excited state of a one-dimensional harmonic oscillator is given in Example 15.4. a) Show that it is normalized (use integral tables). b) Sketch the probability

per unit distance. c) Find the probability that x is positive.

15.19 The ground state wave function of a one-dimensional harmonic oscillator can be written as Ae^{-bx^2}. a) Find the constant b. b) Find the constant A, if it is normalized. c) Show that the energy agrees with Eq. (15.4).

15.20 The wave function $(1/\sqrt{8})(\alpha/\pi)^{1/4}(4\alpha x^2 - 2)e^{-\alpha x^2/2}$ is that of a harmonic oscillator. Prove the statement and find the quantum number n.

15.21 Is the following statement true? "The quantum number of the nth excited state for a one-dimensional harmonic oscillator is n." What is the equivalent statement for the particle in a box? For the Bohr hydrogen atom?

15.22 The wave functions for a harmonic oscillator can be written as $\psi_n = H_n(x)e^{-ax^2}$. Write the differential equation that $H_n(x)$ must satisfy. (The solutions for $H_n(x)$ are called Hermite polynomials.)

15.23 An atom is oscillating in one dimension at a frequency of 2.0×10^{13} Hz. a) If its mass is 4.7×10^{-26} kg, what is the effective force constant? b) What values of energy may it have?

15.24 The semiclassical amplitude, A, of a harmonic oscillator can be obtained from $\frac{1}{2}m\omega_c^2A^2 = E$. Show that A is quantized and find its allowed values. Why can $|x|$ be greater than A?

15.25 The maximum speed, v_{max}, of a harmonic oscillator can be obtained from $\frac{1}{2}mv_{max}^2 = E$. Show that v_{max} is quantized and find its allowed values.

15.26 For small angles, a simple pendulum is a harmonic oscillator with $\omega_c = \sqrt{g/l}$. Find the ground state energy and the energy difference between states for a simple pendulum of length $l = 1.00$ m, where $g = 9.80$ m/s². Are these values detectable?

15.27 A one-dimensional harmonic oscillator is changed by the addition of another force so that $V = b/x^2 + \frac{1}{2}m\omega_c^2x^2$. One resulting wave function is $\psi = Ax^ne^{-ax^2}$ and

$$\frac{d^2\psi}{dx^2} = \left[\frac{n(n-1)}{x^2} - 2a(2n+1) + 4a^2x^2\right]\psi.$$

a) Evaluate a. b) Find E in terms of b, h, and ω_c.

Three

Atomic Physics

CHAPTER 16

The Hydrogen-Like Atom and the SWE

I hope you won't let the mathematics of this chapter intimidate you. Please try to follow the logic of the mathematical arguments. Basically, we're going to take the three-dimensional SWE for the hydrogen-like atom, break it down into three one-dimensional equations, and then discuss separately each of the three solutions. The total wave function is then simply the product of those three one-dimensional solutions. In the process, we'll introduce three quantum numbers and the important rules governing them.

16.1
SEPARATION OF VARIABLES

When the potential energy, V, doesn't depend on time, we can write the three-dimensional, time-independent, nonrelativistic Schrödinger wave equation (SWE) as

$$-\frac{\hbar^2}{2m} \nabla^2\psi + V\psi = E\psi. \tag{A}$$

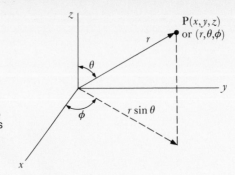

FIGURE 16.1
Spherical coordinates, r, θ, and ϕ. The diagram shows $x = r \sin \theta \cos \phi$, $y = r \sin \theta \sin \phi$, and $z = r \cos \theta$.

For a hydrogen-like atom, m is replaced by the reduced mass m_r, and V is replaced by $-(1/4\pi\epsilon_0)Ze^2/r$. This potential energy depends on r, the separation. It doesn't depend on θ or ϕ in spherical coordinates, shown in Fig. 16.1. Therefore V will have the same value at every point on the surface of a sphere of radius r, and we say that V has spherical symmetry. To take advantage of this symmetry, we use the ∇^2 operator for spherical coordinates:

$$\nabla^2 = \frac{1}{r^2}\frac{\partial}{\partial r}\left(r^2 \frac{\partial}{\partial r}\right) + \frac{1}{r^2}\left[\frac{1}{\sin\theta}\frac{\partial}{\partial\theta}\left(\sin\theta \frac{\partial}{\partial\theta}\right) + \frac{1}{\sin^2\theta}\frac{\partial^2}{\partial\phi^2}\right]. \quad (16.1)$$

Also, we let the wave function $\psi(r, \theta, \phi)$ be the product of a radial function, $R(r)$, and two angular functions, $\Theta(\theta)$ and $\Phi(\phi)$:

$$\psi(r, \theta, \phi) = R(r)\Theta(\theta)\Phi(\phi). \quad (16.2)$$

Because $R(r)$, $\Theta(\theta)$, and $\Phi(\phi)$ are each obvious functions of their small letter variables, we'll often leave off the variables and their parentheses. For example, R is the function, and r is its variable, so we can write Eq. (16.2) in shortened form as $\psi = R\Theta\Phi$.

Substituting Eqs. (16.1) and (16.2) into Eq. (A) gives

$$-\frac{\hbar^2}{2m}\left\{\frac{\Theta\Phi}{r^2}\frac{d}{dr}\left(r^2\frac{dR}{dr}\right) + \frac{1}{r^2}\left[\frac{R\Phi}{\sin\theta}\frac{d}{d\theta}\left(\sin\theta\frac{d\Theta}{d\theta}\right)\right.\right.$$
$$\left.\left. + \frac{R\Theta}{\sin^2\theta}\frac{d^2\Phi}{d\phi^2}\right]\right\} + VR\Theta\Phi = ER\Theta\Phi. \quad (B)$$

We can multiply Eq. (B) through by $-(2m/\hbar^2)(r^2 \sin^2\theta/R\Theta\Phi)$ and then rearrange the result to obtain

$$\frac{1}{\Phi}\frac{d^2\Phi}{d\phi^2} = \frac{2m}{\hbar^2}r^2 \sin^2\theta(V - E) - \frac{\sin^2\theta}{R}\frac{d}{dr}\left(r^2\frac{dR}{dr}\right)$$
$$- \frac{\sin\theta}{\Theta}\frac{d}{d\theta}\left(\sin\theta\frac{d\Theta}{d\theta}\right), \quad (C)$$

where r, θ, and ϕ are three **independent variables.** That means that changing ϕ in the left-hand side of Eq. (C) requires no change in r, θ, or therefore in the right-hand side of Eq. (C). This result is possible only if both sides equal a common constant. We let this constant equal $-m_l^2$. We choose $-m_l^2$ because it will simplify the final results. (The constant m_l isn't a mass.) Therefore we can rearrange the left-hand side of Eq. (C) to get

$$\frac{d^2\Phi}{d\phi^2} + m_l^2\Phi = 0. \qquad (16.3)$$

We can also rearrange the right-hand side of Eq. (C) to obtain

$$\frac{1}{\Theta \sin \theta} \frac{d}{d\theta}\left(\sin \theta \frac{d\Theta}{d\theta}\right) - \frac{m_l^2}{\sin^2 \theta} = \frac{2m}{\hbar^2} r^2(V - E) - \frac{1}{R} \frac{d}{dr}\left(r^2 \frac{dR}{dr}\right). \quad (D)$$

The independence of θ and r allows us, again, to set both sides of Eq. (D) equal to a common constant. This common constant is different from $-m_l^2$. We let it equal $-l(l + 1)$, which will again yield simpler final results. This approach gives us two equations:

$$\frac{1}{\sin \theta} \frac{d}{d\theta}\left(\sin \theta \frac{d\Theta}{d\theta}\right) + \left[l(l + 1) - \frac{m_l^2}{\sin^2 \theta}\right]\Theta = 0 \qquad (16.4)$$

and

$$\frac{1}{r^2} \frac{d}{dr}\left(r^2 \frac{dR}{dr}\right) + \frac{2m}{\hbar^2}\left[E - V - \frac{\hbar^2}{2mr^2} l(l + 1)\right]R = 0. \qquad (16.5)$$

EXAMPLE 16.1   How will Eqs. (16.3), (16.4), and (16.5) be changed, if at all, for a hydrogen-like atom?

Solution For a hydrogen-like atom, $V = -(1/4\pi\epsilon_0)Ze^2/r$ and $m = m_r$. These particular values for V and m are then substituted into Eq. (16.5). However, Eqs. (16.3) and (16.4) won't be changed because V and m don't appear in them. ●

16.2
THE RADIAL SOLUTIONS

Substituting the Coulomb potential energy $V = -(1/4\pi\epsilon_0)Ze^2/r$ and $m = m_r$ into Eq. (16.5) gives a differential equation that is difficult to solve. Courses in quantum mechanics derive the solutions. For example, the solutions to the radial equation, Eq. (16.5), have the general form $R_{nl}(r)$, where n and l are integers. The infinitesimal spherical volume

TABLE 16.1
Some Normalized $R_{nl}(r)$ Solutions

$$R_{10}(r) = \left(\frac{Z}{a_0}\right)^{3/2} 2e^{-Zr/a_0}$$

$$R_{20}(r) = \left(\frac{Z}{2a_0}\right)^{3/2}\left(2 - \frac{Zr}{a_0}\right)e^{-Zr/2a_0}$$

$$R_{21}(r) = \left(\frac{Z}{2a_0}\right)^{3/2}\frac{Zr}{\sqrt{3}a_0}\,e^{-Zr/2a_0}$$

$$R_{30}(r) = \left(\frac{Z}{3a_0}\right)^{3/2}2\left[1 - \frac{2Zr}{3a_0} + \frac{2}{27}\left(\frac{Zr}{a_0}\right)^2\right]e^{-Zr/3a_0}$$

$$R_{31}(r) = \left(\frac{Z}{3a_0}\right)^{3/2}\frac{4\sqrt{2}}{3}\frac{Zr}{a_0}\left(1 - \frac{1}{6}\frac{Zr}{a_0}\right)e^{-Zr/3a_0}$$

$$R_{32}(r) = \left(\frac{Z}{3a_0}\right)^{3/2}\frac{2}{27}\sqrt{\frac{2}{5}}\left(\frac{Zr}{a_0}\right)^2 e^{-Zr/3a_0}$$

element, dv, equals $r^2\,dr\sin\theta\,d\theta\,d\phi$. The radial solutions are normalized so that

$$\int_0^\infty R_{nl}^2(r)r^2\,dr = 1. \tag{16.6}$$

Some of these normalized solutions are presented in Table 16.1.

EXAMPLE 16.2   Evaluate the constant a_0, which appears in the $R_{nl}(r)$ solutions.

Solution We can substitute the simplest solution, $R_{10}(r)$, into Eq. (16.5) along with $m = m_r$, $V = -(1/4\pi\epsilon_0)Ze^2/r$, and $l = 0$. This gives us

$$\frac{1}{r^2}\frac{d}{dr}\left(r^2\frac{d}{dr}e^{-Zr/a_0}\right) + \left(\frac{2m_r}{\hbar^2}\right)\left(E + \frac{1}{4\pi\epsilon_0}\frac{Ze^2}{r} - 0\right)e^{-Zr/a_0} = 0$$

by dividing through by $(Z/a_0)^{3/2}2$. Taking the derivatives, dividing by e^{-Zr/a_0}, and collecting terms, we get

$$\left(\frac{m_r e^2}{\hbar^2 4\pi\epsilon_0} - \frac{1}{a_0}\right)\frac{2Z}{r} + \left(\frac{Z}{a_0}\right)^2 + \frac{2m_r}{\hbar^2}E = 0.$$

Since this relation must be true for all r, the first term in parentheses must equal zero. Setting that term equal to zero gives

$$a_0 = \frac{\hbar^2 4\pi\epsilon_0}{m_r e^2} = \frac{\epsilon_0 h^2}{m_r \pi e^2}.$$

For hydrogen, evaluation of this expression yields $a_0 = 52.9$ pm and is the ground state separation of the nucleus and the electron in Bohr's model. (See Eq. (C) in Section 12.2.) ●

Continuing to solve the equations in Example 16.2, we see that $(Z/a_0)^2 + 2m_rE/\hbar^2$ must also add up to zero, which gives

$$E = \frac{-Z^2 m_r e^4}{8\epsilon_0 h^2} = -Z^2(13.60 \text{ eV})$$

for the hydrogen m_r. This E is the ground state energy of the hydrogen-like atom in the Bohr model.

Example 16.2 and the preceding paragraph indicate that the n in $R_{nl}(r)$ is the principal quantum number from the Bohr theory. Also, for the hydrogen m_r, the constant a_0 in the $R_{nl}(r)$ solutions is *not* the radius of the ground state orbit of the electron in the Bohr atom, but rather the slightly larger separation of the nucleus and the electron.

EXAMPLE 16.3   Prove that $R_{32}(r)$ is normalized.

Solution If $R_{32}(r)$ is normalized, then $\int_0^\infty R_{32}^2(r)r^2 \, dr = 1$. Substituting the function $R_{32}(r)$ from Table 16.1 gives

$$\left[\left(\frac{Z}{3a_0}\right)^{3/2} \frac{2}{27} \sqrt{\frac{2}{5}} \left(\frac{Z}{a_0}\right)^2 \right]^2 \int_0^\infty (r^2 e^{-Zr/3a_0})^2 r^2 \, dr$$

$$= \frac{8Z^7}{98415a_0^7} \int_0^\infty r^6 e^{-2Zr/3a_0} \, dr.$$

Integral tables or repeated integration by parts tell us that

$$\int_0^\infty x^p e^{-ax} \, dx = p!/a^{p+1}$$

when p is a positive integer and $a > 0$. Therefore the integral becomes

$$\frac{8Z^7(6!)}{98415a_0^7(2Z/3a_0)^7} = \frac{8(720)Z^7}{98415a_0^7(128Z^7/2187a_0^7)} = 1.$$ ●

16.3
THE ANGULAR SOLUTIONS

The easiest of the three differential equations to solve is $d^2\Phi/d\phi^2 + m_l^2\Phi = 0$. A normalized solution is

$$\Phi_{m_l}(\phi) = \frac{1}{\sqrt{2\pi}} e^{im_l\phi}, \tag{16.7}$$

so that

$$\int_0^{2\pi} \Phi^*\Phi \, d\phi = 1. \tag{16.8}$$

The wave function can have only one value at a given angle, but ϕ and $\phi + 2\pi$ give the same orientation in space. Therefore

$$e^{im_l\phi} = e^{im_l(\phi+2\pi)} \quad \text{or} \quad 1 = e^{im_l 2\pi} = \cos(m_l 2\pi) + i \sin(m_l 2\pi).$$

The values of the cosine and sine functions make $\cos(m_l 2\pi) = 1$ and $\sin(m_l 2\pi) = 0$ only if m_l equals an integer (positive, negative, or zero).

EXAMPLE 16.4   Prove that $\Phi_{m_l} = (1/\sqrt{2\pi})e^{im_l\phi}$ satisfies Eq. (16.3).

Solution We simply substitute the given expression into Eq. (16.3). The first time we differentiate $(1/\sqrt{2\pi})e^{im_l\phi}$ we get $(1/\sqrt{2\pi})(im_l)e^{im_l\phi}$. Then the second derivative gives

$$\left(\frac{1}{\sqrt{2\pi}}\right)(im_l)^2 e^{im_l\phi} = -\left(\frac{1}{\sqrt{2\pi}}\right)m_l^2 e^{im_l\phi} \quad \text{or} \quad -m_l^2\Phi.$$

Therefore

$$\frac{d^2\Phi}{d\phi^2} + m_l^2\Phi = -m_l^2\Phi + m_l^2\Phi = 0. \qquad \bullet$$

Equation (16.4) is another difficult differential equation. Solutions derived in quantum mechanics courses are given the general form $\Theta_{lm_l}(\theta)$ and are normalized so that

$$\int_0^\pi [\Theta_{lm_l}(\theta)]^2 \sin\theta \, d\theta = 1. \tag{16.9}$$

Some values for the normalized solutions are given in Table 16.2.

TABLE 16.2
Some Normalized $\Theta_{lm_l}(\theta)$ Solutions

$$\Theta_{00} = \frac{1}{\sqrt{2}} \qquad \Theta_{10} = \sqrt{\frac{3}{2}}\cos\theta \qquad \Theta_{1\pm1} = \frac{\sqrt{3}}{2}\sin\theta$$

$$\Theta_{20} = \frac{1}{2}\sqrt{\frac{5}{2}}(3\cos^2\theta - 1) \qquad \Theta_{2\pm1} = \frac{\sqrt{15}}{2}\sin\theta\cos\theta$$

$$\Theta_{2\pm2} = \frac{\sqrt{15}}{4}\sin^2\theta$$

EXAMPLE 16.5   Prove that $r\Theta_{10}$ is a function of z only.

Solution From Table 16.2, $r\Theta_{10} = r\sqrt{3/2}(\cos\theta)$. But $r\cos\theta = z$, so $r\Theta_{10} = \sqrt{3/2}z$. (This fact is useful when you are analyzing the symmetry of the wave function and charge distribution.)   ●

16.4
QUANTUM NUMBERS

Part of the radial solution is a power series in r. The wave function can't approach infinity as r approaches infinity. This means that the power series can't be infinite, but has to end after a certain number of terms. Detailed analysis shows that this requirement gives two results.

The first result is that the n in $R_{nl}(r)$ must be a positive, nonzero integer. The second result is that l can be any integer from 0 to $n-1$. The energy of the hydrogen-like atom, E_n, can be shown to equal $-Z^2(13.60\text{ eV})/n^2$ for the hydrogen m_r. Therefore n is the principal quantum number. For reasons that we'll discuss later, l is called the **orbital angular momentum quantum number,** or sometimes just the **orbital quantum number.**

Further analysis also shows that a solution to Eq. (16.4) that is well-behaved is possible only if $l^2 \geq m_l^2$. You've already seen that m_l can be zero or a positive or negative integer. Therefore the allowed values of m_l range from $-l$ through 0 to $+l$ in steps of one. The quantity m_l is called the **orbital magnetic quantum number,** again for reasons that we'll discuss later. To summarize:

$$\begin{aligned}
&\text{Principal quantum number}\\
&\quad n = 1, 2, 3, 4, \ldots\\
&\text{Orbital angular momentum quantum number}\\
&\quad l = 0, 1, \ldots, n-1.\\
&\text{Orbital magnetic quantum number}\\
&\quad m_l = -l, -l+1, \ldots, 0, \ldots, l-1, l.
\end{aligned} \qquad (16.10)$$

EXAMPLE 16.6   For the second excited energy level of a hydrogen-like atom, what are the allowed principal, orbital angular momentum, and orbital magnetic quantum numbers?

Solution Recall that $n = 1$ gives the ground level, so $n = 2$ gives the first excited level and $n = 3$ gives the second excited level. For $n = 3$, $n - 1 = $

FIGURE 16.2

(a) This amazing assembly of brain power took place in Brussels in 1927. It was the fifth international conference sponsored by the Solvay Institute. Eventually, 19 Nobel prizes were awarded to those present (Marie Curie won two). Ages ranged from 25 (Dirac and Heisenberg) to 75 (Lorentz, the president of the conference). Most of the physicists responsible for the fundamental development of modern physics were in attendance (Sir W. H. Bragg missed the picture). (b) They are: 1. A. Piccard; 2. E. Henriot; 3. P. Ehrenfest; 4. E. Herzen; 5. Th. de Donder; 6. E. Schrödinger; 7. E. Verschaffelt; 8. W. Pauli; 9. W. Heisenberg; 10. R. H. Fowler; 11. L. Brillouin; 12. P. Debye; 13. M. Knudsen; 14. W. L. Bragg; 15. H. A. Kramers; 16. P. A. M. Dirac; 17. A. H. Compton; 18. L. de Broglie; 19. M. Born; 20. N. Bohr; 21. I. Langmuir; 22. M. Planck; 23. M. Curie; 24. H. A. Lorentz; 25. A. Einstein; 26. P. Langevin; 27. C. E. Guye; 28. C. T. R. Wilson; and 29. O. W. Richardson.

Source: Courtesy of Central Scientific Company.

$3 - 1 = 2$. Therefore l can equal 0, 1, or 2. For a given l, m_l can vary from $-l$ to $+l$ in steps of 1. Therefore

for $l = 0$, $m_l = 0$ only;
for $l = 1$, $m_l = -1, 0,$ or $+1$;
for $l = 2$, $m_l = -2, -1, 0, +1,$ or $+2$.

Table 16.3 gives the answers in tabular form.   ●

You can see that for $n = 3$, there are nine different sets of the three quantum numbers. In general, for any l there are $2l + 1$ different m_l's. Also, l can have values ranging from $l = 0$ to $l = n - 1$. Therefore the number of different sets of the three quantum numbers equals

$$\sum_{l=0}^{n-1} (2l + 1) = 1 + 3 + \cdots + (2n - 1).$$

This series contains n terms. The average value of each term in the series is $[1 + (2n - 1)]/2 = 2n/2 = n$. The sum of the series equals the number of terms, n, times the average value per term, n, or n^2. To summarize: For any n, there are n^2 different sets of the three quantum numbers n, l, and m_l. Thus in Example 16.6, $n = 3$ and $n^2 = 9$. The importance of the number of different sets of quantum numbers will become clear later.

In this chapter, you met $R_{nl}(r)$, the radial part of the wave function derived specifically for the hydrogen-like atom. We also introduced you to $\Theta_{lm_l}(\theta)$ and $\Phi_{m_l}(\phi)$, the two angular parts of the wave function. The Θ and Φ solutions are much more general than the R solution in that they are correct for any potential energy, V, which depends only on r. We summarized for you the allowed values of the principal, orbital angular momentum, and orbital magnetic quantum numbers. We'll use these important rules again and again in later chapters.

TABLE 16.3
The Allowed Values of l and m_l for $n = 3$

n	3	3	3	3	3	3	3	3	3
l	0	1	1	1	2	2	2	2	2
m_l	0	-1	0	$+1$	-2	-1	0	$+1$	$+2$

QUESTIONS AND PROBLEMS

16.1 What restrictions did we place on the potential energy when we separated the variables?

16.2 Explain in your own words why both sides of Eqs. (C) and (D) must be set equal to a constant.

16.3 How would $V = A/r^{12} - B/r^6$ change Eqs. (16.3), (16.4), and (16.5)? Would it change their solutions?

16.4 Would Eqs. (16.3), (16.4), and (16.5) be changed by a) $V = -C/x$? b) $V = -D(x^2 + y^2 + z^2)^{1/2}$? c) $V = -Ex \cos \omega t$?

16.5 Will the $R_{nl}(r)$'s given be the same for a) any $V(r)$? b) any V? Explain.

16.6 Prove that $R_{10}(r)$ is normalized as given.

16.7 Prove that $R_{21}(r)$ is normalized as given.

16.8 Find E_2 by substituting $R_{21}(r)$ into the proper equation for a hydrogen atom. Compare the result with Bohr's result.

16.9 Use $R_{32}(r)$ to show that $E_3 = -13.60$ eV for a Li^{2+} ion.

16.10 What does a_0 in the given $R_{nl}(r)$ equal for positronium?

16.11 Does a_0 in the given radial solutions always equal 52.9 pm a) for all Z with the hydrogen m_r? b) for all m_r?

16.12 Sketch the x, y, and z (toward the top of your paper) Cartesian axes. Show r, θ, and ϕ on the sketch. Then illustrate

$$dv = r^2 \, dr \sin \theta \, d\theta \, d\phi = (dr)(r \sin \theta \, d\phi)(r \, d\theta).$$

16.13 Prove that the Φ_{m_l} given is normalized.

16.14 Does m_l have to be an integer for Φ_{m_l} to satisfy Eq. (16.3)? Explain.

16.15 Prove that

$$\int_0^{2\pi} \Phi_{m_l}^* \, \Phi_{m_l'} \, d\phi = 0,$$

if $m_l' \neq m_l$.

16.16 Calculate α for all m_l's, so that

$$\int_0^{\alpha} \Phi_{m_l}^* \, \Phi_{m_l} \, d\phi = 1/2.$$

16.17 Prove that Θ_{00} is normalized.

16.18 Prove that Θ_{10} is normalized.

16.19 Explain what restrictions there are on V to make the given Φ_{m_l} and Θ_{lm_l} solutions correct.

16.20 Prove that the $\Theta_{10}(\theta)$ given is a solution for all θ, if and only if $m_l = 0$.

16.21 Prove that the $\Theta_{2\pm2}(\theta)$ given is a solution.

16.22 Give the allowed principal, orbital angular momentum, and orbital magnetic quantum numbers for the ground state of a hydrogen-like atom.

16.23 In what energy levels of a hydrogen-like atom will you *not* find $l = 3$?

16.24 How many different sets of the three quantum numbers will you find in the first excited energy level of hydrogen? List them.

16.25 a) What values of the orbital angular momentum and orbital magnetic quantum numbers are allowed when the principal quantum number is 5? b) How many different sets are there?

16.26 Consider hydrogen-like atoms. a) When is $l = 4$ not allowed? b) When is $m_l = -2$ not allowed? c) When is $n = 0$ not allowed?

CHAPTER

Wave Functions of the Hydrogen-Like Atom

We know from quantum mechanics that we can't represent the electron as a small particle orbiting in a circle that has an exact radius. Rather, we can only expect to calculate the probability of finding the electron in some volume in space. To do that, we need the wave function. In this chapter we'll simply multiply the three normalized parts of the product wave function for the hydrogen-like atom together to obtain the normalized wave function desired. Knowing the wave functions for various sets of quantum numbers, we'll then discuss and illustrate the probability per volume and probability per radial length for those states. Also, you'll learn the letters associated with different orbital angular momentum quantum number states.

17.1
HYDROGEN-LIKE WAVE FUNCTIONS

In Chapter 16 we showed that $\psi(r, \theta, \phi) = R(r)\Theta(\theta)\Phi(\phi)$, if V depends only on r. We were able to write separate equations for $R(r)$, $\Theta(\theta)$, and $\Phi(\phi)$. The Θ and Φ solutions were true, for any V which depended only on the separation r. The R solutions given were for $V = (1/4\pi\epsilon_0)Ze^2/r$, as in the hydrogen-like atom. For the hydrogen-like atom, therefore,

$$\psi_{nlm_l}(r, \theta, \phi) = R_{nl}(r)\Theta_{lm_l}(\theta)\Phi_{m_l}(\phi).$$

Some normalized examples are given in Table 17.1.

EXAMPLE 17.1   What is ψ_{32-1}? Is it normalized?

Solution Using the solutions from Chapter 16, we see that

$$\psi_{32-1} = R_{32}(r)\Theta_{2-1}(\theta)\Phi_{-1}(\phi)$$

$$= \left[\left(\frac{Z}{3a_0}\right)^{3/2} \frac{2}{27}\sqrt{\frac{2}{5}}\left(\frac{Zr}{a_0}\right)^2 e^{-Zr/3a_0}\right]$$

$$\times \left(\frac{\sqrt{15}}{2}\sin\theta\cos\theta\right)\left(\frac{1}{\sqrt{2\pi}}e^{-i\phi}\right)$$

$$= \frac{1}{81\sqrt{\pi}}\left(\frac{Z}{a_0}\right)^{7/2} r^2 e^{-Zr/3a_0}\sin\theta\cos\theta\, e^{-i\phi}.$$

The functions R, Θ, and Φ were normalized; therefore ψ is normalized. We don't have to prove that $\int_0^{2\pi}\int_0^\pi\int_0^\infty \psi^*\psi r^2\, dr \sin\theta\, d\theta\, d\phi = 1$.   ●

TABLE 17.1
Some Normalized $\psi_{nlm_l}(r, \theta, \phi)$ Solutions

$$\psi_{100} = \frac{1}{\sqrt{\pi}}\left(\frac{Z}{a_0}\right)^{3/2} e^{-Zr/a_0}$$

$$\psi_{200} = \frac{1}{4\sqrt{2\pi}}\left(\frac{Z}{a_0}\right)^{3/2}\left(2 - \frac{Zr}{a_0}\right)e^{-Zr/2a_0}$$

$$\psi_{210} = \frac{1}{4\sqrt{2\pi}}\left(\frac{Z}{a_0}\right)^{3/2}\frac{Zr}{a_0}e^{-Zr/2a_0}\cos\theta$$

$$\psi_{21\pm1} = \frac{1}{8\sqrt{\pi}}\left(\frac{Z}{a_0}\right)^{3/2}\frac{Zr}{a_0}e^{-Zr/2a_0}\sin\theta\, e^{\pm i\phi}$$

$$\psi_{300} = \frac{1}{81\sqrt{3\pi}}\left(\frac{Z}{a_0}\right)^{3/2}\left(27 - 18\frac{Zr}{a_0} + \frac{2Z^2r^2}{a_0^2}\right)e^{-Zr/3a_0}$$

$$\psi_{310} = \frac{1}{81}\sqrt{\frac{2}{\pi}}\left(\frac{Z}{a_0}\right)^{3/2}\left(6 - \frac{Zr}{a_0}\right)\frac{Zr}{a_0}e^{-Zr/3a_0}\cos\theta$$

$$\psi_{31\pm1} = \frac{1}{81\sqrt{\pi}}\left(\frac{Z}{a_0}\right)^{3/2}\left(6 - \frac{Zr}{a_0}\right)\frac{Zr}{a_0}e^{-Zr/3a_0}\sin\theta\, e^{\pm i\phi}$$

17.2
STATE NOTATIONS

For historical reasons involving the appearance of optical spectra, letters are associated with the orbital quantum number l. When we associate the quantum number with an individual electron, we use a lowercase letter. When we use it for an atom, we use an uppercase letter. The notations are shown in Table 17.2. The sequential s, p, d, and f notations stand for **s**harp, **p**rincipal, **d**iffuse, and **f**undamental. From f onward, the letters are in alphabetical order.

The notation 1s stands for a state with $n = 1$ and $l = 0$. The number gives the value of n and the letter the value of l. A 6d state would have $n = 6$ and $l = 2$.

TABLE 17.2
Notations for Orbital Angular Momentum States

Particle \ l	0	1	2	3	4	5	$\cdots$
Electron	s	p	d	f	g	h	$\cdots$
Atom	S	P	D	F	G	H	$\cdots$

EXAMPLE 17.2 Give the principal, orbital angular momentum, and orbital magnetic quantum numbers possible in

a) a 3f state and   b) a 4f state for a hydrogen-like atom.

Solution We're looking for the possible values of n, l, and m_l, respectively.

a) We know that 3f means that $n = 3$ and $l = 3$, but for $n = 3$, l can only equal 0, 1, or 2. Therefore there are *no* possible quantum numbers for a 3f state, as 3f states are not allowed for hydrogen-like atoms.

b) The notation 4f means that $n = 4$ and $l = 3$, which is allowed. For $l = 3$, m_l can equal $-3, -2, -1, 0, +1, +2$, or $+3$.   ●

17.3
PROBABILITY DENSITY

The probability that an electron in a hydrogen-like atom will be found in the volume dv is $\psi^*\psi \, dv = |\psi|^2 \, r^2 \, dr \sin\theta \, d\theta \, d\phi$ (assuming a normalized ψ). The probability per volume, $\psi^*\psi$, is called the **probability density.** All hydrogen-like probability density terms contain $e^{-im_l\phi}e^{+im_l\phi} = 1$, so the probability density doesn't depend on ϕ. That is, all hydrogen-like probability densities for a particular set of n, l, and m_l will be symmetric about the z axis. Or, alternatively, revolution through any angle ϕ about the z axis won't change the probability density of a hydrogen-like atom.

    The probability density is represented by Fig. 17.1 for some states. The s states contain no angular dependence and so are spherically symmetric. For a 2p ($m_l = 0$) state, $|\psi_{210}|^2$ is proportional to $\cos^2\theta$ and so has its maximum value parallel and antiparallel to the z axis (at $\theta = 0°$ and $\theta = 180°$). Also, $|\psi_{210}|^2$ is zero perpendicular to the z axis in the x–y plane (at $\theta = 90°$).

EXAMPLE 17.3   Describe the probability density for a 2p ($m_l = \pm1$) state.

Solution The probability density is $|\psi_{21\pm1}|^2$. We find the wave function in Table 17.1 and see that the probability density is proportional to $r^2 e^{-Zr/a_0}$

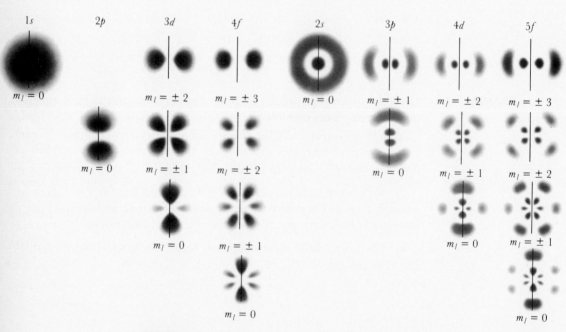

FIGURE 17.1

A photographic representation of the electron probability density, $\psi^*\psi$, for some hydrogen-like states. These are views sectioned along any plane containing the z axis but are not to scale.

Source: Courtesy of Robert B. Leighton, California Institute of Technology.

2p

$m_l = \pm 1$

FIGURE 17.2
Another representation of the electron probability density.
Source: Courtesy of Robert B. Leighton, California Institute of Technology.

times $\sin^2 \theta$. The r dependence will give a zero at $r = 0$ and an approach to zero for large r. Since $\sin 90° = 1$ and $\sin 0 = \sin 180° = 0, |\psi_{21\pm1}|^2$ will be largest perpendicular to the z axis ($\theta = 90°$) and will be 0 along the $+z$ and $-z$ axes ($\theta = 0$ and $180°$). As we explained previously in this section, the probability density has no ϕ dependence and is rotationally symmetric about the z axis. In three dimensions, this configuration looks something like a doughnut with the hole only partially cut out of the center. (See Fig. 17.2.) ●

17.4
THE RADIAL PROBABILITY DENSITY

In the Bohr model of the hydrogen-like atom, the electron is a point particle moving in a circle of definite radius about the nucleus. However, quantum mechanics can give us only the probability of finding the electron in a certain region of space about the nucleus. The probability per unit length along a radius a distance r from the nucleus is called the **radial probability density**.

If we take $|\psi|^2 r^2 \, dr \sin \theta \, d\theta \, d\phi$ for a hydrogen-like atom and integrate the angular portion, only $[R(r)]^2 r^2 dr$ will be left. (The angular integrals will both equal 1 because Θ and Φ are normalized.) The infinitesimal length dr is the thickness of a spherical shell of radius r. Therefore

$$\text{Radial probability density} = [R(r)]^2 r^2. \qquad (17.1)$$

EXAMPLE 17.4 What is the radial probability density for a 3d state of hydrogen?

Solution The radial probability density is $[R(r)]^2 r^2$. For hydrogen, $Z = 1$. Also, 3d means $n = 3$, $l = 2$, and so we want $[R_{32}(r)]^2 r^2$ with $Z = 1$. Using

R_{32} from Table 16.1,

$$\text{Radial probability density} = \left(\frac{1}{3a_0}\right)^3 \left[\frac{4(2)}{729(5)}\right] \left(\frac{r}{a_0}\right)^4 e^{-2r/3a_0}\, r^2$$

$$= \left(\frac{8}{98415a_0^7}\right) r^6 e^{-2r/3a_0}.$$

The r^6/a_0^7 quotient gives the unit of $(\text{length})^{-1}$.   ●

EXAMPLE 17.5   Find the value of r at which the radial probability density is the greatest for a 3d state in hydrogen. Compare this answer with the Bohr model.

Solution This maximum will occur where the derivative with respect to r of the radial probability density equals zero. Differentiating (from Example 17.4),

$$[R_{32}(r)]^2 r^2 = \left(\frac{8}{98415a_0^7}\right) r^6 e^{-2r/3a_0}$$

with respect to r and setting it equal to zero gives us

$$\left(\frac{8}{98415a_0^7}\right)\left[6r^5 - r^6\left(\frac{2}{3a_0}\right)\right] e^{-2r/3a_0} = 0,$$

which reduces to $r = 9a_0$. (The $r = 0$ and $r = \infty$ roots give minima.) The Bohr model of the hydrogen atom gives $r_n = n^2 r_1$ for the orbit radius and $r = n^2 a_0$ for the separation. For $n = 3$ in this problem, $r = 3^2 a_0 = 9a_0$. Therefore the maximum radial probability density of a 3d state occurs at the Bohr separation of that state. The major difference between the two concepts is that in the Bohr model the electron is located precisely a distance $9a_0$ from the nucleus, whereas in quantum mechanics the probability is good that the electron will be found elsewhere. (See Fig. 17.3.)   ●

Note that Fig. 17.3 shows that the distribution for a 3p state has two maxima and that for a 3s state has three maxima. In general, the number of maxima in the radial probability distribution is $n - l$. Only in a highest l state is there one maximum, with that maximum being found at the Bohr separation.

In this chapter we examined various wave functions and their probability densities. You saw that this probability density is always symmetric about the z axis. We pointed out one great difference from the Bohr model for any given state in an atom: In the Bohr model the electron moves in a circle, separated from the nucleus by the exact value of $n^2 a_0$, whereas

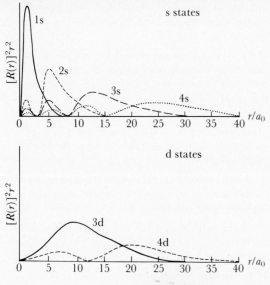

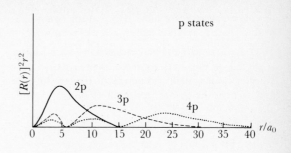

FIGURE 17.3
Graphs of the radial probability density for the ground state and most of the first three excited states of the hydrogen atom.

quantum mechanics can give us only the probabilities of finding the electron in some range of separations. Often the radial probability density has not one, but several maxima, none of which occurs at n^2a_0.

QUESTIONS AND PROBLEMS

17.1 How does ψ_{32-1} differ from ψ_{32+1}?

17.2 Write the wave function for hydrogen when the principal quantum number is 3, the orbital angular momentum quantum number has its largest possible value, and the orbital magnetic quantum number is neither positive nor negative.

17.3 Write the $\psi_{32\pm2}$ for He$^+$. If you ignore the effective mass correction, what is a_0?

17.4 Write the three 3p wave functions for hydrogen.

17.5 What are the possible principal quantum numbers for g states?

17.6 What state has a) $n = 2$ and $l = 0$? b) $n = 2$ and $l = 2$? c) $n = 5$ and $l = 1$?

17.7 Give the possible orbital magnetic quantum numbers for a) an s state b) a d state.

17.8 a) Where will $\psi^*\psi = 0$ for a 2s state in a hydrogen-like atom? b) Make a rough sketch of the probability density for that state of a hydrogen-like atom.

17.9 a) Explain why the 3d ($m_l = 0$) representation in Fig. 17.1 is reasonable. b) At what angle(s) θ is the probability density zero?

17.10 a) At what value of r will the probability

density equal zero for a hydrogen atom 3p state?
b) If $m_l = \pm 1$, at what values of θ will $\psi^*\psi = 0$?

17.11 Explain why the representation of a 3d state ($m_l = \pm 1$) in Fig. 17.1 seems to have zeros parallel, antiparallel, and perpendicular to the z axis. Why is the greatest probability density found along $\theta = 45°$ and $135°$?

17.12 Make a rough sketch of the probability density for a 3d ($m_l = \pm 2$) hydrogen state.

17.13 If the shell is thin enough compared to r, the probability of finding an electron in a thickness Δr is approximately $[R_{nl}(\bar{r})]^2\bar{r}^2\,\Delta r$, where $\bar{r}$ is the average radius of the shell. Calculate the probability of finding the electron between $r = 2.97a_0$ and $3.03a_0$ for a hydrogen 3d state.

17.14 a) Explain why $-e \times \psi^*\psi$ equals the electron charge density. b) Use this idea and the idea in Problem 17.13 to find the approximate charge between $0.99a_0$ and $1.01a_0$ in a hydrogen 1s state. c) Compare the result in part (b) to the Bohr atom result.

17.15 The radius of a proton is approximately $(3 \times 10^{-5})a_0$. Use the idea in Problem 17.13 to obtain the approximate probability of finding the electron between $r = 0$ and $r = 3 \times 10^{-5}a_0$

("inside the proton") for the hydrogen ground energy level. Assume that $R(r)$ is given by a function in Table 16.1.

17.16 a) Write the expression for the probability that the electron in a hydrogen atom 1s state will be found between $r = 0$ and $r = a_0$. b) Evaluate this probability.

17.17 Find the value of r at which the radial probability density is a maximum for a 1s state a) of hydrogen; b) of Li^{2+}; c) of positronium.

17.18 At what separation does the radial probability density have a maximum for the first excited s energy level of hydrogen? Compare this result with the Bohr model.

17.19 For a hydrogen 1s state, calculate the ratios of the radial probability density at $r = a_0$ to those at $0.5a_0$ and $1.5a_0$.

17.20 At what values of r is the radial probability density equal to zero for a 3s state of He^+?

17.21 a) Show that Table 16.1 is consistent with $R_{nl_{max}}(r) = C(r/a_0)^{l_{max}}e^{-r/na_0}$ for hydrogen. b) Show that C, the normalizing constant, equals $(a_0)^{-3/2}(2/n)^{n+(1/2)}[(2n!)]^{-1/2}$. c) Show that the radial probability density for $R_{nl_{max}}(r)$ has one maximum, found at $r = n^2a_0$.

CHAPTER

Quantized Angular Momentum

Magnetic materials have many scientific and technological uses. One of the causes of magnetism is the orbital motion of electrons. In the original Bohr model, the angular momentum of the various orbits equaled exactly $n\hbar$. However, the magnetic effects that this motion would give in hydrogen-like atoms simply aren't there. In this chapter we'll introduce the correct expression for the orbital angular momentum, finding that it is quantized not only in magnitude but also in direction in space. We'll then relate this angular momentum to the orbital magnetic properties of the electron.

18.1
ORBITAL ANGULAR MOMENTUM

Equation (16.5), the radial equation, includes the terms $E - V - (\hbar^2/2mr^2)l(l + 1)$. These are all energy terms, and $(\hbar^2/2mr^2)l(l + 1)$ is a kinetic energy term. Since we can express kinetic energy as $p^2/2m$, let's equate $p^2/2m$ to $(\hbar^2/2mr^2)l(l + 1)$. Multiplying both sides by $2mr^2$ and

203

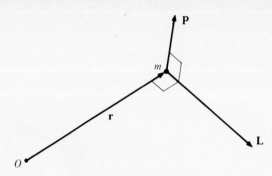

FIGURE 18.1
In classical mechanics the angular momentum of a particle of mass m with respect to the point O is given by $\mathbf{L} = \mathbf{r} \times \mathbf{p}$. Neither the magnitude nor the direction of $\mathbf{L}$ is quantized.

then taking the square roots gives us $rp = \sqrt{l(l+1)}\hbar$. If $\mathbf{r}$ and $\mathbf{p}$ are perpendicular to each other, rp is the magnitude of the angular momentum in classical physics. (See Fig. 18.1.)

The preceding in no way proves that the angular momentum has the magnitude $\sqrt{l(l+1)}\hbar$. However, further analysis shows that it's indeed the correct quantum-mechanics result. That is, the magnitude of the **orbital angular momentum, L,** is

$$L = \sqrt{l(l+1)}\hbar \qquad (18.1)$$

Recall that $\hbar = h/2\pi = 1.0546 \times 10^{-34}$ J · s $= 6.582 \times 10^{-16}$ eV · s.

The relation $L = \sqrt{l(l+1)}\hbar$ agrees with experimental results, but the Bohr assumption $L = n\hbar$, doesn't. The relation $L = \sqrt{l(l+1)}\hbar$ also shows why l is called the orbital angular momentum quantum number: l is the quantum number that *completely* determines the magnitude of the orbital angular momentum.

Different values of l give different values of L, so those values of L are quantized. In classical physics, L can have any value. In quantum physics, the Schrödinger wave equation allows only certain values. For example, $L = 0$ for all s $(l = 0)$ states, $L = \sqrt{2}\hbar$ for all p $(l = 1)$ states, $L = \sqrt{6}\hbar$ for all d $(l = 2)$ states, and so on, as shown in Fig. 18.2.

FIGURE 18.2
In quantum mechanics the magnitude of the angular momentum has only certain allowed values.

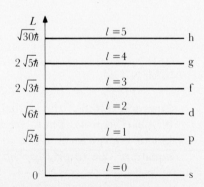

Example 18.1   Calculate the magnitude of the orbital angular momentum of an f state in SI units.

Solution The f state means that $l = 3$. Therefore

$$L = \sqrt{l(l + 1)}\hbar = \sqrt{3(3 + 1)}\hbar = \sqrt{12}\hbar = 2\sqrt{3}\hbar.$$

Numerically, in SI units,

$$L = 2\sqrt{3}(1.0546 \times 10^{-34}\,\text{J} \cdot \text{s}) = 3.653 \times 10^{-34}\,\text{kg} \cdot \text{m}^2/\text{s}.$$

We used the unit conversion $\text{J} \cdot \text{s} = \text{kg} \cdot \text{m}^2/\text{s}$, recalling that $1\,\text{J} = 1\,\text{kg} \cdot \text{m}^2/\text{s}^2$. ●

In the hydrogen atom with only the Coulomb force acting, the energy depends only on the principal quantum number, n. That is, $E_n = -13.60$ eV$/n^2$. Because this energy doesn't depend on l or m_l, different states with the same n but different l or m_l have the same energy. States having the same energy, but which are otherwise different, are called **degenerate states.** The number of different states with the same energy is called the **degeneracy.**

Example 18.2   Zero angular momentum in classical physics would correspond to an orbit in which the electron oscillated back and forth in a straight line through the nucleus. Describe the zero orbital angular momentum situation for hydrogen-like atoms.

Solution Only s $(l = 0)$ states have zero orbital angular momentum $(L = 0)$. These states are spherically symmetric states, not states where the greatest probability would lie along a straight line through the origin. (Their zero orbital angular momentum can be thought of as resulting from the canceling effects of having equal probability of rotation in one direction as in any other.) ●

18.2
THE z COMPONENT OF **L**

In classical physics a system with a potential energy V, a mass m, and a momentum p has a total mechanical energy of $p^2/2m + V = E$. Multiplying this expression by ψ gives us the relation $(p^2/2m)\psi + V\psi = E\psi$. Comparing this to the SWE, $-(\hbar^2/2m)\nabla^2\psi + V\psi = E\psi$, we see that the linear momentum **p** of classical mechanics is replaced by the operator $-i\hbar\nabla$ in quantum mechanics. The classical angular momentum is $\mathbf{L} = \mathbf{r} \times \mathbf{p}$. It has a z component, $L_z = xp_y - yp_x$, from the cross product. In quantum mechanics, p_y is replaced by $-i\hbar\,\partial/\partial y$ and p_x by $-i\hbar\,\partial/\partial x$, or two components of $-i\hbar\nabla$. Converting from Cartesian coordinates to spherical coordinates gives $L_z = -i\hbar\,\partial/\partial\phi$ as the operator representing the z component of the orbital angular momentum in quantum physics.

Just as operating on the wave function ψ with $[-(\hbar^2/2m)\nabla^2 + V]$ yields the energy E times ψ, operating on ψ with $-i\hbar\,\partial/\partial\phi$ gives the z component of L times ψ. When V is a function of r only,

$$\psi = R\Theta\Phi = R\Theta(1/\sqrt{2\pi})e^{im_l\phi}.$$

Therefore

$$-i\hbar\frac{\partial\psi}{\partial\phi} = -i\hbar\frac{R\Theta}{\sqrt{2\pi}}\frac{\partial}{\partial\phi}\,e^{im_l\phi} = -i\hbar(im_l)\frac{R\Theta}{\sqrt{2\pi}}\,e^{im_l\phi} = m_l\hbar\psi,$$

which gives $L_z\psi = m_l\hbar\psi$, or

$$L_z = m_l\hbar. \tag{18.2}$$

In words, the z component of the orbital angular momentum equals the orbital magnetic quantum number times Planck's constant divided by two pi.

Recall that m_l can have values from $-l$ to $+l$ in steps of 1. Therefore the z component of $\mathbf{L}$ can have only certain allowed values. This further means that $\mathbf{L}$ can have only certain allowed orientations in space. (See Fig. 18.3.) This restriction is called the **space quantization** of the orbital angular momentum. We can calculate the angle θ between $\mathbf{L}$ and the z axis from

$$\cos\theta = \frac{L_z}{L} = \frac{m_l\hbar}{\sqrt{l(l+1)}\hbar} = \frac{m_l}{\sqrt{l(l+1)}}. \tag{18.3}$$

Example 18.3   Calculate the orbital angular momentum, the allowed z components of the orbital angular momentum, and the allowed angles with the z axis for an electron in a d state.

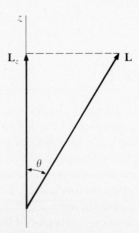

FIGURE 18.3
The orbital angular
momentum vector and its
z component.

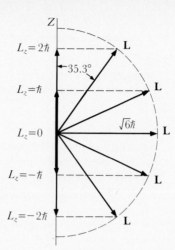

FIGURE 18.4
Space quantization of the
orbital angular momentum
for a d-state electron
($l = 2$).

Solution We begin with $L = \sqrt{l(l + 1)}\hbar$. For a d state, $l = 2$. Therefore the magnitude of the orbital angular momentum is $L = \sqrt{6}\hbar$. We find the allowed values of L_z by recalling that for $l = 2$, m_l can equal $-2, -1, 0, +1$, or $+2$. Therefore the five possible values for L_z are $L_z = m_l \hbar = -2\hbar, -1\hbar$, $0, +1\hbar$, and $+2\hbar$. The corresponding five possible values for $\cos \theta = m_l/\sqrt{l(l + 1)}$ are $-2/\sqrt{6}, -1/\sqrt{6}, 0, +1/\sqrt{6}$, and $+2/\sqrt{6}$, so $\theta = \cos^{-1}(-2/\sqrt{6}) \ldots$ gives us $\theta = 144.7°, 114.1°, 90°, 65.9°$, and $35.3°$. ●

Example 18.4 Sketch the possible directions and z components of the orbital angular momentum vector for an electron in a d state.

Solution Example 18.3 provides the values and angles allowed for $\mathbf{L}$ and L_z in a d state. We first sketch a dashed semicircle, letting its radius be $\sqrt{6}\hbar = 2.45\hbar$, the magnitude of $\mathbf{L}$. Then $\mathbf{L}$ will lie along five radii of the semicircle such that $L_z = +2\hbar, +1\hbar, 0, -1\hbar$, and $-2\hbar$. We show only the $35.3°$ angle for clarity in Fig. 18.4. ●

18.3
THE ORBITAL MAGNETIC DIPOLE MOMENT

A charged object moving in an orbit will have an **orbital magnetic dipole moment**, $\boldsymbol{\mu}_l$, if it has an orbital angular momentum, $\mathbf{L}$. In classical physics, a current loop gives a magnetic dipole moment with $\mu = i\mathbf{A}$, where A is the area enclosed by the loop, and i is the electrical current, or charge per time,

FIGURE 18.5
The infinitesimal area swept out in time dt by the position vector.

passing a point on the loop. An electron has a charge of $-e$. An electron in orbit passes a point on the orbit once every T seconds, where T is the period. Therefore $\boldsymbol{\mu}_l = -(e/T)\mathbf{A} = -e(\mathbf{A}/T)$.

The angular momentum is given by

$$\mathbf{L} = \mathbf{r} \times \mathbf{p} = \mathbf{r} \times m\mathbf{v} = m\mathbf{r} \times \frac{d\mathbf{s}}{dt}.$$

Figure 18.5 shows that the infinitesimal area swept out in time dt is $d\mathbf{A} = \frac{1}{2}(\mathbf{r} \times d\mathbf{s})$ (because the area of a triangle is one-half the height times the base). Therefore $\mathbf{r} \times d\mathbf{s} = 2d\mathbf{A}$ gives $\mathbf{L} = 2m \, d\mathbf{A}/dt$. With no external torque acting, $\mathbf{L}$ is constant, so $d\mathbf{A}/dt$ is also constant, and we can replace it by $\mathbf{A}/T$, giving $\mathbf{L} = 2m(\mathbf{A}/T)$.

If we now use $\boldsymbol{\mu}_l = -e(\mathbf{A}/T)$ and $\mathbf{L} = 2m(\mathbf{A}/T)$ to eliminate $(\mathbf{A}/T)$, we have

$$\boldsymbol{\mu}_l = -\frac{e}{2m} \mathbf{L}, \tag{18.4}$$

which shows that $\boldsymbol{\mu}_l$ and $\mathbf{L}$ are directly proportional. The proportionality constant, $-e/2m$, equals -8.794×10^{10} C/kg. The vectors $\boldsymbol{\mu}_l$ and $\mathbf{L}$ are in opposite directions because the electron has a negative charge. A quantum-mechanics derivation gives the same result.

Since L is quantized at $\sqrt{l(l + 1)}\hbar$, we see that the orbital magnetic dipole moment of the electron is also quantized, or

$$\mu_l = \frac{e\hbar}{2m} \sqrt{l(l + 1)}. \tag{18.5}$$

(We dropped the minus sign because we're giving the expression for the *magnitude* of the orbital magnetic dipole moment vector.) We find that $\mu_l = 0$ for s-state electrons because they have no orbital angular momentum and so no orbital magnetic dipole moment.

We can obtain the z component of the orbital magnetic dipole moment by taking the z component of $\boldsymbol{\mu}_l = -(e/2m)\mathbf{L}$. This relation is simply $\mu_{lz} = -(e/2m)L_z$. Since $L_z = m_l\hbar$,

$$\mu_{lz} = -\frac{e\hbar}{2m} m_l. \tag{18.6}$$

The constant $e\hbar/2m$ is called the **Bohr magneton**, μ_B. Evaluating the Bohr magneton gives

$$\mu_B = 9.274 \times 10^{-24} \, \text{J/T} = 5.788 \times 10^{-5} \, \text{eV/T},$$

where 1 T = 1 tesla = 1 Wb/m^2 = 10^4 gauss. We can rewrite Eqs. (18.5) and (18.6) in terms of the Bohr magneton as

$$\mu_l = \mu_B \sqrt{l(l+1)} \tag{18.5a}$$

or

$$\mu_{lz} = -\mu_B m_l. \tag{18.6a}$$

Example 18.5 The magnitude of the orbital magnetic dipole moment of an electron in a hydrogen-like atom is 1.3115×10^{-23} A · m^2. What state is the electron in? (*Hint:* A · m^2 = J/T.)

Solution We are given μ_l, and Eq. (18.5a) is $\mu_l = \mu_B \sqrt{l(l+1)}$. Substituting μ_l and the Bohr magneton, μ_B, into Eq. (18.5a) gives 1.3115×10^{-23} J/T = 9.274×10^{-24} J/T $\sqrt{l(l+1)}$. Dividing by μ_B and squaring both sides yields $2.000 = l^2 + l$. Solving $l^2 + l - 2 = 0$, or $(l-1)(l+2) = 0$, gives $l = 1$ as the acceptable root. (The root $l = -2$ isn't acceptable because l must be ≥ 0, varying from 0 to $n - 1$.) A value of $l = 1$ means that the electron is in a p state. ●

Example 18.6 What are the allowed z components of the orbital magnetic dipole moment of a p-state electron in units of eV/T?

Solution For a p state, $l = 1$. This allows m_l to equal -1, 0, and $+1$. Then Eq. (18.6a) gives $\mu_{lz} = -\mu_B m_l = -5.788 \times 10^{-5}$ eV/T times (-1, 0, and $+1$). Therefore the allowed values are $\mu_{lz} = +5.788 \times 10^{-5}$ eV/T, 0, and -5.788×10^{-5} eV/T. ●

18.4
MAGNETIC-FIELD EFFECTS

From classical physics, we know that applying a magnetic field **B** to a magnetic dipole causes a torque on the dipole of $\mathbf{T} = \boldsymbol{\mu} \times \mathbf{B}$. However, if a torque is applied perpendicular to the angular momentum **L** of an object (such as a gyroscope), the object begins to precess. The torque from **B** is perpendicular to $\boldsymbol{\mu}$ and also to **L**, which is opposite to $\boldsymbol{\mu}$. Therefore the magnetic field causes the magnetic dipole moment to precess about the z axis, as shown in Fig. 18.6. (Recall that we choose the z axis to be in the direction of the magnetic field.) In quantum physics, this precession results

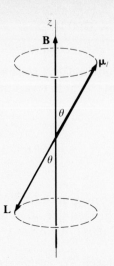

FIGURE 18.6
The orbital magnetic dipole moment vector and the orbital angular momentum vector are in opposite directions for an electron. Both vectors precess about the z axis, which is parallel to the applied magnetic field.

in our inability to determine exactly either L_x or L_y when we have measured L_z.

Example 18.7   Show that although neither L_x nor L_y can be exactly determined, $L_x^2 + L_y^2$ has definite quantized values.

Solution   Since $L^2 = L_x^2 + L_y^2 + L_z^2$, then $(L_x^2 + L_y^2) = L^2 - L_z^2$. With $L = \sqrt{l(l+1)}\hbar$ and $L_z = m_l\hbar$ from Eqs. (18.1) and (18.2), we have $(L_x^2 + L_y^2) = [l(l+1) - m_l^2]\hbar^2$. The allowed values of l and m_l will give the definite quantized values. Similarly, neither μ_{lx} nor μ_{ly} can be determined exactly when μ_{lz} has been measured, but $(\mu_{lx}^2 + \mu_{ly}^2)$ can be.   ●

The magnetic potential energy for a magnetic dipole in a magnetic field changes the total energy by $\delta E_m = -\boldsymbol{\mu} \cdot \mathbf{B}$. Since the z axis is determined by the direction of $\mathbf{B}$, we can write

$$\delta E_m = -\mu B \cos\theta = -\mu \cos\theta\, B = -\mu_z B.$$

Substituting $\mu_{lz} = -\mu_B m_l$ for the orbital magnetic dipole moment yields

$$\delta E_m = \mu_B m_l B. \tag{18.7}$$

Equation (18.7) gives the change in the energy of the system caused by the interaction of the orbital magnetic dipole moment of the electron and the magnetic field.

We know that m_l can have $2l + 1$ different values (from $-l$ to $+l$ in steps of 1). Applying a magnetic field to an orbital state will therefore split the state into $2l + 1$ different energy levels. (In making this statement we assume that the only magnetic dipole moment is the orbital one.)

Example 18.8 Sketch an energy-level diagram to show what happens to a d state as a magnetic field is increased from zero. Assume only an orbital magnetic dipole moment.

Solution With $B = 0$, there is only one degenerate energy level with some energy E_d. A d state has $l = 2$, so m_l can equal $-2, -1, 0, +1$, and $+2$. Using Eq. (18.7), we can see that this energy is changed by $\delta E_m = \mu_B m_l B = -2\mu_B B, -\mu_B B, 0, +\mu_B B$, and $+2\mu_B B$. As B increases, the levels split farther apart. These five energy levels are shown in Fig. 18.7 and correspond to the five orientations in space previously shown in Fig. 18.4.   ●

The maximum value of m_l ($+2$ for the d state) occurs when **L** is closest to parallel to **B** and when $\boldsymbol{\mu}_l$ and **B** are closest to antiparallel. But $\boldsymbol{\mu}_l$ and **B** are closest to being lined up for the lowest energy state, which corresponds to the lowest m_l (-2 for a d state). Since, if possible, any system tends to go to its lowest energy state, $\boldsymbol{\mu}_l$ tends to line up with **B** like a compass needle tends to line up with the earth's magnetic field.

In quantum physics, we can calculate the probability that a transition from one state to another will occur. Quantum mechanics shows that some

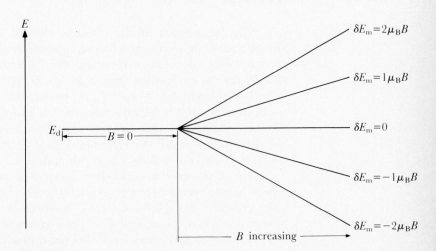

FIGURE 18.7
The energies of a d state in an applied magnetic field, assuming only an orbital magnetic dipole moment.

types of transitions are very improbable under normal conditions. These improbable transitions are called, with some exaggeration, **forbidden transitions.** The more probable transitions are called **allowed transitions.** Allowed transitions occur when certain **selection rules** are followed. These selection rules are obtained from probability calculations, which involve integrals that include both the original state and the final state wave functions. For example, these integrals will be small or zero under normal conditions unless

$$\Delta l = \pm 1 \quad \text{and} \quad \Delta m_l = 0 \quad \text{or} \quad \Delta m_l = \pm 1. \tag{18.8}$$

That is, the selection rules state that transitions from one state to another are allowed only if the orbital angular momentum quantum number, l, increases or decreases by 1 while the orbital magnetic quantum number is either constant or increases or decreases by 1.

Example 18.9   From a d state with $m_l = -1$, what transitions are allowed?

Solution For a d state, $l = 2$. The $\Delta l = \pm 1$ selection rule means that l must change from 2 to either 3 (an f state) or 1 (a p state). No transitions are allowed to s, d, g, h, . . . states. In the transition to the f state, the $\Delta m_l = 0$ or ± 1 selection rule means that m_l can change from -1 to 0 or -2, or m_l can remain -1. So there are only three transitions to each f state. For a p state, the $m_l = -2$ quantum number isn't allowed. Therefore the transitions to the p states are only to the $m_l = -1$ or $m_l = 0$ states.   ●

The frequencies of the photons emitted in transitions between atomic states can be changed by the application of a magnetic field. You can see how this occurs by referring to Fig. 18.8, the energy-level diagram that represents a transition from a d state to a lower energy p state.

The relation $\Delta E_0 = E_d - E_p$ represents the energy change for the transition when $B = 0$. When a magnetic field is applied—as you can see in Fig. 18.8—three energies are now allowed for the photons emitted during the transition: $\Delta E = \Delta E_0$ (when $\Delta m_l = 0$), $\Delta E = \Delta E_0 - \mu_B B$ (when $\Delta m_l = +1$), and $\Delta E = \Delta E_0 + \mu_B B$ (when $\Delta m_l = -1$).

Again, you need to remember that $\mu_B B$ is much less than ΔE_0 for a typical atom, even in strong magnetic fields. The splitting of levels is exaggerated in Fig. 18.8 for clarity. Nevertheless, the ability to change energy levels with a magnetic field does have several applications.

Recall that the photon emitted in the transition will have a frequency $\nu_0 = \Delta E_0 / h$. Then applying a magnetic field will split one spectral line of frequency $\nu_0 = \Delta E_0 / h$ into three lines. For $\Delta m_l = 0, +1$, and -1, respec-

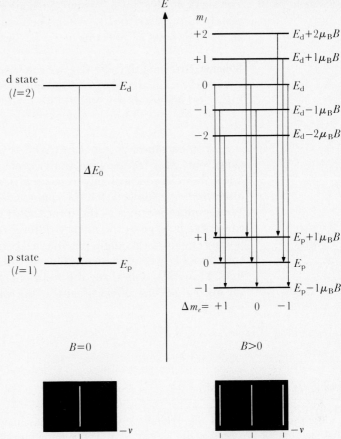

FIGURE 18.8
The cause of the normal Zeeman effect: The magnetic field splits the levels but selection rules allow transitions with only three different energy changes, giving three different frequencies (not to scale).

tively, the frequencies of the three lines will be

$$v = v_0, \qquad v = v_0 - \frac{\mu_B}{h} B, \quad \text{and} \quad v = v_0 + \frac{\mu_B}{h} B, \qquad (18.9)$$

where μ_B/h has the value 1.400×10^{10} Hz/T $= 14.00$ GHz/T.

This splitting of one line into three equally spaced lines by a magnetic field is called the **normal Zeeman effect** and was discovered by Pieter Zeeman in 1894 in Leiden. This effect, however, is "normal" only in that it is based on just the orbital motion of the electron and can even be derived from semiclassical models of the atom. But only a few atoms, such as Ca and Zn, actually exhibit the "normal" Zeeman effect. We'll discuss the reasons for this limited occurrence in Chapter 19.

Example 18.10   For small B, the three normal Zeeman effect frequencies are approximately v_0, $v_0 - dv$, and $v_0 + dv$. Their wavelengths are λ_0, $\lambda_0 - d\lambda$, and $\lambda_0 + d\lambda$. Express $d\lambda$ in terms of λ_0, a constant, and B.

Solution   Knowing that $\lambda = c/v$, we can write $d\lambda = -(c/v^2)dv$. Here $v = v_0$, $c/v_0^2 = \lambda_0^2/c$, and $dv = (\mu_{\rm B}/h)B$, giving $d\lambda = -(\lambda_0^2/c)(\mu_{\rm B}/h)B$. Substituting c, $\mu_{\rm B}$, and h gives $d\lambda = -46.69 \ ({\rm T} \cdot {\rm m})^{-1}\lambda_0^2 B$.   ●

You may have gotten used to the idea of quantized angular momentums, energies, frequencies, and wavelengths. But unless you already knew about it, you probably were surprised to find that angles can also be quantized. We showed that this quantization is with respect to the z axis. We defined the z axis as the direction of the applied magnetic field, because this field gives the energy difference and torque, which make the measurement of the angle possible. Degenerate high l energy levels may split into a large number of sublevels with the application of a magnetic field. However, only three different energy transitions are measured in the normal Zeeman effect because of the selection rules governing allowed transitions.

QUESTIONS AND PROBLEMS

18.1 Calculate the magnitude of the orbital angular momentum of an electron in   a) an s state and   b) a g state in terms of Planck's constant divided by 2π.

18.2 The orbital angular momentum of an electron has a magnitude of 2.583×10^{-34} kg $\cdot$ m^2/s. In which state is the electron?

18.3 Calculate the maximum orbital angular momentum for an electron (in magnitude) in a hydrogen atom for states with principal quantum numbers of 1, 10, and 100. Compare each with the Bohr atom postulated value of $n\hbar$. What trend do you see?

18.4 The z component of the orbital angular momentum is $-3\hbar$. For which states, if any, could this statement be false?

18.5 Calculate the fractional change in the magnitude of the orbital angular momentum in going from a state with one orbital angular

momentum quantum number to the next one. The minimum classical fractional change is zero (classically, L can have any value). Show that the quantum physics fractional value approaches zero in the limit of large quantum numbers.

18.6 Calculate the minimum angle between the orbital angular momentum and its z component for states with principal quantum numbers of 1, 100, and 10,000. Does this angle approach the minimum classical value in the limit of large quantum numbers?

18.7 All states have a possible orientation in which the z component of the orbital angular momentum equals zero, even though L may not equal zero. Prove this statement and explain.

18.8 The maximum angle between the orbital angular momentum and the z axis is 153.435°. What state is the electron in?

18.9 For a d state the z component of the orbital

angular momentum is 81.65% of the magnitude of **L**. Calculate the orbital magnetic quantum number of the state.

18.10 Calculate the allowed angles and then sketch the possible directions and z components of the orbital angular momentum of an electron in a) an s state,   b) a p state, and   c) an f state.

18.11 Calculate the allowed z components and the magnitude of the orbital angular momentum for s, d, and f electrons.

18.12 For the Bohr hydrogen atom in a ground state, $\mu_l = \mu_B$. Compare this to the quantum physics result.

18.13 Classical charges that don't move in straight-line orbits have magnetic dipole moments. Even though they don't have straight-line charge distributions, s states have no magnetic dipole moment due to orbital motion. Can you explain this apparent contradiction?

18.14 The minimum angle between the magnetic dipole moment and its z component in a given state (due to orbital motion) is 24.095°. What is that state?

18.15 Express the torque in terms of **L**, **B**, and constants.

18.16 The torque exerted is also quantized. Derive an expression for the magnitude of the torque in terms of the Bohr magneton, the magnetic field, the orbital angular momentum quantum number, and the orbital magnetic quantum number.

18.17 For a given l, how many orientations of **L** and $\boldsymbol{\mu}_l$ are there?

18.18 Sketch energy-level diagrams to show what happens to the s and f states as a magnetic field is turned on. (Consider orbital magnetic dipole moment effects only.)

18.19 For an f state, what are the orientations of **L** and $\boldsymbol{\mu}_l$ for $\delta E_m = +3\mu_B B$, 0, and $-3\mu_B B$?

18.20 When a 0.60-T magnetic field acts on the orbital magnetic dipole moment of a d state, by how many joules is the d state spread?

18.21 The energy difference between the most parallel and most antiparallel orientations of the orbital magnetic dipole moment with a magnetic field is 9.953×10^{-5} eV for a p state. What is the magnitude of the magnetic field?

18.22 The energy difference between the most parallel orientation and the perpendicular orientation of $\boldsymbol{\mu}_l$ with **B** is 1.67×10^{-23} J for a 0.45-T field. What orbital state is this?

18.23 From an s state, what three transitions are allowed to a state of a given n?

18.24 Denoting a state as (n, l, m_l), which of the following transitions are allowed?   a) $(5, 1, -1)$ to $(4, 0, 0)$.   b) $(4, 3, -3)$ to $(3, 2, 0)$.   c) $(3, 2, 1)$ to $(2, 0, 0)$.

18.25 Eq. (18.9) gives the three frequencies. What are the three wavelengths in the normal Zeeman effect?

18.26 Prove that the small wavelength change in the normal Zeeman effect is $\Delta\lambda \approx \pm\lambda_0^2 eB/(4\pi mc)$.

18.27 Calculate the frequencies and wavelengths of the three normal Zeeman spectral lines for a 1.000-T magnetic field if $\lambda = 422.6728$ nm when $B = 0$.

18.28 What magnetic field will give the normal Zeeman effect lines at 636.4055 nm, 636.4209 nm, and 636.4363 nm?

18.29 Explain how spectral lines from stars can reveal the stars' magnetic fields.

18.30 For a given B, the frequency and energy shifts for all normal Zeeman effect spectral lines are the same, but the wavelength shifts aren't. Explain.

18.31 Is the word "normal" in *normal Zeeman effect* a misnomer? Explain.

18.32 Which type of angular momentum underlies the normal Zeeman effect?

18.33 Draw a diagram similar to Fig. 18.8 for an f to d state transition and show that there are still only three spectral lines.

19

Electron Spin

The magnetic properties of matter are only partially the result of the orbital motion of electrons. Another cause is the magnetic effects of a different type of angular momentum, the spin angular momentum, which we present in this chapter. Spin and orbital effects often interact with each other. We'll discuss results of these interactions, including the spin–orbit interaction, total angular momentum, total magnetic dipole moment, and the anomalous Zeeman effect. Finally, you'll learn how to interpret and write spectroscopic notation, a standard method of identifying electron states in atoms.

19.1
SPIN QUANTUM NUMBERS

The angular momentum of the earth has two components. One is its orbital angular momentum as the earth moves around the sun every year. The other is a spin angular momentum as the earth spins about its own axis every day.

We will also discover that many small particles have a spin angular momentum. Although objects in classical physics can have any values for their spin angular momentums, every electron that has ever been measured has the same magnitude of **spin angular momentum, S.** This spin angular momentum of electrons was first suggested by S. A. Goudsmit and G. E. Uhlenbeck in 1925 to explain experimental results. Unfortunately, R. de L. Kronig had the same idea at the same time. We say unfortunately, because Kronig asked W. Pauli for his opinion of the idea before publication, and Pauli convinced him the idea had no merit.

Spin angular momentum doesn't result from the solutions of the nonrelativistic Schrödinger wave equation (which gave us the quantum numbers n, l, and m_l). Rather, as P. A. M. Dirac showed in 1928, spin angular momentum comes from solving a correct relativistic wave equation. The solutions to Dirac's relativistic equation involve a fourth quantum number, s, which is called the **spin angular momentum quantum number,** or sometimes, just the **spin.**

For electrons, s has only one value. Contrary to all the nonrelativistically derived quantum numbers you have encountered, s is not an integer for electrons. For electrons, $s = \frac{1}{2}$. Because $s = \frac{1}{2}$, an electron is called a "spin-$\frac{1}{2}$ particle."

The terms S and s are related in the same way that L and l are related. That is, just as $L = \sqrt{l(l + 1)}\hbar$, $S = \sqrt{s(s + 1)}\hbar$. For electrons then

$$S = \sqrt{s(s + 1)}\hbar = \frac{\sqrt{3}}{2}\hbar. \tag{19.1}$$

Substituting the value of $\hbar$ into Eq. (19.1), we see that all electrons have an intrinsic spin angular momentum of magnitude $(\sqrt{3}/2)\hbar = 9.133 \times 10^{-35}$ kg · m²/s.

As we discussed in Section 18.2, L_z is the z component of the orbital angular momentum; that is, $L_z = m_l\hbar$, where m_l varies from $-l$ to $+l$ in steps of 1, and m_l is the orbital magnetic quantum number. Similarly, S_z, the z component of the spin angular momentum, equals $m_s\hbar$, where m_s varies from $-s$ to $+s$ in steps of 1; that is, $m_s = -\frac{1}{2}$ and $+\frac{1}{2}$ for electrons. The term m_s is called the **spin magnetic quantum number.** Therefore, for electrons,

$$S_z = m_s\hbar = \pm\tfrac{1}{2}\hbar. \tag{19.2}$$

The z component of the spin angular momentum of all electrons is $\pm\frac{1}{2}\hbar = \pm 5.273 \times 10^{-35}$ kg · m²/s. The component parallel to the z axis is often called **spin up** ($m_s = +\frac{1}{2}$), and the component antiparallel to the z axis is then called **spin down** ($m_s = -\frac{1}{2}$).

EXAMPLE 19.1   Sketch the possible spin angular momentum vectors and their z components for an electron. Calculate the angles they make with the z axis.

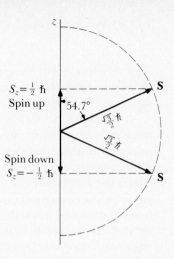

FIGURE 19.1
Space quantization of the spin angular momentum of a spin-$\frac{1}{2}$ particle, such as an electron.

Solution From Eq. (19.1), $S = (\sqrt{3}/2)\hbar = 0.866\hbar$. We first draw a dashed semicircle in Fig. 19.1, letting its radius be $0.866\hbar$. Then **S** will lie along the two radii of the circle such that, from Eq. (19.2), $S_z = +\frac{1}{2}\hbar$ or $-\frac{1}{2}\hbar$. We can see that $\cos\ \theta = S_z/S = \pm\frac{1}{2}\hbar/[(\sqrt{3}/2)\hbar] = \pm 1/\sqrt{3}$. Therefore $\theta = 54.7356°$ and $125.2644°$. $\qquad\qquad\qquad\bullet$

19.2
MAGNETIC EFFECTS, SPIN ONLY

Recall that the relation between the orbital magnetic dipole moment and the orbital angular momentum is $\mu_l = -(e/2m)\mathbf{L}$. Experimentally and from relativistic quantum theory, the **spin magnetic dipole moment,** $\boldsymbol{\mu_s}$, is found to be related to the spin angular momentum by

$$\mu_s \approx -2\,\frac{e}{2m}\,\mathbf{S}, \qquad (19.3)$$

where μ_s and **S** are opposite because of the negative charge on the electron. The 2 multiplying the right-hand side of the equation is not exactly 2; it is 2.0023193044. Since that number differs from 2 by only slightly over one-tenth of one percent, we can use the value 2 for many problems. Using $(\sqrt{3}/2)\hbar$ as the magnitude of **S**, we have

$$\mu_s = (2.00232)(e/2m)(\sqrt{3}/2)\hbar = (1.00116)[(e\hbar/2m)\sqrt{3} = \mu_B\sqrt{3}]$$
$$= 1.6082 \times 10^{-23} \text{J/T} = 1.0037 \times 10^{-4} \text{ eV/T}.$$

In a similar manner

$$\mu_{sz} = \mp(1.00116)\mu_{\mathrm{B}} \approx \mp\mu_{\mathrm{B}}. \tag{19.4}$$

The z component of the spin magnetic dipole moment is only about a tenth of one percent greater than the Bohr magneton. Using 2.00232. . . instead of 2, we get $\mu_{sz} = \mp 9.285 \times 10^{-24}$ J/T $= \mp 5.795 \times 10^{-5}$ eV/T.

The SWE doesn't give the electron spin in its solutions, and Dirac's relativistic wave equation gives exactly 2, not 2.00232. . . . . The part of quantum physics that predicts the magnitude of greater than 2 is called quantum electrodynamics (QED). To date, there's no disagreement between the extremely precise value predicted by QED and the measured value (within experimental error). Quantum electrodynamics was developed in the late 1940s. Four of the physicists who had much to do with its development were Richard Feynman, Julian Schwinger, Sin-itiro Tomonaga, and Freeman Dyson.

EXAMPLE 19.2   What are the orientations of μ_s for the spin-up and spin-down states?

Solution Since μ_s is in the opposite direction from **S**, the orientation of μ_s in the spin-up state is such that μ_{sz} is antiparallel to the z axis, while in the spin-down state μ_{sz} is parallel to the z axis. (See Fig. 19.2.)   ●

When we apply an external magnetic field, there will be a torque on the spin magnetic dipole moment of the electron. This torque will cause precession of **S** and μ_s about the z axis. (The direction of the magnetic field, again, is the z direction.) The energy will also change because of **spin magnetic potential energy,** or

$$\delta E_{\mathrm{s}} = -\mu_{sz}B = \pm(9.285 \times 10^{-24}\ \mathrm{J/T})B$$
$$= \pm(5.795 \times 10^{-5}\ \mathrm{eV/T})B \approx m_s 2\mu_{\mathrm{B}}B. \tag{19.5}$$

For example, in an s state (this "s" refers to $l = 0$) the orbital angular momentum, **L**, is zero, but there is spin angular momentum for a single

FIGURE 19.2
The spin magnetic dipole moment vectors and their spin angular momentum vectors have opposite directions for an electron.

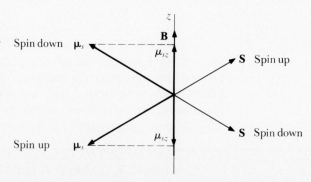

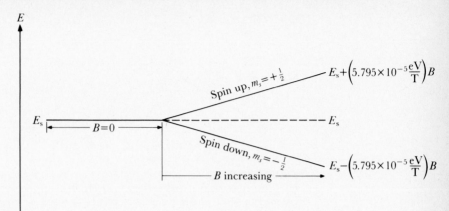

FIGURE 19.3
An s state of a single electron is split by the application of a magnetic field.

electron. When $B = 0$, the s state is two-fold degenerate with spin-up and spin-down states having the same energy. For $B > 0$, the two states are split apart and the degeneracy is removed, as shown in Fig. 19.3.

EXAMPLE 19.3 Calculate the spin magnetic potential energy for a single electron in a 1.600-T magnetic field in electron volts.

Solution From Eq. (19.5),

$$\delta E_s = \pm (5.795 \times 10^{-5} \text{ eV/T})(1.600 \text{ T}) = \pm 9.272 \times 10^{-5} \text{ eV}.$$

The spin-up state increases its energy by this amount, while the spin-down state decreases its energy. Even this strong magnetic field changes the energy of the atomic energy states less than 10^{-4} eV. ●

19.3
SOME APPLICATIONS

A force may be exerted on a magnetic dipole if it is placed in a nonuniform magnetic field. If **B** is set up as shown in Fig. 19.4, we can think of a spin-up electron as having its south pole in a stronger part of the field than its north pole, giving a net force downward. A spin-down electron will have its north pole in a stronger part of the field than its south pole, giving a net force upward.

Therefore a beam of atoms having all their electron angular momentums canceling, except for one "left-over" spin angular momentum unit, will be separated into spin-up and spin-down beams, with no other beams

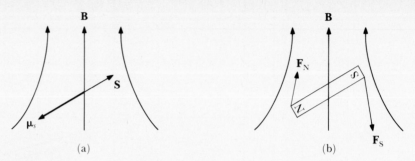

FIGURE 19.4

(a) A spin-up electron in a nonuniform magnetic field. (b) The electron is replaced by a tiny bar magnet. The S pole is in a stronger part of the magnetic field than the N pole.

between. This separation was first demonstrated by O. Stern and W. Gerlach in 1921, four years before the intrinsic spin angular momentum of electrons was suggested.

Years before that, however, researchers had noticed that the very strong yellow spectral line in Na (the "sodium D line") was in fact composed of two lines at 589.0 nm and 589.6 nm. The transition involves a 3p electron dropping into a 3s state. (We'll discuss why the 3s state is so much lower in energy than the 3p state in Na, while they are degenerate in H, in Chapter 20.) The split into two lines suggests an interaction between the electron's spin magnetic dipole moment and a magnetic field, but the split is found in the absence of an external magnetic field. Why? The reason could be an internal magnetic field within the atom. What could cause this field?

Suppose that you were able to seat yourself on a swivel chair on an orbiting electron. (See Fig. 19.5.) From your frame of reference, the nucleus would seem to be orbiting about you (just as the sun seems to rotate about the earth). The rotating positively charged nucleus would

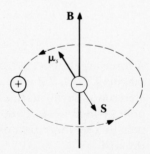

FIGURE 19.5

The nucleus rotates relative to the electron. The resulting magnetic field interacts with the spin magnetic dipole moment of the electron to give the spin–orbit interaction.

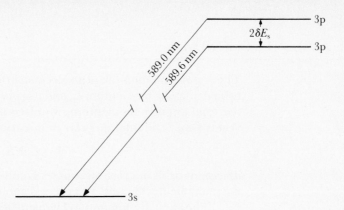

FIGURE 19.6
Spin–orbit splitting of the sodium D lines.

create a current loop and therefore a magnetic field. Of course, if the electron had no net orbital angular momentum (s state), this effect wouldn't occur. This interaction is called the **spin–orbit interaction.** It is found in all states except s states. The spin–orbit interaction causes spectral lines to split, energy bands in solids to split, and magnetic anisotropy to occur in solids, among other phenomena. (See Fig. 19.6.)

EXAMPLE 19.4 Calculate the effective magnetic field for the Na atom 3p state.

Solution The splitting of the sodium D lines isn't caused by any splitting of the final 3s state by a spin–orbit interaction, because there's no spin–orbit interaction in s states. Rather, the two lines result from the spin–orbit splitting of the initial 3p state. The 589.0-nm photon has an energy of

$$\frac{1.240 \times 10^{-6}\ \text{eV} \cdot \text{m}}{589.0 \times 10^{-9}\ \text{m}} = 2.1053\ \text{eV} \qquad \text{from } E = \frac{hc}{\lambda}.$$

The 589.6-nm photon has an energy of

$$\frac{1.240 \times 10^{-6}\ \text{eV} \cdot \text{m}}{589.6 \times 10^{-9}\ \text{m}} = 2.1031\ \text{eV}.$$

The energy difference is $(2.1053 - 2.1031)$ eV $= 2.2 \times 10^{-3}$ eV. The spin-up state has its energy increased by δE_s, and the spin-down state has its energy decreased by δE_s. Therefore the energy difference between the spin-down state and the spin-up state is $2\delta E_s$, with δE_s given by Eq. (19.5). That is

$$2(5.795 \times 10^{-5}\ \text{eV/T})B = 2.2 \times 10^{-3}\ \text{eV} \quad \text{or} \quad B = 19\ \text{T},$$

which is a very strong magnetic field. ●

19.4
TOTAL ANGULAR MOMENTUM

The angular momentum of an electron in an atom has two components: its orbital angular momentum **L**, and its spin angular momentum **S**. To find the total angular momentum, we often simply add **L** and **S**, which is a vector sum. The **total angular momentum** then is **J**, where

$$\mathbf{J} = \mathbf{L} + \mathbf{S}. \tag{19.6}$$

The spin–orbit interaction couples **L** and **S**, so an external torque from a magnetic field will cause **J** to precess about the z axis. (See Fig. 19.7.)

The angular momentum **J** is quantized, just as **L** and **S** were, and **J** has a magnitude of

$$J = \sqrt{j(j+1)}\,\hbar, \tag{19.7}$$

where j is the **total angular momentum quantum number.** It can have either of two values for a given l: $l + s$, or $l - s$. Since s is $\frac{1}{2}$ for an electron,

$$j = l \pm \tfrac{1}{2}. \tag{19.8}$$

For a single electron, the one exception to Eq. (19.8) is for an $l = 0$ state. Since $j = -\frac{1}{2}$ would give an imaginary J, only $j = +\frac{1}{2}$ is allowed for $l = 0$.

The z component of **J** is

$$J_z = m_j \hbar, \tag{19.9}$$

where m_j is the **total magnetic quantum number** and can vary from $-j$ to $+j$ in steps of 1.

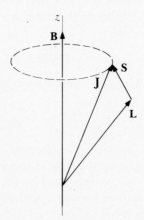

FIGURE 19.7
The total angular momentum as the vector sum of the orbital angular momentum and the spin angular momentum.

EXAMPLE 19.5 a) Write the different possible combinations of l, m_l, and m_s in the form (l, m_l, m_s) for a p state.   b) Write the different possible combinations of the orbital angular momentum, total angular momentum, and total magnetic quantum numbers for a p state.

Solution For a p state, $l = 1$.

a) Then $m_l = -1, 0$, and $+1$; m_s can equal $+\frac{1}{2}$ or $-\frac{1}{2}$. The $1 \times 3 \times 2 = 6$ possibilities are: $(1, -1, -\frac{1}{2})$, $(1, -1, \frac{1}{2})$, $(1, 0, -\frac{1}{2})$, $(1, 0, \frac{1}{2})$, $(1, 1, -\frac{1}{2})$, and $(1, 1, \frac{1}{2})$. There are six different p states.

b) For $l = 1$, $j = l \pm \frac{1}{2}$ gives $j = \frac{1}{2}$ and $j = \frac{3}{2}$. For $j = \frac{1}{2}$, $m_j = -\frac{1}{2}$ and $+\frac{1}{2}$. For $j = \frac{3}{2}$, $m_j = -\frac{3}{2}, -\frac{1}{2}, +\frac{1}{2}$, and $+\frac{3}{2}$. The $2 + 4 = 6$ possibilities, in the form (l, j, m_j), are: $(1, \frac{1}{2}, -\frac{1}{2})$, $(1, \frac{1}{2}, \frac{1}{2})$, $(1, \frac{3}{2}, -\frac{3}{2})$, $(1, \frac{3}{2}, -\frac{1}{2})$, $(1, \frac{3}{2}, \frac{1}{2})$, and $(1, \frac{3}{2}, \frac{3}{2})$. Again, there are six different p states for a given n. ●

We can also determine the angle between $\mathbf{J}$ and the z axis by using $J_z = J \cos \theta$. Since $m_j \hbar / [\sqrt{j(j+1)}\hbar] = m_j / \sqrt{j(j+1)}$, from Eqs. (19.7) and (19.9), we have

$$\cos \theta = \frac{J_z}{J} = \frac{m_j}{\sqrt{j(j+1)}}. \qquad (19.10)$$

EXAMPLE 19.6 The minimum angle between the total angular momentum vector and the z axis is $32.31°$. Calculate the total angular momentum quantum number.

Solution Since $\cos \theta = m_j / \sqrt{j(j+1)}$, the minimum θ will occur for the maximum m_j. (The smallest θ's give the largest cosines between $0°$ and $180°$.) The maximum value allowed for m_j is j. Therefore $\cos \theta_{\min} = j / \sqrt{j(j+1)}$. Solving this expression, we get $j = 1/[(1/\cos^2 \theta) - 1] = $ ctn$^2 \theta$. For $\theta = 32.31°$, $j = 2.500$ or $j = \frac{5}{2}$. ●

Recall that $\boldsymbol{\mu}_l = -(e/2m)\mathbf{L}$ and $\boldsymbol{\mu}_s \approx -2(e/2m)\mathbf{S}$, with each magnetic dipole moment directed antiparallel to its angular momentum. We can't write any such simple general relation between the **total magnetic dipole moment**, $\boldsymbol{\mu}_j$, and $\mathbf{J}$. The reason is that the spin angular momentum is about twice as effective as the orbital angular momentum in producing a magnetic dipole moment. Figure 19.8 shows, therefore, that the vector addition triangles won't generally be similar. Thus $\boldsymbol{\mu}_j$ won't generally be antiparallel to $\mathbf{J}$, so $\boldsymbol{\mu}_j$ and $\mathbf{J}$ will make different angles with the z axes.

EXAMPLE 19.7 Calculate the magnitude of the total magnetic dipole moment in terms of the total angular momentum quantum number, the orbital angular momentum quantum number, the spin angular momentum quantum number, and the Bohr magneton.

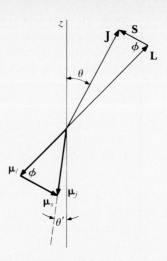

FIGURE 19.8

In this diagram, μ_l is opposite L and μ_s is opposite S, but μ_j isn't opposite J. Therefore $\theta' \neq \theta$.

Solution Applying the law of cosines to the two vector triangles in Fig. 19.8 we have

$$J^2 = L^2 + S^2 - 2LS \cos \phi \quad \text{and} \quad \mu_j^2 = \mu_l^2 + \mu_s^2 - 2\mu_l\mu_s \cos \phi.$$

Solving both equations for $-2 \cos \phi$ and then equating them gives $(J^2 - L^2 - S^2)/LS = (\mu_j^2 - \mu_l^2 - \mu_s^2)/\mu_l\mu_s$. Using $\mu_l = (e/2m)L$ and $\mu_s \approx 2(e/2m)S$ yields (after some algebra) $\mu_j^2 \approx (2J^2 - L^2 + 2S^2)(e/2m)^2$. Then $J^2 = j(j+1)\hbar^2$, $L^2 = l(l+1)\hbar^2$, $S^2 = s(s+1)\hbar^2$, and $\mu_B = (e\hbar/2m)$ give $\mu_j \approx \sqrt{2j(j+1) - l(l+1) + 2s(s+1)}\mu_B$.   ●

To check the solution to Example 19.7: For s states, $l = 0$, which gives $j = s$. Then

$$\mu_s \approx \sqrt{2s(s+1) - 0 + 2s(s+1)}\mu_B = 2\sqrt{s(s+1)}\mu_B.$$

With $s = \frac{1}{2}$, we then have $\mu_s \approx \sqrt{3}\mu_B$, which agrees with the calculation preceding Eq. (19.4). Ignoring spin, or letting $s = 0$, we have $j = l$. Then

$$\mu_l = \sqrt{2l(l+1) - l(l+1) + 0}\mu_B = \sqrt{l(l+1)}\mu_B,$$

which agrees with Eq. (18.5a).

The interaction of the total magnetic dipole moment with a magnetic field will split the energy levels. These splits, along with selection rules, result in the so-called **anomalous Zeeman effect.** Despite its name, the anomalous Zeeman effect is more common than the normal Zeeman effect—and also much more complicated to explain. The story is told that

W. Pauli once was asked why he looked so depressed. "How can one avoid despondency if one thinks of the anomalous Zeeman effect?" he replied. To avoid depression, we'll investigate this effect no further in this text.

In many cases the total angular momentum won't be simply a sum of **L** and **S**. For example, the spin–orbit interaction coupling **L** and **S** may be overwhelmed by strong external fields acting on the atom. In those cases, other methods are used to find the total angular momentum.

19.5
SPECTROSCOPIC NOTATION

Suppose that we wanted to discuss a single atomic electron having a principal quantum number of 5, an orbital angular momentum quantum number of 2, and a total angular momentum quantum number of $\frac{3}{2}$. We can easily reduce all these words to the term $5\ ^2D_{\frac{3}{2}}$, which is an example of **spectroscopic notation.** We pronounce $5\ ^2D_{\frac{3}{2}}$ as "five doublet D three halves." The general notation is

$$n\ ^{2s+1}(\text{capital letter for } l)_j. \qquad (19.11)$$

EXAMPLE 19.8   Show that $5\ ^2D_{\frac{3}{2}}$ is correct for the electron state given.

Solution The principal quantum number, n, is given as 5. For a single electron, $s = \frac{1}{2}$, which gives $2s + 1 = 2(\frac{1}{2}) + 1 = 2$. When $l = 2$ the atomic state is D. We know that j could be $2 + \frac{1}{2} = \frac{5}{2}$ or $2 - \frac{1}{2} = \frac{3}{2}$, and here j is given as $\frac{3}{2}$. Substituting into Eq. (19.11) we get $5\ ^2D_{\frac{3}{2}}$.   ●

In sodium, the $3\ ^2P_{\frac{1}{2}}$ and $3\ ^2P_{\frac{3}{2}}$ levels are split apart about 2.2 meV by the spin–orbit interaction. Transitions from these two levels to a $3\ ^2S_{\frac{1}{2}}$ ground state give the sodium D line doublet. Thanks to spectroscopic notation, quite a lot of specific information was given in the two preceding sentences.

In this chapter we've shown that altogether we need an orbital angular momentum, orbital magnetic dipole moment, and quantum numbers l and m_l; spin angular momentum, spin magnetic dipole moment, and quantum numbers s and m_s; and total angular momentum, total magnetic dipole moment, and quantum numbers j and m_j. There are quantizations of both direction and magnitude. You've learned that the spin motion is about twice as effective as the orbital motion in causing a magnetic dipole moment. One result is that μ_j isn't opposite **J**, except in special cases. We've discussed the spin–orbit interaction and the anomalous Zeeman effect, and we showed how to interpret or write $6\ {}^2\mathrm{F}_{\frac{7}{2}}$. If you're having trouble keeping all these quantum numbers and the effects associated with them straight, they're summarized in Table 20.4, which appears at the end of Chapter 20.

QUESTIONS AND PROBLEMS

19.1 As precisely as possible in your own words, state the difference between S and s.

19.2 In an experiment, can we change the a) magnitude or   b) direction of the spin angular momentum of an electron?

19.3 Calculate the spin angular momentum, the spin magnetic quantum number(s), and the z component(s) of the spin angular momentum of a) a "spin-zero" particle ($s = 0$) and   b) a "spin-one" particle ($s = 1$).

19.4 All electrons have exactly the same s. What else do they have that is exactly the same?

19.5 Which is closer to the Bohr magneton, the electron's spin magnetic dipole moment or the z component of the electron's spin magnetic dipole moment? How close?

19.6 In electron spin resonance (ESR) electrons in the lower of two spin states in a magnetic field can absorb a photon of the right frequency and move to the higher state.
a) Why would this transition be called a "spin flip"?
b) Does this transition tend to align or anti-align μ_s with $\mathbf{B}$?
c) Does this transition tend to align or anti-align $\mathbf{S}$ with $\mathbf{B}$?
d) Prove that the photon frequency is $(2.802 \times 10^{10}\ \mathrm{Hz/T})B = (28.02\ \mathrm{GHz/T})B$.
e) Calculate the magnetic field needed for resonance with 3.00-cm microwaves.

19.7 Calculate the energy difference in an s state between the spin-up and spin-down states with a magnetic field of 0.600 T.

19.8 The Boltzmann distribution (which we will discuss in Chapter 22) gives the number of electrons in the spin-up state, N_u, as equaling the number of electrons in the spin-down state, N_d, times $e^{-\Delta E/kT}$. In the exponent, ΔE is the energy difference between the two states, and $kT \approx 1/40$ eV at room temperature. For every million electrons in the spin-down state, how many are in the spin-up state for   a) $B = 5 \times 10^{-5}$ T (the earth's field)?   b) $B = 0.10$ T?   c) $B = 10$ T?

19.9 There's no spin–orbit splitting of the hydrogen atom ground energy level. Explain why.

19.10 In which orbital angular momentum state is the spin–orbit interaction zero? Why?

19.11 a) Write the different possible combinations of (n, l, m_l, m_s) for the first excited level of the hydrogen atom.   b) Repeat for (n, l, j, m_j).

19.12 Write the 10 different (j, m_j) combinations for a d state.

19.13 If j can be $\frac{5}{2}$ or $\frac{7}{2}$, is the state an s, p, d, f, or g state?

19.14 For a single atomic electron, one of the two possible angles closest to 90° between the total angular momentum vector and the z axis is 97.238°.   a) What's the other angle?   b) What's the total angular momentum quantum number?   c) What values could the orbital angular momentum have?

19.15 What are the six different angles possible between the z axis and $\mathbf{J}$ for a p state?

19.16 For a single electron state in an atom, m_j will never be zero. Explain.

19.17 In two certain special cases, μ_j and $\mathbf{J}$ will be antiparallel. What are these two special cases?

19.18 Is $m_j = m_l \pm m_s$?

19.19 Is μ_{jz} approximately equal to $(m_l + 2m_s)\mu_B$? Explain.

19.20 a) Find an advanced text with a discussion

of the Lande g factor. Show that

$$g = 1 + \frac{j(j+1) + s(s+1) - l(l+1)}{2j(j+1)} \quad \text{and}$$

$$\delta E_j = g m_j \mu_B B.$$

b) Calculate the g factor for $s = 0$ and also for $l = 0$. Explain your answers.

c) Why is this expression slightly in error? (*Hint:* What exactly is g for $l = 0$?)

19.21 Write the spectroscopic notation for all possible single electron states if

a) $n = 1$ d) $n = 2$ and $l = 2$
b) $n = 3$ and $l = 2$ e) $n = 6$ and $l = 3$.
c) $n = 4$ and $l = 1$

19.22 Give all the quantum numbers you can for the states

a) $4\,{}^2S_{\frac{1}{2}}$ d) $8\,{}^2G_{\frac{7}{2}}$
b) $2\,{}^2P_{\frac{3}{2}}$ e) $7\,{}^2F_{\frac{7}{2}}$.
c) $2\,{}^2P_{\frac{1}{2}}$

19.23 Calculate the angles possible between the total angular momentum and the z axis, as well as the magnitude of $\mathbf{J}$ and possible J_z's, for a $5\,{}^2P_{\frac{3}{2}}$ state.

19.24 Calculate J, J_z, and the minimum θ for a $6\,{}^2F_{\frac{7}{2}}$ state.

19.25 For what reasons isn't a $2\,{}^2D_{\frac{1}{2}}$ state possible?

19.26 For a given orbital angular momentum quantum number, the most negative z component of the total angular momentum occurs at $151.874°$. The total energy also has its most negative possible value. Give the spectroscopic notation for this single electron state.

CHAPTER *20*

Many-Electron Atoms

In this chapter we'll derive — to an extent — the periodic table of the elements. We can solve exactly the SWE for one electron around a nucleus, but for more than one electron only approximate solutions exist. However, we can use the one-electron solutions to find out about atoms that have more than one electron. Important tools in doing so are the Pauli exclusion principle and the quantum number rules, which we'll discuss in this chapter. You'll see why 2s levels are below 2p levels and when the outer electron of the alkaline metals and singly ionized alkaline earths have a hydrogen-like binding energy. We'll end the chapter by deriving an approximate relation for the frequency of those characteristic x-ray lines that we skipped in Section 8.1.

20.1
THE PAULI EXCLUSION PRINCIPLE

We can specify the state of any electron by using four quantum numbers. Examples are (n, l, m_l, m_s) and (n, l, j, m_j). We can speak of an electron in an $n = 3$, $l = 1$, $m_l = -1$, and $m_s = +\frac{1}{2}$ state. A particular wave function will then be associated with this electron. Experimental evidence indicates, however, that if one electron with that wave function and those quantum numbers exists in an atom, no other electrons will exist in the same state in that atom.

This condition is a special case of the generalization arrived at by Wolfgang Pauli in 1925. The generalization is called the **Pauli exclusion principle,** which we state in either of two forms:

No two electrons can occupy the same state in a given system.

or

No two electrons can have the same set of quantum numbers in a given system.

The Pauli exclusion principle has been verified for many different types of systems and holds for any particle having a spin angular momentum quantum number of $\frac{1}{2}, \frac{3}{2}, \frac{5}{2}, \ldots$.

EXAMPLE 20.1   In a given atom, can two electrons have $(1, 0, 0, \frac{1}{2})$ and $(1, 0, 0, -\frac{1}{2})$, respectively, as their quantum numbers (n, l, m_l, m_s)? If so, why?

Solution Yes. First, the numbers agree with the rules for the quantum numbers. For $n = 1$, l must equal 0; then m_l must also equal 0, and m_s can be $+\frac{1}{2}$ or $-\frac{1}{2}$. Second, the two sets of quantum numbers are *different* (corresponding to two different states), so they don't violate the Pauli exclusion principle.   ●

EXAMPLE 20.2   Can two nearby atoms both have $(1, 0, 0, \frac{1}{2})$ electrons? If so, why?

Solution Yes. The Pauli exclusion principle applies to one unconnected system at a time (for example, either atom). Therefore two $(1, 0, 0, \frac{1}{2})$ electrons *couldn't* be in the *same* atom (=the same system), but these two electrons *could* be in two *different* atoms (=two different unconnected systems).   ●

20.2
ALKALI METALS

One of the phenomena that the Pauli exclusion principle explains is the formation of closed shells and subshells in atoms. You may have studied these previously in a chemistry class. For instance, a sodium atom has 11

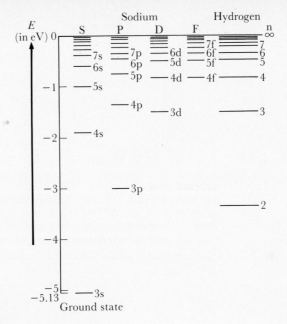

FIGURE 20.1
Energy levels for Na and H atoms.

electrons. Ten of these electrons form two closed shells (one with $n = 1$, the other with $n = 2$). The remaining electron is in an $n = 3$ ground state.

The $2n^2 = 18$ different $n = 3$ states of Na aren't all degenerate. Experiments show that the 3s states are about 2.1 eV lower than the 3p states, with the 3p states about 1.5 eV lower than the 3d states. Moreover, the energy of the 3d states is quite close to the energy of the $n = 3$ states in hydrogen, as seen in Fig. 20.1. Also, the 4s states are more than 0.4 eV below the 3d states in energy, despite the fact that they have a higher n.

We can explain these results reasonably well by using Gauss's law and the radial probability density. Gauss's law tells us that the magnitude of the electric field is $(1/4\pi\epsilon_0) \, q_{enclosed}/r^2$ outside a spherical charge distribution, where $q_{enclosed}$ is the net charge enclosed by a spherical Gaussian surface of radius r. If a spherically symmetric collection of 10 electrons and 11 protons is enclosed, $q_{enclosed}$ equals $-10e + 11e = +1e$. Thus, an eleventh electron *completely* outside this collection of charges would be attracted by an effective charge of $+1e$. Therefore

$$E_n = -\frac{Z_{eff}^2}{n^2}(13.6 \text{ eV}) \tag{20.1}$$

would have $Z_{eff} = 1$, so $E_n = -13.6 \text{ eV}/n^2$, just as for hydrogen (ignoring

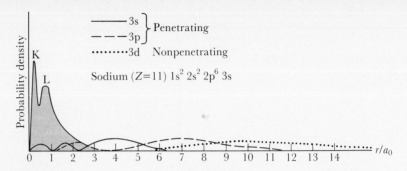

FIGURE 20.2

The probability density for the electrons in sodium. The terms K, L, and $1s^2 2s^2 2p^6 3s$ will be explained in Section 20.3. The shaded area represents the probability for the 10 inner electrons, so a 3d electron spends almost 100 percent of its time outside those 10 electrons. A 3p electron spends less time outside and a 3s electron even less.

Source: Adapted with permission from H. E. White, *Introduction to Atomic Spectra.* New York, McGraw-Hill, 1934, p. 103.

the small reduced mass corrections). We say that the 10 electrons **screen** 10 of the 11 protons, leaving an effective net charge of $+1e$, or an effective Z, Z_{eff}, of 1.

The radial probability density for an $l = n - 1$ state (for example, a 3d state), has one peak (see Fig. 17.3), which puts its most probable position outside the positions of the electrons with smaller n values. (Those electrons also are pulled closer to the nucleus than in hydrogen because they are less effectively screened from the positive nucleus.) In Na, a 3d electron spends most of its time well outside the $n = 1$ and $n = 2$ states (labeled K and L in Fig. 20.2). The 10 electrons in the $n = 1$ and $n = 2$ states then screen off almost all but one of the 11 protons, leaving a net charge of nearly $+1e$ and an energy of nearly $-13.6 \text{ eV}/3^2 = -1.5 \text{ eV}$ (the $n = 3$ energy for hydrogen).

The radial probability density for an $l = n - 2$ state has two peaks, for an $l = n - 3$ state has three peaks, and so on. The small separation 3p peak in Na gives a 3p electron a good chance of being well inside the high probability positions for the $n = 2$ states. A 3p electron, therefore, spends some of its time within the inner closed shells, so the 11 protons aren't always screened by all 10 inner electrons. This behavior means that Z_{eff}, on average, will be greater than 1 and E_{3p} less than -1.5 eV. A 3s electron spends even more time within the inner electron shells than a 3p electron, giving an even larger Z_{eff} and an even more negative energy.

EXAMPLE 20.3   The energy of the 3s state of Na is -5.138 eV. Calculate the effective charge.

Solution We use $E_n = -Z_{\text{eff}}^2(13.6 \text{ eV})/n^2$, with $E_n = -5.138 \text{ eV}$ and $n = 3$. Solving for Z_{eff}, we get $3\sqrt{5.138/13.6} = Z_{\text{eff}} = 1.84$. The effective charge is $+1.84e$. ●

In Example 20.3, the 11 protons are screened by an average of $11 - 1.84 = 9.16$ electrons instead of 10 electrons because of the penetration of the inner shells by the 3s electron. In this example we assume that the only reason for the change in E_3 is the effective Z. We ignore any differences in E_3 due to the changes given the 3s wave function by other effects.

The alkali metals Li, Na, K, Rb, and Cs have one more electron than the corresponding noble gases He, Ne, Ar, and Xe. This extra electron is outside electrons that are closer to the nucleus. Therefore all the alkali metals will behave much like Na.

EXAMPLE 20.4 The outer electron in K has a 4s ground state. Discuss the relative energies of its 4s, 4p, 4d, and 4f states. Calculate the approximate energy of the state having the smallest Z_{eff}.

Solution The 4f state is the largest orbital angular momentum state and is one for which the electron will spend the most time outside the closed shells and subshells. This condition will make Z_{eff} close to 1, which gives $E_4 \approx -1^2(13.6 \text{ eV})/4^2 = -0.85 \text{ eV}$ as the approximate energy of the 4f state. The electron in the 4d state will spend a bit more time within the inner shells, and the energy will therefore be a bit more negative. For the same reason, the 4p state will have an even lower energy, and the 4s state will have the lowest (experimentally, the energy of the 4s state is -4.339 eV). ●

We can extend this analysis by considering the singly ionized alkaline earths: Be^+, Mg^+, Ca^+, Sr^+, and Ba^+. For any n, their highest l state will have an effective charge of almost $+2e$, so $Z_{\text{eff}} \approx 2$. The 3d state for Mg^+, for example, will have an energy of about $-2^2(13.6 \text{ eV})/3^2 = -6.0 \text{ eV}$.

20.3
THE PERIODIC TABLE

The Pauli exclusion principle has many important applications in areas as diverse as nuclear physics, solid state physics, and astrophysics. In this section, we'll apply the Pauli exclusion principle to derive, in effect, the periodic table of the elements. First, recall that for any n, l can vary from 0 to $n - 1$ in steps of 1, for any l, m_l can vary from $-l$ to $+l$ in steps of 1, and for any m_l, m_s can be $-\frac{1}{2}$ or $+\frac{1}{2}$.

Also recall that $2n^2$ different combinations of n, l, m_l, and m_s exist for any n. Thus there are $2n^2$ different states, which can hold up to $2n^2$ electrons. For a given n in an atom, the

$$\text{Number of states} = \text{Maximum number of electrons} = 2n^2. \qquad (20.2)$$

For $n = 1$ there are 2 states, for $n = 2$ there are 8 states, for $n = 3$ there are 18 states, and so on.

EXAMPLE 20.5   How many electrons can be found in an $n = 4$ level in an atom?

Solution The number of states is $2n^2 = 2 \times 4^2 = 32$. Therefore any number of electrons from 0 to 32 (but no more than 32) can be found in those states.   ●

Let's first consider the $n = 1$ level. Hydrogen, $_1$H, in its ground state, will have one electron with $n = 1$, $l = 0$, $m_l = 0$, and $m_s = -\frac{1}{2}$. Helium, $_2$He, is the next element. It will have 2 electrons. We can assign to one electron in He, in its ground state, the (n, l, m_l, m_s) quantum numbers $(1, 0, 0, -\frac{1}{2})$. The other electron will then have to have $(1, 0, 0, +\frac{1}{2})$. The $n = 1$ level has only two states, so the two $_2$He electrons have completely filled these states.

The two $n = 1$ states are said to form a shell, called the **K shell.** A **shell** is a collection of states having the same principal quantum number, n. Based on our preceding discussion, we say that the He atom has a **closed shell.** That is, all the states in its K shell are full, and no more electrons can be added to that shell.

Next let's move on to $_3$Li with its three electrons. Two of its electrons can have the same two sets of quantum numbers as the two He electrons. The third electron can't go in the K shell, because that would be forbidden by the Pauli exclusion principle. The two $n = 1$ states are full, so it must go to an $n = 2$ state. The eight $n = 2$ states form the **L shell.** The letters are assigned alphabetically:

$$n = 1\ 2\ 3\ 4\ 5\ \ldots$$
$$\text{Shell} = K\ L\ M\ N\ O\ \ldots$$

Our discussion of the alkali metals showed that for $n = 2$, the $l = 0$, or s, states have lower energy than the $l = 1$, or p, states. Since all systems go to the lowest energy state they can reach, the third electron in the ground state in Li would have $(2, 0, 0, -\frac{1}{2})$ as its four quantum numbers.

The element $_4$Be with four electrons would add another electron with $(2, 0, 0, +\frac{1}{2})$ quantum numbers, thus filling the $l = 0$, or s, subshell. Within a shell, a **subshell** is a collection of states, all of which have the same

TABLE 20.1
Quantum Numbers for the First 18 Electrons

Electron	n	l	m_l	m_s	
1	1	0	0	$-\frac{1}{2}$	
2	1	0	0	$+\frac{1}{2}$	
					Filled K shell
3	2	0	0	$-\frac{1}{2}$	
4	2	0	0	$+\frac{1}{2}$	
					Filled s subshell of the L shell
5	2	1	$+1$	$-\frac{1}{2}$	
6	2	1	0	$-\frac{1}{2}$	
7	2	1	-1	$-\frac{1}{2}$	
8	2	1	$+1$	$+\frac{1}{2}$	
9	2	1	0	$+\frac{1}{2}$	
10	2	1	-1	$+\frac{1}{2}$	
					Filled p subshell and L shell
11	3	0	0	$-\frac{1}{2}$	
12	3	0	0	$+\frac{1}{2}$	
					Filled s subshell of the M shell
13	3	1	$+1$	$-\frac{1}{2}$	
14	3	1	0	$-\frac{1}{2}$	
15	3	1	-1	$-\frac{1}{2}$	
16	3	1	$+1$	$+\frac{1}{2}$	
17	3	1	0	$+\frac{1}{2}$	
18	3	1	-1	$+\frac{1}{2}$	
					Filled p subshell of the M shell

orbital quantum number, l. It will take another six electrons to fill the $l = 1$, or p, subshell within the L shell. That would give another **filled subshell,** as well as another filled shell.

Quantum numbers for the first 18 electrons are shown in Table 20.1. The actual order in which the m_l and m_s quantum numbers occur for a given l is unimportant for our purposes, because we'll not discuss the effects that certain rules of ordering explain.

EXAMPLE 20.6   What are the quantum numbers of the electron in the ground state of $_6$C?

Solution The subscript preceding C tells us that C has six electrons. In the ground state, the six electrons will have the quantum numbers of the first six electrons given in Table 20.1.   ●

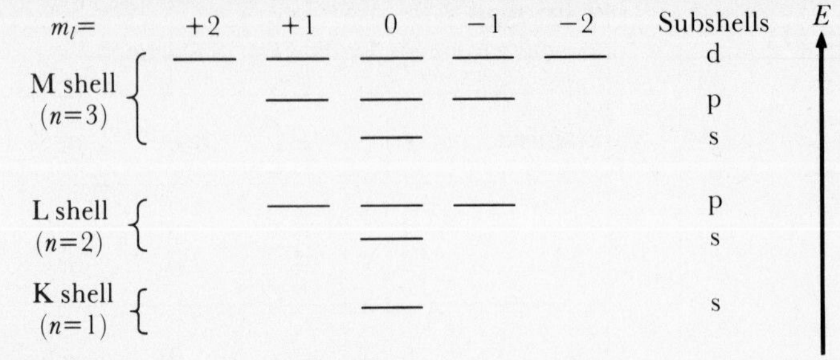

FIGURE 20.3
The different m_l levels of s, p, and d subshells of K, L, and M shells in an atom. Each m_l level can contain 0, 1, or 2 electrons.

Another way of representing what's happening is with an energy-level diagram such as that presented in Fig. 20.3. Then some of the elements would be represented as shown in Fig. 20.4.

EXAMPLE 20.7 Represent the ground state electron distribution in $_{14}$Si with an energy-level diagram.

Solution The quantum numbers $(1, 0, 0, \pm\frac{1}{2})$ give two electrons in the K shell; $(2, 0, 0, \pm\frac{1}{2})$ give two electrons in the s subshell of the L shell; $(2, 1, [\pm 1 \text{ or } 0], \pm\frac{1}{2})$ give six electrons in the p subshell of the L shell; $(3, 0, 0, \pm\frac{1}{2})$ give two electrons in the s subshell of the M shell. So far we have $2 + 2 + 6 + 2 = 12$ electrons. We can place the remaining two with $(3, 1, 1, -\frac{1}{2})$ and $(3, 1, 0, -\frac{1}{2})$, as shown in Fig. 20.5.   ●

Since the actual order in which the orbital magnetic and spin magnetic quantum numbers occur is unimportant for our purposes, we can use

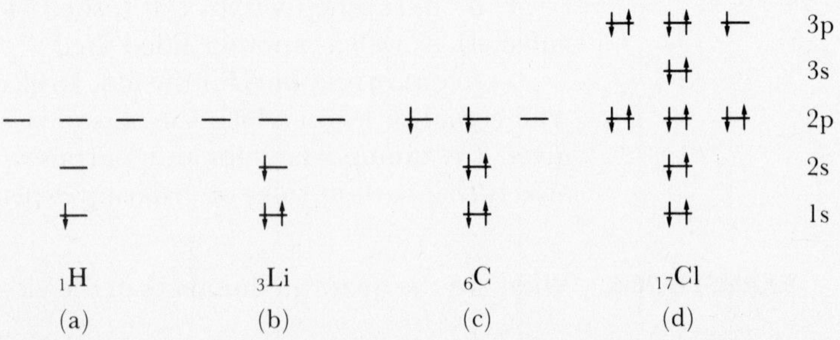

FIGURE 20.4
The lowest energy levels are filled first, as shown for the ground states of four elements.
Note: ↓ means a spin-down electron in the level, and ↑ means a spin-up electron.

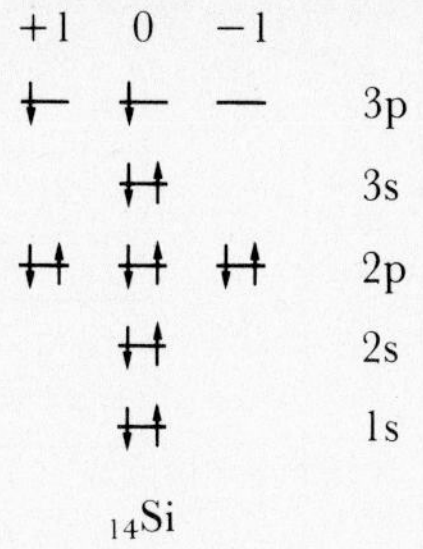

$$
\begin{array}{ccc}
+1 & 0 & -1 \\
\end{array}
$$

+1	0	−1	
⊢	⊢	—	3p
	⊢⊣		3s
⊢⊣	⊢⊣	⊢⊣	2p
	⊢⊣		2s
	⊢⊣		1s

$_{14}$Si

FIGURE 20.5
The ground state electron distribution in $_{14}$Si.

a third way involving just the principal and orbital quantum numbers. The n and l values of each electron are given by the corresponding number and letter. A superscript on the letter also shows the number of electrons in that subshell (the number with that n and l value). Recall that for a given l, there are $2l + 1$ possible values of m_l and 2 values of m_s. Therefore the maximum number for the superscript (which is the number of states in the subshell) is $2(2l + 1)$. This number is 2 for the s subshells, 6 for the p subshells, 10 for the d subshells, and so on. This manner of representing all electrons in an atom is called the **electron configuration** of the atom. Table 20.2 gives some examples of these configurations.

Note that both He and Ne have closed shells. This configuration makes it energetically difficult to add or subtract electrons from He and

TABLE 20.2
Ground State Electron Configurations
for Some of the First 18 Elements

Element	Configuration
$_1$H	$1s$
$_2$He	$1s^2$
$_3$Li	$1s^2 2s$
$_4$Be	$1s^2 2s^2$
$_5$B	$1s^2 2s^2 2p$
$_6$C	$1s^2 2s^2 2p^2$
$_7$N	$1s^2 2s^2 2p^3$
$_9$F	$1s^2 2s^2 2p^5$
$_{10}$Ne	$1s^2 2s^2 2p^6$
$_{11}$Na	$1s^2 2s^2 2p^6 3s$
$_{12}$Mg	$1s^2 2s^2 2p^6 3s^2$
$_{13}$Al	$1s^2 2s^2 2p^6 3s^2 3p$
$_{17}$Cl	$1s^2 2s^2 2p^6 3s^2 3p^5$
$_{18}$Ar	$1s^2 2s^2 2p^6 3s^2 3p^6$

TABLE 20.3
The Periodic Table of the Elements

Group I	II		III	IV	V	VI	VII	VIII
H 1.0079 1s¹								2 He 4.003 1s²
3 Li 6.941 2s¹	4 Be 9.012 2s²		5 B 10.81 2p¹	6 C 12.011 2p²	7 N 14.007 2p³	8 O 15.999 2p⁴	9 F 18.998 2p⁵	10 Ne 20.18 2p⁶
11 Na 22.99 3s¹	12 Mg 24.31 3s²		13 Al 26.98 3p¹	14 Si 28.09 3p²	15 P 30.97 3p³	16 S 32.07 3p⁴	17 Cl 35.45 3p⁵	18 Ar 39.95 3p⁶
19 K 39.10 4s¹	20 Ca 40.08 4s²	(3d)	31 Ga 69.72 4p¹	32 Ge 72.59 4p²	33 As 74.92 4p³	34 Se 78.96 4p⁴	35 Br 79.90 4p⁵	36 Kr 83.80 4p⁶
37 Rb 85.47 5s¹	38 Sr 87.62 5s²	(4d)	49 In 114.82 5p¹	50 Sn 118.71 5p²	51 Sb 121.75 5p³	52 Te 127.60 5p⁴	53 I 126.90 5p⁵	54 Xe 131.29 5p⁶
55 Cs 132.91 6s¹	56 Ba 137.33 6s²	(5d)	81 Tl 204.4 6p¹	82 Pb 207.2 6p²	83 Bi 209.0 6p³	84 Po (210) 6p⁴	85 At (210) 6p⁵	86 Rn 222.0 6p⁶
87 Fr (223) 7s¹	88 Ra 226.0 7s²	(6d)						

3d transition series

21 Sc 44.96 4s²3d¹	22 Ti 47.88 4s²3d²	23 V 50.94 4s²3d³	24 Cr 52.00 4s¹3d⁵	25 Mn 54.94 4s²3d⁵	26 Fe 55.85 4s²3d⁶	27 Co 58.93 4s²3d⁷	28 Ni 58.69 4s²3d⁸	29 Cu 63.55 4s¹3d¹⁰	30 Zn 65.39 4s²3d¹⁰

4d transition series

39 Y 88.91 5s²4d¹	40 Zr 91.22 5s²4d²	41 Nb 92.91 5s¹4d⁴	42 Mo 95.94 5s¹4d⁵	43 Tc (99) 5s²4d⁵	44 Ru 101.07 5s¹4d⁷	45 Rh 102.91 5s¹4d⁸	46 Pd 106.42 5s⁰4d¹⁰	47 Ag 107.87 5s¹4d¹⁰	48 Cd 112.41 5s²4d¹⁰

5d transition series

57-71	72 Hf 178.49 6s²5d²	73 Ta 180.95 6s²5d³	74 W 183.85 6s²5d⁴	75 Re 186.21 6s²5d⁵	76 Os 190.2 6s²5d⁶	77 Ir 192.2 6s²5d⁷	78 Pt 195.1 6s¹5d⁹	79 Au 197.0 6s¹5d¹⁰	80 Hg 200.6 6s²5d¹⁰

6d transition series

89-103	104 * (261) 7s²6d²	105 * (262) 7s²6d³	106 * (263) 7s²6d⁴	107 * (262)				109 * (266)	

4f (Lanthanides)

57 La 138.91 6s²5d¹	58 Ce 140.12 6s²4f²	59 Pr 140.91 6s²4f³	60 Nd 144.24 6s²4f⁴	61 Pm (147) 6s²4f⁵	62 Sm 150.36 6s²4f⁶	63 Eu 151.96 6s²4f⁷	64 Gd 157.25 6s²4f⁷5d¹	65 Tb 158.93 6s²4f⁹	66 Dy 162.50 6s²4f¹⁰	67 Ho 164.93 6s²4f¹¹	68 Er 167.26 6s²4f¹²	69 Tm 168.93 6s²4f¹³	70 Yb 173.04 6s²4f¹⁴	71 Lu 174.97 6s²4f¹⁴5d¹

5f (Actinides)

89 Ac 227.0 7s²6d¹	90 Th 232.0 7s²6d²	91 Pa 231.0 7s²5f²6d¹	92 U 238.0 7s²5f³6d¹	93 Np (237) 7s²5f⁴6d¹	94 Pu (239) 7s²5f⁶	95 Am (243) 7s²5f⁷	96 Cm (247) 7s²5f⁷6d¹	97 Bk (247) 7s²5f⁹	98 Cf (251) 7s²5f¹⁰	99 Es (252) 7s²5f¹¹	100 Fm (257) 7s²5f¹²	101 Md (258) 7s²5f¹³	102 No (259) 7s²5f¹⁴	103 Lr (260) 7s²5f¹⁴6d¹

* Not named or names disputed.

FIGURE 20.6
A scheme to help you remember the approximate overall order for the ground state electron configuration of atoms up to $_{88}$Ra.

Ne. Both are noble gases, and both are virtually chemically inert. Both Li and Na have one outer s electron and are chemically similar. Because they both lack one electron to form a closed p subshell, F and Cl are alike chemically. Thus you see how the periodic table, which was originally made up by comparing chemical reactions, so far can be derived by using the Pauli exclusion principle and the rules for allowed quantum numbers.

You might think that the electron configuration of $_{19}$K would simply be that of $_{18}$Ar plus one 3d electron. However, the 4s states are screened so much less by the inner electron shells than are the 3d states that the 4s states usually have lower energy than the 3d states. The two 4s states are filled before the ten 3d states (with two exceptions). Therefore $_{19}$K has the $_{18}$Ar electron configuration plus a $4s^1$ (or just 4s) electron; $_{20}$Ca adds $4s^2$, $_{21}$Sc adds $4s^2 3d^1$, and so on to $_{30}$Zn which adds $4s^2 3d^{10}$. The exceptions are $_{24}$Cr with $4s^1 3d^5$ and $_{29}$Cu with $4s^1 3d^{10}$ added to the electron configuration of $_{18}$Ar. (See Table 20.3.)

Roughly, the overall order is

$$1s^2 2s^2 2p^6 3s^2 3p^6 4s^2 3d^{10} 4p^6 5s^2 4d^{10} 5p^6 6s^2 4f^{14} 5d^{10} 6p^6 7s^2,$$

which is the ground state electron configuration of $_{88}$Ra. (See Fig. 20.6.) Mixtures of 6d and 5f states follow the $7s^2$. For instance, $_{92}$U adds $6d^1 5f^3$ to the $_{88}$Ra electron configuration.

For any atom or ion, there is one ground state electron configuration and an infinite number of possible excited state configurations. For example, for He in the ground state we have $1s^2$. For He in the first excited state we have 1s2s. (Singly ionized He^+ would have 1s for its ground state.)

EXAMPLE 20.8 Give the ground state and an excited state electron configuration for $_{26}$Fe.

Solution We start with the given order $1s^2 2s^2 2p^6 3s^2$ . . . , adding the superscripts until we reach a number less than or equal to 26. This occurs at $4s^2$, at which point we have 20 electrons. We then add 6 of the possible 10 3d electrons to arrive at the total of 26. The answer is

$$1s^2 2s^2 2p^6 3s^2 3p^6 4s^2 3d^6.$$

One of the possible excited states would have the same configuration as the ground state with the exception of $4s^1 3d^7$. This new configuration still adds to 26 electrons in allowed states.   ●

20.4
CHARACTERISTIC X-RAYS

Figure 8.3 showed that an x-ray tube could produce a continuous spectrum plus peaks. We explained that the continuous spectrum is caused by bremsstrahlung. The peaks are at different places for different target materials. We say, therefore, that the peaks are characteristic of the target element in the x-ray tube.

Suppose that the target element is $_{29}$Cu. We can draw an energy-level representation of $_{29}$Cu as shown in Fig. 20.7. When the energetic electron hits the Cu target material in the x-ray tube, one of the inner electrons in a $_{29}$Cu atom may be excited to a higher energy state (if that state is empty). Or that inner electron may be knocked completely free of the $_{29}$Cu atom.

In either case the result would be a missing electron, or **hole** in a subshell. As a more energetic electron drops down to fill the hole, energy

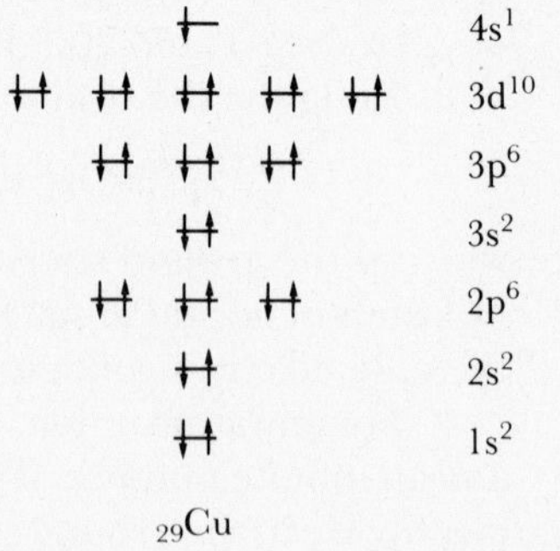

FIGURE 20.7
The energy level diagram for $_{29}$Cu.

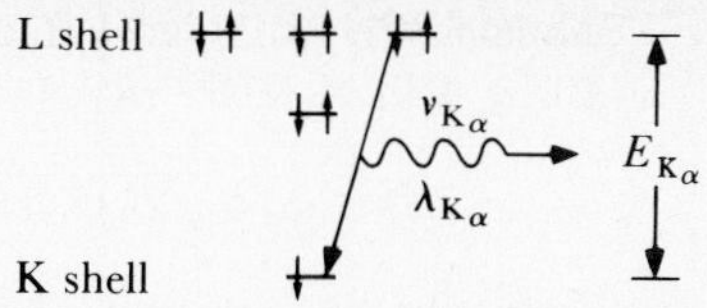

FIGURE 20.8

An L-shell electron makes the transition to fill the empty state (hole) in the K shell, causing the emission of a K_α photon.

will be released, just as energy was released when the electron in the hydrogen atom dropped to a lower energy state. Sometimes this energy will be transferred to an electron loosely enough bound that it can be ejected from the atom. This ejection is called the **Auger effect;** the ejected electron, an **Auger electron.**

When a more energetic electron drops down to fill a hole, the energy may also be released in the form of an x-ray photon. (See Fig. 20.8.) A whole set of different x-ray spectral line series can result. The K_α, K_β, K_γ, and so on, lines come from electrons dropping to the K shell from the L shell, the M shell, the N shell, and so on. The L lines result from electrons falling into holes in the L shell.

We can estimate the energy of a K_α **characteristic x-ray** photon. A K_α **photon** is emitted when an L electron drops down to fill a hole in the K shell. Inside the orbit of the L electron are Z protons and the one electron left in the K shell. The effective charge experienced by the L electron is therefore approximately $(Z - 1)e$. Using $E_{K_\alpha} = E_2 - E_1$ along with $E_n \approx -(Z - 1)^2 (13.6 \text{ eV})/n^2$, we obtain

$$E_{K_\alpha} \approx (Z - 1)^2 (1/1^2 - 1/2^2)(13.6 \text{ eV}),$$

or

$$E_{K_\alpha} \approx (Z - 1)^2 (10.2 \text{ eV}). \tag{20.3}$$

Then $\nu_{K_\alpha} = E_{K_\alpha}/h$ and $\lambda = c/\nu$ give

$$\nu_{K_\alpha} \approx (Z - 1)^2 (2.47 \times 10^{15} \text{ Hz}) \tag{20.4}$$

and

$$\lambda_{K_\alpha} \approx \frac{1.21 \times 10^{-7} \text{ m}}{(Z - 1)^2}. \tag{20.5}$$

EXAMPLE 20.9 The K_α wavelength for a particular element is 154 pm. What is the element?

Solution We are given that $\lambda_{K_\alpha} = 154$ pm $= 154 \times 10^{-12}$ m. From Eq. (20.5), 154×10^{-12} m $\approx (1.21 \times 10^{-7}$ m$)/(Z - 1)^2$, so

$$Z \approx 1 + \sqrt{(1.21 \times 10^{-7})/(154 \times 10^{-12})} = 1 + 28.03 = 29.$$

(*Z must* be an integer!) The result, $Z = 29$, refers to an element with 29 protons and 29 electrons in its neutral atom. This element is $_{29}$Cu.   ●

The frequency is

$$\frac{3.00 \times 10^8 \text{ m/s}}{154 \times 10^{-12} \text{ m}} = 1.95 \times 10^{18} \text{ Hz.}$$

We could have used this frequency in Eq. (20.4) to obtain Z. The charac-

TABLE 20.4
Summary of Quantum Numbers for Atoms

Name	Symbol	Applications
Principal	n	The quantum number n principally determines the energy, as in $E_n = -Z^2(13.60 \text{ eV})/n^2$. $n = 1 \ 2 \ 3 \ 4 \ 5 \ \ldots$ Shell $= $ K L M N O $\ldots$
Orbital angular momentum	l	The quantum numbers n and l determine the radial probability density, $R_{nl}^2 r^2$; l determines the magnitude of the orbital angular momentum, $L = \sqrt{l(l+1)}\hbar$; l can vary from 0 to $n - 1$ in steps of one. $l = 0 \ 1 \ 2 \ 3 \ 4 \ 5 \ \ldots$ Subshell $= $ s p d f g h $\ldots$ The orbital magnetic dipole moment is $\mu_l = \mu_B \sqrt{l(l+1)}$, directed antiparallel to **L**. In many-electron atoms, the lowest l subshells have the lowest energy for a given shell and fill up first. In alkali metals the $l = n - 1$ outer electron has an energy very near -13.6 eV$/n^2$; in singly ionized alkali earths, it is -54.4 eV$/n^2$.
Orbital magnetic	m_l	The quantum numbers n, l, and m_l determine the probability density, $\psi_{nlm_l}^* \psi_{nlm_l}$; m_l determines the z component of the orbital angular momentum, $L_z = m_l \hbar$; m_l can vary from $-l$ to $+l$ in steps of 1. The z component of the orbital magnetic dipole moment is $\mu_{lz} = -\mu_B m_l$. The interaction of just μ_l with **B** gives $\delta E_m = \mu_B m_l B$ and the normal Zeeman effect.

teristic x-ray photon energy is

$$\frac{hc}{\lambda} = \frac{1.240 \times 10^{-6} \text{ eV} \cdot \text{m}}{154 \times 10^{-12} \text{ m}} = 8.05 \times 10^3 \text{ eV} = 8.05 \text{ keV}.$$

We could have used 8.05×10^3 eV in Eq. (20.3) to find Z.

In this chapter we applied the Pauli exclusion principle to many-electron atoms. We'll use this very important principle again in the solid state and nuclear physics chapters yet to come. Although we can't solve the SWE exactly for many-electron atoms, we may still use the single-electron model for some atoms and for K_α x-rays. Some of you may have been given a lot of information about the periodic table in chemistry class, but with little explanation of the rules on which it's based. That information should be more understandable now. We present Table 20.4 to help tie together many of the ideas of these last five chapters.

Name	Symbol	Applications
Spin angular momentum	s	The quantum number s determines the magnitude of the spin angular momentum. For any electron, $s = \frac{1}{2}$, giving $S = \sqrt{3}\hbar/2$. The spin magnetic dipole moment is $\mu_s \approx \mu_B \sqrt{3}$, directed antiparallel to $\mathbf{S}$ for any electron.
Spin magnetic	m_s	The quantum number m_s determines the z component of the spin angular momentum; $m_s = \pm\frac{1}{2}$ for any electron, giving $S_z = \pm\hbar/2$. The z component of the spin magnetic dipole moment is $\mu_{sz} \approx \mp\mu_B$ for any electron. This interaction of only μ_s with $\mathbf{B}$ gives $\delta E_s \approx \pm\mu_B B$.
Total angular momentum	j	The quantum number j determines the magnitude of the total angular momentum, $\mathbf{J}$; $\mathbf{J} = \mathbf{L} + \mathbf{S}$, $j = |l \pm \frac{1}{2}|$ for an electron, and $J = \sqrt{j(j+1)}\hbar$. The total magnetic dipole moment is given by a complicated expression and is opposite $\mathbf{J}$ only in special cases. The interaction of μ_j with $\mathbf{B}$ gives the anomalous Zeeman effect, except for special cases.
Total magnetic	m_j	The quantum number m_j determines the z component of the total angular momentum, $J_z = m_j\hbar$; m_j varies from $-j$ to $+j$ in steps of 1. The z component of the total magnetic dipole moment is not a simple function of m_j, except for special cases.

QUESTIONS AND PROBLEMS

20.1 If the Pauli exclusion principle didn't hold true, in what ground state would all the electrons be found for all atoms?

20.2 Is it true that if one electron is in a 1s state in an atom, no other electron in that atom can be found in a 1s state because of the Pauli exclusion principle?

20.3 A coupled pair of electrons can have a total spin quantum number of zero. Can two or more coupled pairs then have the same set of quantum numbers in a given system, despite the Pauli exclusion principle?

20.4 In an electron gas at $T = 0$ K, electrons fill all states up to an energy E_F. Why don't all the electrons drop down to the lowest state?

20.5 "The number of states, not including spin, is N. Therefore as many as $2N$ electrons can be found in these states." Explain as much of the preceding statement as you can.

20.6 Protons and neutrons both have $s = \frac{1}{2}$. Are they also bound by the Pauli exclusion principle?

20.7 Why doesn't the spin–orbit splitting show up on the scale of Fig. 20.1?

20.8 The energy needed to ionize Li by removing its outer electron from its 2s ground state is 5.39 eV. Assuming that the wave function is otherwise unchanged, what is the effective charge?

20.9 a) Explain why the 2s level in Li is below the 2p level. b) What is the approximate energy of the 2p level?

20.10 Calculate the approximate energy of one of the $n = 5$ states of Rb.

20.11 The effective charge is $2.771e$ for the outer 5s electron in Rb. Calculate the minimum energy needed to remove the 5s electron from Rb, assuming that the wave function is otherwise unchanged.

20.12 Calculate the approximate energy of the least penetrating of the a) $n = 2$ states of Be^+ and b) $n = 4$ states of Ca^+.

20.13 The energy needed to remove the second 5s electron from the ground state to obtain Sr^{2+} is

11.027 eV. Calculate the effective Z, assuming that the wave function is otherwise unchanged.

20.14 Calculate the maximum number of electrons that could be found in each of the following shells:
  a) K   d) N
  b) L   e) O.
  c) M

20.15 Which shell contains 50 states?

20.16 Is it possible to find a shell that will accept a maximum of 36 electrons?

20.17 How many subshells are there in each of the following shells?
  a) K   d) N
  b) L   e) O.
  c) M

20.18 Without looking at Table 20.1, what are the quantum numbers of the ground state electrons in the following atoms?
  a) $_5B$ c) $_{18}Ar$
  b) $_8O$ d) $_{20}Ca$.
How would these quantum numbers differ for an excited state?

20.19 Represent the distribution of the ground state electrons with an energy-level diagram for
  a) $_5B$ d) $_{18}Ar$
  b) $_8O$ e) $_{20}Ca$.
  c) $_{15}P$
How would these diagrams differ for excited states?

20.20 Give the ground state electron configuration of
  a) $_8O$ c) $_{83}Bi$
  b) $_{15}P$ d) $_{87}Fr$.
How would the configurations change for excited states?

20.21 Give the ground state electron configuration for $_{92}U$ if the Pauli exclusion principle no longer held.

20.22 The outer ground state electron configuration in $_{23}V$ is $4s^2 3d^3$, while in $_{24}Cr$ it's $4s^1 3d^5$. What must be the reason?

20.23 What is the origin of the L_α, L_β, L_γ, . . . series of x-ray lines?

20.24 The K_α characteristic x-ray line is actually split into a K_{α_1} and a K_{α_2} line. Why?

20.25 Will any more radiation be emitted after a K_α x-ray is emitted? Explain.

20.26 Explain how an atom could emit *no* x-ray when an electron falls into a hole in the K shell.

20.27 How can one atom emit both K_α and L x-ray photons almost simultaneously?

20.28 What is the other name we use in this chapter for the internal photoelectric effect?

20.29 In 1913, H. G. J. Moseley determined experimentally that $\sqrt{v_{K_\alpha}}$ is linearly related to Z. Explain why. If you were to graph $\sqrt{v_{K_\alpha}}$ versus Z, what would be the slope and intercept of the line?

20.30 For the L_α x-ray photon, we have $E_{L\alpha} \approx (Z - 7.4)^2 \, (5/36)(13.6 \text{ eV})$. Derive the 5/36. What must be the origin of the 7.4?

20.31 Compare the calculated value of the K_α x-ray wavelength for $_{74}W$ with the experimental value of 20.9 pm.

20.32 What element will have a v_{K_α} of about one-fourth that of $_{47}Ag$?

20.33 A K_α x-ray is emitted from a sample with a frequency of 5.3×10^{18} Hz. What is the element?

20.34 Calculate the energy, frequency, and wavelength of the K_α x-ray for   a) $_{13}Al$,   b) $_{25}Mn$, and   c) $_{37}Rb$.

20.35 A K_α x-ray is emitted from a sample with an energy of 9.89 keV. What is the target element?

20.36 Why does the Pauli exclusion principle require that an inner electron be removed before a characteristic x-ray is produced?

20.37 Can an x-ray photon of energy $E = hv_{K_\alpha}$ be absorbed by an atom? Explain.

20.38 Use the *Handbook of Chemistry and Physics* to find out how well the approximate expressions for the K_α wavelength and energy work for   a) low Z,   b) medium Z, and   c) high Z.

20.39 Rydberg atoms are atoms having an outermost electron in an excited state with a very large principal quantum number.

a) Why do all Rydberg atoms with the same n value have essentially the same ionization energy?

b) What would cause some minor differences in the ionization energy for different Rydberg atoms with the same n?

c) Why can a Rydberg atom be described as "fragile"?

d) What is the ionization energy for a Rydberg atom with a principal quantum number of 290?

e) What is the approximate size of the Rydberg atom with $n = 732$?

(Note: 290 was obtained in an earth laboratory, and 732 was detected in interstellar space.)

PART Four

Molecular, Statistical, and Solid State Physics

CHAPTER *21*

Energy in Molecules

Astronomers examine the skies with their instruments and then announce that the "empty space" between the stars isn't empty at all, but contains an amazing number of different types of molecules. How can they tell? They have not brought interstellar probes back to their labs. Rather, they have examined the missing frequencies in the spectrum of those electromagnetic waves that have traveled through interstellar space to their spectroscopes. Each type of molecule in space has absorbed its own characteristic frequencies. In this chapter we will show why by examining the vibrational and rotational energy levels of diatomic molecules.

21.1
MOLECULES

Atoms often join together to form molecules. For instance, the air we breathe is primarily composed of N_2, O_2, and H_2O molecules, along with some Ar atoms. The attractive forces that bond molecules together are basically electrostatic. We will discuss the types of binding in Chapter 23.

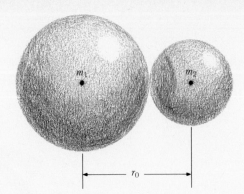

FIGURE 21.1
A diatomic molecule.

A **diatomic molecule** is made up of two atoms. They may be the same, as in H_2, or they may be different, as in CO. Since electrons are so light, we can consider the entire mass of each atom to be concentrated at its nucleus. The two nuclei in a diatomic molecule will be separated by a distance r_0, as illustrated in Fig. 21.1. Called the **equilibrium separation,** r_0 will vary for different molecules.

EXAMPLE 21.1   The basically electrostatic binding energy of H_2 is 4.52 eV. Show that 4.52 eV is much greater than the absolute value of the gravitational potential energy, Gm_1m_2/r, where $G = 6.67 \times 10^{-11}$ N $\cdot$ m^2/kg^2. The equilibrium separation is 74.1 pm, and $m_1 = m_2 = 1.67 \times 10^{-27}$ kg for H.

Solution Substituting the given values, the absolute value of the gravitational potential energy is

$$G\frac{m_1m_2}{r} = 6.67 \times 10^{-11} \frac{\text{N} \cdot \text{m}^2}{\text{kg}^2} \frac{(1.67 \times 10^{-27} \text{ kg})^2}{74.1 \times 10^{-12} \text{ m}}$$

$$= 2.51 \times 10^{-54} \text{ J} \frac{1 \text{ eV}}{1.60 \times 10^{-19} \text{ J}} = 1.57 \times 10^{-35} \text{ eV}.$$

Certainly, 4.52 eV $\gg 1.57 \times 10^{-35}$ eV. Therefore this gravitational potential energy is negligible.   ●

21.2
VIBRATIONAL ENERGY

The atoms in a molecule won't remain at rest, but will vibrate about their equilibrium positions. A typical graph of the potential energy V versus separation r is shown in Fig. 21.2 for a diatomic molecule. For small

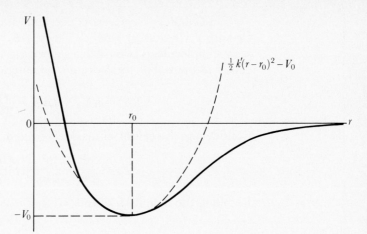

$\frac{1}{2}k'(r-r_0)^2 - V_0$

FIGURE 21.2
The potential energy as a function of separation for a typical diatomic molecule. The dashed curve represents the harmonic oscillator potential energy.

displacements from equilibrium, we can closely approximate the actual potential energy curve by using the relation $V' = \frac{1}{2}k'(r - r_0)^2 = \frac{1}{2}k'x^2$. Here V' is the potential energy above the equilibrium value, $-V_0$. Remember that $V = \frac{1}{2}k'x^2$ is the potential energy of a one-dimensional harmonic oscillator. In Chapter 15, we rewrote $\frac{1}{2}k'x^2$ as $\frac{1}{2}m\,\omega_c^2 x^2$ and found that the allowed energies above the equilibrium energy are

$$E_v = (n + \tfrac{1}{2})\hbar\omega_c = (n + \tfrac{1}{2})h\nu_c. \qquad (21.1)$$

In Eq. (21.1), n is the **vibrational quantum number** and may equal 0, 1, 2, 3, ... The ground state energy with respect to the equilibrium value isn't 0. The energy levels are evenly spaced $\hbar\omega_c = h\nu_c$ apart in energy, as shown in Fig. 21.3.

FIGURE 21.3
The ground state and first three excited state energy levels for vibration in the one-dimensional harmonic oscillator approximation. The levels are equally spaced.

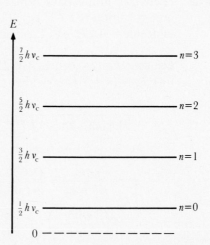

In Chapter 15 we were concerned with the motion of only one particle with a vibrational kinetic energy, $KE_v = p^2/2m$. Now we treat a diatomic molecule as two particles vibrating with respect to their common center of mass. The total vibrational kinetic energy then is

$$KE_v = KE_{v1} + KE_{v2} = \frac{p_1^2}{2m_1} + \frac{p_2^2}{2m_2}.$$

Considering the center of mass to be fixed, we have $|\mathbf{p}_1| = |-\mathbf{p}_2| = p$, so

$$KE_v = \frac{p^2}{2m_1} + \frac{p^2}{2m_2} = \frac{p^2(m_1 + m_2)}{2m_1 m_2}.$$

Recalling the definition of the reduced mass, m_r,

$$m_r = \frac{m_1 m_2}{m_1 + m_2}, \tag{21.2}$$

we have $KE_v = p^2/2m_r$.

This result means that we can write the SWE for the vibrational energies of a diatomic molecule the same way that we can write the expression for a one-dimensional harmonic oscillator, but with m replaced by m_r. Then

$$E_v = (n + \tfrac{1}{2})\hbar\omega_c = (n + \tfrac{1}{2})h\nu_c$$

will be the energies, where

$$\omega_c = 2\pi\nu_c = \sqrt{\frac{k'}{m_r}}, \tag{21.3}$$

and k' is the force constant.

EXAMPLE 21.2   Hydrogen atoms have a mass of 1.008 u (1 u = 1 unified atomic mass unit = 1.6605×10^{-27} kg). The ground state vibrational energy of an H_2 molecule is 0.273 eV. Calculate the force constant for the H_2 molecule.

Solution We're given $E_v = 0.273$ eV for $n = 0$ (the ground state). Therefore $E_v = (n + \tfrac{1}{2})\hbar\omega_c$ yields

$$\frac{0.273 \text{ eV}}{(0 + \tfrac{1}{2})(6.582 \times 10^{-16} \text{ eV} \cdot \text{s})} = 8.295 \times 10^{14} \text{ rad/s} = \omega_c.$$

To determine k', we need m_r. Here $m_1 = m_2 = 1.008$ u, so

$$m_r = \frac{m_1 m_2}{m_1 + m_2} = \frac{(1.008 \text{ u})^2}{2(1.008 \text{ u})}$$

$$= \left(\frac{1.008 \text{ u}}{2}\right)\left(\frac{1.6605 \times 10^{-27} \text{ kg}}{1 \text{ u}}\right) = 8.369 \times 10^{-28} \text{ kg}.$$

Finally, we solve $\omega_c = \sqrt{k'/m_r}$ for k', which gives

$$k' = m_r\omega_c^2 = (8.369 \times 10^{-28} \text{ kg})(8.295 \times 10^{14} \text{ rad/s})^2 = 576 \text{ N/m},$$

because $1\text{N/m} = (1 \text{ kg} \cdot \text{m/s}^2)/\text{m} = 1 \text{ kg/s}^2$.   ●

21.3
ROTATIONAL ENERGY

The atoms of a molecule not only vibrate with respect to their center of mass, they also rotate about it. A principal axis of rotation passes through the center of mass of a diatomic molecule; this axis is perpendicular to the line joining the two nuclei. The rotational inertia of a point mass m rotating in a circle of radius r is $I = mr^2$. (I is also called the moment of inertia.) The nuclei are close enough to being point masses that their total rotational inertia, I, equals $m_1 r_1^2 + m_2 r_2^2$, where r_1 and r_2 are the distances from the center of mass to the nuclei of atoms 1 and 2. Figure 21.4 shows that $r_0 = r_1 + r_2$. We'd like to find an expression for I in terms of m_r and r_0. Multiplying $m_1 r_1^2 + m_2 r_2^2$ by $(m_1 + m_2)/(m_1 + m_2)$ gives us

$$I = \frac{m_1 r_1^2 m_2 + m_2 r_2^2 m_1 + m_1^2 r_1^2 + m_2^2 r_2^2}{m_1 + m_2}. \tag{A}$$

By definition of the center of mass, $m_1 r_1 = m_2 r_2$, so $m_1^2 r_1^2 + m_2^2 r_2^2 = 2m_1 r_1 m_2 r_2$. Substituting into Eq. (A) gives $I = [m_1 m_2/(m_1 + m_2)](r_1 + r_2)^2$. With $m_r = m_1 m_2/(m_1 + m_2)$ and $r_0 = r_1 + r_2$, we finally obtain

$$I = m_r r_0^2. \tag{21.4}$$

EXAMPLE 21.3   Calculate the rotational inertia of an H_2 molecule.

Solution From Example 21.1 we have $r_0 = 74.1 \times 10^{-12}$ m and from Example 21.2 we have $m_r = 8.369 \times 10^{-28}$ kg for H_2. Therefore

$$I = m_r r_0^2 = (8.369 \times 10^{-28} \text{ kg})(74.1 \times 10^{-12} \text{ m})^2 = 4.60 \times 10^{-48} \text{ kg} \cdot \text{m}^2.$$
●

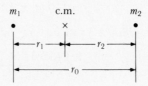

FIGURE 21.4
The atoms in a diatomic molecule rotate about an axis through their center of mass.

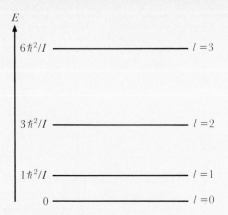

FIGURE 21.5
The ground state and first three excited rotational energy levels for a diatomic molecule. The levels aren't equally spaced.

The rotational angular momentum is given by $\mathbf{L} = I\omega$. Since we can write the SWE of a diatomic molecule with V as a function of r only, the angular momentum is quantized in the same way as for the hydrogen atom. That is, $L = \sqrt{l(l+1)}\hbar$. The rotational energy is all kinetic. The expression for rotational kinetic energy here is $E_r = \frac{1}{2}I\omega^2 = I^2\omega^2/2I = L^2/2I$, which is the rotational equivalent of $p^2/2m$. Substituting $L^2 = l(l+1)\hbar^2$ gives

$$E_r = \frac{l(l+1)\hbar^2}{2I}. \tag{21.5}$$

This l term is the **rotational angular momentum quantum number** and equals 0, 1, 2, 3, ... The allowed energies are illustrated in Fig. 21.5.

EXAMPLE 21.4 Give the ground energy level and first two excited level rotational energies of an H_2 molecule in eV.

Solution Example 21.3 gave us $I = 4.60 \times 10^{-48}$ kg · m², an answer in SI units. One way to use $E_r = l(l+1)\hbar^2/2I$ is to substitute $\hbar^2$ in SI units, then convert the energy in joules to eV. Another way is to write

$$\hbar^2 = \hbar \times \hbar = (1.0546 \times 10^{-34} \text{ J} \cdot \text{s})(6.582 \times 10^{-16} \text{ eV} \cdot \text{s})$$
$$= 6.941 \times 10^{-50} \text{ kg} \cdot \text{m}^2 \cdot \text{eV},$$

using 1 J = 1 kg · m²/s². Then

$$E_r = \frac{l(l+1)(6.941 \times 10^{-50} \text{ kg} \cdot \text{m}^2 \cdot \text{eV})}{2(4.60 \times 10^{-48} \text{ kg} \cdot \text{m}^2)} = l(l+1)(7.54 \times 10^{-3} \text{ eV}).$$

In the ground level $l = 0$, and therefore $E_r = 0$. In the first excited level

$l = 1$ and

$$E_r = 1(1 + 1)(7.54 \times 10^{-3}\ eV) = 1.51 \times 10^{-2}\ eV.$$

In the second excited level $l = 2$ and

$$E_r = 4.53 \times 10^{-2}\ eV. \qquad \bullet$$

We showed that the ground level rotational energy is zero. It is quite possible for molecules to have *no* rotational energy and so *no* rotational motion. Note that the energy levels aren't evenly spaced as they are in the vibrational case.

21.4
MOLECULAR SPECTRA

The total energy of a molecule is made up of three main parts: $E = E_e + E_v + E_r$. The term E_e gives the energy due to the electrons. These electronic energy levels are much more complicated in molecules than in a hydrogen atom. The outer electron levels still typically have energies with absolute values in the eV range. The term E_v gives the vibrational energy. The 0.273 eV value given in Example 21.2 for H_2 is one of the largest values known because H_2 has the lowest reduced mass of any molecule. Therefore E_v is considerably less than $|E_e|$. The E_r values for H_2 calculated in Example 21.4 were in the 10^{-2} eV range, or much less than the E_v values. (Again, the small reduced mass and rotational inertia of H_2 means that its E_r values are greater, often much greater, than the E_r values for other molecules.)

All these behaviors lead to a complicated energy-level diagram, sketched in Fig. 21.6. We can start to draw it with two electronic levels, A and B. Then we add evenly spaced vibrational levels to both of the electronic levels. Typically, the separation between adjacent vibrational levels will be far less than the separation between A and B. Finally, we add rotational levels to each vibrational level. Ordinarily, the separation between adjacent rotational levels will be the smallest separation of all.

Photons can be absorbed by molecules to cause transitions to higher energy states. Also, photons can be emitted when the molecules make transitions to lower energy states. Selection rules again limit the number of high probability transitions under normal conditions. If the two atoms are the same (as in H_2, N_2, O_2, for example), the electric dipole moment of the molecule is zero and transitions involving photons improbable. The selection rules for vibrational and rotational levels both require that the transition involve an increase or decrease of 1 in the quantum number. That is $\Delta n = \pm 1$ and $\Delta l = \pm 1$.

A different electronic level will result in different electron distributions with corresponding changes in k', r_0, I, and other factors. To eliminate this and other complications, we'll concern ourselves with those transitions in which only E_v and E_r change. The Pauli exclusion principle will change the selection rules if the two nuclei in the diatomic molecule are identical. Therefore we'll further limit ourselves to molecules with two different nuclei.

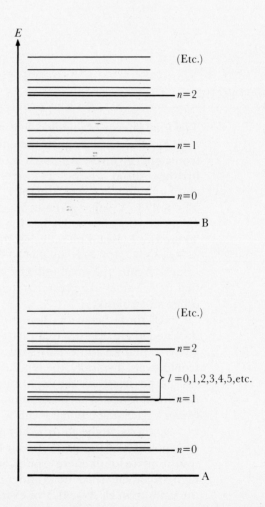

FIGURE 21.6
Energy levels in a diatomic molecule. Electronic levels A and B have vibrational levels ($n = 0, 1, 2, \ldots$) and rotational levels ($l = 0, 1, 2, 3, 4, 5, \ldots$) between the vibrational levels.

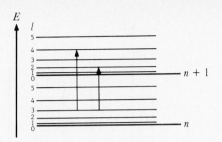

FIGURE 21.7
The two allowed absorption transitions, which start from $n = n$ and $l = 3$, within a specific electronic level.

Suppose that an initial level has the quantum numbers n and l. For absorption, n must increase by 1 to $n + 1$ for the larger vibrational energy change. At the same time, l may either increase to $l + 1$ or decrease to $l - 1$ (unless $l = 0$) for the smaller rotational energy change, as shown in Fig. 21.7. We use $h\nu = \Delta E$ to find the photon frequency. With $E = E_e + E_v + E_r$, $\Delta E_e = 0$, $E_v = (n + \frac{1}{2})h\nu_c$, and $E_r = l(l + 1)\hbar^2 / 2I$, we have

$$h\nu = (n + 1 + \tfrac{1}{2})h\nu_c + \frac{(l \pm 1)(l \pm 1 + 1)\hbar^2}{2I} - (n + \tfrac{1}{2})h\nu_c - \frac{l(l + 1)\hbar^2}{2I}.$$

Dividing both sides by h, recalling that $\hbar = h/2\pi$, we find two solutions: $\nu = \nu_c + (l + 1)\hbar/2\pi I$, when l increases to $l + 1$, and $\nu = \nu_c - l\hbar/2\pi I$, when l decreases to $l - 1$. Remember that l is the quantum number of the *initial* level. Since there is no $l = -1$ level, the transition from $l = 0$ to $l = -1$ isn't possible. Therefore $\nu \neq \nu_c$. We can express all of this more simply by saying that the possible frequencies are

$$\nu = \nu_c \pm \frac{(l' + 1)\hbar}{2\pi I}, \tag{21.6}$$

where l' is 0, 1, 2, 3, . . .

Emission merely involves a downward, rather than an upward, transition in energy. The same selection rules would apply. Therefore the probable emission frequencies are the same as the probable absorption frequencies expressed by Eq. (21.6).

The spectrum has no line at ν_c, but does contain lines at frequencies of $\nu_c \pm \hbar/2\pi I$, $\nu_c \pm 2\hbar/2\pi I$, $\nu_c \pm 3\hbar/2\pi I$, and so on. That is, the lines would form two branches in this model of a diatomic molecule. As shown in Fig. 21.8, each branch would be symmetric about ν_c. Also, each line would be separated by $\hbar/2\pi I$ from its neighbors in the branch. The lines are in the infrared portion of the electromagnetic spectrum.

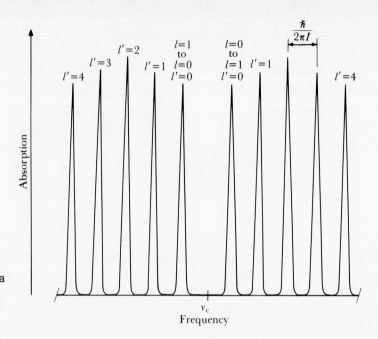

FIGURE 21.8
A portion of an idealized
absorption spectrum for a
diatomic molecule
containing two different
atoms.

EXAMPLE 21.5   The two peaks on either side of the central frequency are at 8.657×10^{13} Hz and 8.483×10^{13} Hz in the absorption spectrum of a gas of CH molecules. The mass of C is exactly 12 u and the mass of H is 1.008 u. Find the equilibrium separation and the force constant of CH.

Solution We are given the two $l' = 0$ frequencies; that is

$$8.657 \times 10^{13} \text{ Hz} = v_c + (0 + 1)\hbar/2\pi I$$

and

$$8.483 \times 10^{13} \text{ Hz} = v_c - (0 + 1)\hbar/2\pi I.$$

Subtracting the two equations, we get

$$0.174 \times 10^{13} \text{ Hz} = 0 + 2(0 + 1)\hbar/2\pi I.$$

Solving for I yields

$$I = \frac{\hbar}{\pi(0.174 \times 10^{13}\,\text{Hz})} = \frac{1.0546 \times 10^{-34}\,\text{J} \cdot \text{s}}{0.174\pi \times 10^{13}\,\text{Hz}} = 1.929 \times 10^{-47}\,\text{kg} \cdot \text{m}^2,$$

using $\text{J} \cdot \text{s} = \text{kg} \cdot \text{m}^2/\text{s}$ and $\text{Hz} = \text{s}^{-1}$.

Since $I = m_r r_0^2$, we need m_r before we can calculate r_0:

$$m_r = \frac{m_1 m_2}{m_1 + m_2} = \frac{(12 \text{ u})(1.008 \text{ u})}{12 \text{ u} + 1.008 \text{ u}} = (0.9299 \text{ u})(1.6605 \times 10^{-27} \text{ kg/u})$$
$$= 1.544 \times 10^{-27} \text{ kg}.$$

Then

$$r_0 = \sqrt{\frac{I}{m_r}} = \sqrt{\frac{1.929 \times 10^{-47} \text{ kg} \cdot \text{m}^2}{1.544 \times 10^{-27} \text{ kg}}} = 1.12 \times 10^{-10} \text{ m} = 0.112 \text{ nm}.$$

Adding the two frequency equations yields 1.7140×10^{14} Hz $= 2\nu_c + 0$, or 8.570×10^{13} Hz $= \nu_c$. Then $\nu_c = \omega_c/2\pi = \sqrt{k'/m_r}/2\pi$ becomes

$$k' = 4\pi^2 \nu_c^2 m_r = 4\pi^2 (8.570 \times 10^{13} \text{ Hz})^2 (1.544 \times 10^{-27} \text{ kg}) = 448 \text{ N/m}.$$

●

21.5
VARIATIONS FROM THE MODEL

The actual spectra of a diatomic molecule may vary to a greater or lesser degree from the predictions of the simple model used in Section 21.4. We'll consider three reasons for the variation: isotopes, anharmonic vibrations, and rotation.

First, isotopes of the same atom have different masses (because they have different numbers of neutrons in the nucleus). The neutral neutron isn't expected to cause much change in the force constant. But isotopes having different masses will result in different reduced masses, vibrational energies, rotational inertia, and rotational energies—and therefore in different emission and absorption spectra. This difference won't be as great between adjacent heavy atom isotopes, such as ^{234}U ($m = 234.04$ u) and ^{235}U ($m = 235.04$ u), as it will between isotopes of lighter atoms.

EXAMPLE 21.6   Where will we find the two peaks on either side of the central frequency in the absorption spectrum of a gas of CH molecules if the carbon is ^{12}C and the hydrogen is ^{3}H? A radioactive isotope, ^{3}H has a mass of 3.016 u.

Solution For ^{12}C^{3}H,

$$m_r = \frac{m_1 m_2}{m_1 + m_2} = \frac{(12 \text{ u})(3.016 \text{ u})}{12 \text{ u} + 3.016 \text{ u}} (1.6605 \times 10^{-27} \text{ kg/u})$$

$$= 4.002 \times 10^{-27} \text{ kg}.$$

Assuming that the k' we obtained in Example 21.5 is unchanged, we have

$$\nu_c = \frac{\sqrt{k'/m_r}}{2\pi} = \frac{\sqrt{(448 \text{ N/m})/4.002 \times 10^{-27} \text{ kg}}}{2\pi} = 5.325 \times 10^{13} \text{ Hz}.$$

We further assume that the same forces will also result in the same equilib-

rium separation, or $r_0 = 0.112$ nm, from Example 21.5.
Then

$$I = m_r r_0^2 = (4.002 \times 10^{-27} \text{ kg})(0.112 \times 10^{-9} \text{ m})^2$$
$$= 5.020 \times 10^{-47} \text{ kg} \cdot \text{m}^2$$

and

$$\frac{\hbar}{2\pi I} = \frac{1.0546 \times 10^{-34} \text{ J} \cdot \text{s}}{2\pi(5.020 \times 10^{-47} \text{ kg} \cdot \text{m}^2)} = 3.3 \times 10^{11} \text{ Hz}.$$

Finally, our answer is that the two $l' = 0$ lines are at

$$\nu = \nu_c \pm \frac{\hbar}{2\pi I} = 5.325 \times 10^{13} \text{ Hz} \pm 3.3 \times 10^{11} \text{ Hz}$$

or

$$\nu = 5.358 \times 10^{13} \text{ Hz} \quad \text{and} \quad \nu = 5.292 \times 10^{13} \text{ Hz}. \qquad \bullet$$

The two frequencies calculated for $^{12}C^3H$ in Example 21.6 differ greatly from the corresponding two frequencies for $^{12}C^1H$ given in Example 21.5. This large difference occurs because the isotope ^3H has almost three times the mass of the isotope ^1H.

Second, with regard to anharmonic vibrations, we replaced the actual potential energy with $V = \frac{1}{2}k'x^2 - V_0$. This expression will yield poorer approximations for higher energy states. Therefore the effective force constants and the vibrational energies will differ from what we've shown. Also, the shape of the typical potential energy curve (Figure 21.2) shows us that it's usually much harder for atoms to approach closer than r_0 than it is to move away from r_0. (Remember that the slope of the potential energy curve gives the negative of the force.) Therefore the average separation of the nuclei usually increases for higher vibrational energies (which is why most substances expand when heated). Increasing r_0 increases I and changes the rotational energy.

And third, as a diatomic molecule rotates faster, more of the attractive force is needed to provide centripetal force. Therefore not as much attractive force is available to balance the repulsive force, and the nuclei are pushed farther apart. That is, r_0 increases, which increases I and changes k'. This effect is relatively small for light molecules and negligible for heavy molecules.

In this chapter we presented a basis for understanding molecular energy levels and spectra by examining a simple quantum model of a diatomic molecule. We found the usual harmonic oscillator energy levels for vibration of the diatomic molecule. The levels were evenly spaced, but we used the reduced mass in the expressions for the angular frequency. The reduced mass also appeared in the relation for the rotational inertia.

The rotational kinetic energy had low quantum number levels that were much more closely spaced than the vibrational levels. Another difference was that the rotational energy levels were not evenly spaced. The total allowed energy levels then resulted from the sum of the electron energies in the molecules, the vibrational energies, and the rotational energies. We determined the allowed emission and absorption frequencies by using selection rules. Finally, we discussed some reasons for experimentally measured variations from the predictions of the model.

QUESTIONS AND PROBLEMS

21.1 The atom having the smallest fraction of its mass in the nucleus is ^1H. Use data from Chapter 12 to calculate this fraction for ^1H.

21.2 Will the gravitational forces holding a molecule together still be negligible a) for molecules with atoms having hundreds of times the mass of ^1H? b) on planets with much higher g's?

21.3 For vibrational energy in a stationary state, $KE + V =$ constant. Draw horizontal lines on Fig. 21.2 for two stationary states. Approximate the average separation for these states. How does the average separation change with increasing kinetic energy?

21.4 The force constant for the most common form of HCl is 516 N/m. The masses of H and Cl are 1.008 u and 34.969 u, respectively. Calculate the ground state and first excited state vibrational energies.

21.5 The second most common form of HCl has a Cl isotope with a mass of 36.966 u. How does this change the answers to Problem 21.4?

21.6 A table shows that the difference between the vibrational energy of the ground state and the first excited state for I_2 is 26.60 meV and that the force constant is 172 N/m. What would be the mass of an I atom in unified atomic mass units (u)?

21.7 The force constant for CO is 1902 N/m. Take the atomic masses of C and O to be 12 u (exactly) and 15.995 u, respectively. Calculate the minimum vibrational energy of a CO molecule.

21.8 Explain each step in the derivation of $I = m_r r_0^2$.

21.9 Derive an expression for the angular speed of rotation of a diatomic molecule in terms of the rotational angular momentum quantum number.

21.10 For the ground and first excited rotational levels of H_2, calculate a) the angular speed of rotation and b) the tangential speed (from $v = \omega r$) of the protons.

21.11 The equilibrium separation for N_2 is 0.1098 nm. The most common isotope of N is ^{14}N, with a mass of 14.003 u.

a) Calculate the energy of the third excited rotational level of $^{14}N_2$.

b) Some 0.37% of N is ^{15}N, with a mass of 15.000 u. Calculate the energy of the third excited rotational levels of $^{15}N_2$ and $^{14}N^{15}N$, assuming the same equilibrium separation for all N_2.

21.12 For common HCl, the atomic masses of H and Cl are 1.008 u and 34.969 u, respectively, and the energy of the first excited rotational level of the molecule is 2.62 meV. Calculate the equilibrium separation.

21.13 Consider the ground state vibrational and rotational energies. Which may be zero? Which may not be zero?

21.14 Are the values of the rotational angular momentum quantum number discussed in this chapter limited to 0, 1, 2, . . . , $n - 1$, where n is the vibrational quantum number?

21.15 Find the expression for the rotational energy in terms of the rotational angular momentum quantum number, Planck's constant, the force constant, the classical vibrational frequency, the equilibrium separation, and constants.

21.16 Consider E_e, E_v, and E_r in the harmonic oscillator model of a diatomic molecule. Which energy levels are evenly spaced and which aren't? Explain.

21.17 Explain why H_2 has one of the highest classical vibrational frequencies.

21.18 Explain why the minimum, nonzero rotational energy for H_2 is the largest of all diatomic molecules.

21.19 a) For what type(s) of diatomic molecules will Eq. (21.6) not hold true? b) Equation (21.6) is based on the assumption that one of the E terms is constant. Which one?

21.20 How is l' in Eq. (21.6) related to l for absorption when a) E_r increases? b) E_r decreases?

21.21 Equation (21.6) was derived for absorption, assuming no change in E_e, an increase in E_v, and either an increase or a decrease in E_r. Why are these three E terms treated differently?

21.22 What selection rule forbids a) a vibration only transition? b) a rotation only transition? c) an absorption or emission line at v_c?

21.23 Is there any way the spectra of H_2, N_2, or O_2 can look like that in Fig. 21.8?

21.24 Why is the percentage isotope variation for an adjacent isotope the greatest in hydrogen?

21.25 a) Why are the equilibrium separation and force constant treated as unaffected by variations in isotope mass?

b) Would the corrections for anharmonic vibrations and rotation in excited states be the same for molecules made of different isotopes of the same atoms?

21.26 Calculate the position of the two peaks on either side of the central frequency in the absorption spectrum of a gas of $H^{35}Cl$ molecules using data from Problems 21.4 and 21.12.

21.27 Why does the rotational correction become smaller as the masses of the atoms in the molecules become larger?

21.28 Consider NaCl molecules with the masses of the atoms being 22.99 u and 34.97 u or 36.97 u, respectively. The equilibrium separation is 0.2361 nm, and the force constant is 109 N/m. Because of the two Cl isotopes, the absorption spectrum shows closely spaced lines on either side of two central frequencies. Calculate the frequency separation of those lines closest to each of the two central frequencies.

21.29 The increase in r_0 due to molecular rotation is approximately $m_r r_0 \omega^2 / k'$. Assume that this increase is small compared to r_0 and show that

$$E_r = \frac{l(l+1)\hbar^2}{2I_0}\left[1 - \frac{2l(l+1)\hbar^2}{k' m_r r_0^4}\right].$$

21.30 The absorption spectrum of $^{28}Si^{16}O$ has peaks on either side of the central frequency at 3.7191×10^{13} Hz and 3.7278×10^{13} Hz. The atomic masses are 27.977 u and 15.995 u, respectively. Calculate the equilibrium separation and the force constant.

21.31 Write the numerical description of the idealized vibration–rotation spectrum of $^{12}C^{16}O$. The equilibrium separation is 112.8 pm. Use other data from Problem 21.7.

CHAPTER *22*

Maxwell–Boltzmann Statistics

Lasers and their applications are some of the great developments of the last half of this century. In a laser, huge numbers of electrons are changing energy states. It would be impossible to figure out what each electron is doing at some instant of time. Instead, we describe the system by using statistics, which give us averages and probabilities that a particular electron will be in a particular state. The kind of statistics we'll use in this chapter is Maxwell–Boltzmann statistics. You may already know something about these statistics from a previous physics course — when you learned about the Maxwell speed distribution of the particles of an ideal gas. Since you may not have covered it, we'll discuss that distribution. We also referred briefly to a similar distribution in Section 6.2.

22.1
THE MAXWELL–BOLTZMANN DISTRIBUTION

Let's consider a very large number of particles—for example, the molecules in a gas or the free electrons in a metal. If you could examine them all at a single instant, it's highly unlikely that they'd all have the same speed or momentum or energy. In fact, we do find that there is a distribution of these speeds, momentums, and energies. This distribution depends on a number of factors, such as the type of particle and the temperature of the system.

There are three main **probability distributions:** the Maxwell–Boltzmann, the Fermi–Dirac, and the Bose–Einstein. We represent a *distribution* by the symbol $f(E)$, which stands for the average number of particles with a particular state of energy E. Suppose there are $G(E)$ states that have this energy and the average number of particles with this energy is $N(E)$. Then the average number of particles equals the number of states times the average number of particles per state, all at E. That is,

$$N(E) = G(E)f(E), \tag{22.1}$$

where $G(E)$ is called the *degeneracy* of the energy level E.

EXAMPLE 22.1   Let's say that we have an average of 4.4×10^{20} atoms, each with a total energy of -3.4 eV. The degeneracy of this level is 8 (the first excited level of the hydrogen atom). Calculate the average number of atoms for each -3.4-eV state.

Solution We know that $N(E)$, the average number of particles, is 4.4×10^{20} atoms and that $G(E)$, the degeneracy, is 8 states. Therefore from $N(E) = G(E)f(E)$ we can calculate the average number of particles per state:

$$f(E) = \frac{N(E)}{G(E)} = \frac{4.4 \times 10^{20} \text{ atoms}}{8 \text{ states}} = 5.5 \times 10^{19} \text{ atoms/state.} \quad \bullet$$

The **Maxwell–Boltzmann distribution** is derived by considering the number of ways that a total number of distinguishable, but otherwise identical, particles can be distributed. A macroscopic example of this type of particle would be the balls used in the game of pool. Each ball is identical in terms of mass, radius, and elasticity, but is distinguishable by the markings on it. Systems that are well-described by classical physics obey Maxwell–Boltzmann statistics. Examples include virtually all gases over a large range of temperatures. The **Maxwell–Boltzmann distribution function,** $f_{MB}(E)$, is

$$f_{MB}(E) = Ce^{-E/kT}, \tag{22.2}$$

where C is a constant, E is the energy of the state, k is Boltzmann's constant, that is,

$$k = 1.381 \times 10^{-23} \text{ J/K} = 8.62 \times 10^{-5} \text{ eV/K}$$

and T is the absolute temperature.

EXAMPLE 22.2 A gas of hydrogen atoms is at a temperature of 4150 K (the minimum temperature within our sun). The ground energy level of hydrogen has a degeneracy of 2 and an energy of -13.60 eV. The first excited level has a degeneracy of 8 and an energy of -3.40 eV. Using Maxwell–Boltzmann statistics, calculate the ratio of the average number of atoms with an electron in the first excited level to the average number of atoms with an electron in the ground level.

Solution We want the ratio $N(-3.40 \text{ eV})/N(-13.60 \text{ eV})$. Remember that $N(E) = G(E)f(E)$. We are given that $G(-13.60 \text{ eV}) = 2$, that $G(-3.40 \text{ eV}) = 8$. Here $f(E) = f_{MB}(E) = Ce^{-E/kT}$. Since T is given as an absolute temperature (in K, not °C or °F), $kT = (8.62 \times 10^{-5} \text{ eV/K})(4150 \text{ K}) = 0.3577$ eV. Therefore the ratio is

$$\frac{N(-3.40 \text{ eV})}{N(-13.60 \text{ eV})} = \frac{G(-3.40 \text{ eV})f_{MB}(-3.40 \text{ eV})}{G(-13.60 \text{ eV})f_{MB}(-13.60 \text{ eV})} = \frac{8Ce^{-(-3.40/0.3577)}}{2Ce^{-(-13.60/0.3577)}}$$

$$= 4e^{(3.40-13.60)/0.3577} = 4e^{-28.51} = 1.66 \times 10^{-12}. \qquad \bullet$$

Example 22.2 showed that there are, on average, only about five hydrogen atoms in their first excited level for every 3×10^{12} hydrogen atoms in the ground level, even at this elevated temperature. We have to apply quantum concepts to find the allowed energies of *each* atom. However, the collection of atoms makes up a classical gas, so we can apply Maxwell–Boltzmann statistics to the *system* of hydrogen atoms.

Example 22.2 showed us that for Maxwell–Boltzmann particles,

$$\frac{N_2}{N_1} = \frac{G_2}{G_1} e^{-\Delta E/kT}, \tag{22.3}$$

where $\Delta E = E_2 - E_1$. As illustrated in Fig. 22.1, N_2, G_2, and E_2 are the average number of particles, the degeneracy, and the energy of the upper energy level. Similarly, N_1, G_1, and E_1 are those terms for the lower level. And ΔE is the energy separation of the levels.

The ratio N_2/N_1 approaches zero as T approaches zero. That is, at absolute zero all particles are in their lowest possible energy levels. The other extreme is as T approaches infinity. Then N_2/N_1 approaches G_2/G_1. Particles are thermally excited from their lowest energy levels to higher

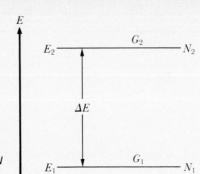

FIGURE 22.1
Two energy levels, with energy E, degeneracy G, and number of particles N separated by ΔE.

energy levels as the temperature increases. If $G_2 = G_1$, there can be no more particles in the upper level than there are in the lower level (when the system is in thermal equilibrium).

EXAMPLE 22.3   Derive the expression for the energy of N identical particles spread over two levels of equal degeneracy at a temperature T. The levels are separated by ΔE and the particles are distinguishable.

Solution With distinguishable particles, we can use $f_{MB}(E)$. With equal degeneracy, $G_2 = G_1$. Therefore $N_2/N_1 = (G_2/G_1)e^{-\Delta E/kT}$ becomes $N_2 e^{\Delta E/kT} = N_1$. The N particles are divided, with N_1 having energy E_1 and N_2 having energy E_2. Therefore $N = N_1 + N_2$, or $N - N_2 = N_1$. To get the total energy, we add the average number of particles with energy E_1 times E_1 to the average number of particles with energy E_2 times E_2. That is, $E = N_1 E_1 + N_2 E_2$. Since $N_1 = N - N_2$,

$$E = (N - N_2)E_1 + N_2 E_2 = NE_1 + N_2(E_2 - E_1) = NE_1 + N_2 \Delta E.$$

Since $N_1 = N_2 e^{\Delta E/kT}$, we have $N = N_1 + N_2 = N_2(e^{\Delta E/kT} + 1)$, or $N/(e^{\Delta E/kT} + 1) = N_2$. Finally, we substitute this N_2 into $E = NE_1 + N_2 \Delta E$ to obtain

$$E = N\left(E_1 + \frac{\Delta E}{e^{\Delta E/kT} + 1}\right). \qquad \bullet$$

22.2
THE MAXWELL SPEED DISTRIBUTION

One important application of the Maxwell–Boltzmann distribution is finding the distribution of molecular speeds in an ideal gas. The first step is to move from $N(E) = G(E)f(E)$, the expression for discrete quantized energy

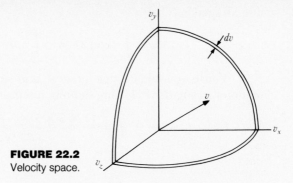

FIGURE 22.2
Velocity space.

levels, to

$$n(v) = g(v)f(v). \tag{22.4}$$

Equation (22.4) is the expression for a continuous distribution of allowed speeds, v, where $n(v)$ is the average number of particles per unit speed range and $g(v)$ is the average number of states per unit speed range. Therefore $n(v)\,dv$ will be the average number of particles with speeds between v and $v + dv$, and $g(v)\,dv$ will be the number of states in that same range of speeds. The relation $n(v) = g(v)f(v)$ will also be accurate for discrete quantized levels that are very close to one another.

We first need to determine $g(v)$. Suppose that we set up a "velocity space," such as that shown in Fig. 22.2. Assuming that one velocity is as attainable as any other, we can say that the number of speed states between v and $v + dv$ is proportional to the volume of a spherical shell of radius v and thickness dv. This volume is $4\pi v^2\,dv$. Therefore $g(v) = Bv^2$, where B is a proportionality constant that includes the 4π.

EXAMPLE 22.4   Recall that $E = \frac{1}{2}mv^2$ for the translational kinetic energy of ideal gas molecules. What is an expression for the average number of particles per unit speed range?

Solution   Substituting $E = \frac{1}{2}mv^2$ into $f_{MB}(E) = Ce^{-E/kT}$ gives $f(v) = Ce^{-mv^2/2kT}$. With $g(v) = Bv^2$, $n(v) = g(v)f(v)$ becomes $n(v) = Av^2e^{-mv^2/2kT}$, where $A = BC$.   ●

The answer in Example 22.4 immediately leads us to determine the constant A. Since $n(v)\,dv$ gives the average number of molecules with speeds between v and $v + dv$, the integral of $n(v)\,dv$ from v_1 to v_2 would give us the average number of molecules in that speed range. Let N be the total number of molecules. To make sure that we get all N molecules, we could integrate $n(v)\,dv$ from zero to infinity. This upper limit doesn't mean that any molecule has infinite speed; it's just a mathematical way of including all speeds. That is,

$$N = \int_0^\infty n(v)\, dv = A \int_0^\infty v^2 e^{-mv^2/2kT}\, dv.$$

You may evaluate this integral, and similar ones that you will encounter, using the following definite integrals from math tables:

$$\int_0^\infty x^{2n} e^{-ax^2}\, dx = \frac{[1 \times 3 \times 5 \times \cdots \times (2n-1)]\sqrt{\pi}}{(2^{n+1})(a^{n+(1/2)})}$$

and

$$\int_0^\infty x^{2n+1} e^{-ax^2}\, dx = \frac{n!}{2a^{n+1}},$$

where $a > 0$ and $n =$ an integer. Using the first definite integral with $x = v$, $n = 1$, and $a = m/2kT$, we get $N = A[1]\sqrt{\pi}/[(2^2)(m/2kT)^{3/2}]$. Solving for A and substituting into the answer in Example 22.4, we arrive at the **Maxwell speed distribution** for an ideal gas:

$$n(v) = \left(\frac{m}{kT}\right)^{3/2} \sqrt{\frac{2}{\pi}}\, Nv^2 e^{-mv^2/2kT}. \tag{22.5}$$

This distribution, divided by N, is plotted in Fig. 22.3 for two different temperatures.

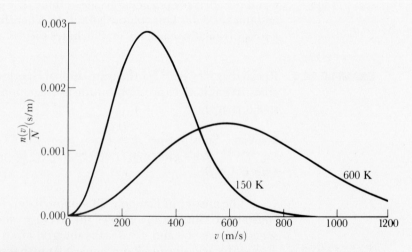

FIGURE 22.3
The Maxwell speed distribution divided by N for air molecules ($m \approx 30\ \mu$) at 150 K and 600 K. Note that increasing T by a factor of 4 makes the curve "twice as wide and half as tall."

TABLE 22.1

The Average Fraction of Ideal Gas Molecules with Speeds between Zero and the Speeds Shown

$v/\sqrt{kT/m}$	Fraction	$v/\sqrt{kT/m}$	Fraction
0.00	0.000	1.75	0.618
0.25	0.004	2.00	0.739
0.50	0.031	2.25	0.833
0.75	0.095	2.50	0.900
1.00	0.199	3.00	0.971
1.25	0.332	3.50	0.993
1.50	0.478	4.00	0.999

EXAMPLE 22.5   What is the physical meaning of   a) $n(v)\, dv/N$?   b) $\int_0^v n(v)\, dv/N$?

Solution

a) The term $n(v)\, dv$ gives the average number of molecules with speeds between v and $v + dv$, and N is the total number of molecules. Thus $n(v)\, dv/N$ is the average fraction of the molecules with speeds between v and $v + dv$.

b) The integral of $n(v)\, dv$ from v_1 to v_2 gives the average number of molecules in that speed range, so $\int_0^v n(v)\, dv/N$ gives the average fraction of molecules with speeds between 0 and v.   ●

Table 22.1 shows values for $\int_0^v n(v)\, dv/N$, numerically evaluated on a pocket calculator with an integral button, for various speeds.

We commonly determine three special speeds for the Maxwell speed distribution. The first is the **most probable speed,** v_p, the speed at which the average number of molecules per unit speed range is the greatest. To obtain it, all you need to do is set the derivative of $n(v)$ with respect to v equal to zero and solve for the correct root. Doing so, you get

$$v_p = \sqrt{\frac{2kT}{m}}. \tag{22.6}$$

EXAMPLE 22.6   The typical air molecule has a mass of about 30 u (or a molecular mass of 30 g/mol or kg/kmol). For typical air molecules at 300 K (about room temperature)   a) what is the most probable speed?   b) what fraction of molecules have speeds of less than v_p?

Solution

a) We solve $v_p = \sqrt{2kT/m}$, with $k = 1.38 \times 10^{-23}$ J/K, $T = 300$ K, and $m = (30 \text{ u})(1.66 \times 10^{-27} \text{ kg/u}) = 4.98 \times 10^{-26}$ kg. The result is $v_p = 0.41$ km/s.

b) Again, we use $v_p = \sqrt{2kT/m}$, or $1.414\sqrt{kT/m}$. In terms of $v/\sqrt{kT/m}$, Table 22.1 gives 0.332 for 1.25 and 0.478 for 1.50. Using linear interpolation to find the fraction for 1.414, we obtain

$$0.332 + \frac{0.478 - 0.332}{1.50 - 1.25}(1.414 - 1.25) = 0.428. \qquad \bullet$$

The answer in Example 22.6 (b) is true for any ideal gas at any temperature. Some 42.8 percent, or three of every seven molecules, will have speeds of less than the most probable value.

The other two special speeds are the average speed, $\bar{v}$, and the rms speed, v_{rms}. We find the **average speed** from $\int_0^\infty vn(v)\,dv/N$ and the average (mean) of the square of the speed, similarly, from $\int_0^\infty v^2 n(v)\,dv/N$. The **root-mean-square (rms) speed,** then, is the square root of the mean of the square of the speed. Using the definite integrals in the paragraph before Eq. (22.5), we obtain

$$\bar{v} = \sqrt{\frac{8kT}{\pi m}} \qquad (22.7)$$

and

$$v_{rms} = \sqrt{\frac{3kT}{m}}. \qquad (22.8)$$

The square of the rms speed is the average value of v^2, so the average translational kinetic energy of an ideal gas molecule is $\frac{1}{2}m(v_{rms})^2 = \frac{1}{2}m(3kT/m) = \frac{3}{2}kT$. Figure 22.4 shows the relative positions of these three special speeds on a plot of the Maxwell speed distribution.

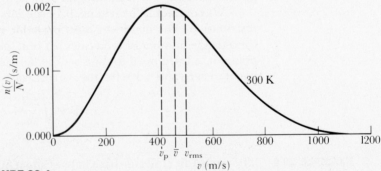

FIGURE 22.4
The Maxwell speed distribution divided by N for air molecules at room temperature. The most probable speed is 410 m/s, the average speed is 460 m/s, and the rms speed is 500 m/s. Half the molecules have speeds below 440 m/s.

EXAMPLE 22.7 What is the ratio of the rms speed to the average speed to the most probable speed of air molecules at room temperature?

Solution We want the ratio $v_{rms} : \bar{v} : v_p$. Equations (22.8), (22.7), and (22.6) tell us that this ratio is

$$\sqrt{3kT/m} : \sqrt{8kT/\pi m} : \sqrt{2kT/m} = 1.732 : 1.596 : 1.414$$

or

$$1.225 : 1.128 : 1.000. \qquad \bullet$$

Note that in the solution to Example 22.7, kT/m canceled out in the ratio. Therefore the ratio obtained applies to any ideal gas molecules at any temperature, not just air molecules at room temperature.

22.3
APPLICATION TO DIATOMIC MOLECULES

The height of the peaks in the absorption spectra of diatomic molecules depends on the temperature. The reason is that the number of molecules in the initial state determines the probability that there will be a transition from the initial to the final state. We can use the Maxwell–Boltzmann distribution to determine this number for molecules which are well separated from one another.

There's one wave function for each energy for the one-dimensional harmonic oscillator. Therefore the degeneracy, G, of each vibrational level is 1. Recall that $E_v = (n + \frac{1}{2})h\nu_c$. Then

$$\Delta E = E_2 - E_1 = (n_2 - n_1)h\nu_c = \Delta n h\nu_c.$$

Therefore $N_2/N_1 = (G_2/G_1)e^{-\Delta E/kT}$ becomes (with $G_2 = 1 = G_1$)

$$\frac{N_2}{N_1} = e^{-\Delta n h\nu_c/kT} \qquad (22.9)$$

for the vibrational levels of a gas of diatomic molecules in the harmonic oscillator approximation.

EXAMPLE 22.8 The absorption spectrum of $^{28}Si^{16}O$ has peaks on both sides of the central frequency at 3.7191×10^{13} Hz and 3.7278×10^{13} Hz. Calculate the number of molecules in their first excited vibrational state at a room temperature of 20 °C if 100,000 molecules are in their ground state.

Solution We solve for v_c, the central frequency, or the average of the two given frequencies. As in Example 21.5, we have

$$v_c = \tfrac{1}{2}(3.7191 + 3.7278) \times 10^{13} \text{ Hz} = 3.72345 \times 10^{13} \text{ Hz}.$$

Since $n = 0$ for the ground state and $n = 1$ for the first excited state, $\Delta n = 1 - 0 = 1$. We need T to be the absolute temperature, so $T = (20 + 273) \text{ K} = 293 \text{ K}$. Therefore

$$\frac{\Delta n h v_c}{kT} = \frac{1(4.136 \times 10^{-15} \text{ eV} \cdot \text{s})(3.72345 \times 10^{13} \text{ Hz})}{(8.62 \times 10^{-5} \text{ eV/K})(293 \text{ K})}$$

$$= \frac{0.154}{0.02526} = 6.097,$$

which gives $N_2/N_1 = e^{-6.097} = 2.25 \times 10^{-3}$. Letting $N_1 = 100,000$ we have $N_2 = (100,000)(2.25 \times 10^{-3}) = 225$. Therefore on average at room temperature, there are 225 $^{28}\text{Si}^{16}\text{O}$ molecules in their first excited vibrational state for every 100,000 in their ground state.   ●

The degeneracy of each vibrational level of a diatomic molecule is simply 1, but 1 is not the degeneracy for each of the rotational levels. There are $2l + 1$ wave functions for each value of the rotational quantum number l. These wave functions correspond to the $2l + 1$ orientations of the rotational angular momentum vector (given by $m_l = -l$ to $+l$ in steps of 1). Therefore the degeneracy of each rotational level equals $2l + 1$.

Recalling that $E_r = l(l + 1)\hbar^2/2I$, we have

$$\Delta E = E_2 - E_1 = [l_2(l_2 + 1) - l_1(l_1 + 1)]\hbar^2/2I.$$

Therefore $N_2/N_1 = (G_2/G_1)e^{-\Delta E/kT}$ becomes

$$\frac{N_2}{N_1} = \frac{2l_2 + 1}{2l_1 + 1}\, e^{-[l_2(l_2+1)-l_1(l_1+1)]\hbar^2/2IkT} \tag{22.10}$$

for the rotational levels of diatomic molecules (assuming that I remains constant).

EXAMPLE 22.9   Calculate the relative numbers of molecules in the ground rotational level and the first two excited rotational levels of $^{28}\text{Si}^{16}\text{O}$ at 20 °C from the data in Example 22.8.

Solution As in Example 21.5, we subtract the two frequencies to get $8.7 \times 10^{10} \text{ Hz} = (\hbar/\pi I)$. Next, we use Eq. (22.10). Since we know $(\hbar/\pi I)$, but not I, we can write the $\hbar^2/2I$ in the exponent as

$$\left(\frac{\hbar}{\pi I}\right)\left(\frac{\pi\hbar}{2}\right) = (8.7 \times 10^{10} \text{ Hz})\left[\frac{\pi(6.582 \times 10^{-16} \text{ eV} \cdot \text{s})}{2}\right]$$

$$= 8.99 \times 10^{-5} \text{ eV}.$$

As in Example 22.8, $kT = 0.02526$ eV, so $\hbar^2/2IkT = 0.00356$. Then for the ground level, $l = 0$. For the first excited level, $l = 1$. For the second excited level, $l = 2$. Therefore

$$\frac{N_1}{N_g} = \left[\frac{2(1) + 1}{2(0) + 1}\right] e^{-[1(1+1)-0(0+1)](0.00356)} = 3e^{-0.0071} = 3.0.$$

Also,

$$\frac{N_2}{N_g} = \left[\frac{2(2) + 1}{2(0) + 1}\right] e^{-[2(2+1)-0(0+1)](0.00356)} = 5e^{-0.0214} = 4.9.$$

Therefore for every 10 $^{28}\text{Si}^{16}\text{O}$ molecules in the rotational ground level at 20 °C, another 30 will be in the first excited rotational level and 49 in the second excited rotational level, on average. ●

   Continuing this solution to the higher excited rotational levels gives numbers of molecules that continually decrease following the 144 molecules in the $l = 11$ level. Therefore the $l = 11$ level is the most highly occupied, and the $l = 11$ to $l = 10$, or 12, transitions will be the most probable. Thus the lines in the spectrum that correspond to these transitions will show the greatest emission or absorption of all.

22.4
LASERS

A **laser** is a device that emits highly monochromatic, well-collimated beams of coherent light. In other words, the light waves from a laser have only a small variation about a single frequency or wavelength, the waves do not spread out much in space, and the waves retain their phase differences with one another.

   To see how a laser works, let's consider two electron energy levels in equilibrium. The relation $f = Ce^{-E/kT}$ gives $f_2/f_1 = e^{-\Delta E/kT}$. If $\Delta E = 2$ eV (corresponding to $\lambda = 620$ nm, which is red light) and $kT = 0.20$ eV (corresponding to $T = 2300$ K, a glowing gas temperature), then

$f_2 = e^{-10}f_1 = 45 \times 10^{-6}f_1$. That is, there will be, on average, only 45 electrons per state in the higher level for every million electrons per state in the lower level.

Photons can be absorbed by electrons, exciting them from the lower states (if $h\nu = \Delta E$). This process is called **stimulated absorption.** Once an electron is excited, it can naturally return to a lower state in a random process with an average characteristic lifetime. This natural return is called **spontaneous emisson.**

The excited electron can also be influenced to return to a lower state in a shorter time. In **stimulated emission** a photon of energy $h\nu = \Delta E$ passes by and, because of a resonance effect, stimulates transition to a state ΔE lower. This transition yields the emission of a second photon of the same energy, frequency, and wavelength. In short, as illustrated in Fig. 22.5, one photon goes in but two photons come out in the process of stimulated emission. Therefore the light's energy has been amplified, which is the basis for operation of the laser. The very word *laser* is an acronym for **l**ight **a**mplification by the **s**timulated **e**mission of **r**adiation.

In addition to all his other triumphs, Einstein is also linked to laser theory. In 1916 he studied the relative probabilities involved in emission and absorption. He did so not by using quantum mechanics (which hadn't yet been invented), but by considering a gas of photons in dynamic equilibrium with a gas of N atoms. The result from Einstein's study that we're interested in here is that the probabilities of stimulated emission or absorption are proportional to the Maxwell–Boltzmann distribution function, $f_{MB}(E)$. (Problem 22.34 can guide you to his result.) Therefore the probability of a stimulated transition is directly proportional to the average number of electrons per state in the initial level.

The reason that all materials don't exhibit laser action can be seen in our previous calculation for f_2 near the beginning of this section. Even at 2300 K, the great majority of the electrons are in their lower energy states. Therefore a photon interacting with an electron is much more likely to find an electron in a lower state and be absorbed than it is to find an electron in an excited state and to stimulate an emission. In our

FIGURE 22.5

Transitions between two energy levels involving photons.

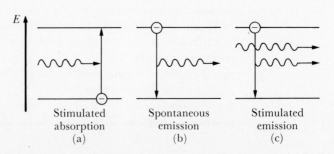

Stimulated absorption
(a)

Spontaneous emission
(b)

Stimulated emission
(c)

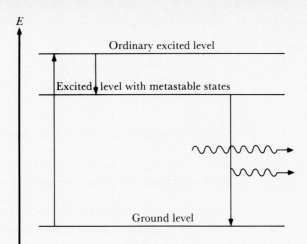

FIGURE 22.6
Stimulated emission in a
three-level laser.

$f_2 = (45 \times 10^{-6})f_1$ example, stimulated absorption is more probable than stimulated emission by a ratio of one million to forty-five.

What's needed then are more electrons per state in the higher levels than in the lower levels, or $f_2 > f_1$. This is called a **population inversion,** where *population* means the average number of electrons per state. One way that it can be obtained is in a **three-level laser,** as shown in Fig. 22.6. In a three-level laser, electrons are excited (pumped) from their ground levels to an ordinary excited level by absorbing a photon, or through collisions or other processes. The electrons then spontaneously decay (within a time of 10^{-8} s or so) to a much longer lived excited level. A long lifetime state in this level is called a **metastable state.** An electron has a long lifetime in a metastable state because its transition to a ground state is a "forbidden" one.

This long lifetime causes the number of electrons to "pile up" in the metastable states until there are more electrons per state in the metastable states than in the ground states. (People waiting in a series of lines pile up in the line where the transaction takes the longest.) With more electrons per state in the metastable states, a population inversion has been obtained. Then stimulated emission is more probable than stimulated absorption, and laser action is possible.

There are also other types of lasers. For example, the population inversion in a semiconductor laser is obtained by driving electrons and holes to a p–n junction (which we'll discuss in Chapters 26 and 27) with a dc electric field. In a chemical laser, a chemical reaction forms molecules in metastable excited states. In a CO_2 gas dynamic laser, the population inversion results from a rapid expansion of the gas. There are also free electron lasers, apparent natural lasers in interstellar space and the Mar-

FIGURE 22.7
A laser tube. Photons that escape through the sides have had little chance of stimulating photon emission.

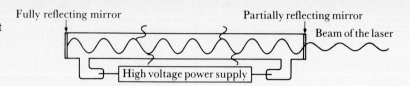

Fully reflecting mirror   Partially reflecting mirror

Beam of the laser

High voltage power supply

tian atmosphere, and even x-ray lasers pumped by small nuclear explosions.

Laser light is highly monochromatic because the stimulated emission is a resonance effect, occurring only when $h\nu = \Delta E$. Laser light is coherent because the stimulating and emitted photons must be in phase (otherwise they destructively interfere). Then these two photons can each stimulate two more photons, also in phase, and so on. Laser light doesn't diverge much because it usually comes from a resonant cavity with parallel mirrors (at least one only partially reflecting) at both ends, as shown in Fig. 22.7.

The laser that you are probably most familiar with is the **He–Ne gas laser,** with its 632.8-nm red light. In this laser, an electrical current is passed through a mixture of He and Ne gases, exciting He atoms to their metastable 1s2s energy level. This excitation energy, plus a little kinetic

FIGURE 22.8
Some energy levels in the helium–neon (He–Ne) laser.

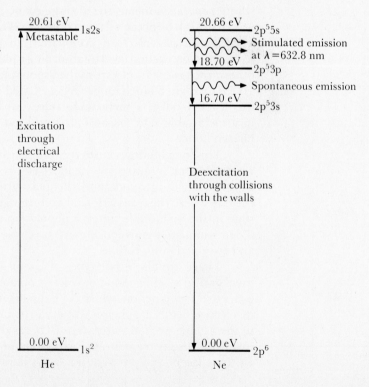

energy, can be transferred through inelastic collisions to Ne atoms to give a population inversion in Ne $2p^55s$ excited levels, although these levels are not metastable. The 632.8-nm laser transition is shown in Fig. 22.8. There are other collision transfer and deexcitation paths. Even those few in the figure indicate the low efficiency of a He–Ne gas laser. Only about 10^{-3} percent of the input energy actually is radiated out as 632.8-nm laser light.

A **maser** works like a laser, except that the energy levels are much closer together so that the frequency emitted is in the microwave range. (*Maser* is an acronym for **m**icrowave **a**mplification by the **s**timulated **e**mission of **r**adiation.) Research is underway to attempt to extend the concept to other frequency ranges, for instance the gamma-ray region of the spectrum (to be called a graser?).

EXAMPLE 22.10   Calculate the apparent absolute temperature required for a population inversion.

Solution For a population inversion, $f_2/f_1 > 1$. Since $f_2/f_1 = e^{-\Delta E/kT}$, this means that $(>1) = e^{-\Delta E/kT}$. Taking the natural log of both sides of the equation gives $\ln(>1) = -\Delta E/kT$, or $T = -\Delta E/[k \ln(>1)]$. Since ΔE, k, and the ln of any number greater than 1 are all positive, the absolute temperature T *appears* to be negative!   ●

But absolute zero is the lowest possible temperature, and as $T \to \infty$, $f_2/f_1 \to 1$. Therefore we could say that this apparent negative absolute temperature is "hotter than infinity."

Actually, there are *no* negative absolute temperatures. The problem here is that we tried to apply the equilibrium concepts of temperature and the Maxwell–Boltzmann distribution function to a nonequilibrium situation. Metastable states are long-lived, but they'll eventually decay to the ground state. As they do, the distribution will approach the Maxwell–Boltzmann distribution for the real temperature of the system.

Fluorescence and phosphorescence are somewhat related to our discussion of lasers. In **fluorescence,** high frequency photons are absorbed to bring electrons to high energy states. The electrons then drop back to the ground states in a series of steps, emitting a corresponding series of lower frequency photons. For example, the high frequency ultraviolet radiation emitted by the mercury gas inside the tube of a fluorescent light is absorbed by the coating on the tube. Light is then reemitted by the coating in the lower frequency visible portion of the spectrum. In **phosphorescence,** electrons are excited to metastable states having unusually long lifetimes. Therefore the phosphorescent material is able to continue to emit light long after excitation to any metastable state has ceased.

In this chapter we introduced the classical Maxwell–Boltzmann distribution and explained why it can be used with some systems having energies determined by quantum mechanics. We applied that distribution to the Maxwell speed distribution for the molecules of an ideal gas, to the vibrational and rotational energies of diatomic molecules, and to the operation of a three-level laser. We discussed the requirements for laser operation and the properties of laser light. Finally, we briefly described fluorescence and phosphorescence.

QUESTIONS AND PROBLEMS

22.1 Is there a situation in which all of the particles in a Maxwell-Boltzmann distribution have the same energy?

22.2 The average number of hydrogen atoms in each ground state is 26×10^{26}. What is the average number of hydrogen atoms having the ground state energy?

22.3 If the average number of molecules per state at an energy of -3.752 eV is 2.2×10^{22} and the average number of molecules with an energy of -3.752 eV is 1.1×10^{23}, what is the degeneracy of this energy level?

22.4 At what temperature are the populations of the ground level and the first excited level equal in a collection of hydrogen atoms?

22.5 The energy levels of the hydrogen atom can be explained only by quantum physics. Why then can classical Maxwell–Boltzmann statistics be applied to a gas of hydrogen atoms?

22.6 a) Derive the expression for the fraction of particles in the upper of two levels, N_2/N, as a function of temperature.

b) Graph that function for equal degeneracies as a function of the temperature in units of $\Delta E/k$.

22.7 Continue Example 22.3 to find the heat capacity per particle, $(dE/dT)/N$. At what temperature does this heat capacity have its maximum? How does the heat capacity depend on temperature when $T \gg \Delta E/k$?

22.8 The degeneracies of the levels for a particle in a one-dimensional box all equal 1. Give the ratios of the populations of two levels in terms of the quantum numbers, particle mass, and box width.

22.9 Consider a system of N atoms that exhibit a normal Zeeman effect. For a p state, calculate the number of those N atoms in each of the three orbital magnetic quantum number levels as a function of the magnetic field.

22.10 The spin-up and spin-down levels of an s state are separated by $(1.159 \times 10^{-4}$ eV/T$)B$, where B is the magnetic field. For a 0.350-T magnetic field, what is the ratio of the populations of the two levels at   a) 300 K? b) 77 K? c) 4.2 K?

22.11 Spin paramagnetism: In a magnetic field B, the net magnetic dipole moment of a collection of atoms that have only a spin magnetic dipole moment is approximately $(N_1 - N_2)\mu_B$, and the difference in energy is approximately $2\mu_B B$.

a) Find the expression for the net magnetic dipole moment of N atoms as a function of temperature.

b) Graph that expression as a function of B/T from $B/T = 0$ to $B/T = 4$.

c) Show that for small B/T, the expression is proportional to $1/T$ (which leads to Curie's law).

d) Show that for large B/T, the expression approaches $N\mu_B$ (saturation).

22.12 Adiabatic demagnetization: Two energy levels of equal degeneracy are split apart by an energy proportional to B.

a) Explain why the average energy is less than $(E_1 + E_2)/2$ for any finite temperature.

b) Explain why the average energy increases toward $(E_1 + E_2)/2$ as B approaches zero.

c) Explain why the system will cool if the energy increases adiabatically (no heat added or removed).

d) If the two states are pure spin $\frac{1}{2}$ states, find the expression for the average energy change per atom for adiabatic demagnetization (see Problem 22.10).

22.13 Consider the temperature to be constant so that the atmospheric pressure, p, is proportional to the number of atoms. Derive the relationship $p = p_0 e^{-mgy/kT}$.

22.14 Find the median speed of the molecules of an ideal gas in terms of k, T, and m.

22.15 The magnitude of the escape velocity at the surface of the moon is 2.4 km/s. a) At what temperature would 10% of all H_2 molecules ($m = 2.0$ u) have a speed greater than 2.4 km/s? b) Repeat part (a) for O_2 molecules ($m = 32.0$ u).

22.16 Derive Eq. (22.6).

22.17 Derive Eq. (22.7).

22.18 Derive Eq. (22.8).

22.19 Consider a gas of four molecules, with speeds of 100 m/s, 200 m/s, 200 m/s, and 400 m/s. Find the most probable, average, and rms speeds of this gas.

22.20 Find the fraction of molecules of an ideal gas with speeds greater than the average speed.

22.21 At sufficiently low temperatures, almost all the molecules in a gas are found in the lowest vibrational level. Therefore almost none of the heat added to the gas will go to increasing the vibrational energy, making the vibrational contribution to the heat capacity practically zero. To get an idea of the region in which the vibrational contribution to the heat capacity of H_2 gas begins to rise much above zero, calculate the temperature at which the ratio of the number of molecules in the first excited vibrational state to the number of molecules in the vibrational ground state equals 1%. Use the data from Example 21.2.

22.22 From the data of Example 21.5, determine the relative populations of the vibrational ground state and first two excited states of a gas of $^{12}C^{1}H$ molecules at 500 K. Assume that the one-dimensional harmonic oscillator model holds.

22.23 Repeat Problem 22.22 for $^{12}C^{3}H$. (See Example 21.6.)

22.24 The force constant for $^{1}H^{35}Cl$ is 516 N/m, and the masses of the atoms are 1.008 u and 34.969 u, respectively. Calculate the temperature that would be required so that there would be one $^{1}H^{35}Cl$ molecule in the first excited vibrational state for every four in the vibrational ground state.

22.25 Prove that if the ratio of the number of molecules in the first excited vibrational state of a diatomic molecule to the number in the ground state is r, then the ratio of the number in the nth excited state to the ground state number is r^n. What assumption or assumptions underlie your answer?

22.26 Repeat Problem 22.21 for the rotational energy levels of H_2, using data from Example 21.4.

22.27 Let N_l and N_g refer to excited and ground rotational levels respectively.
  a) Show that $N_l/N_g = (2l + 1)e^{-[l(l+1)]\hbar^2/2IkT}$.
  b) Use standard calculus techniques to show that the most populated level, l_{mp}, at temperature T is $l_{mp} = \sqrt{IkT}/\hbar - \frac{1}{2}$.
  c) What significance does l_{mp} have for the absorption spectrum?

22.28 Continue Example 22.9 for the 10th, 11th, and 12th excited levels.

22.29 At what temperature do the ground and second rotational excited levels have the same population in $^{28}Si^{16}O$? (See Example 22.9.)

22.30 Relate the discussion of Section 21.5 to the relative populations of vibrational and rotational levels in diatomic molecules.

22.31 Using probability concepts, explain why a population inversion is needed for laser operation.

22.32 Why isn't a laser perfectly monochromatic?

22.33 We discussed stimulated and spontaneous emission, but only stimulated absorption. Is there spontaneous absorption? Explain.

22.34 Einstein's A and B coefficients: Einstein let a gas of photons (energy density per unit frequency, u_ν) be in dynamic equilibrium with a gas of N atoms. Then A_{21} was the probability/time per volume of spontaneous emission, $B_{21}u_\nu$ of stimulated emission, and $B_{12}u_\nu$ of stimulated absorption. The B's and A's are independent of T and u_ν, and u_ν is evaluated at $\nu = \Delta E/h$.
  a) Explain why "dynamic equilibrium" means that $B_{12}u_\nu N_1 = (B_{21}u_\nu + A_{21})N_2$ and solve the equation for N_1/N_2.
  b) Solve Eq. (22.3) for N_1/N_2.

c) Explain why $N_1/N_2 \rightarrow (G_1/G_2)$ and $B_{21}u_v \gg A_{21}$ at very high temperatures.

d) Then prove that $B_{12} = (G_2/G_1)B_{21}$.

e) Continue, showing that $u_v = (A_{21}/B_{21})/[\exp(hv/kT) - 1]$.

f) Use the information in Chapter 6 to find A_{21}/B_{21} in terms of h, v, and c; then find the same ratio in terms of h, ΔE, and c.

g) Using your answer to part (d), show that the ratio of the probability/time of stimulated emission to stimulated absorption is $\exp(-\Delta E/kT)$.

22.35 Strictly speaking, the requirement for a population inversion is that $f_2/f_1 > 1$. Books often state this requirement as $N_2 > N_1$, which in general, is incorrect. Under what special condition is $f_2/f_1 > 1$ the same as $N_2 > N_1$?

22.36 Example 22.10 showed that $f_2/f_1 > 1$ seems to infer a negative absolute temperature for a population inversion. Under what conditions is $N_2 > N_1$ with a positive absolute temperature? ($N_2 > N_1$ is not, in general, the condition for population inversion; see Problem 22.35.)

22.37 A particular metastable state has a 10^{-5} s lifetime.

a) Show that 10^{-5} s is a long lifetime compared to 10^{-8} s by finding a comparable lifetime proportional to an average human lifespan.

b) Use the same ratio to find out how long a second transaction would take if the first normal transaction took 1 min. Describe how the "pile-up" would occur if people first arrived one per minute for the 1-min transaction, then went to a line for the longer transaction.

22.38 What's the difference between lasers having a continuous output and those having a pulsed output?

22.39 What is the purpose of mirrors or polished surfaces at the ends of laser cavities?

22.40 Is pumping to a normal excited state the only way to get electrons to a metastable state?

22.41 Why is laser light usually a) highly monochromatic? b) coherent? c) well-collimated?

22.42 A three-level laser also emits monochromatic light that's not coherent. Why?

22.43 a) How are phosphorescence and fluorescence related to the operation of a three-level laser?

b) Does your answer mean that the light emitted in phosphorescence and fluorescence is monochromatic, coherent, and well-collimated?

22.44 Why can it be said that fluorescent dyes added to a detergent make shirts "whiter than white"?

CHAPTER

Crystals and Binding

Many years ago I wrote a computer program for a physics class (it did a Simpson's rule integration of the Maxwell speed distribution). To run it, I needed a computer that occupied most of a room and cost a couple of decades of a professor's wages. I punched cards, the computer compiled the program and it punched cards, the computer was reconfigured to run the compiled program, and hours after I started I found my inevitable programming mistake. A few years ago I wrote the same program. To run it, I used my programmable pocket calculator, which cost a couple of hours' wages. And I was able to find my mistake in a matter of minutes.

Transistors, large-scale integrated circuits (as pictured in Fig. 23.1), smaller, faster, and cheaper electronic devices—all are fruits of research into the properties of condensed matter, especially crystalline solids. We'll begin the study of solid state physics in this chapter by considering the descriptions and regularities of some types of crystal structure. We'll also discuss the forces that hold solids together. An acquaintance with crystal structure and binding is essential to understanding the fabrication and operation of some types of solid state devices.

FIGURE 23.1

A view of riverfront development from high atop a granite outcrop? No. This view of a silicon microprocessor chip was taken through a scanning electron microscope with a grain of salt in the foreground for scale. The actual chips used are salt free.

Photo courtesy of Xerox Microelectronics Center.

23.1
SOLIDS

The main states of matter are often cited as solid, liquid, gas, and plasma. (A **plasma** is a highly ionized gas.) The study of solids and liquids is called **condensed matter physics.** In a **crystalline solid,** the atoms have a regular, long-range order. Because of this regularity, any crystal has a considerable amount of microscopic symmetry. This microscopic symmetry shows up in macroscopic properties. For example, a particular class of crystals will tend to grow, or to be easily broken, into certain shapes. Also the ways in which electromagnetic, thermal, or mechanical properties change with angle relate to the atomic symmetry of the crystal. Figure 23.2 gives an example of a variation with angle.

Many seemingly solid materials don't have long-range order — only short-range order. (By short range we mean a distance of only a few atoms or less.) Examples are glass and cold butter, which we call **amorphous**

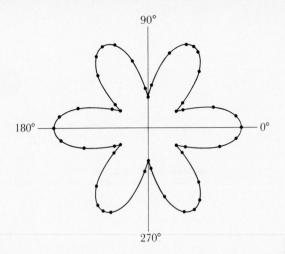

FIGURE 23.2
Experimental plot of the magnetization as a function of angle for a manganese–bismuth (MnBi) crystal at low temperature in a constant magnetic field. The MnBi crystal structure has a hexagonal atomic symmetry in a plane parallel to the magnetic field.

Data used in W. E. Stutius, Tu Chen, and T. R. Sandin, *AIP Conf. Proc.* **18**, 1222 (1974).

solids. If either glass or butter is heated, it doesn't melt at a certain temperature. It just gets softer and more "runny." That is, it gets less viscous. For this reason, some scientists don't class amorphous materials as solids at all, but rather as highly viscous liquids.

Amorphous solids have many properties that are interestingly different than their crystalline counterparts and are often easier to manufacture. For instance, there is considerable interest in making solar cells of cheap amorphous silicon rather than of expensive single crystals. Another ambiguous class of material is that of liquid crystals, which are liquids that possess some degree of long-range order.

You can mentally construct the ideal crystalline solid by imagining the continuous repetition in space of some basic structural unit. This basic structural unit may consist of just one atom (as in copper), or it may be made up of as many as thousands of atoms (as in crystalline protein). The **crystal structure** is made up of a lattice and a basis, or

$$\text{Crystal structure} = \text{Lattice} + \text{Basis}, \qquad (23.1)$$

where the **lattice** — a mathematical abstraction — is a regular set of points in space and the **basis** is a group of atoms associated with *each* lattice point. Figure 23.3 is a two dimensional representation of Eq. (23.1).

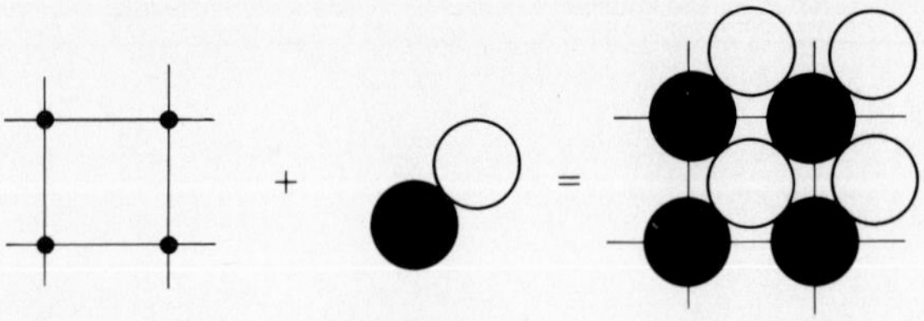

Part of a
two-dimensional
lattice

A basis of two
different atoms

Part of a
two-dimensional
crystal structure

FIGURE 23.3
The lattice plus the basis equals the crystal structure. Note that the basis is associated with
each lattice point.

EXAMPLE 23.1   How does Eq. (23.1) apply to an amorphous solid?

Solution The short-range order of an amorphous solid can group atoms
together in a basis. But since there's no long-range order, there's no
regular set of mathematical points in space to which the basis can associate.
That is, the amorphous solid doesn't have a regular crystal structure because it has no lattice.   ●

23.2
BRAGG REFLECTION

Suppose that you draw one plane through one set of atoms in a crystal and
then draw a parallel plane through the next set of atoms. The two planes
will be separated by some perpendicular distance d. Now suppose that you
send waves into that space with rays that make an angle θ with the planes,
are reflected from the planes, and have equal angles of incidence and
reflection. (We customarily measure θ with respect to the crystal planes,
rather than with respect to the normal to the surface, as you probably did
for reflection and refraction in a previous physics course.) Figure 23.4
shows that the bottom ray travels an extra distance $d \sin \theta + d \sin \theta =
2d \sin \theta$. If the extra distance equals a whole number of wavelengths, $n\lambda$,
the reflected rays will be in phase with each other, and a diffraction maximum will result. The integer n is called the **order number.** The resulting equation is called the **Bragg equation,** after the father–son team of
Sir W. H. Bragg and Sir W. L. Bragg:

$$2d \sin \theta = n\lambda, \quad \text{with } n = 1, 2, 3, \ldots \tag{23.2}$$

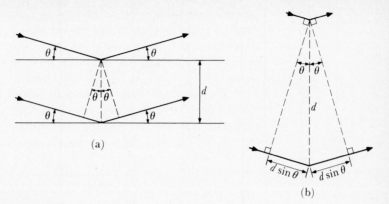

FIGURE 23.4
(a) Rays reflecting from two adjacent parallel planes of atoms in a crystal. (b) Magnification of part (a).

EXAMPLE 23.2 What is the distance between the planes if the first-order diffraction maximum occurs at $\theta = 15°$ for 150-pm x-rays?

Solution We solve $2d \sin \theta = n\lambda$ to obtain $d = n\lambda/(2 \sin \theta)$. We know that $n = 1$ for the first order, $\lambda = 150$ pm, and $\sin 15° = 0.2588$. Therefore $d = 1(150 \text{ pm})/(2 \times 0.2588) = 290$ pm. ●

We can use electrons, neutrons, and other particles, as well as x-ray photons, to determine crystal structures experimentally. Figures 10.1 and 23.5 are examples of maxima measured at different angles from different

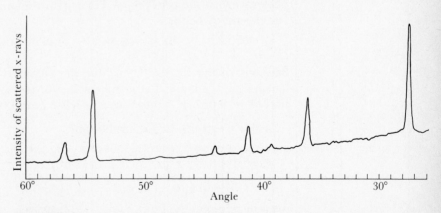

FIGURE 23.5
X-ray diffraction data using Cu $K_{\alpha 1}$ x-rays and a powdered sample. Different sets of crystal planes have different separations and so give maxima at different angles. The intensities of the maxima depend on factors such as the number of electrons in the atom, the density of atoms in the plane, and the angle.

planes of atoms in crystals. The Bragg equation, Eq. (23.2), will give the distance between planes, and the symmetry of the overall diffraction pattern will relate to the symmetry of the crystal. Finding the crystal structure from diffraction data can be tricky; there are more unknowns than equations.

We may need to obtain the wavelength for the Bragg equation from the de Broglie relationship, $\lambda = h/p$. For nonrelativistic neutrons and electrons, $KE = p^2/2m$ and for photons, $KE = pc$. In terms of the kinetic energy, we can then show the wavelengths to be

$$\lambda = \frac{1240 \text{ keV} \cdot \text{pm}}{KE} \qquad \text{for photons;} \qquad (23.3)$$

$$\lambda = \frac{28.6 \text{ eV}^{1/2} \cdot \text{pm}}{\sqrt{KE}} \qquad \text{for neutrons;} \qquad (23.4)$$

$$\lambda = \frac{1226 \text{ eV}^{1/2} \cdot \text{pm}}{\sqrt{KE}} \qquad \text{for electrons.} \qquad (23.5)$$

For example, the wavelength equals 100 pm for 12.40-keV photons, for 0.0818-eV neutrons, and for 150-eV electrons. Since the sizes of all atoms and the consequent spacing of crystal planes are typically of the order of 10^{-10} m, Å units are often used, where 1 Å $= 10^{-10}$ m. However, for consistency with the rest of the text and with SI, we'll use pm or nm. Conversions are easy: 1 Å $= 0.1$ nm $= 100$ pm.

EXAMPLE 23.3 The total scattering angle, 2θ, is to be 10.0° for a material in which crystal planes are separated by 138 pm. For a first-order maximum, what must the kinetic energy be for

a) electrons,   b) x-rays, and   c) neutrons?

Solution Solving $2d \sin \theta = n\lambda$ for the required wavelength gives $\lambda = 2d \sin \theta/n$. We know that $n = 1$ for first order, $d = 138$ pm, and $\sin \theta = \sin 5.0° = 0.08716$. Therefore $\lambda = 2(138 \text{ pm})(0.08716)/1 = 24.05$ pm.

a) For electrons, Eq. (23.5) becomes

$$KE = \left(\frac{1226 \text{ eV}^{1/2} \cdot \text{pm}}{\lambda}\right)^2 = \left(\frac{1226 \text{ eV}^{1/2} \cdot \text{pm}}{24.05 \text{ pm}}\right)^2 = 2600 \text{ eV}$$
$$= 2.60 \text{ keV}.$$

b) For x-ray photons, Eq. (23.3) becomes

$$KE = \frac{1240 \text{ keV} \cdot \text{pm}}{\lambda} = \frac{1240 \text{ keV} \cdot \text{pm}}{24.05 \text{ pm}} = 51.5 \text{ keV}.$$

c) For neutrons, Eq. (23.4) becomes

$$KE = \left(\frac{28.6 \text{ eV}^{1/2} \cdot \text{pm}}{\lambda}\right)^2 = \left(\frac{28.6 \text{ eV}^{1/2} \cdot \text{pm}}{24.05 \text{ pm}}\right)^2 = 1.41 \text{ eV.} \quad \bullet$$

Because electrons are light, charged particles, they tend to interact more with atoms and so mostly give information about the surface planes of atoms in a crystal. X-rays tend to interact more with the bulk of the sample and most may even pass completely through thin samples. Neutrons interact more with the nuclei of atoms and are especially useful for investigating magnetic interactions or for obtaining information from scattering with kinetic energy changes (called inelastic scattering).

23.3
CRYSTAL STRUCTURES

Fourteen distinct lattices are allowed by symmetry. We'll consider only the three cubic lattices. To illustrate them, we first draw the edges of three cubes in space, as in Fig. 23.6. The **simple cubic (sc) lattice** has one lattice point at each of the cube's eight corners. The **body-centered cubic (bcc) lattice** has one lattice point at each corner plus one lattice point in the center of the cubical volume. The **face-centered cubic (fcc) lattice** has one lattice point at each corner plus one lattice point in the center of each of the six faces of the cube.

But this is just the lattice. To get the entire crystal, we must add the basis to the lattice, repeating the cube over and over to make up the entire solid. The simplest basis would be one atom at each lattice point. No element has the sc crystal structure of the sc lattice plus one basis atom

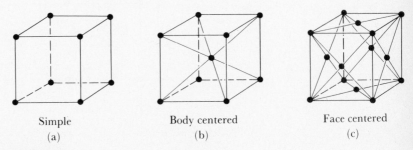

Simple (a)   Body centered (b)   Face centered (c)

FIGURE 23.6
The conventional cells of the three cubic lattices.

(except possibly under exotic laboratory conditions). However, many elements have the bcc crystal structure, with one atom at each bcc lattice point. Examples include Li and the other alkali metals, Cr, Fe, and W. Also, many elements have the fcc crystal structure, with one atom at each fcc lattice point. Examples include Al, Ca, Cu, Ag, and Au.

EXAMPLE 23.4   For Cu, the side of the cube is 361 pm. Show that $d = 361/2$ pm for planes parallel to the cube faces.

Solution The crystal structure of Cu metal is composed of an fcc lattice plus a basis of one atom at each lattice point. Therefore we see in Fig. 23.7 that we could draw one plane through the atoms at the top of the cube. The next closest parallel plane would be through the atoms in the centers of the sides, half way down the cube. These two planes are separated by 361/2 pm.   ●

Two important, related crystal structures have the face-centered cubic lattice: the diamond structure and the cubic zinc sulphide (also called the zincblende) structure. In the **diamond structure** the basis has two *identical* atoms. One of the two atoms is placed at each lattice point. The other atom is displaced by $\frac{1}{4}$ of the cube-edge distance in each of the x, y, and z directions. (The x, y, and z axes are set up parallel to the cube edges.) So if we place the origin at a lattice point, one of the two atoms of the basis would be at $(0, 0, 0)$ and the other atom at $(\frac{1}{4}, \frac{1}{4}, \frac{1}{4})$. Therefore we say that the diamond structure is composed of an fcc lattice plus a basis of one atom at $(0, 0, 0)$ and an identical atom at $(\frac{1}{4}, \frac{1}{4}, \frac{1}{4})$. The group IV elements C (in the diamond form), Si, Ge, and a form of Sn have this structure. For these elements, the cube-edge distances are 356 pm, 543 pm, 565 pm, and 646 pm, respectively.

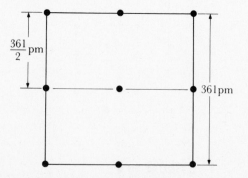

FIGURE 23.7
A side view of a conventional cell of Cu. The atoms actually fill most of the volume.

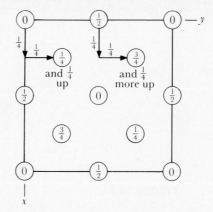

FIGURE 23.8

Looking down at the atoms in a conventional cell of diamond, with the top layer removed. The fractions are distances above the bottom layer (up, or toward you) expressed in terms of the cube-edge distance. The atoms actually occupy most of the volume of space. Silicon (Si), germanium (Ge), and gray tin (Sn) have this same crystal structure. The cubic zinc–sulfide (ZnS) structure is similar to this.

EXAMPLE 23.5   How far above the bottom face of the cube are Si atoms located?

Solution The cube-edge distance for Si is 543 pm. Figure 23.8 shows atoms in the bottom face (0); atoms $(\frac{1}{4})$543 pm $= 136$ pm above it; atoms $(\frac{1}{2})$543 pm $= 272$ pm above it; atoms $(\frac{3}{4})$543 pm $= 407$ pm above it; and atoms in the top layer 543 pm above the bottom. Then there are more layers of atoms above that as the entire crystal is built up by repetition of the cube.   ●

The **cubic zinc sulphide structure** is similar to the diamond structure. Its lattice is also fcc, and its basis is also two atoms in the same positions as in diamond. This crystal structure differs from the diamond structure by having two *different* atoms in its basis. In ZnS, one atom is Zn at $(0, 0, 0)$ and the other atom is S at $(\frac{1}{4}, \frac{1}{4}, \frac{1}{4})$, as shown in Fig. 23.9. The ZnS, SiC, and CuCl crystals and the "III–V compounds," such as GaAs, InSb, and GaP, have this structure.

EXAMPLE 23.6   If the bottom face of the cubic cell contains Ga atoms, where are the As atoms in GaAs? The cube-edge distance is 565 pm.

Solution The Ga atoms are in the bottom face of a face-centered cubic structure. Therefore a Ga atom is located at each lattice point throughout the crystal. The As atoms are then located at a displacement of $\frac{1}{4}$ cube-edge

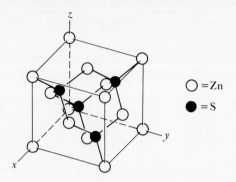

FIGURE 23.9
The cubic ZnS structure. Comparing this structure to that in Fig. 23.8, you can see that the Zn atoms have distances of 0 and $\frac{1}{2}$ above the bottom layer (as well as 1), whereas the S atoms are at the $\frac{1}{4}$ and $\frac{3}{4}$ distances.

distance in each of the x, y, and z directions from each lattice point. Therefore in Figs. 23.8 and 23.9, the As atoms will be found at $(\frac{1}{4})565$ pm $= 141$ pm and at $(\frac{3}{4})565$ pm $= 424$ pm above the bottom face. They'll also be at equivalent positions in other cubes throughout the entire crystal.   ●

23.4
BINDING

The forces that hold crystals together are mainly electrostatic. The five main types of binding are called ionic binding, covalent binding, metallic binding, hydrogen binding, and van der Waals binding. The word *bond* may also be used in these terms, as in *metallic bond*.

Ionic binding is caused by the Coulomb's law attraction of the plus and minus ions that make up an ionic crystal, as shown in Fig. 23.10. This bond is very strong. Since the potential energy of two point charges equals $(1/4\pi\epsilon_0)(q_1q_2/r)$, a charge of $+e$ and $-e$ separated by 282 pm (as in Na$^+$Cl$^-$) gives

$$PE = 9.0 \times 10^9 \frac{\text{N} \cdot \text{m}^2}{\text{C}^2} \frac{(-)(1.60 \times 10^{-19}\,\text{C})^2}{282 \times 10^{-12}\,\text{m}} \left(\frac{1\,\text{eV}}{1.60 \times 10^{-19}\,\text{J}}\right)$$

$$= -5.1\,\text{eV}.$$

Because each ion attracts more than one other ion, the average binding energy per ion pair in solid NaCl is even higher than 5.1 eV. Ionic binding

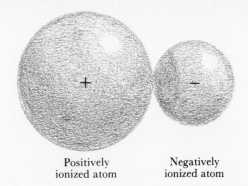

Positively ionized atom   Negatively ionized atom

FIGURE 23.10
The ionic binding of a molecule is caused by the attraction of unlike charges.

is especially important in the alkali halides (made up of positive alkali metal ions and negative halogen ions), such as NaCl, LiF, and RbF.

Covalent binding is caused by a continuous exchange of electrons between adjacent atoms, as illustrated in Fig. 23.11. This exchange results in a lowering of the total energy and therefore in binding. This bond also is very strong. In C (diamond), Si, and Ge, each atom has four electrons outside a closed shell and needs four electrons to close a p subshell. So these atoms continually share one electron with each of their four nearest neighbors.

The simplest covalent bond is in the hydrogen molecule, made up of two hydrogen atoms. The two hydrogen atoms have identical, and therefore interchangeable, nuclei and identical electrons. Let's write two possible wave functions for separated hydrogen atoms:

$$\psi_1 = \psi_{Aa}\psi_{Bb} \quad \text{and} \quad \psi_2 = \psi_{Ab}\psi_{Ba},$$

where ψ_{Aa} is the ground energy level wave function, ψ_{100}, for nucleus A surrounded by the electron probability cloud of electron a, ψ_{Bb} for nucleus B and electron b, and so on. As the atoms are brought together, the electron probability clouds begin to overlap; there's some probability then

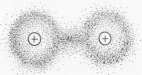

FIGURE 23.11
Covalent binding: The H_2 molecule.

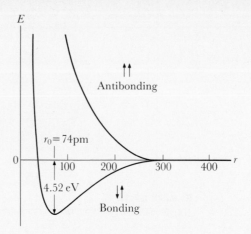

FIGURE 23.12
The energy change as two hydrogen atoms are brought together to form an H_2 molecule; $r_0 = 74.1$ pm.

that the electron from nucleus A will spend time nearer nucleus B than nucleus A. A new wave function must be constructed to recognize this condition.

Since the atoms are identical, the equally weighted linear combinations $\psi_+ = \psi_1 + \psi_2$ and $\psi_- = \psi_1 - \psi_2$ are the new wave functions. When the energy is calculated for the equilibrium separation of the nuclei, $\psi_1 + \psi_2$ gives an energy 4.52 eV lower than for the separated atoms (bonding) as shown in Fig. 23.12 but $\psi_1 - \psi_2$ gives a higher energy (antibonding).

Since both electrons are represented by $\psi_1 + \psi_2$, one electron must have spin up and the other spin down to obey the Pauli exclusion principle. Figure 23.13 shows that $|\psi_1 + \psi_2|^2$ is large between the nuclei. In general, the covalent bond between a pair of atoms usually involves one electron from each atom. The two electrons have opposite spin magnetic quantum numbers. The electrons spend just as much time near one atom as the other if the atoms are identical, but they don't do so if the atoms are different. In any case, most of their time is spent between the atoms, and it's the attraction of the positively charged remainder of the atoms to the negative electron cloud between them that results in the bonding.

EXAMPLE 23.7   The bonding of GaAs is said to be 31% ionic and 69% covalent. The reason is that an unbalanced sharing leaves a positive Ga ion and a negative As ion. Explain how this condition could occur.

Solution Evidently, the probability is such that the shared electrons will spend more time near the As than the Ga, leaving, on average, the As more

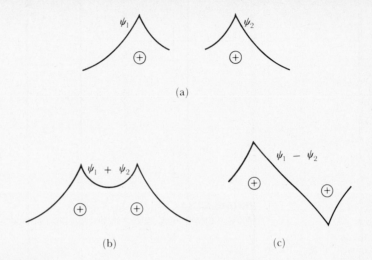

FIGURE 23.13
(a) A representation of the wave functions when the atoms are well separated. (b) The bonding wave function, for which the probability is large that the electrons are between the two nuclei. (c) The antibonding wave function, which results in a repulsion of the two atoms.

minus and the Ga more plus. Therefore ionic binding between the negative As and the positive Ga occurs, as does the covalent bonding caused by the partial sharing of the electrons.

Actually, a completely ionic bond is an ideal. Some other binding is always going on as well.

Metals are good conductors of electricity and heat because their outer (valence) electrons are free to move throughout the crystal. If an outer electron leaves, a positive ion core remains. One cause of **metallic binding** is the attraction of the negative free-electron "gas" and the positive ion cores illustrated in Fig. 23.14. Another cause is the lowering of the allowed energies of particles when they are delocalized. For example, for a particle in a one-dimensional box the allowed energies are $E_n = n^2 h^2 / 8mL^2$. Delocalizing the particle means making L larger so that the particle isn't restricted to as small a locality. Making L larger decreases E_n. This decrease in energy results in binding.

The metallic bond per pair of nearest neighbor atoms isn't as strong as ionic or covalent bonds. Nevertheless, ordinary metals are quite strong because metal crystals typically have structures with large numbers of nearest neighbors. Also, because the metallic bond isn't as localized as other types, we can say that it's more forgiving of the local defects that are always present in real solids.

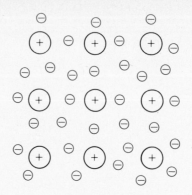

FIGURE 23.14
Metallic binding: A negative free-electron gas permeating a structure of positive ion cores.

EXAMPLE 23.8   An atom is missing in a crystal (this type of crystal defect is called a *vacancy*). Discuss the effect on the binding and mechanical strength of the crystal if the binding is mainly a) ionic,   b) covalent,   and c) metallic.

Solution

a) An ion will be missing here and therefore the surrounding ions will be missing some attraction. This decrease in attraction will lower the binding energy and leave a mechanically weak place in the crystal.

b) There will be no atom here to swap electrons with. The result again will be a lowering of the binding energy and a weak place in the crystal.

c) An ion core will be missing here, which will lower the binding energy somewhat. However, the metallic binding process isn't as localized as the ionic or covalent processes, so the "weak place" will be much more spread out and less serious.   ●

Hydrogen binding occurs when a hydrogen atom forms a bond with other atoms. In its most extreme form, the hydrogen atom loses its electron completely. Then the bare positive proton attracts negative ions, as in Fig. 23.15. Hydrogen binding is of medium energy, a few tenths of an eV per bond. For example, ice is held together primarily by hydrogen binding. The hydrogen bonds in DNA can be "zippered" open for replication of gene structure, then closed again until needed. The hydrogen bond is also important in certain ferroelectric and polymerization processes.

The weakest form of binding is **van der Waals binding,** but it occurs with all atoms. Only in the inert gas crystals (which solidify at temperatures of only a few degrees from absolute zero), though, is the van der Waals binding the predominant form of binding. Inert gas atoms are closed shell,

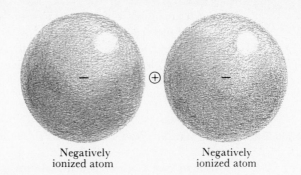

Negatively   Negatively
ionized atom   ionized atom

FIGURE 23.15
The hydrogen binding of two negative ions.

neutral atoms with strongly held valence electrons. With no hydrogen about, their closed shells rule out the other types of binding.

We can understand how two neutral atoms can attract each other by realizing that the Heisenberg uncertainty principle, $\Delta p_x \Delta x \geq h$, doesn't allow the center of the negative charge distribution (the center of the electron cloud) to be at exactly the same place as the center of the positive charge distribution (the nucleus) all the time. Therefore, at some instant of time, the plus and minus charge centers of atom 1 will be separated by some distance. This separation gives atom 1 an electric dipole moment and sets up an electric dipole field centered at atom 1. As illustrated in Fig. 23.16, atom 2 in that field will be slightly polarized and attracted toward atom 1.

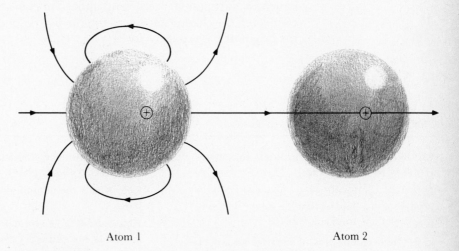

Atom 1   Atom 2

FIGURE 23.16
Atom 1 has an instantaneous electric dipole moment. The resulting dipole electric field instantaneously polarizes and attracts atom 2, resulting in the weak van der Waals binding.

The dipole electric field from atom 1 is proportional to r^{-3}, the induced electric dipole moment of atom 2 is proportional to the dipole electric field, and the electric potential energy is proportional to the product of the electric field and the electric dipole moment. Therefore the weak van der Waals binding energy is proportional to r^{-6} (r is the separation of the centers of the atoms).

EXAMPLE 23.9   Will van der Waals binding be the most important binding in Cu?

Solution   No. Copper is a metal and is held together mainly by metallic binding. However, there is some van der Waals binding of the positive ion core in metals. This weak binding contributes only a small fraction of the total binding energy.   ●

The five types of binding involve attractive forces between atoms. When atoms get close together, repulsion occurs. The equilibrium position is where the attractive and repulsive forces balance.

The **repulsive force** arises because the electron clouds begin to overlap as atoms approach one another. However, the electrons from one atom can't occupy the low energy states already occupied in another atom. (The Pauli exclusion principle forbids it.) Therefore the electrons are forced into higher energy states. If moving atoms closer together increases the energy, a repulsive force must be involved.

This chapter introduced you to amorphous and crystalline solids. You saw how to determine the separation of crystal planes, using the Bragg equation. We identified the relationship of lattice and basis to crystal structure and applied it to several important cubic crystal structures. We completed the chapter by describing the five types of crystal binding and the origin of the repulsive force.

QUESTIONS AND PROBLEMS

23.1 Why is the free-electron gas sometimes referred to as a plasma?

23.2 Find out the difference between a "single crystal" and a "polycrystalline solid." Is the polycrystalline solid amorphous?

23.3 The magnetization of a crystal has a maximum every 60° as the crystal is rotated. Do you think that the crystal structure is cubic? Explain.

23.4 Must there be an atom at each lattice point in a perfect crystal? Explain.

23.5 What's the minimum number of atoms in a basis?

23.6 Considerable symmetry is associated with the lattice. Is any symmetry associated with the basis?

23.7 The crystal structure may not have as much symmetry as the lattice. Why?

23.8 The distance between planes in an NaCl crystal is 282.0 pm. a) If the x-ray wavelength is 20.90 pm, what is the total scattering angle 2θ for a first-order diffraction maximum? b) Is this answer correct only for x-rays?

23.9 Calculate the wavelength of a) an electron, b) a photon, and c) a neutron with a kinetic energy of 10 eV.

23.10 What potential difference should electrons be accelerated through if they're to give a first-order Bragg diffraction maximum at an angle of 5.6° with crystal planes that are 326.5 pm apart?

23.11 Thermal neutrons with a kinetic energy of 0.025 eV have a first-order diffraction maximum at a total scattering angle of $2\theta = 42.4°$. Calculate the separation of the planes.

23.12 Calculate the energy of x-ray photons that give a first-order diffraction maximum at a total scattering angle of $2\theta = 9.74°$ from a crystal with a spacing of 303.6 pm between planes.

23.13 The NaCl structure has an fcc lattice, and a basis of one Na at each lattice point and one Cl at a displacement of $\frac{1}{2}$ cube-edge distance in each of the x, y, and z directions from each lattice point. Sketch a conventional cube of NaCl. Why isn't it an sc lattice?

23.14 The CsCl structure has an sc lattice, a basis of one Cs at each lattice point, and one Cl at a displacement of $\frac{1}{2}$ cube-edge distance in each of the x, y, and z directions from each lattice point. Sketch a conventional cube of CsCl. Why isn't it a bcc lattice?

23.15 In terms of Eq. (23.1), how are the diamond and cubic ZnS crystals similar? How are they different?

23.16 Why can it be said that the diamond structure is formed of two interpenetrating fcc structures separated by $(\frac{1}{4}, \frac{1}{4}, \frac{1}{4})$?

23.17 The "filling factor" is the percent of a solid that's occupied or filled by atoms (considered as spheres that touch their nearest neighbors). Calculate the filling factor for crystals that have a basis of one atom and a lattice that is a) sc, b) bcc, and c) fcc.

23.18 a) Calculate the nearest neighbor distances (nucleus to nucleus of closest atoms) in Cu. b) Calculate the next nearest neighbor distance.

23.19 a) Calculate the distance between the center of a Ga atom and the center of the nearest As atom in GaAs. b) Calculate the smallest Ga–Ga separation (center to center) in GaAs.

23.20 The electron sharing in GaAs is unbalanced. Will it also be unbalanced in Si?

23.21 In the diamond structure, one of the four covalent bonds extends from an atom in the $\hat{x} + \hat{y} + \hat{z}$ direction and another in the $\hat{x} - \hat{y} - \hat{z}$ direction ($\hat{x}$, $\hat{y}$, and $\hat{z}$ are unit vectors). Recalling that $\mathbf{A} \cdot \mathbf{B} = AB \cos \theta$, calculate the angle between the two bonds.

23.22 Each atom in the diamond structure has four nearest neighbors. Why is it reasonable that covalently bonded Si and Ge are found in the diamond structure?

23.23 Why are ordinary metals mechanically stronger than ordinary ionic or covalently bonded materials?

23.24 Quartz (crystalline SiO_2) filaments that are stronger than metal filaments of the same size are now available. How is this possible?

23.25 Graphite is said to be slippery because it forms in covalently bonded planes, with atoms in one plane attracting those in parallel planes by van der Waals forces. How does this binding explain why the planes slip across one another easily?

23.26 How can the binding be ionic, covalent, and van der Waals simultaneously?

23.27 Does the hydrogen bond hold the H_2 molecule together? Explain.

23.28 When ^2H is used in a hydrogen bonded material in place of ^1H, how does this affect the vibrational frequencies of the atoms? (The material has been "deuterated.")

23.29 The potential energy of a van der Waals bond between two atoms can be represented by $PE = Ar^{-12} - Br^{-6}$, where A and B are constants that depend on the atoms. Calculate the equilibrium separation in terms of A and B. (Remember: $F = -d(PE)/dr$. What does F equal at equilibrium?)

23.30 We can represent the potential energy of an ionic bond between two atoms by $PE = \lambda e^{-r/\rho} + (1/4\pi\epsilon_0)(-e^2/r)$, where λ and ρ aren't wavelength and some kind of density, but constants with units of energy and length, respectively. Prove that at the equilibrium separation, r_0, $PE = (\rho/r_0 - 1) \cdot (1/4\pi\epsilon_0)(e^2/r_0)$. *Hint:* Reread Problem 23.29.

23.31 In arriving at the total potential energy of a crystal, you have to sum both the attractive energy and the repulsive energy.

a) Consider each form of binding and state whether you would sum the attractive energy over the entire crystal or only over nearest neighbors.

b) Why would you sum the repulsive energy only over nearest neighbors?

CHAPTER

Quantum Statistics

The design of many devices requires knowledge of the electrical and thermal properties of the materials used to make them. For a solid, these properties are largely determined by the behavior of huge numbers of electrons and quanta called phonons. Neither electrons nor phonons follow Maxwell–Boltzmann statistics, except in the correspondence-principle limit. In this chapter, we'll define phonon, describe the distribution function for phonons, and then present a thermal application. We'll also give another distribution function for electrons and show some of its thermal and electrical consequences in the free-electron gas model.

24.1
TYPES OF QUANTUM STATISTICS

Maxwell–Boltzmann statistics give the probability distributions for systems of distinguishable particles that are well-described by the laws of classical physics. However, when the quantum nature of the particles becomes important, we have to use other types of statistics. For instance, suppose that we have a dense gas of electrons with each electron limited by the Pauli exclusion principle and "smeared out" by the Heisenberg uncertainty principle. This gas of electrons would be a quantum system of indistinguishable identical particles, so we couldn't describe it well by the Maxwell–Boltzmann distribution function. However, we can use Maxwell–Boltzmann statistics with accuracy if the average separation between particles is much greater than the de Broglie wavelength, h/p, of the particles.

Quantum systems of indistinguishable, identical particles will be distributed differently depending on whether only one particle can be put into each state (fermions) or there is no such limitation (bosons). **Bose–Einstein statistics** are derived for indistinguishable, identical particles of integral spin angular momentum quantum number, s. (That is, particles called **bosons** for which $s = 0, 1, \ldots$) Examples of bosons include photons, He-4, and paired electrons. The **Bose–Einstein distribution function,** $f_{BE}(E)$, is

$$f_{BE}(E) = \frac{1}{e^{\alpha}e^{E/kT} - 1},$$

(24.1)

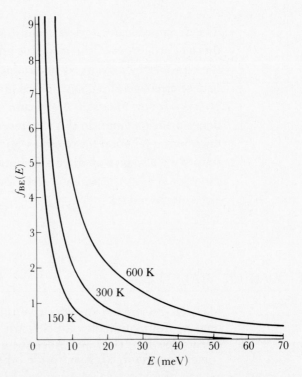

FIGURE 24.1
The Bose–Einstein distribution function when the total number of particles isn't constant.

where α equals zero (giving $e^\alpha = 1$) if the total number of particles is not constant within the system (such as the photons in the cavity in blackbody radiation). Figure 24.1 shows the Bose–Einstein distribution function when α equals zero.

For systems having a constant total number of particles, α decreases to zero as the temperature is lowered. Near a particular low temperature, this behavior causes the particles to drop rapidly into the ground energy level. This "condensation" of particles is called a **Bose–Einstein condensation** and is part of the explanation for the superfluidity (zero viscosity) of liquid helium and for superconductivity.

Another "particle" that follows Bose–Einstein statistics is the phonon. As you will recall, a photon is a quantum of electromagnetic wave energy, with $E = h\nu$. Similarly, in a solid, a **phonon** is a quantum of mechanical wave energy, with

$$E = h\nu, \tag{24.2}$$

where E is the phonon energy and ν is the phonon frequency.

EXAMPLE 24.1 The total number of phonons increases with temperature. How does the average number of phonons per energy state increase with temperature for frequencies that are much less than $(k/h)T = (2.08 \times 10^{10} \text{ Hz/K})T$? *Hint:* $e^x \approx 1 + x$ for small x.

Solution The distribution function, $f(E)$, gives the average number per energy state, and phonons follow Bose–Einstein statistics. Since the total number of phonons isn't conserved, $e^\alpha = 1$ in Eq. (24.1). Also, $E = h\nu$ gives the energy of each phonon, or $E/kT = h\nu/kT$ in Eq. (24.1). For $\nu \ll (k/h)T$, $h\nu/kT \ll 1$, so that $f_{\text{BE}} = 1/(e^{h\nu/kT} - 1)$ accurately equals $1/(1 + h\nu/kT - 1) = kT/h\nu$. That is, the average number of phonons per energy state is directly proportional to the temperature for the frequencies given. ●

At the other extreme (high frequencies and low temperatures) from Example 24.1, the average number of phonons per energy state will be approximately proportional to $e^{-h\nu/kT}$, rapidly approaching zero as T approaches zero.

In the classical model of a monatomic solid, each atom has $\frac{1}{2}kT$ of energy per degree of freedom (on average), with six degrees of freedom. Therefore N atoms have a total energy of $E_t = N(6)(\frac{1}{2}kT) = 3NkT$. The heat capacity associated with these atoms is

$$C = dE_t/dT = d(3NkT)/dT = 3Nk$$

in the classical physics model. This relation is called the law of Dulong and Petit. The value $3Nk$ agrees with experimental results fairly well for most

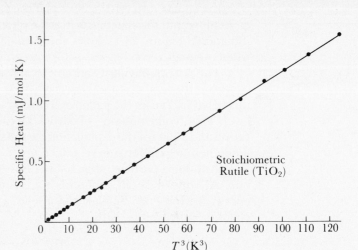

FIGURE 24.2
The specific heat of a solid caused by the motion of its atoms is proportional to the cube of the absolute temperature at sufficiently low temperatures.

Source: T. R. Sandin and P. H. Keesom, *Phys. Rev.* **177**, 1370 (1969).

elemental solids at high temperatures; at low temperatures C is much lower than the constant $3Nk$ value. Experimentally, C is found to be proportional to T^3, approaching zero as $T \to 0$, as in Fig. 24.2.

To explain this result, Einstein proposed a model (the **Einstein model**) in which each atom vibrates with a single frequency v, so that all phonons have the same frequency and therefore the same energy, hv. Increasing the temperature increases the vibrational energy and therefore the number of phonons. We get this number by evaluating Eq. (24.1), with $\alpha = 0$. In three dimensions, the total phonon energy becomes $E_t = 3Nhv/(e^{hv/kT} - 1)$. From this expression we obtain

$$C = dE_t/dT = 3Nk(hv/kT)^2/(e^{hv/kT} - 1)^2.$$

EXAMPLE 24.2   Show that the Einstein model agrees with the law of Dulong and Petit as the absolute temperature approaches infinity and that it gives $C \to 0$ as $T \to 0$.

Solution Evaluating $C = 3Nk(hv/kT)^2/(e^{hv/kT} - 1)^2$ as $T \to \infty$ gives $C = 3Nk0/0$, so let's rewrite it as

$$C = \lim_{x \to 0} 3Nkx^2/(e^{2x} - 2e^x + 1).$$

Using L'Hopital's rule once gives

$$C = \lim_{x \to 0} 3Nk2x/(2e^{2x} - 2e^x),$$

and, again, yields

$$C = \lim_{x \to 0} 3Nk2/(4e^{2x} - 2e^x) = 3Nk,$$

or the law of Dulong and Petit. However, as $T \to 0$, we get $C \to 3Nk\infty/\infty$. So we again use the L'Hopital result, this time with $x \to \infty$, which gives $C \to 3Nk2/\infty = 0$. ●

The Einstein model agrees with experimental results at the high temperature limit, but at low temperatures C isn't proportional to T^3 in the Einstein model. The **Debye model** of a solid is based on the assumption that the frequency varies—and in such a way that the wave speed is constant up to a certain cutoff frequency. This model gives the desired T^3 dependence.

Fermi–Dirac statistics are quantum statistics derived for indistinguishable, identical particles of half odd integral spin angular momentum quantum number, s (that is, particles called **fermions** for which $s = \frac{1}{2}$, $\frac{3}{2}, \ldots$). Electrons, protons, and neutrons are examples of fermions. The **Fermi–Dirac distribution function,** $f_{FD}(E)$, is

$$f_{FD}(E) = \frac{1}{e^{(E - E_F)/kT} + 1}, \tag{24.3}$$

where E_F is called the **Fermi energy.** The Fermi energy is the energy at which the Fermi–Dirac distribution function equals $\frac{1}{2}$, because

$$\frac{1}{e^{(E_F - E_F)/kT} + 1} = \frac{1}{e^0 + 1} = \frac{1}{1 + 1} = \frac{1}{2}.$$

The Fermi energy may vary with temperature for a given system.

The function $f(E)$ gives the average number of particles in a particular energy state. However, for fermions the Pauli exclusion principle limits us to a maximum of one particle per individual quantum state. This condition means that $f_{FD}(E)$ varies from 0 to 1 and therefore is also the probability that an individual state of energy E will be occupied. At the Fermi energy, one-half the states with an energy E_F will be occupied (on average); alternatively, we can say that the probability is one-half that an individual state will be occupied.

EXAMPLE 24.3   At what energy is the probability of occupying a particular state
          a) 99% and   b) 1% for a group of fermions?

Solution The function $f_{FD}(E)$ also gives the probability desired, so we use Eq. (24.3).

a) Substituting, we get

$$0.99 = \frac{1}{e^{(E-E_F)/kT} + 1} \quad \text{or} \quad \frac{1}{0.99} - 1 = e^{(E-E_F)/kT}.$$

Then

$$\ln\left(\frac{1}{0.99} - 1\right) = \frac{E - E_F}{kT} \quad \text{or} \quad E = E_F - 4.6kT,$$

skipping a few algebraic steps.

b) Replace all the 0.99s by $1\% = 0.01$ to get $E = E_F + 4.6kT$.   ●

In general, if the probability of occupation is P for an energy that is ΔE above E_F, the probability of occupation will be $1 - P$ for an energy that is ΔE below E_F.

24.2
ELECTRON DISTRIBUTION

In the **free-electron gas model** of a metal, the interactions of individual electrons with ionized atoms are ignored because they are relatively constant throughout the solid. We consider the potential energy term, V, in the Schrödinger wave equation to be zero except at the boundaries of the solid. Since the chance that electrons will spontaneously appear outside the boundaries of a solid at normal temperatures is virtually zero, we allow the potential energy at the boundaries of the solid to be infinite. Thus we have a three-dimensional particle in a box problem.

For simplicity, let's choose a cube of side L for our box. Therefore $V = 0$ for x, y, and z between 0 and L; $V = \infty$ elsewhere. The three-dimensional, time-independent, nonrelativistic SWE becomes

$$-\frac{\hbar^2}{2m}\left(\frac{\partial^2}{\partial x^2} + \frac{\partial^2}{\partial y^2} + \frac{\partial^2}{\partial z^2}\right)\psi + 0\psi = E\psi. \tag{24.4}$$

The normalized solution to this SWE is the product of the normalized one-dimensional solutions (see Eq. (14.1)), or

$$\psi = \sqrt{\frac{8}{L^3}} \sin\frac{n_x \pi}{L} x \sin\frac{n_y \pi}{L} y \sin\frac{n_z \pi}{L} z, \tag{24.5}$$

where n_x, n_y, and n_z are three positive, nonzero, integer quantum numbers. The possible wave functions are all zero if x, y, or z equals 0 or L. The normalization integral is $\int_0^L\int_0^L\int_0^L \psi^*\psi \, dx \, dy \, dz = 1$. Substituting the wave function, Eq. (24.5), back into the SWE, Eq. (24.4), yields the allowed energies for a free electron:

$$E = (n_x^2 + n_y^2 + n_z^2)\frac{h^2}{8mL^2}. \tag{24.6}$$

EXAMPLE 24.4   Find the minimum electron energy in a free-electron gas.

Solution The smallest possible value of the positive integers n_x, n_y, and n_z is 1. (If any of the n's is given a value of zero, $\sin 0 = 0$ makes the wave function zero also, showing that the probability of that state existing is zero.) Therefore, substituting into Eq. (24.6), we get

$$E = (1^2 + 1^2 + 1^2)h^2/(8mL^2) \quad \text{or} \quad E = 3h^2/(8mL^2). \qquad \bullet$$

At absolute zero ($T = 0$ K), electrons are in their lowest possible energy states. There are two energy states for n_x, n_y, and $n_z = 1$: one with spin up ($m_s = +\frac{1}{2}$) and one with spin down ($m_s = -\frac{1}{2}$). But all the electrons can't crowd into these two lowest energy states because of the Pauli exclusion principle. Instead, the electrons will fill all the states up to E_{F0}, the Fermi energy at $T = 0$ K. All states above E_{F0} will be empty. The Fermi–Dirac distribution function shows us this result.

EXAMPLE 24.5   Find the probability that at absolute zero an energy state will be occupied for

a) $E < E_{F0}$ and   b) $E > E_{F0}$.

Solution The probability is given by $f_{FD}(E) = 1/[e^{(E-E_F)/kT} + 1]$. At $T = 0$, $E_F = E_{F0}$.

a) If E is less than E_{F0}, then $(E - E_{F0})$ must be negative. Therefore $(E - E_{F0})/kT = (-)/0 = -\infty$. Since $e^{-\infty} = 1/e^{\infty} = 0$, at absolute zero $f_{FD}(E) = 1/(0 + 1) = 1$. So the probability that states below the Fermi energy are occupied is 1; that is, they are *all* occupied.

b) If E is greater than E_{F0}, then $E - E_{F0}$ must be positive. This condition gives $e^{+\infty}$, and $f_{FD}(E) = 1/(\infty + 1) = 0$. So the probability that states above the Fermi energy are occupied is 0; that is, they are *all* empty. $\qquad \bullet$

At temperatures well above absolute zero, some of the states slightly below E_F will be empty and some of the states slightly above E_F will be full as a small fraction of the electrons gain some thermal energy. This dependence is shown in Fig. 24.3.

We can use $f_{FD}(E)$ to find the probability that an energy state is occupied by an electron. To find the number of electrons in the region of that energy, however, we also need to know how many states there are

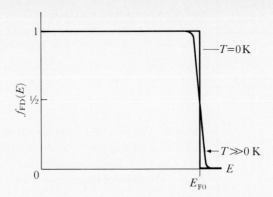

FIGURE 24.3
The Fermi–Dirac distribution function. Because of the Pauli exclusion principle, it gives both the average number of particles in a particular state and the probability that a state is occupied at that energy. This graph is based on the assumption that E_F doesn't depend on T.

near that energy. At $T = 0$ the situation is simplest: N states have energies less than or equal to the Fermi energy, and those states are fully occupied with N electrons. If we plot n_x along one Cartesian axis and n_y and n_z along the other two axes, as in Fig. 24.4, we have a description of the energy states in "n space." All states with n_x, n_y, and n_z greater than zero are filled to a maximum of $E_{F0} = (n_x^2 + n_y^2 + n_z^2)_{max} h^2/(8mL^2)$. If we let $(n_x^2 + n_y^2 + n_z^2)_{max} = n_{max}^2$, all the states in n space are filled to n_{max}, a radius in n space. The n's are positive integers, so the volume filled in n space is only one-eighth of a sphere.

The quantum numbers n_x, n_y, and n_z change in steps of 1. Therefore in a volume of $1 \times 1 \times 1 = 1$ in n space, there are exactly two electron

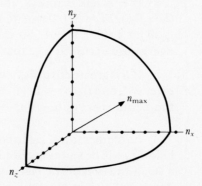

FIGURE 24.4
Allowed electron states in n space for the free-electron gas model. At absolute zero, the electrons occupy the one-eighth of the sphere shown, filling all states up to n_{max}. Therefore n_{max} corresponds to E_{F0}.

states (one spin up, the other spin down). For example, in a cube of side 1 and volume 1 in n space centered at $n_x = 1066$, $n_y = 1492$, and $n_z = 1776$, there are two states having quantum numbers of $(1066, 1492, 1776, +\frac{1}{2})$ and $(1066, 1492, 1776, -\frac{1}{2})$.

We can obtain the number of states from the product of the number of states per volume and the volume. With N states, two states per unit volume in n space, and a volume of one-eighth of a sphere of radius n_{max}, we have $N = 2(1/8)(4\pi n_{max}^3/3)$. Using $E_{F0} = n_{max}^2 h^2/(8mL^2)$ to eliminate n_{max}, letting $L^3 = V$, the volume of the solid, and solving for E_{F0} we obtain

$$E_{F0} = \frac{h^2}{8m}\left(\frac{3N}{\pi V}\right)^{2/3} = (3.646 \times 10^{-19} \text{ eV} \cdot \text{m}^2)\left(\frac{N}{V}\right)^{2/3}, \quad (24.7)$$

where N/V equals the number of free electrons per volume (in m^{-3}) and is called the **electron concentration.** We obtain the 3.646 . . . term by substituting the values of h, π, and the mass of the electron.

EXAMPLE 24.6 At absolute zero, Cu has a free-electron concentration of $8.5 \times 10^{28} \text{ m}^{-3}$. Using the free-electron gas model,

   a) calculate the Fermi energy at absolute zero for copper and

   b) calculate the maximum radius of the occupied $1/8$ sphere in n space for 1 cm^3 of Cu at $T = 0$ K.

Solution

   a) With $N/V = 8.5 \times 10^{28} \text{ m}^{-3}$, the easiest solution is

$$E_{F0} = (3.646 \times 10^{-19} \text{ eV} \cdot \text{m}^2)(8.5 \times 10^{28} \text{ m}^{-3})^{2/3} = 7.0 \text{ eV}.$$

The harder solution is to use the middle of Eq. (24.7), substituting $h = 6.626 \times 10^{-34} \text{ J} \cdot \text{s}$, $m = 9.11 \times 10^{-31}$ kg, the constants, and the given value of N/V—then converting the answer from joules to the more customary electron volts.

   b) Using $E_{F0} = n_{max}^2 h^2/(8mL^2)$ to solve for n_{max} and $L = 1$ cm $= 1 \times 10^{-2}$ m for a 1-cm^3 cube, we have

$$n_{max} = \frac{\sqrt{8mE_{F0}}\, L}{h}$$

$$= \frac{\sqrt{8(9.11 \times 10^{-31} \text{ kg})(7.0 \text{ eV})(1.60 \times 10^{-19} \text{ J/eV})}(1 \times 10^{-2} \text{ m})}{6.626 \times 10^{-34} \text{ J} \cdot \text{s}}$$

$$= 4.3 \times 10^7. \qquad\qquad\qquad\qquad\qquad \bullet$$

Note that the quantum numbers of the occupied states in Example 24.6 may vary from 1 to about 43 million. With such a huge number of states, we can treat them as being virtually continuous in energy. That is, the energy is quantized, but the energy difference between adjacent states is practically zero.

In the first part of Eq. (24.7), let's replace E_{F0} by E and N by $\mathcal{N}$. Then $\mathcal{N}$ is the number of states with energies between zero and E. Solving for $\mathcal{N}$ gives $\mathcal{N} = (\pi V/3)(8mE/h^2)^{3/2}$. Differentiating with respect to E, we get

$$d\mathcal{N}/dE = (\pi V/2)(8m/h^2)^{3/2}E^{1/2}.$$

The term $d\mathcal{N}/dE$ is the number of states per unit energy range and is called the **density of states,** or $g(E)$. Since we can solve the first part of Eq. (24.7) to obtain

$$(\pi V/2)(8m/h^2)^{3/2} = (3N/2)(E_{F0})^{-3/2},$$

we have

$$g(E) = \frac{\pi V}{2}\left(\frac{8m}{h^2}\right)^{3/2}E^{1/2} = \frac{3N}{2E_{F0}^{3/2}}E^{1/2}. \tag{24.8}$$

EXAMPLE 24.7   Show that the energy may be treated as being essentially continuous by calculating the density of states for the free electrons at the Fermi energy in a mole of Cu at $T = 0$. Copper has one free electron per atom.

Solution With one free electron per atom and Avogadro's number of atoms of Cu in a mole, we have $N = N_A = 6.02 \times 10^{23}$. For Cu we have shown that $E_{F0} = 7.0$ eV. We want $g(E)$ at $E = E_{F0} = 7.0$ eV. Substituting into the right-hand side of Eq. (24.8), we get

$$g(E_{F0}) = \frac{3(6.02 \times 10^{23})}{2(7.0\text{ eV})^{3/2}}(7.0\text{ eV})^{1/2} = 1.3 \times 10^{23}\text{ states/eV.} \qquad \bullet$$

With this huge number of states per unit energy range, it's clear that we may consider the energy to be virtually continuous.

We can now apply another variation of $N(E) = G(E)f(E)$ to obtain $n(E)$, the number of electrons per unit energy range. It is equal to the number of states per unit energy range times the average number of electrons per state. That is, $n(E) = g(E)f(E)$. Therefore for the free-electron gas, using the expression for the density of states, Eq. (24.8), and the Fermi–Dirac distribution function, Eq. (24.3), we have

$$n(E) = \frac{\pi V}{2}\left(\frac{8m}{h^2}\right)^{3/2}\frac{E^{1/2}}{e^{(E-E_F)/kT}+1} = \frac{3N}{2E_{F0}^{3/2}}\frac{E^{1/2}}{e^{(E-E_F)/kT}+1}. \tag{24.9}$$

Thus $n(E)$ varies as $E^{1/2}$ and with the Fermi–Dirac distribution function. Its graph is shown in Fig. 24.5.

The integral of $n(E)\,dE$ from E_1 to E_2 gives the number of electrons between E_1 and E_2, and $N = \int_0^\infty n(E)\,dE$ determines the Fermi energy at

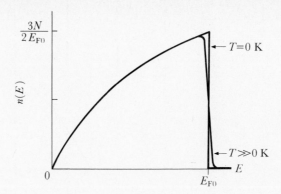

FIGURE 24.5
The electron distribution with energy for a free-electron gas. The total number of electrons equals the area under either curve. The graphs are based on the assumption that the Fermi energy is the same at the two temperatures. For most metals, the $T \gg 0$ K curve *as drawn* would be for a temperature nearer the melting point than to room temperature.

temperatures above absolute zero. For metals at room temperature, $E_{F0} = E_F$ because the Fermi energy changes very little between absolute zero and that temperature.

EXAMPLE 24.8   Check Eq. (24.9) by calculating N at $T = 0$ K.

Solution At $T = 0$ K, $E_F = E_{F0}$, $(E - E_{F0})/kT = -\infty$ for $E < E_{F0}$, and $(E - E_{F0})/kT = +\infty$ for $E > E_{F0}$. Therefore

$$N = \int_0^\infty n(E)\, dE = \left(\frac{3N}{2E_{F0}^{3/2}}\right)\left[\int_0^{E_{F0}} 1\, E^{1/2}\, dE + \int_{E_{F0}}^\infty 0\, E^{1/2}\, dE\right]$$

$$= \left(\frac{3N}{2E_{F0}^{3/2}}\right)\left(\frac{2E_{F0}^{3/2}}{3}\right) = N.$$

Remember: $1/(e^{-\infty} + 1) = 1$ and $1/(e^{+\infty} + 1) = 0$.   ●

EXAMPLE 24.9   What fraction of the free electrons in Cu have a kinetic energy between 3.95 eV and 4.05 eV at room temperature?

Solution A room temperature of 20 °C = 293 K gives $kT = 0.025$ eV = $1/40$ eV. The exact answer to this question would be given by the integral $\int_{3.95\,\text{eV}}^{4.05\,\text{eV}} n(E)\, dE/N$. However, $\Delta E = (4.05 - 3.95)$ eV = 0.10 eV is reasonably small, so we can approximate the integral by $n(\overline{E})\Delta E/N$, with $\overline{E}$ being the average energy, or 4.00 eV in this case. As previously stated, we can accurately allow E_F to equal E_{F0}, which is 7.0 eV for Cu. Using the right-

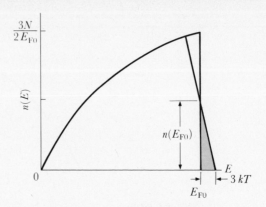

FIGURE 24.6
An approximation to Fig. 24.5 but not to scale ($3kT \ll E_{F0}$).

hand side of Eq. (24.9) for $n(\overline{E})$, we have

$$\frac{n(\overline{E})\Delta E}{N} = \left[\frac{3}{2(7.0 \text{ eV})^{3/2}}\right]\left[\frac{(4.00 \text{ eV})^{1/2}(0.10 \text{ eV})}{e^{(4.00-7.0)/(1/40)} + 1}\right]$$

$$= \frac{0.0162}{e^{-120} + 1} = 0.016. \qquad\bullet$$

The total energy of the free-electron gas equals $\int_0^\infty En(E)\,dE$. We can evaluate this integral exactly only at $T = 0$ K, where the total energy is $(3/5)NE_{F0}$. At low temperatures, we can approximate part of Fig. 24.5 ($T \gg 0$) by a straight line. Since $1/(e^3 + 1) < 0.05$, we let the straight line intercept the E axis at $E_{F0} + 3kT$ in Fig. 24.6. The shaded area under the curve equals the number of electrons promoted above the Fermi energy by the addition of heat to warm the electron gas above absolute zero. The shaded area is a triangle (area = one-half the base times the height). The base here is $3kT$. The height is the value of $n(E)$ at E_{F0}, which, from the right-hand side of Eq. (24.9) is

$$n(E_{F0}) = \left(\frac{3N}{2E_{F0}^{3/2}}\right)\left(\frac{E_{F0}^{1/2}}{e^0 + 1}\right) = \frac{3N}{4E_{F0}}.$$

Therefore the number of electrons promoted equals the shaded area, or

$$\frac{1}{2}(3kT)\left(\frac{3N}{4E_{F0}}\right) = \frac{9NkT}{8E_{F0}}.$$

Since the centroid of the triangle is one-third of $3kT$, or kT, above E_{F0}, the average electron was promoted from kT below E_{F0} to kT above E_{F0}, thereby being given $2kT$ of heat. The total heat added then is $2kT$ average per electron times $9NkT/8E_{F0}$ electrons, or $9Nk^2T^2/4E_{F0}$. Finally, the heat capacity of the free-electron gas is

$$C_{\text{el}} = \frac{dE_t}{dT} = \frac{d}{dT}\left(\frac{3NE_{\text{F0}}}{5} + \frac{9Nk^2T^2}{4E_{\text{F0}}}\right) = \frac{9Nk^2}{2E_{\text{F0}}}T.$$

You can see that the low temperature heat capacity of the free-electron gas is directly proportional to the absolute temperature.

The heat capacity of a classical electron gas is constant at $\frac{3}{2}Nk$, but the heat capacity of the quantum free-electron gas is much smaller at most temperatures, approaching zero as T approaches zero. If we had performed a more exact analysis by expanding the integral for the energy in a series form, we would have almost the same result. The difference would be that C_{el} would have a $\pi^2/2 = 4.9$ term instead of our $9/2 = 4.5$ estimation.

The heat capacity of the free-electron gas at low temperatures is far below the constant classical value because of the Pauli exclusion principle. With a rise in temperature, dT, only those few electrons near the Fermi energy can gain energy. For $E \ll E_{\text{F}}$, all the states are nearly full, so there are almost no empty states for the electrons to move into if they were to gain energy. For $E \gg E_{\text{F}}$, all the states are nearly empty, so there are almost no electrons to gain energy.

24.3
THE ELECTRICAL CONDUCTIVITY OF THE FREE-ELECTRON GAS

Suppose that we sketch two new graphs for the free-electron gas. This time let's plot the linear momentum components p_x, p_y, and p_z allowed for the electrons on the three Cartesian axes (in momentum space). The two-dimensional versions are shown in Figs. 24.7 and 24.8.

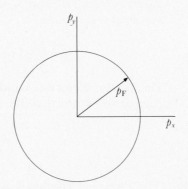

FIGURE 24.7
The Fermi sphere at $t = 0$ (or with $\mathbf{E} = 0$).

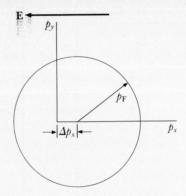

FIGURE 24.8
The Fermi sphere in equilibrium with $\mathbf{E} \neq 0$ (Δp_x grossly exaggerated).

At absolute zero, all states are filled up to the momentum that corresponds to the Fermi energy. The result is a full sphere called the **Fermi sphere.** Example 24.3 shows that the sphere won't change much at higher temperatures. Its surface will just be a bit "fuzzier" or "smeared out" over that small range of momentum and energy where $f_{FD}(E)$ isn't either almost 1 or almost 0.

Newton's second law of motion is $\mathbf{F} = d\mathbf{p}/dt$, where $\mathbf{F} = (-e)\mathbf{E}$ for electrons in an electric field $\mathbf{E}$. Therefore, applying a constant electric field to all electrons for a time τ will change their momentum by $\Delta\mathbf{p} = (-e)\mathbf{E}\tau$. If we turn this electric field on at $t = 0$, the electrons will begin to scatter off impurities, other defects, and phonons after the **collision time** τ. The average momentum of the electrons in the Fermi sphere increases from zero to $\Delta\mathbf{p}$, where it remains in equilibrium. The **drift velocity** is the average velocity of the electrons. Its symbol is $\mathbf{v_d}$ and, for the negative electrons, its direction is opposite that of the electric field. Then $\mathbf{v_d} = \Delta\mathbf{p}/m = -e\mathbf{E}\tau/m$, and $\mathbf{j}$, the current density (the current per cross-sectional area) equals $(N/V)q\mathbf{v_d}$. For the electrons,

$$\mathbf{j} = \left(\frac{N}{V}\right)(-e)\left(\frac{-e\mathbf{E}\tau}{m}\right) = \left(\frac{N}{V}\right)\left(\frac{e^2\mathbf{E}\tau}{m}\right).$$

Finally, the **electrical conductivity,** σ, is defined by $\mathbf{j} = \sigma\mathbf{E}$, so simple comparison shows that the electrical conductivity of a free-electron gas is

$$\sigma = \frac{N}{V}\frac{e^2\tau}{m}. \tag{24.10}$$

EXAMPLE 24.10   The electrical conductivity of Cu at room temperature is 5.9×10^7 (ohm $\cdot$ m)$^{-1}$, and the electron concentration is 8.5×10^{28} m^{-3} for its free-electron gas. Calculate the collision time.

Solution We use $\sigma = (N/V)e^2\tau/m$ to obtain $\tau = \sigma m/e^2(N/V)$. With $\sigma = 5.9 \times 10^7 \ (\Omega \cdot m)^{-1}$, $m = 9.11 \times 10^{-31}$ kg, $e = 1.60 \times 10^{-19}$ C, and $(N/V) = 8.5 \times 10^{28}$ m^{-3}, we have $\tau = 2.5 \times 10^{-14}$ s. ●

If we were to multiply the short time calculated in Example 24.10 by the drift velocity (even for $j = 1$ A/mm$^2 = 10^6$ A/m^2, $v_d \approx 10^{-4}$ m/s for Cu), it would seem that electrons would be scattered before they even move the width of a nucleus. But this isn't the case. Most of the electrons are below the surface of the Fermi sphere. Thus there are no empty states for most of the electrons to scatter into, with the small energy changes per electron involved. Instead of the magnitude of the drift velocity, we should use the speed of the electrons on the surface of the Fermi sphere (where there are some empty states for them to scatter into).

This speed is called the **Fermi speed**, v_F, and is defined by $\frac{1}{2}mv_F^2 = E_F$. Therefore the **mean free path** (average distance between collisions), l, of an electron in the free-electron gas model of a metal is

$$l = v_F\tau, \qquad (24.11)$$

where $v_F = \sqrt{2E_F/m}$.

EXAMPLE 24.11 Calculate the mean free path of electrons in Cu at room temperature.

Solution For Cu, $E_F = (7.0 \ \text{eV})(1.60 \times 10^{-19} \ \text{J/eV}) = 1.12 \times 10^{-18}$ J. Then

$$v_F = \sqrt{\frac{2(1.12 \times 10^{-18} \ \text{J})}{9.11 \times 10^{-31} \ \text{kg}}} = 1.57 \times 10^6 \ \text{m/s}.$$

Taking the value of τ from Example 24.10 and substituting, we get

$$l = v_F\tau = (1.57 \times 10^6 \ \text{m/s})(2.5 \times 10^{-14} \ \text{s}) = 3.9 \times 10^{-8} \ \text{m} = 39 \ \text{nm}.$$
●

In Example 24.11, the Cu electrons get past an average of about 200 atoms before being scattered. At low temperatures, this value may be 10^5 times larger for extraordinarily pure and defect-free single crystals of Cu. The distance may be many centimeters long in other ultra-high purity single crystal metals. (If so, the size of the sample itself may limit the mean free path.) The mean free path is very long because these extraordinary samples have very few impurities and defects to scatter the electrons. Also, at low temperatures the atoms have low vibrational energies, which means that there are very few phonons to scatter the electrons.

In real solids, the "free" outer electrons are not completely free. Also, V is not a constant in the SWE. These deviations from the free-electron gas model assumptions can often be compensated for by replacing the mass of the electron, m, by an **effective mass, $m*$**, in the equations of this chapter. The value of $m*$ may be greater or less than m.

This chapter introduced you to both the Bose–Einstein and the Fermi–Dirac distribution functions. We used phonons as examples of bosons and electrons as examples of fermions. We presented the assumptions of the free-electron gas model of conduction electrons in a metal. We also found relations for the Fermi energy, density of states, and number of electrons per unit energy range in that model. We further showed how to determine quantities such as numbers of electrons and average energies. We then moved to the behavior of the free-electron gas under the influence of an electric field. We found relations between the macroscopic electrical conductivity and microscopic quantities such as electron concentration, charge, mass, collision time, and mean free path. We described the effects of the Pauli exclusion principle on thermal and electrical properties. Finally, we mentioned a parameter called the effective mass, which often can be used to bring experimental results for a real metal into agreement with the predictions of the free-electron gas model.

QUESTIONS AND PROBLEMS

24.1 For a "phonon gas" in a cavity of volume V, $\alpha = 0$ and the frequency density of states, $g(\nu)$, equals $8\pi\nu^2 V/c^3$. Write an expression for the energy between ν and $\nu + d\nu$ at a temperature T.

24.2 In the Debye model for phonons in a solid, the density of states, $g(E)$, equals $12\pi VE^2/(\hbar^3 v_s)$, where v_s is the speed of sound (assumed to be independent of frequency, direction, or polarization). This relation holds up to $E_D = k\theta_D$, where θ_D is called the Debye temperature. Show that at low temperatures the heat capacity, dE_t/dT, is proportional to T^3. (*Hint:* Let $E/kT = x$ and note that $E_D/kT \to \infty$ as $T \to 0$.)

24.3 In "optical modes" of vibration in crystals, the frequency is found experimentally to be relatively independent of the wavelength. Therefore does the classical model, the Einstein model, or the Debye model best represent the heat capacity resulting from the optical mode phonons? Explain.

24.4 a) Explain why the probability of "phonon-assisted processes" in which phonons are absorbed (for example, to conserve momentum) increases with temperature.

b) Thermal conductivity caused by atomic vibrations can be explained as the result of the imbalance in the number of phonons at the hot end compared to the cold end of a sample. This imbalance causes the phonons to diffuse, carrying energy. Which end (hot or cold) has the most phonons per volume? Explain.

24.5 The Fermi energy of Al at room temperature is 11.6 eV. What is the probability of finding a specific 11.7-eV state occupied at room temperature in Al?

24.6 Prove that if the probability of occupying a state ΔE above E_F is P, the probability of occupation ΔE below E_F is $1 - P$.

24.7 The probability of occupying a specific electron state of a given energy equals P at a temperature T. In terms of P, T, and E_F, what is the energy of the state?

24.8 Prove that the trial wave function

$$\psi = (8/L^3)^{1/2} \cos(n_x\pi/L)x \, \cos(n_y\pi/L)y \, \cos(n_z\pi/L)z$$

satisfies the free-electron gas SWE and yields the correct energy but doesn't satisfy the boundary conditions.

24.9 Show that the wave function given for the free-electron gas is correctly normalized.

24.10 Calculate the lowest possible electron energy in a 1.00-cm cube for a free-electron gas. What speed does this correspond to?

24.11 a) Calculate the maximum n_x for the free-electron gas in a 1.00-cm cube of Al at $T = 0$ K. The Fermi energy of Al is 11.6 eV at absolute zero.

b) Calculate the speed for this case.

24.12 Calculate and graph the Fermi–Dirac distribution function from $5kT$ below the Fermi energy to $5kT$ above the Fermi energy. What are its approximate values below and above these limits?

24.13 If Li has a Fermi energy of 4.75 eV at absolute zero, calculate its corresponding electron concentration.

24.14 The electron concentration in Au at $T = 0$ K is 5.9×10^{28} m^{-3}. Calculate the Fermi energy of Au at absolute zero.

24.15 Calculate the density of states for 5.00×10^{22} free electrons in Na at 3.00 eV. The Fermi energy at $T = 0$ K in Na is 3.25 eV.

24.16 Use two different methods to calculate the density of states at 8.50 eV for 6.02×10^{23} free electrons in In. Indium has an electron concentration of 1.15×10^{29} m^{-3} at $T = 0$ K.

24.17 Prove by integration that the average energy of a free-electron gas electron at absolute zero is 60% of the Fermi energy at that temperature.

24.18 Calculate the fraction of free electrons below 6.8 eV at room temperature in Cu by integrating. (*Hint:* Why can you ignore the exponential term?)

24.19 Calculate the total energy of a gas of 6.02×10^{23} free electrons in Cu at room tempera-

ture carried by just those electrons with energies of less than 6.8 eV. (See Problem 24.18.)

24.20 Calculate the ratio of the number of free electrons per unit energy range in a metal at T_2 to that at T_1. For what energies is this ratio less than one? Equal to one? Greater than one?

24.21 Calculate the net energy of those electrons between 5.95 eV and 6.05 eV in a free-electron gas composed of 6.02×10^{23} electrons at 77 K in Mg. The electron concentration equals 8.6×10^{28} m^{-3}.

24.22 Calculate the probability of finding a free electron in Ag in the range from 4.95 eV to 5.05 eV at $T = 100$ °C if there are 5.88×10^{28} free electrons in a cubic meter of Ag at absolute zero. Assume the Fermi energy to be constant with T.

24.23 Calculate the average energy of those free electrons that have energies of less than half the Fermi energy at absolute zero.

24.24 Calculate the probability of finding an electron with an energy of less than $3E_{F0}/5$ at $T = 0$ K in a free-electron gas.

24.25 Calculate the median energy for the free-electron gas at absolute zero.

24.26 How does the Pauli exclusion principle affect the electron specific heat and the mean free path?

24.27 On Fig. 24.8, sketch a possible electron scattering event that would transfer energy from the electron to the crystal.

24.28 Aluminum has an electron concentration of 1.81×10^{29} m^{-3} and an electrical conductivity of 3.65×10^7 (ohm · m)$^{-1}$ at room temperature. Calculate a) the collision time, b) the Fermi energy (assuming negligible change with temperature), and c) the Fermi speed and the mean free path.

24.29 Zinc has an electron concentration of 1.31×10^{29} m^{-3}. If the collision time is 4.2×10^{-11} s at low temperatures, calculate a) the conductivity, b) the Fermi energy, and c) the Fermi speed and the mean free path.

24.30 If the resistance in pure metals is mainly the result of electron–phonon scattering, why does resistance increase with temperature?

24.31 In Eq. (24.10), what is temperature dependent and to what extent?

24.32 Which numbered equations in this chapter would change if the effective mass weren't equal to the mass of the free electron? In which direction would the equations change if the effective

mass were smaller than the mass of the free electron?

24.33 Why is m^* often not equal to m in actual materials?

24.34 Is m^* different from m for relativistic reasons?

24.35 If m^* is greater than m, does this mean that the electrons weigh more?

CHAPTER **25**

The Band Theory of Solids

March 18, 1987, New York City: a jammed auditorium — those who
waited in line for hours getting the choice seats — hundreds of others
standing along the sides and back — more outside watching on monitors
and waiting for a chance to get in — some still left in the auditorium at
6 A.M. the next morning. Who were these people? Enthusiastic fans of a
rock group or a famous performer? No. They were physicists anxious
to hear the latest research results in the field of superconductivity, one
of the topics that we'll discuss in this chapter.

In a typical modern circuit, conductors carry currents to and from
semiconductor devices. Insulators keep the currents from going where
they shouldn't. But when atoms are collected to form a solid, what
determines whether that solid will be a conductor, a semiconductor, or
an insulator? We'll discuss this question in this chapter and find out
what the superconductivity excitement was all about. And we'll also explain the physics behind such common solid state concepts as energy
gap, hole, doping; and intrinsic, *n*-type, *p*-type, and compensated
semiconductors.

25.1
ENERGY BANDS

A lithium atom in its ground state has two electrons filling the 1s states and one electron in one of the two 2s states. An isolated Li atom has 1s, 2s, 2p, 3s, and so on, energy levels, which have their energies smeared out only slightly by the uncertainty principle and by small interactions within the atom. However, when Li atoms (or any type of atoms) are brought closer and closer together, the wave functions of their outer electrons begin to overlap. (The outer electrons are called **valence electrons**.) For a given state in two equal atoms, instead of two equal and independent wave functions, ψ_1 and ψ_2, the Pauli exclusion principle will require two different wave functions for the overlap situation. We relate one of these two wave functions, ψ_+, to the sum of wave functions and the other wave function, ψ_-, to the difference.

Since ψ_+ and ψ_- will have different energy states, one energy level has been split into two levels for two atoms. In a solid, N atoms interact and the energy splits into N different levels. The typical solid has more than 10^{22} atoms in each cubic centimeter, so N is a huge number. The energy difference between the highest and lowest of the N levels is only a few eV. Therefore the difference in energy between any two adjacent levels is negligible. In the solid, the original single allowed energy level of the isolated atom has been spread into an essentially continuous band of allowed energy levels called an **energy band.**

The greatest overlap occurs for the valence electron states (and states above the valence electron states). As a result, the greatest widening of levels into bands occurs for those states. Also, the probability of finding a valence electron has been spread over the entire solid instead of being localized to one atom. Therefore these outer electron energy bands extend throughout the solid. However, for the inner electron states, the overlap of the wave functions between adjacent atoms ranges from very little to essentially zero. Therefore, as illustrated in Fig. 25.1, these inner states are only slightly broadened or virtually unaffected as the atoms are brought together to form a solid.

In an isolated atom of Li, there is one electron in the two possible states of the 2s sublevel. In solid Li, with N atoms of Li, there are N electrons in the $2N$ possible states of the 2s band. These electrons are free to gain energy within the band, perhaps from an external electric field or from some added heat. Since they are free to move within the extended band throughout the solid—and are therefore free to conduct electricity and heat—the 2s electrons in solid Li behave like a free-electron gas in what we can call a **conduction band.** Lithium belongs to the class of metals that are good conductors because they have partially filled conduction bands, as shown in Fig. 25.2(a).

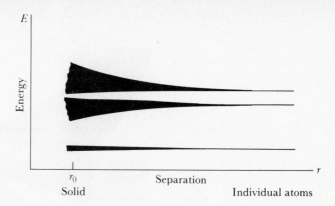

FIGURE 25.1
A representation of the change from discrete localized energy levels (on right of sketch) to energy bands with decreasing separation of the atoms.

In other materials the **valence band** may be completely full of electrons at $T = 0$ K. The electrons in the valence band are involved in the binding of the atoms. When the valence band is full, all its electrons are so involved, and they don't contribute to the conduction process.

However, suppose that the spreading of energy levels causes the next highest empty energy band to overlap the valence band in energy, as in Fig. 25.2(b). Then there will still be a continuous range of possible energies for the valence electrons. The outer electrons can therefore easily gain energy and momentum to conduct. The solid will be a good conductor because the two bands have overlapped to form a conduction band.

In two types of materials the valence band and the conduction band don't overlap, remaining separated: **semiconductors** and **insulators.** In

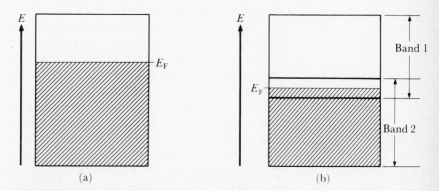

FIGURE 25.2
Solids may be conductors because of (a) a partially filled conduction band, or (b) overlapping valence and conduction bands.

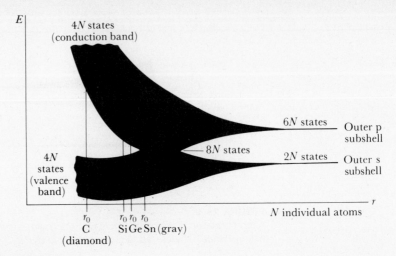

FIGURE 25.3
Gray tin, germanium, silicon, and diamond have two s and two p outer electrons and the same crystal structure. In gray Sn, the valence and conduction bands overlap, making it a metal. In Ge and Si, an energy gap of about 1 eV between those bands makes them semiconductors. In diamond, that forbidden energy gap is greater than 5 eV, making diamond an insulator.

both types at absolute zero, all the valence band states are full of electrons, and all the conduction band states are empty. In **intrinsic** (pure) semiconductors and insulators there is an energy region between the two bands that has no allowed states, as shown in Fig. 25.3. This energy region is called the **forbidden energy gap.** From solutions to the SWE, electrons can't be found in the forbidden energy gap in an ideal crystal, just as the SWE doesn't allow an electron in a -10-eV state in a hydrogen atom (where the allowed energies are -13.60 eV$/n^2$).

At absolute zero in both pure insulators and semiconductors, all the outer, or valence, electrons are "frozen" in the binding process (that is, they are all in the valence band) and none is free to move throughout the solid and conduct (that is, none is in the conduction band). Therefore the electrical conductivity is zero at $T = 0$ K. However, as the temperature is raised, a few electrons can gain enough thermal energy to be excited across the forbidden gap to the conduction band. This excitation happens much more readily with semiconductors than with insulators because of the relative "widths" of the forbidden energy gaps, E_g. For a semiconductor, E_g is on the order of 1 eV. For example, at room temperature the energy gap for Ge is 0.66 eV, for Si it's 1.12 eV, and for GaAs it's 1.42 eV. As indicated in Fig. 25.4, insulators have larger energy gaps of several eV. For example, values are 3.0 eV for both hexagonal SiC and TiO_2 (rutile) and 5.5 eV for diamond.

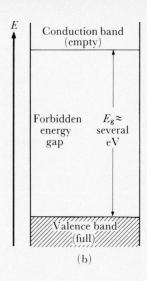

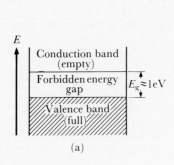

FIGURE 25.4

Energy band diagrams of intrinsic (a) semiconductors and (b) insulators at absolute zero. Electrons can be excited from the filled valence band to the empty conduction band if they receive enough energy to cross the forbidden energy gap.

EXAMPLE 25.1   The conductivity, σ, of conductors decreases as the temperature increases because of the increasing number of electron–phonon interactions. How does σ change with the temperature for intrinsic semiconductors and insulators?

Solution   As stated in this section, at $T = 0$ K all the electrons in intrinsic semiconductors and insulators are in the full valence band and there is no conduction. Therefore the electrical conductivity is zero at absolute zero. As T increases above absolute zero, electrons are thermally excited into the conduction band where they conduct. Therefore σ increases with T.

●

The increase described in Example 25.1 is much greater for semiconductors than for insulators, because semiconductors have a narrower energy gap across which the electrons must be excited. The variation of σ with T (or the corresponding reciprocal variation of the resistivity) is used to design resistance thermometers, thermistors, and other devices.

The electrons in the conduction band aren't considered to be the only charge carriers in semiconductors and insulators. In intrinsic material, when an electron jumps across the energy gap into the conduction band, it leaves an empty state behind in the valence band. As Fig. 25.5 shows, the excited electron leaves an empty bond, or hole, which can be

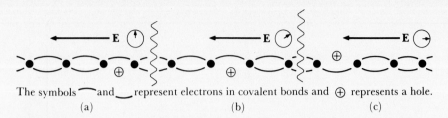

The symbols ⌢ and ＿ represent electrons in covalent bonds and ⊕ represents a hole.
(a)　　　　　　　　　(b)　　　　　　　　　(c)

FIGURE 25.5
The positions of a hole at three consecutive times, (a), (b), and (c), with an electric field to the left.

filled by a nearby electron. Movement of that electron would leave another hole to be filled by yet another electron, and so on.

Under the influence of an electric field, **E**, to the left, the electrons are pushed to the right (because of their negative charge). Therefore the hole moves to the left, or in the same direction as the electric field. That is, the hole acts as if it had a positive charge, $q = +e$. Rather than talk about the motions of the many electrons involved, it's much easier to talk about the motion of one hole.

Similarly, in describing a fluid full of bubbles, it's much easier to describe the motion of a bubble — the absence of fluid molecules — than it is to describe the motions of all the surrounding molecules in the fluid. A huge number of the molecules move down under the influence of the gravitational field while a single bubble moves up. A hole, that is, the absence of an electron, is a positive "quasi-particle" (it acts like a particle), which gives electrical conduction in the valence band.

EXAMPLE 25.2　Sufficiently energetic photons can be absorbed to create separated electron-hole pairs in a semiconductor.

a) For what light frequency would this absorption occur in GaAs if $E_g = 1.42$ eV at room temperature?

b) If 8.0×10^{21} free electrons/m³ resulted, what would be the hole concentration for an intrinsic semiconductor?

c) What would a graph of σ versus v look like (roughly) for a thin slab of GaAs at constant temperature?

Solution

a) To give the electrons at the top of the valence band just enough energy to jump the gap, the photon energy $h v$ would have to equal E_g.

Therefore

$$v_{min} = \frac{E_g}{h} = \frac{1.42 \text{ eV}}{4.136 \times 10^{-15} \text{ eV} \cdot \text{s}} = 3.43 \times 10^{14} \text{ Hz}.$$

The answer, then, is $v \geq 3.43 \times 10^{14}$ Hz.

b) In pure semiconductors, every electron moved from the valence band to the conduction band leaves one hole behind. Therefore the electron and hole concentrations are equal. In this example that means 8.0×10^{21} holes/m³. (These holes also take part in the conduction process.)

c) The conductivity would remain constant with increasing frequency up to 3.43×10^{14} Hz. Then σ would begin to increase rapidly as more and more electrons are excited across the forbidden energy gap to the conduction band (with the holes they left behind also conducting, but in the valence band). The slab must be thin for light to reach into most of the material. ●

The visible spectrum ranges from 4.3×10^{14} Hz to 7.5×10^{14} Hz, so the 3.43×10^{14} Hz obtained in Example 25.2 is in the infrared. Thus all higher infrared, as well as visible and ultraviolet, frequencies can also create electron–hole pairs (the basis of photoconductive devices, solar cells, and the like). The electron–hole pairs could also be created by even higher energy photons (x-rays or γ rays) as well as by bombardment by charged particles (the basis of semiconductor radiation detectors).

It's possible for the electron and hole to become incompletely separated, forming a positronium-like structure called an **exciton.** In that case, the energy to create an exciton would equal E_g minus the exciton binding energy. Typical exciton ground state binding energies range from a few meV to hundreds of meV. Excitons can collect to form microscopic droplets within certain regions of the solid, but the lifetimes of these electron–hole drops are usually considerably less than a millisecond.

25.2
SUPERCONDUCTIVITY

In 1908 Kamerlingh Onnes was the first to liquefy helium (naturally occurring helium boils at 4.2 K at 1 atm pressure). Three years later a research program he had established in Leiden showed that mercury apparently loses all its electrical resistance at a temperature somewhat below 4.2 K. The difference between the temperature dependence of the elec-

Conductor

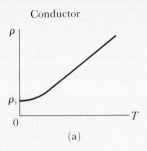

(a)

Superconductor

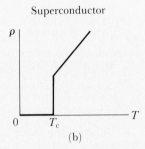

(b)

FIGURE 25.6
The low temperature behavior of the resistivity for (a) a conductor (the value of ρ_i depends on impurities and crystal defects), and (b) a superconductor.

trical resistivity of a normal conductor and that of a superconductor is shown in Fig. 25.6.

Superconductivity is a phenomena that entails more than just zero dc resistance. A superconductor cooled below a certain temperature in a sufficiently low applied magnetic field will also expel all or part of the magnetic flux present in the bulk of the material. This expulsion of magnetic flux is called the **Meissner effect.**

Superconductivity can be destroyed by raising the temperature above a certain **critical temperature,** T_c, and/or by raising the applied external magnetic field beyond a certain **critical field** B_c. The critical field decreases as the temperature increases. In fact, as T approaches T_c, B_c approaches zero. Above the critical temperature and/or the critical field, the resistance isn't zero, magnetic flux penetrates into the bulk of the sample, and the material is said to be **normal.**

For 75 years after the discovery of superconductivity, the highest T_c attained was just over 20 K. Thus superconductivity was possible only if the sample were cooled by using refrigerants such as somewhat rare and costly liquid helium or explosive liquid hydrogen. Then, in January 1986, Karl Müller and Johannes Bednorz found an oxide of barium, lanthanum, and copper that has a T_c almost double the previous high, and the race was

Cooper pairs tunnel through the insulating barrier from one superconductor to the other.

If we increase the direct current above a certain critical value, i_c, two things happen. First, a resistance develops, giving a dc voltage across the Josephson junction. Second, an alternating current is generated and has a frequency of

$$v = \frac{2e}{h} V, \qquad (25.1)$$

where V is the applied dc voltage and $2e/h = 4.836 \times 10^{14}$ Hz/V. Equation (25.1) expresses the **ac Josephson effect.** We could use measurements of the frequency and voltage in the ac Josephson effect to determine precisely the ratio e/h or, equivalently, to measure Planck's constant in units of eV · s.

Multiplying both sides of Eq. (25.1) by h gives $hv = (2e)V$ and sheds some light on this effect. The work done in moving a Cooper pair of charge $2e$ through a potential difference of V is $(2e)V$. This work is used to create a photon of energy hv.

EXAMPLE 25.5 What is the dc potential difference across a Josephson junction if

a) the direct current is less than i_c?   b) a 6.00 GHz alternating current is produced?

Solution

a) For $i < i_c$, the potential difference (voltage drop) across the junction is zero (the dc Josephson effect).

b) This is the ac Josephson effect (so the direct current must be greater than i_c). Evaluating $v = (2e/h)V$ gives

$$V = \frac{v}{2e/h} = \frac{6.00 \times 10^9 \text{ Hz}}{4.836 \times 10^{14} \text{ Hz/V}} = 12.41 \ \mu\text{V}.$$

Alternatively,

$$V = \frac{hv}{2e} = \frac{(4.136 \times 10^{-15} \text{ eV · s})(6.00 \times 10^9 \text{ Hz})}{2e} = 12.41 \ \mu\text{V}. \qquad \bullet$$

The term $2e/h$ has units of Hz/V or $(\text{V · s})^{-1} = (\text{Wb})^{-1}$. Therefore its reciprocal, $h/2e$, equals 2.068×10^{-15} Wb and has units of magnetic flux. In fact, wave-function phase considerations associated with the $B = 0$

superconducting state require that the magnetic flux, Φ_B, through the hole of a superconducting ring must be quantized in units of $h/2e$:

$$\Phi_B = n\,\frac{h}{2e}, \tag{25.2}$$

where $n = 0, 1, 2, 3, \ldots$, and Φ_B is the sum of the magnetic flux from external sources (such as current-carrying coils and permanent magnets) and the magnetic flux from the superconducting current of the ring. Equation (25.2) is a statement of **flux quantization.**

EXAMPLE 25.6   Only six quanta of magnetic flux pass through a Type I superconducting ring. The flux through the ring from an external source is 7.3×10^{-15} Wb. What is the magnetic flux in the hole due to the superconducting currents on the surface of the ring?

Solution   The sum of the external flux and the superconducting current's flux is

$$\Phi_B = n\left(\frac{h}{2e}\right) = 6(2.068 \times 10^{-15}\text{ Wb}) = 12.4 \times 10^{-15}\text{ Wb}.$$

Therefore the superconducting current's magnetic flux is

$$12.4 \times 10^{-15}\text{ Wb} - 7.3 \times 10^{-15}\text{ Wb} = 5.1 \times 10^{-15}\text{ Wb.} \qquad \bullet$$

The quantum 2.068×10^{-15} Wb is an extremely small amount of magnetic flux. Extraordinarily precise measurements of magnetic effects can be made using devices based on the concept of flux quantization. As an example, the magnetic fields from currents in our bodies have been studied (giving magnetocardiograms and magnetoencephalograms). One such device, which also uses Josephson junctions, is armed with the name SQUID, the acronym for **s**uperconducting **qu**antum **i**nterference **d**evice.

We could fill pages with examples of applications of superconductivity (but we won't). Besides zero resistance (and therefore zero i^2R power loss) applications and magnetic levitation (a result of the Meissner effect or induced currents), superconducting devices have provided improvements in sensitivity, accuracy, and/or speed of 10, 100, and more times over the best normal devices. If all these devices can be made to operate at room temperature, the result may be a new electronics revolution as far reaching as that brought about by the invention of the transistor. No wonder those physicists mentioned at the beginning of the chapter were so excited!

25.3
DEFECT SEMICONDUCTORS

Real semiconductors can be very pure, but none are perfect crystals. Defects include missing atoms (vacancies), atoms in noncrystalline sites (interstitials), and impurity atoms somewhere in the crystal. These defects may result in extra electrons, missing electrons, or no change at all in the number of electrons. A semiconductor that has many more than a minimum amount of defects is called a defect semiconductor or an **extrinsic semiconductor.**

An important type of defect is the **substitutional impurity,** an impurity atom that replaces a regular atom in a solid (for example, a P atom replacing a Si atom in a Si crystal). The tremendous advances in solid state electronics are largely the result of our ability to selectively add various amounts of impurity atoms to create different extrinsic semiconductor regions. The process of adding the impurities is called **doping.**

The most common elemental semiconductor materials are Si and Ge. They crystallize in the diamond structure, forming covalent bonds by sharing one electron per atom with each of their four nearest neighbors. Since Si and Ge have four electrons outside a closed shell (they're Group IV atoms in the periodic table), all four of their valence electrons are used for binding. They have a full valence band and an empty conduction band at absolute zero.

Suppose that we now substitute some Group V atoms (such as N, P, As, Sb, or Bi) for the Ge or Si. Group V atoms have five valence electrons instead of the four of Ge or Si. After four of these five valence electrons have been used for the covalent bonds with the four nearest neighbors, the fifth electron is left over, as shown in Fig. 25.11. This fifth electron is still attracted to the Group V atom, but its energy isn't being lowered by the covalent binding process. Therefore this fifth electron is only weakly attracted to the impurity atom. Just a small amount of energy will allow this electron to break free, that is, to move into the conduction band. Thus the Group V impurity atom is called a **donor** because it can easily donate an electron to the conduction band.

On an energy-band diagram (see Fig. 25.12) a representation of a donor ground state must therefore be placed just slightly below the bottom of the conduction band. It's in the forbidden energy gap (which contains energies forbidden for the electrons of the intrinsic material). When the donor electron is freed to move in the conduction band, no hole is left in the valence band, because that donor electron wasn't previously involved in the binding process. Therefore the conduction process with donors is carried out mostly by *negative* charge carriers (the donated electrons), so the material is called an **n-type semiconductor.**

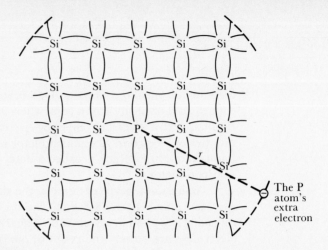

FIGURE 25.11
A schematic representation in two dimensions of a phosphorous (P) substitutional impurity in silicon (Si). Each curved line between adjacent atoms represents an electron being shared in a covalent bond. A phosphorous atom acts as a donor in Si because it can easily donate its extra electron to the conduction process.

Single donor impurity atoms have one easily freed electron. The probability density for this electron is greatest well outside the Z protons and the $Z - 1$ remaining electrons of the donor atom, giving $q_{enclosed,net} = 1e$, or a hydrogen-like atom. Equation (11.3) for the radius of a Bohr orbit has ϵ_0 in the numerator and m in the denominator. Equation (11.4) for the energy of a hydrogen-like atom has m in the numerator and

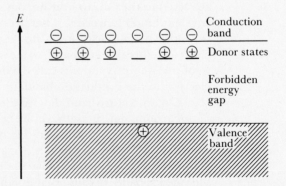

FIGURE 25.12
The energy-band diagram for an n-type semiconductor at $T > 0$ K. All but one of the donor states have been ionized, donating their easily removed electrons to the conduction band. One electron has been thermally excited from the valence band to the conduction band, leaving a hole.

ϵ_0^2 in the denominator. In a semiconductor, we replace the mass m with the conduction band effective mass $m*$ (by multiplying m by the ratio $m*/m$). We also replace the permittivity of a vacuum, ϵ_0, with the permittivity of the semiconductor, $\kappa\epsilon_0$ (by multiplying ϵ_0 by κ, the dielectric constant). Therefore, in this **hydrogenic model** of the donor,

$$r = \frac{\kappa}{m*/m} \, (53 \text{ pm}) \qquad (25.3)$$

and the **donor ionization energy**, E_d, is

$$E_d = \frac{m*/m}{\kappa^2} \, (13.6 \text{ eV}). \qquad (25.4)$$

As you recall the Bohr model was just a first step. We can now better define r as the position of the maximum in the ground state radial probability density of the donor electron about the remaining impurity ion.

EXAMPLE 25.7   The effective mass of an electron in the Ge conduction band is about $0.1m$, and the Ge dielectric constant is 15.8. Calculate the approximate

   a) most probable radius of the donor electron's orbit and   b) donor ionization energy.

Solution The hydrogenic model will give an approximate solution.

   a) We are given $\kappa = 15.8$ and $m*/m = 0.1m/m = 0.1$. Substituting into $r = \kappa(53 \text{ pm})/(m*/m)$, we obtain $r = (15.8)(53 \text{ pm})/(0.1) = 8$ nm.

   b) Substituting the given quantities into $E_d = (m*/m)(13.6 \text{ eV})/\kappa^2$, we have $(0.1)(13.6 \text{ eV})/(15.8)^2 = 0.005$ eV $= 5$ meV.   ●

Germanium atoms are tenths of a nm apart. Part (a) of Example 25.7 shows that so many Ge atoms lie between the donor electron and the remaining donor ion that we were justified in using the macroscopic dielectric constant. That part of the example also shows that if impurity concentrations are sufficiently large, impurity wave functions can begin to overlap one another appreciably, leading to the formation of impurity conduction bands (instead of isolated impurity levels) in the crystal. Actual values of E_d in Ge are about double that calculated in part (b) of Example 25.7 (12 meV for P, 13 meV for As, 10 meV for Sb). More complicated models include the fact that m* varies with crystal direction and give values closer to 10 meV. However, the hydrogenic model agrees reasonably well with experimental results. Also note that Fig. 25.12 isn't drawn to scale. The width of the forbidden energy gap is about 1 eV, but the donor levels are actually only about 1 percent of the gap width (about 0.01 eV) below the conduction band.

Now suppose that we were to substitute some Group III atoms (B, Al, Ga, In, Tl) for intrinsic Group IV Ge or Si. Group III atoms have only three valence electrons compared to the four valence electrons of Ge or Si. Therefore each Group III atom will be missing one valence electron.

It's relatively easy for nearby electrons to move over and fill up the bond vacancy caused by this missing valence electron. If an electron does so, the Group III impurity atom will have accepted an extra electron. The Group III atom in a Group IV material is called an **acceptor** because it can easily accept an electron from the valence band.

The acceptor ground state energy level must therefore be placed slightly above the valence band within the forbidden energy gap. The accepted electron takes part in the covalent binding process and so doesn't move throughout the crystal. However, the hole it leaves behind in the valence band is free to take part in the conduction process. For this type of material the conduction process is carried out mostly by *positive* charge carriers (the holes in the valence band) and so the material is called a **p-type semiconductor.**

EXAMPLE 25.8   Sketch an energy band diagram like that in Fig. 25.12 for a *p*-type semiconductor at $T > 0$ K. Let a circled minus sign also represent an ionized acceptor state (one that has accepted an electron).

Solution See Fig. 25.13.

Step 1. Plot the energy on the vertical axis and the position (roughly) on the horizontal axis.

Step 2. Draw two horizontal lines with the conduction band above the top

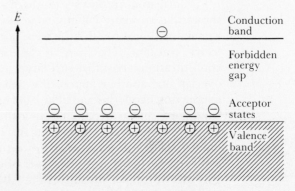

FIGURE 25.13

line, the valence band below the bottom line, and the forbidden energy gap between them.

Step 3. Sketch in some horizontal dashes just above the top of the valence band to represent the acceptor states. We'll use seven of them.

Step 4. Since $T > 0$ K and the acceptor levels are so close to the valence band, let's assume that most of the levels have accepted electrons. Let's put six $\ominus$'s in six of the acceptor levels to represent the six electrons that have been accepted into six impurity binding states and six $\oplus$'s in the valence band to represent the six holes they left behind.

Step 5. Although there are many more valence band and conduction band states than acceptor states, the width of the forbidden energy gap is much greater than the energy needed to move an electron from the top of the valence band to the acceptor state. Therefore we can expect at ordinary temperatures and acceptor concentrations that only a small number of electrons will be excited across the energy gap. To illustrate, let's put one $\ominus$ in the conduction band and one more $\oplus$ in the valence band to represent an electron excited from the valence band to the conduction band, and the hole it left behind.   ●

Note that we have seven $\oplus$'s and seven $\ominus$'s in Fig. 25.13, showing that exciting electrons from one state to another in the solid doesn't change the net charge of the solid. If it was initially neutral, it remains neutral. Another example of maintaining charge neutrality is Fig. 25.12, which has six $\oplus$'s and six $\ominus$'s.

We can also use the hydrogenic model to calculate the acceptor ionization energy and the position of the maximum in the radial probability density of the hole. For this use of the hydrogenic model, m^* is the effective mass of the hole in the valence band.

A **compensated semiconductor** has both donor and acceptor states. If there are more donors than acceptors, the material is *n*-type compensated. With more donors than acceptors, many of the acceptor states would be filled by electrons dropping down from the donor states. (Recall that the electrons will tend to go to the lowest possible energy states.)

If there are more acceptors than donors, the material is *p*-type compensated. At absolute zero, all the donor states would be ionized in this type of material, with the electrons being accepted by the more plentiful acceptor levels. Much of the actual semiconductor material used in devices is compensated material.

In this chapter you saw why energy bands form in solids and how the bands determine whether a material is a conductor, a semiconductor, or an insulator. You learned about the Cooper pairs in the BCS theory, which condense to form a superconducting energy gap, resulting in zero dc resistance below the critical temperature, magnetic field, and current. You also found out about the Meissner effect in type I and type II superconductors. You were introduced to the dc and ac Josephson effects and to magnetic flux quantization. We showed how conduction results from both electrons in the conduction band and holes in the valence band for semiconducting and insulating materials. We described how doping can be used to form n-type, p-type, and compensated semiconductors, the positions of the donor and acceptor levels in the energy gap, and how to use a hydrogenic model. Much of the material presented in this chapter is necessary for an understanding of the next two, quite practical chapters.

QUESTIONS AND PROBLEMS

25.1 Sketch the equivalent of Fig. 25.1 for Na.

25.2 The electron configuration (ground state) for Zn is that of filled 1s, 2s, 2p, 3s, 3p, and 4s subshells. How then can Zn be a metal if its valence subshell is full?

25.3 Some materials make a semiconductor-to-metal transition as the temperature or pressure varies. Using the band model, explain how this is possible.

25.4 An electron combines with a hole in InP, which has an energy gap of 1.35 eV at room temperature. What is the maximum wavelength that could be emitted in the process? Why is it the *maximum* wavelength? How are smaller wavelengths possible?

25.5 In intrinsic material, the electron concentration is approximately proportional to $T^{3/2}e^{-E_g/2kT}$ for various materials. Assume that the proportionality constant is about the same for Ge, Si, GaAs, TiO$_2$, and diamond and find the ratios of the electron concentrations for these five materials at room temperature. What are the equivalent ratios of the hole concentrations?

25.6 State generally how resistance changes with temperature for a) a Pt resistance thermometer b) a Ge resistance thermometer. Explain, using the band model.

25.7 In a semiconductor, what process or processes cause resistance to increase with T? to decrease with T?

25.8 In a previous physics course, you may have found that a long straight wire of radius R carrying a current i uniformly spread over its circular cross-section has a magnetic field of $\mu_0 ir/2\pi R^2$ inside the wire. Why, then, must all the current be on the surface of a long straight type I superconducting wire?

25.9 In a superconductor, the magnetization **M** opposes the applied magnetic field, just as in a diamagnetic material.
   a) Why, then, is a type I superconductor called a perfect diamagnet?
   b) Why will a small permanent magnet float above a horizontal superconducting disk? See Fig. 25.14.

FIGURE 25.14

Source: Photograph by George Stickles, courtesy of Fluoramics, Inc., 103 Pleasant Avenue, Upper Saddle River, NJ 07458.

25.10 The three magnetic vectors, **B**, **H**, and **M**, are related by $\mathbf{B} = \mu_0(\mathbf{H} + \mathbf{M})$, where $\mu_0 = 4\pi \times 10^{-7}$ T · m/A. For a long thin rod with its axis parallel to a uniform magnetic field, **H** inside the rod and **H** just outside the side of the rod are the same. For a superconductor, **M** is almost zero in the normal state. Represent an **M** vector that is opposite **H** by $-M$.

    a) For a type I superconductor, graph $-M$ versus H from $H = 0$ to $H = 1.1B_c/\mu_0$.

    b) For a type II superconductor with $B_{c2} = 3B_{c1}$, sketch $-M$ versus H from $H = 0$ to $H = 1.1B_{c2}/\mu_0$, making an informed guess for the region from B_{c1}/μ_0 to B_{c2}/μ_0.

25.11 The superconducting energy gap in Pb approaches 2.73 meV as T approaches zero.

    a) In this limit, what minimum photon frequency will break up a Cooper pair in Pb?

    b) What can be said about this minimum photon frequency as the temperature increases?

25.12 Microwaves of decreasing wavelength are being directed at a superconductor. Suddenly, at 0.58 mm, the superconductor begins to absorb much more microwave power. Explain why and calculate the width of the superconducting energy gap.

25.13 What is the frequency of the ac component of the current through a Josephson junction if the dc voltage is 3.29 μV?

25.14 A current flowing through a Josephson junction generates 1.24×10^{-5}-eV photons. What is the dc potential difference across the junction?

25.15 The magnetic flux within an N-turn toroid of square cross-section with inner radius a and outer radius b is $[\mu_0 Ni(b - a)/2\pi]\ln b/a$. If the closely wound wires of the toroid are superconducting with negligible flux penetration, what quantized currents are allowed if $b = 3.0$ cm, $a = 2.0$ cm, and $N = 150$?

25.16 Suppose that a magnetic monopole were located just below the center of a superconducting ring. Its magnetic field would be $(\mu_0/4\pi)(q_m/r^2)$, where q_m is its pole strength. The magnetic flux through the hemisphere subtended by the ring would equal this magnetic field times $\frac{1}{2}(4\pi r^2)$. What is the smallest allowed value of the pole strength of a magnetic monopole?

25.17 Adding more donor states to an n-type semiconductor greatly increases its conductivity. Why? This large increase easily overwhelms a small decrease in the conductivity caused by doping. Explain.

25.18 A p-type material carries a net positive charge, and an n-type material carries a net negative charge. True or false? Explain.

25.19 The container in Fig. 25.15 has three separate compartments, A, B, and C. Originally C is filled with marbles, while B and A are empty. Then compartment B is sealed off, several marbles are moved from C to A, and the container is tilted slightly.

    a) What happens?

    b) What does this demonstration have to do with intrinsic semiconductors?

    c) How could you modify the demonstration to make it relate to extrinsic semiconductors?

    d) Is there an analogy to the Pauli exclusion principle here?

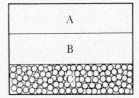

FIGURE 25.15

25.20 There are several ways of removing electrons from the surface of a solid. Can holes be removed from the surface?

25.21 GaSb has an energy gap of 0.72 eV at room temperature. Light of increasing wavelength is shone on GaSb. At what wavelength does the resistivity of a thin intrinsic GaSb film change dramatically? Does it increase or decrease at this wavelength?

25.22 GaAs has an energy gap of 1.521 eV at low temperature. The optical absorption is virtually zero as the photon energy increases to 1.517 eV, when the optical absorption begins to increase. Why is the 1.517-eV value associated with excitons? What is the exciton binding energy?

25.23 Why don't excitons give an electric current as they migrate through a semiconductor?

25.24 The effective mass of conduction band electrons in Si is about $0.2m$ and the dielectric constant is 11.7. Calculate the approximate position of the radial probability density maximum of a single donor electron and the approximate donor ionization energy.

25.25 The single donor ground state for the Bohr hydrogenic model in GaAs has a probability density maximum at approximately 10 nm, and the dielectric constant is 13.13. Calculate the corresponding donor ionization energy.

25.26 Use the results of Problem 25.24 and the fact that there are 4.99×10^{28} Si atoms/m³ to find the approximate percentage of completely uniform donor level concentration that will result in the formation of an impurity band. Then explain why the band will actually form at a lower donor concentration.

25.27 The acceptor ionization energies in Ge are 10 to 11 meV. Calculate the approximate position of the maximum in the radial probability density of the holes in Ge and their approximate valence band effective mass.

25.28 a) Give some possible combinations of dopants (elements) that might be found in compensated Si or Ge.

b) How does the addition of some acceptor levels "compensate" for the donor levels previously present in n-type material?

c) Sketch the energy band diagrams for both p-type and n-type compensated semiconductors at absolute zero, showing ionized donor and acceptor levels and charge neutrality.

d) Repeat part (c) for a compensated semiconductor that is neither n-type nor p-type.

25.29 A piece of Si has adjacent regions A, B, C, and D. Region A is undoped. Regions B and C have equal dopings of element x. Regions C and D have equal dopings of element y. Region D has a higher doping of element x than any other region. Give some possible elements and some relative doping levels if region A is intrinsic, region B is p-type, region C is compensated n-type, and region D is compensated p-type.

25.30 What kind of a donor would you get if you doped Si with a Group VI atom? Could you estimate the two donor ionization energies? Explain.

25.31 An atom from what group could accept two electrons when doped into Ge?

25.32 At absolute zero, where would you expect the Fermi energy to be in   a) an intrinsic semiconductor?   b) an n-type semiconductor?
c) a p-type semiconductor?

CHAPTER

Extrinsic Semiconductors

The p–n junction is the basis of the semiconductor industry. You begin your study of this type of junction in this chapter. You'll also learn about mobility, so that you can start to explain statements such as: "GaAs devices are being developed because the higher electron mobility of GaAs makes them intrinsically faster than Si devices." Conductivity and mobility are electric-field effects, but you'll also be introduced to two useful magnetic field phenomena: the Hall effect and magnetoresistance.

26.1
SOME SEMICONDUCTOR PROPERTIES

One very important electrical conduction property of electrons and holes is their mobility. The **mobility,** μ, is defined as the magnitude of the drift velocity, v_d, divided by the magnitude of the electric field, E, which drives the current. That is,

$$\mu = \frac{v_d}{E}. \tag{26.1}$$

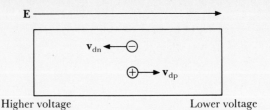

FIGURE 26.1
Holes drift with the electric field (to a lower voltage), while electrons drift against the electric field (to a higher voltage).

Recall that the magnitude of a vector is always positive. Therefore, μ is always a positive number despite the fact that holes drift in the direction of the electric field, while electrons drift opposite to **E**, as illustrated in Fig. 26.1.

Let's start now to use n instead of N/V for the electron concentration. The term n is more convenient to use and shouldn't be confused with a quantum number in the following sections. So the relations $\sigma = j/E$ for the electrical conductivity and $\mathbf{j} = nq\mathbf{v_d}$ for the current density give us $\sigma = n|q|v_d/E = n|q|\mu$.

Suppose that both electrons and holes carry the current. First, we define p to be the number of holes per volume, called the **hole concentration.** Then the total conductivity is the sum of the parallel electron and hole conductivities, or

$$\sigma = e(n\mu_n + p\mu_p). \tag{26.2}$$

In this relation $e = |q| = 1.60 \times 10^{-19}$ C for electrons or holes; n is the free-electron concentration (in m^{-3}); p is the free-hole concentration (in m^{-3}); μ_n and μ_p are the electron and hole mobility, respectively (in m^2/V $\cdot$ s); and σ has units of $(\Omega \cdot m)^{-1}$.

EXAMPLE 26.1   In Si at room temperature the electron mobility is 0.135 m^2/V $\cdot$ s, and the hole mobility is 0.048 m^2/V $\cdot$ s. At room temperature:

a) Calculate the free-electron and free-hole concentrations of intrinsic Si if its conductivity is 3.8×10^{-4} $(\Omega \cdot m)^{-1}$.

b) Calculate the conductivity of n-type Si with 1.3×10^{19} free electrons/m^3 and 1.3×10^{13} free holes/m^3.

c) What are the electron and hole drift velocities if a very large field $\mathbf{E} = 2.0 \times 10^6$ V/m is applied in the $+x$ direction?

Solution

a) Since this is an intrinsic material, the number of electrons in the conduction band equals the number of holes in the valence band. That is, $n = p$. We then have

$$\sigma = e(n\mu_n + p\mu_p) = en(\mu_n + \mu_p)$$

or

$$n = \frac{\sigma}{e(\mu_n + \mu_p)} = \frac{3.8 \times 10^{-4} \ (\Omega \cdot m)^{-1}}{(1.60 \times 10^{-19} \ C)(0.135 + 0.048)m^2/V \cdot s}$$
$$= 1.3 \times 10^{16} \ m^{-3} = p.$$

(The units work out because $\Omega = V/A$ and $C = A \cdot s$).

b) Direct substitution into Eq. (26.2) gives

$$\sigma = (1.60 \times 10^{-19} \ C)[(1.3 \times 10^{19} \ m^{-3})(0.135 \ m^2/V \cdot s)$$
$$+ (1.3 \times 10^{13} \ m^{-3})(0.048 \ m^2/V \cdot s)] = 0.28 \ (\Omega \cdot m)^{-1}.$$

c) The definition of the mobility gives $v_d = \mu E$. For the electrons in room temperature Si,

$$v_d = (0.135 \ m^2/V \cdot s)(2.0 \times 10^6 \ V/m) = 2.7 \times 10^5 \ m/s.$$

Electrons are negative and so are pushed opposite the direction of **E**. Therefore the electron drift velocity is 2.7×10^5 m/s in the $-x$ direction. For the positive holes,

$$v_d = (0.048 \ m^2/V \cdot s)(2.0 \times 10^6 \ V/m) = 9.6 \times 10^4 \ m/s.$$

Holes are positive, so they'll drift in the $+x$ direction. ●

There are 4.99×10^{28} Si atoms/m³, and $n = p = 1.3 \times 10^{16}$ m⁻³ in part (a) of Example 26.1. Therefore the fraction of electrically active impurities must be less than $1.3 \times 10^{16}/4.99 \times 10^{28} = 2.6 \times 10^{-13}$ for Si to be intrinsic at room temperature. That fraction is less than three parts per ten million million. Making such pure material is a very difficult task. In part (b), note that we could neglect the holes' contribution to this conductivity because $n \gg p$. Also note that when n was increased, p was decreased. That occurred because many of the donor electrons dropped down to combine with holes in the valence band. In fact, over large variations of n and p, the product np remains constant at a given temperature for a particular semiconductor.

If a magnetic field **B** is applied perpendicular to a current density **j** flowing in a slab of material, the charge carriers will experience a magnetic force $\mathbf{F} = q\mathbf{v_d} \times \mathbf{B}$. For the directions shown in Fig. 26.2, you can see that this force pushes both negative and positive charge carriers to the top of the slab. This charging will cause an electric field $\mathbf{E_H}$, which at equilibrium will produce a force to balance the magnetic force; that is, $qE_H = qv_d B$. Since $v_d = j_x/n(-e)$ for electrons and $v_d = j_x/p(+e)$ for holes, we can write

$$E_H = \frac{j_x B}{n(-e)} \ \text{(electrons)} \quad \text{and} \quad E_H = \frac{j_x B}{p(+e)} \ \text{(holes)}. \qquad (26.3)$$

FIGURE 26.2
The Hall effect: For the magnetic field and current density directions shown, both electrons and holes are pushed toward the top of the sample by the magnetic force.

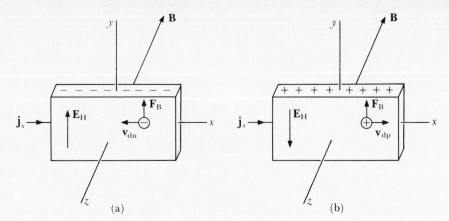

(a)   (b)

The creation of this electric field (and its resulting potential difference) by a transverse magnetic field acting on a current-carrying material is called the **Hall effect.** The term $\mathbf{E}_H$ is the **Hall electric field,** and the resulting potential difference is the **Hall voltage.**

Since the Hall electric field and voltage are directly proportional to the magnetic field, Hall-effect probes are used to measure and control magnetic fields and currents that cause magnetic fields. The Hall voltage is directly proportional to the product of the current and the magnetic field (possibly caused by another current), so a Hall-effect device can be used as an analog multiplier. The Hall voltage is inversely proportional to n or p, so it is used to determine the carrier concentration, with its sign giving the sign of the majority carriers in most situations.

You must take some care in applying the simple model we've used here. More subtle effects may introduce drastic changes in the relations that we derived. For instance, the charge carriers in Al are electrons, but the Hall voltage is positive for Al.

The Hall electric field per current density per magnetic field is called the **Hall coefficient,** R_H. That is,

$$R_H = \frac{E_H}{j_x B}. \tag{26.4}$$

Despite the symbol R, R_H is *not* a resistance; R_H has units of m^3/C in SI. Substituting Eq. (26.3) into Eq. (26.4) gives (for our simple model):

$$R_H = -\frac{1}{ne} \text{ (electrons)} \quad \text{and} \quad R_H = +\frac{1}{pe} \text{ (holes)}. \tag{26.5}$$

If both electrons and holes are present, a more complicated derivation gives

$$R_{\mathrm{H}} = \frac{1}{e} \frac{(p\mu_p^2 - n\mu_n^2)}{(p\mu_p + n\mu_n)^2}. \tag{26.6}$$

EXAMPLE 26.2 Calculate the Hall coefficient for the

a) intrinsic and   b) n-type Si samples of Example 26.1.

Solution

a) For an intrinsic semiconductor, $n = p$, so we must use Eq. (26.6), which simplifies to

$$R_{\mathrm{H}} = \frac{1}{e} \frac{(n\mu_p^2 - n\mu_n^2)}{(n\mu_p + n\mu_n)^2} = \frac{1}{ne} \frac{\mu_p - \mu_n}{\mu_p + \mu_n}.$$

Substituting the quantities $n = 1.3 \times 10^{16}$ m^{-3}, $e = 1.6 \times 10^{-19}$ C, $\mu_p = 0.048$ m^2/V $\cdot$ s, and $\mu_n = 0.135$ m^2/V $\cdot$ s gives $R_{\mathrm{H}} = -2.3 \times 10^2$ m^3/C.

b) For the n-type Si, $n \gg p$ and $\mu_n > \mu_p$, so Eq. (26.6) reduces to the first expression of Eq. (26.5):

$$R_{\mathrm{H}} = -\frac{1}{ne} = -\frac{1}{(1.3 \times 10^{19}\ \mathrm{m}^{-3})(1.60 \times 10^{-19}\ \mathrm{C})} = -0.48\ \mathrm{m}^3/\mathrm{C}. \quad \bullet$$

EXAMPLE 26.3 Calculate the Hall electric field of the

a) intrinsic and

b) n-type Si samples of Examples 26.1 and 26.2 for a current density of 5.0×10^3 A/m^2 and a magnetic field of 0.30 T.

Solution

a) Equation (26.4) gives $E_{\mathrm{H}} = R_{\mathrm{H}} j_x B$. Substituting the value of R_{H} from part (a) of Example 26.2 gives

$$E_{\mathrm{H}} = -(2.3 \times 10^2\ \mathrm{m}^3/\mathrm{C})(5.0 \times 10^3\ \mathrm{A/m}^2)(0.30\ \mathrm{T}) = -3.4 \times 10^5\ \mathrm{V/m}.$$

(For the units, C $=$ A $\cdot$ s and T $=$ V $\cdot$ s/m^2.)

b) Substituting the value of R_{H} from part (b) of Example 26.2 gives

$$E_{\mathrm{H}} = -(0.48\ \mathrm{m}^3/\mathrm{C})(5.0 \times 10^3\ \mathrm{A/m})(0.30\ \mathrm{T}) = -7.2 \times 10^2\ \mathrm{V/m}. \quad \bullet$$

EXAMPLE 26.4 Calculate the Hall coefficient and Hall electric field for Cu (which has $n = 8.5 \times 10^{28}$ m^{-3}) for the same current density and magnetic field as in Example 26.3.

Solution The charge carriers in Cu are all electrons. Therefore

$$R_H = -\frac{1}{ne} = -\frac{1}{(8.5 \times 10^{28}/\text{m}^3)(1.60 \times 10^{-19}\ \text{C})} = -7.4 \times 10^{-11}\ \text{m}^3/\text{C}.$$

Substituting the calculated R_H gives

$$E_H = -(7.4 \times 10^{-11}\ \text{m}^3/\text{C})(5.0 \times 10^3\ \text{A/m}^2)(0.30\ \text{T})$$
$$= -1.1 \times 10^{-7}\ \text{V/m}. \qquad \bullet$$

The actual experimental value of R_H for Cu is $-5.4 \times 10^{-11}\ \text{m}^3/\text{C}$, not the value obtained in Example 26.4. This difference points up some deficiencies in our simple model.

The reason that the Hall field dropped as the number of charge carriers increased in Examples 26.2–26.4 is found from the relation $\mathbf{j}_x = nq\mathbf{v}_d$. More charge carriers result in a lower v_d for a given j_x. Therefore a lower E_H is necessary to balance the decreased $q\mathbf{v}_d \times \mathbf{B}$ force.

The Hall electric field divided by the electric field driving the current, E_H/E_x, is an important ratio. Since $E_x = j_x/\sigma$, $E_H/E_x = E_H\sigma/j_x$ and

$$\frac{E_H\sigma}{j_x} = \frac{(-3.4 \times 10^5)(3.8 \times 10^{-4})}{5.0 \times 10^3} = -2.6 \times 10^{-2} \qquad \text{for intrinsic Si;}$$

$$= \frac{(-7.2 \times 10^2)(0.28)}{5.0 \times 10^3} = -4.0 \times 10^{-2} \qquad \text{for } n\text{-type Si;}$$

$$= \frac{(-1.1 \times 10^{-7})(5.9 \times 10^7)}{5.0 \times 10^3}$$

$$= -1.3 \times 10^{-3} \qquad \text{for Cu (our simple model).}$$

This ratio will be related to the ratio of the Hall voltage to the resistive voltage drop. Thus the Hall voltage will be much lower than the already small voltage drop along a metal sample, making the Hall voltage difficult to measure. Therefore semiconductors are used in Hall-effect devices.

For electrons in our simple model, $j_x/E_H = n(-e)/B$ decreases continuously with increasing B. However, for special geometries that give essentially a two-dimensional electron gas, the ratio j_x/E_H turns out to be quantized at low temperatures and very high magnetic fields. This quantization results from the allowed orbits of the electrons in the magnetic field. Instead of changing continuously with B, the ratio changes in very precise steps. This **quantized Hall effect** has been used to measure e^2/h to an accuracy of less than one part per million.

Our simple model considered every electron to be moving at one drift velocity and every hole to be moving at another drift velocity. Actually, drift velocities will be distributed about these average values. Electrons or holes moving at their drift velocities move in a straight line between collisions with the force from the applied magnetic field balanced by the force from the Hall electric field. But charge carriers with speeds of

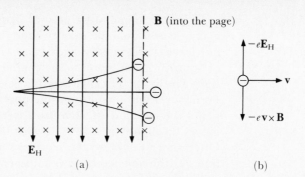

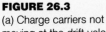

FIGURE 26.3
(a) Charge carriers not moving at the drift velocity have curved paths, which results in magnetoresistance. (b) Vector diagram for an electron.

less than v_d will have $qv_dB < qE_H$ and will curve in one direction. Those with speeds of more than v_d will curve in the opposite direction, as illustrated in Fig. 26.3 for electrons. The charge carriers with a curved mean free path won't move quite as far down the material between collisions. The current will decrease slightly, and therefore the resistance will increase, giving one cause of **magnetoresistance,** the extra resistance due to a magnetic field. Magnetoresistance is also caused by the change in the electron wave functions with **B**. Therefore a piece of material with a large magnetoresistance effect can be used to determine the presence or absence of magnetization in a magnetic memory device.

26.2
THE FERMI LEVEL IN SEMICONDUCTORS

In Section 24.2 we discussed the Fermi energy, E_F, for the free-electron gas and made the following observations about it:

1. At $T = 0$ K, E_{F0} was the maximum electron kinetic energy. All states below E_{F0} were full and all states above E_{F0} were empty.

2. At $T > 0$ K, E_F was the energy where the probability of occupation of an electron state was 50 percent. Below E_F the probability was between 100 percent and 50 percent; above E_F the probability decreased from 50 percent to 0 percent.

However, in a semiconductor, things aren't quite so simple. Quantum statistical methods give the following results. In an intrinsic semiconductor at $T = 0$ K, the outer electrons are all in the valence band. The probability is 100 percent that the states in the valence band are full and 0 percent that the states in the conduction band are occupied. The Fermi energy therefore should be somewhere between the two bands. In fact, for simple bands it's exactly in the center of the forbidden energy gap.

At any temperature greater than absolute zero, electrons are thermally excited from the valence band to the conduction band. This excitation leaves states in the valence band with less than 100 percent probability of occupation and states in the conduction band with greater than 0 percent. The lighter effective mass particles spread over a wider band, and the Fermi level shifts slightly with temperature toward that band. For example, if the electron effective mass is less than the hole effective mass, E_F moves up slightly in energy toward the conduction band as T increases. Often there will be so few electrons in the conduction band and so few holes in the valence band (compared to the huge number of available states) that the electron and hole gases won't be much affected by the Pauli exclusion principle. Therefore we can describe them accurately as classical gases in those bands.

EXAMPLE 26.5   Suppose that energy gaps of 0.2 eV, 1.0 eV, and 5.0 eV exist in intrinsic materials, with the Fermi level at the center of the energy gap. Calculate the probabilities of occupation of electron states at the bottom of the conduction band for these materials at 27 °C.

Solution The probability of occupation of an electron state in a band is most accurately given by the Fermi–Dirac distribution function, $f_{FD}(E) = 1/[e^{(E-E_F)/kT} + 1]$. At 27 °C, $T = (27 + 273)$ K = 300 K, and

$$kT = (8.62 \times 10^{-5} \text{ eV/K})(300 \text{ K}) = 2.586 \times 10^{-2} \text{ eV}.$$

Here $E - E_F$ is the energy difference between the top of the energy gap and its middle, and so $E - E_F = E_g/2$. Substituting the 0.2-eV, 1.0-eV, and 5.0-eV values of E_g gives $(E - E_F)/kT$ of 3.87, 19.3, and 96.7. The values of $f_{FD}(E)$ are then 2.0×10^{-2}, 4.0×10^{-9}, and 1.0×10^{-42}.   ●

Two percent of all states were occupied for the 0.2-eV (narrow gap) semiconductor in Example 26.5. Electrons had more difficulty jumping the more typical 1.0-eV semiconductor energy gap, and only four states of every thousand million were occupied. You should easily see why the material with a 5.0-eV energy gap is an insulator. Only one state of every 10^{42} was occupied at the bottom of the conduction band.

The probability that a hole occupies a state in the valence band is the probability that there's no electron in that state, or $1 - f_{FD}(E)$. For holes at the top of the valence band for the materials of Example 26.5, $E - E_F = -E_g/2$. Since $1 - 1/(e^{-x} + 1) = 1/(e^x + 1)$, the answers in Example 26.5 are also the fractions of those hole states that are occupied.

For an extrinsic semiconductor, you might be tempted to make E_F the 50 percent probability energy for the defect states in the forbidden energy gap, but that's incorrect. For instance, a single donor ground state level may be doubly degenerate (spin up and spin down). But only one electron can occupy this ground state level. Therefore the maximum probability of occupation per state is 0.5. The expression $f_{FD} = 1/[e^{(E-E_F)/kT} + 1]$ is derived for a maximum probability of 1, not 0.5, so f_{FD} isn't correct for the donor and acceptor states in the energy gap. Another complication occurs because of the excited defect states.

Although the probability that a donor level or acceptor level is occupied may not equal 0.5 when the Fermi energy equals the energy of that level, this probability must lie somewhere between 0 and 1. At $T = 0$ K, an n-type semiconductor with no acceptor levels will have a 100 percent probability that there is an electron in each donor and a 0 percent probability that there are any electrons in the conduction band. Therefore E_F is near the top of the energy gap (between the donor ground state levels and the bottom of the conduction band).

Similarly, in a p-type semiconductor at $T = 0$ K with no compensation, the Fermi level is near the bottom of the energy gap (between the top of the valence band and the acceptor levels). The position of the Fermi level will change with temperature, but can generally be found near the top of the energy gap for n-type material and near the bottom of the gap for p-type material. The exception is lightly doped material at elevated temperatures (where the material would act like intrinsic material).

EXAMPLE 26.6 Sketch the energy band diagram at $T = 0$ K, including the Fermi level, for

a) pure Si,   b) Si doped with Al, and   c) Si doped with As.

Solution Pure Si is intrinsic Si, Si doped with Al is p-type Si, and Si doped with As is n-type Si. Based on our previous discussion, the answers are as shown in Fig. 26.4(a), (b), and (c).   ●

FIGURE 26.4

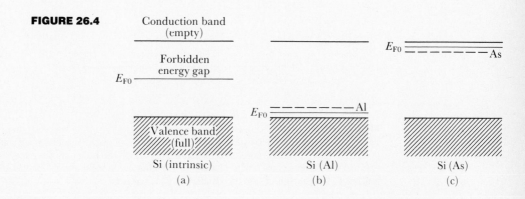

26.3
THE p–n JUNCTION

Suppose that we have a piece of p-type Si and another piece of n-type Si that are separated from each other at room temperature, as in Fig. 26.5. The p-type material would conduct mostly by means of its "hole gas" in its valence band. The n-type material would conduct mostly by means of its "electron gas" in the conduction band. Now if the p-type material were in some way formed in contact with the n-type material, we'd have a **p–n junction.** (We couldn't just stick the two pieces of material together and expect the junction to work properly because of the impossibility of matching their surfaces at the atomic level.)

Suppose that you had a container with a barrier across its center. On one side of the barrier you place oxygen gas and on the other side nitrogen gas. What would happen if you then removed the barrier? Some oxygen molecules would immediately begin to diffuse to the nitrogen side, and some nitrogen molecules would begin to diffuse to the oxygen side. You could then say that this diffusion is a result of the initial concentration difference of either gas between the two parts of the container.

A similar event occurs when the p and n sides of the p–n junction are formed in contact. Some of the hole gas from the p side diffuses into the n side, which gives the n side a positive charge. At the same time, some of the electron gas from the n side diffuses into the p side, which gives the p side a negative charge. Therefore the n side becomes positively and the p side negatively charged! An electron diffusing into the p side finds many holes to fall into, and a hole diffusing into the n side soon recombines with an electron.

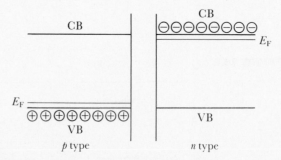

FIGURE 26.5
Separated p-type and n-type pieces of the same semiconductor. The letters CB mean *conduction band,* and the letters VB mean *valence band.* The donor states, acceptor states, and minority carriers were omitted for clarity. (The few electrons in the p-type CB and the few holes in the n-type VB are the omitted minority carriers.)

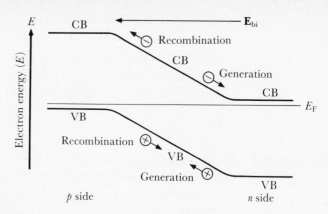

FIGURE 26.6

The *p–n* junction at equilibrium, with no external voltage being applied. Remember, **E** is the electric field (a vector) and *E* is the electron energy (a scalar).

At equilibrium, some of the results of these diffusions include:

1. An electric field, called the **built-in field** ($\mathbf{E_{bi}}$) is set up in the region of the junction. The field is directed from the side having a plus charge to the side having a minus charge. That is, $\mathbf{E_{bi}}$ is directed from the *n* side to the *p* side.

2. A **built-in voltage** results from this built-in electric field.

3. The force on an electron acts opposite to the electric-field direction. Therefore to move an electron from the *n* side to the *p* side requires external work. Consequently, an electron's energy is raised as it travels from the *n* side to the *p* side, giving an "energy hill," as Fig. 26.6 shows.

4. The *p* side energy is raised and the *n* side energy is lowered until the Fermi energy is constant across the junction. This gives a constant probability across the junction for equilibrium (at a given energy in a particular band).

5. The recombination leaves the junction region depleted of charge carriers compared to the remainder of the material.

6. The electrons and holes diffusing across the junction because of the concentration difference (and then recombining) give a **recombination current.** For electrons at zero external voltage, this current is $i_{nr}(0)$. For holes, it is $i_{pr}(0)$.

7. The electron–hole pairs constantly being generated in the junction region by thermal processes are separated and swept out of this region by the built-in field. The result is what's called a **generation**

current. For electrons at zero external voltage, this current is i_{ng} (0). For holes, it is i_{pg} (0).

8. At equilibrium the recombination and generation currents are equal and opposite. If we let i stand for the absolute value of the current, we have i_{nr} (0) $= i_{ng}$ (0) for the electrons and i_{pr} (0) $= i_{pg}$ (0) for the holes.

EXAMPLE 26.7   Explain why a capacitance must be associated with a p–n junction. Recall that the capacitance is defined by $C = q/V$, which means that equal and opposite charges are separated by some distance and have a potential difference V between them.

Solution In the p–n junction the equal and opposite separated charges result from the diffusions of the electron and hole gases. This diffusion makes the p side negative and the n side positive, giving a built-in electric field. The potential difference, V, is then the built-in voltage which results from this built-in field.   ●

In p–i–n junction structures, the p and n sides are separated by an intrinsic layer. You may remember from a previous physics course that $C = \kappa \epsilon_0 A / d$ for a parallel-plate capacitor. The extra intrinsic layer increases the separation, d, and therefore decreases the capacitance.

Incidentally, the junction's built-in voltage can't be used as the emf in a constant temperature circuit. Other voltages will also be developed at the two contacts to the device when it's placed in a complete circuit. The total voltage across two similar contacts and the p–n junction will then add to zero, so a meter across the sample won't measure the built-in voltage. Since potential difference equals work per charge, any other result would mean a constant energy output with no energy input!

We began this chapter by discussing effects connected with an applied electric field. The mobility and electric field effects enable us to determine the drift velocity of the charge carriers. Also, the conductivity (which we related to mobilities and carrier concentrations) and electric field effects allow us to calculate current densities. We then saw how the addition of an applied magnetic field can result in the Hall effect and magnetoresistance. Previously, we had considered the Fermi level in metals; in this chapter we extended our consideration to intrinsic and extrinsic semiconductors. Finally, we completed the chapter with an introduction to the basic properties of the commonly used p–n junction. We will use many of these properties in our descriptions of semiconductor devices in Chapter 27.

QUESTIONS AND PROBLEMS

26.1 For an electron gas, how is the mobility related to the collision time?

26.2 In Ge at room temperature, the electron mobility is $0.36 \text{ m}^2/\text{V} \cdot \text{s}$ and the hole mobility is $0.18 \text{ m}^2/\text{V} \cdot \text{s}$. At room-temperature:

a) The resistivity of intrinsic Ge is $0.42 \ \Omega \cdot \text{m}$. Calculate the concentrations of free electrons and free holes.

b) Calculate the conductivity of p-type Ge with $6.0 \times 10^{21}/\text{m}^3$ free holes and very few free electrons.

c) What electric field will make holes drift at 100 m/s to the left?

d) If holes drift at 100 m/s to the left, at what velocity will electrons drift?

26.3 In GaAs at room temperature, a 10^4 V/m electric field will cause electrons and holes to drift at speeds of 8000 m/s and 300 m/s, respectively. Calculate the conductivity of extrinsic GaAs if it is   a) highly p type with 10^{23} carriers/m^3 and b) highly n type with 10^{23} carriers/m^3.

26.4 For PbTe, a ratio of conductivity to electron concentration for strongly n-type materials has its greatest value at 4 K where it equals 8.0×10^{-17} C $\cdot$ m^2/V $\cdot$ s. What is the corresponding electron drift velocity for a 2.0×10^{-2} V/m field in the $+y$ direction?

26.5 In our examples and problems, we have assumed that the mobility is the same for intrinsic and extrinsic samples of a given semiconductor. Will the mobility actually be greater for intrinsic material or extrinsic material? Explain.

26.6 How will the Hall field be related to the Hall voltage through a dimension of the sample?

26.7 a) From the results of Problem 26.2, calculate the Hall coefficient of intrinsic Ge.
b) What current density is needed for a Hall field of -2.5×10^3 V/m in a 0.60 T magnetic field?

26.8 In a particular semiconductor, the electron mobility is twice the hole mobility. It is p type with 10 times as many holes as electrons. The Hall coefficient equals 2.5×10^{-3} m^3/C. The conductivity equals $3.7 \ (\Omega \cdot \text{m})^{-1}$. Calculate the electron and hole concentration and mobilities.

26.9 Show that $E_H/E_x = R_H \sigma B$.

26.10 Show that the difference between the hole mobility and the electron mobility equals the product of the conductivity and the Hall coefficient for intrinsic semiconductors.

26.11 For a highly n-type semiconductor, the Hall field is -27 V/m for a current density of 1.8×10^4 A/m^2 and a magnetic field of 0.50 T. Calculate the electron concentration.

26.12 For a highly p-type semiconductor with a hole concentration of 1.2×10^{24} m^{-3} what current density and magnetic field will give a Hall field of 100 V/m?

26.13 The terms *transverse magnetoresistance* and *longitudinal magnetoresistance* refer to the direction of **B** relative to $\mathbf{v}_d$. Both of the reasons for magnetoresistance given at the end of Section 26.1 apply to transverse magnetoresistance, but only one of the reasons applies to longitudinal magnetoresistance. Explain why.

26.14 A resistance thermometer is to be used in a magnetic field. How should the thermometer be oriented with respect to the field to minimize errors due to magnetoresistance?

26.15 A completely compensated semiconductor has as many single donors as single acceptors. Sketch the energy band diagram at $T = 0$ K for a completely compensated semiconductor, including the Fermi level.

26.16 At 77 K an intrinsic semiconductor has only 7.5×10^{-48} of all electron states near the bottom of the conduction band and hole states near the top of the valence band occupied. Calculate the energy gap.

26.17 GaSb has an energy gap of 0.68 eV at room temperature ($kT = 1/40$ eV). Assume that the Fermi level is at the center of the energy gap for intrinsic GaSb at this temperature.

a) Find the probability that an electron state at the bottom of the conduction band is occupied.

b) Find the probability that a hole state at the top of the valence band is occupied.

c) In which direction would your answers to (a) and (b) change for p-type GaSb?

26.18 As the temperature rises, a lightly doped semiconductor soon exhausts all its donors or acceptors (the defect levels become nearly all ionized). Then the material begins to behave more and more like intrinsic material. Discuss what happens to the Fermi level as the temperature rises from absolute zero if this type of material is a) n type and b) p type.

26.19 A semiconductor is so highly doped that the impurity band formed overlaps the conduction band. Discuss the position of the Fermi level at low temperature.

26.20 a) Sketch the net charge per volume for a p–n junction at equilibrium as a function of distance from the center of the junction.

b) Make a similar sketch showing both the concentration of free electrons and the concentration of holes.

26.21 Do some outside reading to find methods of fabricating p–n junctions. Describe two.

26.22 If the p material and the n material are the same (except for different dopants), the junction is occasionally called a homojunction. What would a heterojunction be? (Yes, the words homosexual and heterosexual do contain a clue to the answer.)

26.23 How would Fig. 26.6 change as the doping of the sides decreased toward zero?

26.24 How can the n (for *n*egative) side of a p–n junction be positive and the p (for *p*ositive) side be negative?

26.25 The electric field between the plates of a parallel-plate capacitor is constant (neglecting the fringing at the edges). However, the built-in electric field in the p–n junction is not constant. Consider the distributions of charge in the two cases and explain why.

26.26 The hole energy is raised as it travels from the p side to the n side of a p–n junction. Explain.

26.27 Why is the p–n junction region a region of higher resistance than the surrounding semiconductor material?

26.28 a) If the electron generation current across a p–n junction is 73 μA, what is the electron recombination current at zero external voltage?

b) In which directions do each of these currents flow?

26.29 a) If the hole generation current across a p–n junction is 49 μA, what is the hole recombination current at zero external voltage?

b) In which directions do each of these currents flow?

26.30 Structures like the following are now being constructed. A wide-gap semiconductor and a narrow-gap semiconductor with the same crystal structure and atomic spacing are formed in layers in the following order: thick layer of wide gap, thin layer of narrow gap, thick layer of wide gap. Assume that these semiconductors are intrinsic and have equal Fermi energies. Sketch the equivalent of Fig. 26.6 and explain why any available electrons would tend to be trapped in the thin layer of narrow-gap semiconductor.

CHAPTER

Some Semiconductor Devices

In many electronics courses semiconductor devices are explained mainly in terms of their characteristic curves and applications. The scientific basis of how they work is often minimized or ignored. I believe that by learning the physics behind some of these basic devices, you can better understand their operation and applications. You will then be better able to apply what you know to analyzing or designing new semiconductor devices.

27.1
THE JUNCTION DIODE

Let's connect a battery to a piece of semiconductor material containing a p–n junction. Suppose that we connect the plus battery terminal to the p side and the negative battery terminal to the n side, as shown in Fig. 27.1. Such a connection gives the p–n junction a **forward bias,** with an external potential difference V applied to the junction. (This V equals the battery emf minus all iR drops external to the junction.) The doped semiconductor material is a reasonably good conductor, except in the depleted region

355

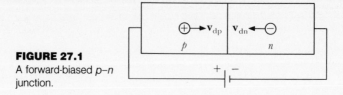

FIGURE 27.1

A forward-biased p–n junction.

of the p–n junction. There the battery sets up a strong electric field $\mathbf{E_B}$, which opposes the built-in electric field, $\mathbf{E_{bi}}$. The total electric field $\mathbf{E}$ is the sum of the oppositely directed vectors $\mathbf{E_{bi}}$ and $\mathbf{E_B}$. Therefore $\mathbf{E}$ is smaller than before the battery was connected. This smaller field results in a smaller force on the electrons; Fig. 27.2 therefore shows a smaller energy difference between the p and n sides of the junction for forward bias.

The smaller electric field and the smaller energy hill don't greatly affect the rate at which electron and hole pairs are thermally generated in the junction region (assuming constant temperature). Therefore the generation current with a potential difference V is approximately the same as with no potential difference for both electrons and holes. Algebraically,

$$i_{ng}(V) \approx i_{ng}(0) \quad \text{and} \quad i_{pg}(V) \approx i_{pg}(0).$$

But now the electron and hole gases don't have to climb as high an energy hill to diffuse (electrons into the p side, holes into the n side). Thus the number diffusing into the same states, which have been lowered by an amount of energy $\Delta E = (-e)V$, can be expected to change by the Boltzmann factor, $e^{-\Delta E/kT} = e^{eV/kT}$. We can use the classical Boltzmann factor here; the Pauli exclusion principle limitation becomes unimportant because there are so many available states and so comparatively few electrons

FIGURE 27.2

The height of the "energy hill" is decreased by forward bias. The vector E's are electric fields (bias, built-in, and resultant), and the scalar E is the electron energy.

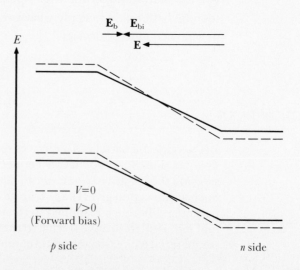

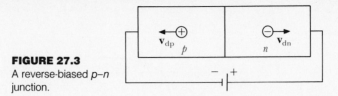

FIGURE 27.3
A reverse-biased *p–n* junction.

and holes in the energy bands. The recombination currents, then, increase to

$$i_{nr}(V) = i_{nr}(0)e^{eV/kT} \quad \text{and} \quad i_{pr}(V) = i_{pr}(0)e^{eV/kT}$$

for forward bias with a potential difference V.

The total forward current for the electrons with a forward bias becomes

$$i_{nr}(V) - i_{ng}(V) = i_{nr}(0)e^{eV/kT} - i_{nr}(0) = i_{nr}(0)(e^{eV/kT} - 1).$$

(We used $i_{ng}(V) = i_{ng}(0)$ and $i_{ng}(0) = i_{nr}(0)$ to derive this expression.) A similar relation holds for the holes: Just replace the n subscripts with p subscripts. Therefore the total current through the p–n junction is

$$i = i_s(e^{eV/kT} - 1). \tag{27.1}$$

If we had connected the plus battery terminal to the n side of the p–n junction and the minus battery terminal to the p side, as in Fig. 27.3, we'd have had a **reverse bias.** Figure 27.4 shows that the field due to the battery then adds to the built-in field. As a result, the energy difference is increased by $\Delta E = (-e)(-V) = eV$. Repeating the derivation for forward

FIGURE 27.4
The height of the "energy hill" is increased by reverse bias. The generation currents hardly change, but the recombination currents decrease by the Boltzmann factor.

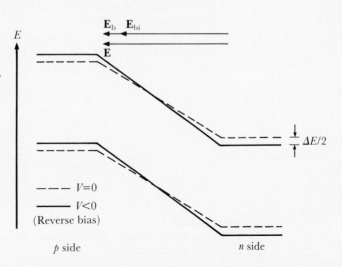

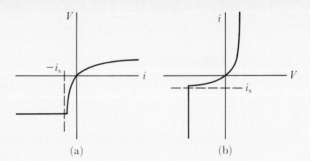

FIGURE 27.5
Junction diode curves:
(a) V versus i; (b) i versus V.

(a)   (b)

bias, we again arrive at $i = i_s(e^{eV/kT} - 1)$, except that V and i are negative for reverse bias.

As V approaches minus infinity, $e^{eV/kT}$ approaches zero and i approaches $-i_s$. Therefore i_s is the maximum theoretical reverse current, or the **saturation current.** As V approaches minus infinity, the height of the energy hill approaches infinity, and the recombination currents approach zero (the electrons and holes can't diffuse up an infinitely high energy hill). Therefore i_s is the sum of the electron and hole generation currents.

Graphs of $i = i_s(e^{eV/kT} - 1)$ in the forms V versus i and i versus V are shown in Fig. 27.5(a) and (b), respectively. Since the resistance R equals V/i, the forward resistance becomes small for large forward currents, while the reverse resistance becomes very large for large reverse voltages. A p–n junction with two terminals that is used to provide a small forward resistance and a large reverse resistance is called a **junction diode.** Allowing current to flow easily in the forward direction while almost stopping the current in the reverse direction gives the junction diode an obvious application: a rectifier, as shown in Fig. 27.6.

We've ignored many effects that will make the current through real p–n junctions deviate from $i = i_s(e^{eV/kT} - 1)$. For instance, before V gets too negative, the junction will "break down" and begin to conduct much more easily in the reverse direction. There are two main types of reverse breakdown. The first is **avalanche breakdown,** which occurs when the

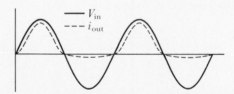

FIGURE 27.6
When a sine wave voltage is applied across a junction diode, the current is mostly rectified (except for very low voltages).

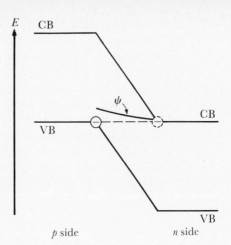

FIGURE 27.7
Tunneling across the *p–n* junction in Zener breakdown first occurs at the precise reverse voltage that brings the top of the *p*-type valence band above the bottom of the *n*-type conduction band.

electric field in the junction becomes so strong that the carriers can gain sufficient energy between collisions to create electron–hole pairs in inelastic collisions. One charge carrier can then create two more charge carriers in a collision, giving one carrier in, three carriers out. Each of these three carriers can then create three more, and so on, and a large current quickly builds up. The heating effect of this large current can damage unprotected devices.

The second main type of reverse breakdown is **Zener breakdown.** Suppose that the reverse bias drops the bottom of the conduction band of the *n*-type material just below the top of the valence band of the *p*-type material, as in Fig. 27.7. If the junction region is thin enough, electrons can then tunnel directly from the *p*-type valence band to the *n*-type conduction band. Increased currents therefore are possible with no further increase in voltage across the junction. (Recall that tunneling is a zero energy loss process, so there's no resistance associated with the tunneling portion of the current.) Zener breakdown is utilized in Zener diodes for such applications as voltage regulation and protection against overloads.

EXAMPLE 27.1   The saturation current for a junction diode is $10\ \mu A$. Calculate the resistance at room temperature for forward and reverse biases of 0.20 V.

Solution At room temperature, $kT = 1/40$ eV so we have $eV/kT = e(\pm 0.20\text{ V})/(1/40\text{ eV}) = \pm 8.0$. Using Eq. (27.1) and $i_s = 10 \times 10^{-6}$ A,

$$R = \frac{V}{i} = \frac{V}{i_s(e^{eV/kT} - 1)} = \frac{\pm 0.20\text{ V}}{(10 \times 10^{-6}\text{ A})(e^{\pm 8.0} - 1)}$$

$$= 6.7\ \Omega \quad \text{and} \quad 20{,}000\ \Omega. \qquad\qquad\bullet$$

The small forward resistance and large reverse resistance in Example 27.1 are typical of junction diodes, if the absolute value of *eV* is *not* much smaller than *kT*. You can see why the forward resistance is so small and the reverse resistance so large from Figs. 27.1 and 27.3. Forward bias drives holes in the *p* side and electrons in the *n* side toward the *p–n* junction. There the electrons can easily fall into the holes (recombine), and the forward resistance is low. However, reverse bias pulls electrons and holes in the opposite direction, or away from the junction. For the current to continue, electrons must be constantly pulled out of holes at the *p–n* junction—a hard thing to do, requiring the expenditure of energy. Therefore a large reverse resistance is created.

27.2
SEMICONDUCTOR OPTOELECTRONIC DEVICES

Semiconductor **optoelectronic devices** emit or absorb light as electrons change energies. Electrical energy is fed into light-emitting diodes and semiconductor lasers to produce light energy. In contrast, light energy is fed into solar cells to produce electrical energy.

Whenever electrons and holes recombine in a semiconductor, energy is released. If an electron falls from near the bottom of the conduction band into a hole near the top of the valence band, the energy released will be approximately E_g, the width of the forbidden energy gap. This recombination energy may be divided among a number of phonons, it may be used to increase the energy of another free carrier, or it may produce a photon with or without some associated phonons. When a photon is produced, the process is called **radiative recombination.**

In a **light-emitting diode (LED),** the diode is forward biased, so the electrons and holes both travel toward the *p–n* junction region. There they undergo radiative recombination and light is emitted. Since $E = h\nu$, we can obtain different frequencies of light by varying the energy difference through which the electron falls. We can vary the energy difference by using materials with different energy gaps, or cause the transitions to be made to defect states within the forbidden energy gap. Using different dopants, we can then obtain different light frequencies.

The **semiconductor laser** operates in much the same way as the LED, except that the transitions are stimulated instead of spontaneous. We can make the sides of the junction into mirrors, thereby providing light that travels parallel to the *p–n* junction in coherent waves. We usually obtained the flat mirrored side by cleaving the crystal and the reflection by the large change in the index of refraction between the semiconductor and its surroundings.

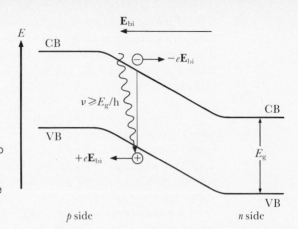

FIGURE 27.8
The photovoltaic effect:
The photon is absorbed to
create an electron–hole
pair, and the built-in
electric field separates the
electron and hole in the
junction region.

The **photovoltaic effect** is the basis for the operation of the solar cell. The photovoltaic effect occurs when photons are absorbed to create electron–hole pairs. If they are created close enough to the p–n junction region, the pairs can migrate to it. There the built-in electric field sweeps the electrons to the n side and the holes to the p side of the junction, as illustrated in Fig. 27.8. If we then connect the junction to an external circuit, we can use the resulting energy, voltage, and current in that circuit. This device is called a **solar cell.** (Despite the word *solar,* sunlight isn't required. Any light source of sufficiently high frequency can provide the photons.)

One interesting area of applied research in condensed matter physics combines solids and conducting liquids. For instance, photoelectrochemical cells are being developed that both produce and store electric energy. The stored energy can be released when there are no incoming photons. The result is a seeming contradiction: a solar cell that works in the dark! Photoelectrochemical cells can also produce useful chemical products, such as fuels.

EXAMPLE 27.2 At what frequencies will a GaAs solar cell at room temperature produce electrical energy? Relate the result to a GaAs LED or laser.

Solution At room temperature GaAs has a band gap of 1.42 eV (see Example 25.2). Therefore photons of energy greater than or equal to 1.42 eV are needed to produce electron–hole pairs. If $h\nu \geq 1.42$ eV, then

$$\nu \geq \frac{1.42 \text{ eV}}{4.136 \times 10^{-15} \text{ eV} \cdot \text{s}} = 3.43 \times 10^{14} \text{ Hz.}$$

Therefore GaAs solar cells will convert light energy into electrical energy for frequencies above 3.43×10^{14} Hz. Conversely, GaAs LEDs and lasers that operate by recombination between the bottom of the conduction band and the top of the valence band will emit 3.43×10^{14} Hz infrared radiation. ●

The 3.43×10^{14} Hz frequency obtained in Example 27.2 can vary several percent because of variations in E_g with temperature and/or pressure. From $\lambda = c/v$ or $E = hc/\lambda = 1.240$ eV $\cdot \mu$m$/\lambda$, 3.43×10^{14} Hz and 1.42 eV correspond to 0.87 μm. Therefore GaAs solar cells will work for wavelengths below 0.87 μm — for example, in the 0.40 μm to 0.70 μm visible light region.

A **photodiode** operates much like a solar cell. A photodiode is constructed so that it can absorb light to create electron–hole pairs, thereby greatly increasing the generation current. If reverse biased, the current will be very small when the photodiode is in the dark and much larger when light shines on it. A photon can also be absorbed to create an electron–hole pair in a **charge-coupled device (CCD).** The electron is then collected in a potential well in n-type material, while holes are allowed to recombine in a substrate. Charge-coupled devices can be used to detect very faint light.

27.3

BIPOLAR JUNCTION TRANSISTORS

A ***pnp* bipolar junction transistor** is formed with a thin ($< 10^{-4}$ m) lightly doped, n-type region between two more heavily doped p-type regions. As Fig. 27.9 shows, this configuration gives two p–n junctions back to back. Suppose that we connect a battery across the *pnp* transistor, as shown in Fig. 27.10. The **base** (center n portion) is so lightly doped that most (typically, > 98 percent) of the holes from the usually heavily doped **emitter** (the left p portion) diffuse right across the base to the **collector** (the

FIGURE 27.9

(a) A *pnp* transistor.
(b) Equilibrium conditions. For purposes of clarity, no donor or acceptor levels are shown, and only the majority carriers are indicated for each region.

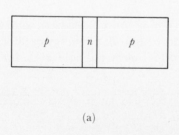

(a)

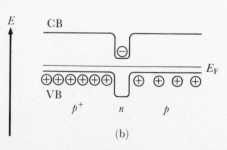

(b)

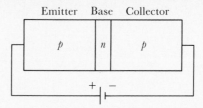

FIGURE 27.10
The current is shut off almost immediately in this circuit.

right p portion). Although typically fewer than 2 percent of the holes from the emitter recombine with electrons in the base, this quantity is enough to quickly make the base positive, thereby setting up a strong electric field to repel any further hole current from the emitter.

You can see that the hole current will be cut off quickly unless the positive charge in the base portion of the pnp transistor is somehow neutralized. We can do so by adding electrons from an external source to the base (which means a conventional electric current away from the base). In summary, a small current from the base will allow a large emitter–collector current to flow. If this small current doesn't flow, the emitter–collector current will be stopped quickly. This explanation shows how you can use a transistor to switch electrical currents.

EXAMPLE 27.3 In a particular pnp bipolar junction transistor, 99.0 percent of the holes can diffuse from the emitter to the collector. What current from the base will allow a current of 2.0 A to flow to the collector?

Solution The 99.0 percent figure means that for every 100 holes flowing from the emitter, 99 get to the collector and 1 must be neutralized by adding an electron to the base. Therefore the current ratio is

$$\frac{99 \text{ charge carriers/time}}{1 \text{ charge carrier/time}} = \frac{2.0 \text{ A}}{i_b}.$$

Solving for the base current gives $i_b = 2.0$ A$(1/99) = 0.020$ A $= 20$ mA. Thus a small current of 20 mA turned on or off from the base can control a large current of 2.0 A to the collector.   ●

Suppose that we now connect the pnp transistor as shown in Fig. 27.11. The emitter–base p–n junction is forward biased and therefore has a small resistance. The base–collector p–n junction, however, is reverse biased and therefore has a large resistance. If the ac source then gives us a small change in voltage, ΔV_{in}, the emitter current will change by Δi. Most of this Δi will also pass through the base–collector junction, but since $\Delta V = \Delta iR$, the large R of this junction will give a large ΔV_{out}. Thus a small

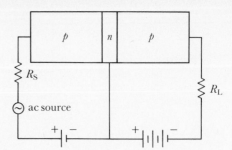

FIGURE 27.11
A possible circuit for
amplification of ac voltage.

voltage change in gives a large voltage change out. This explanation shows how you can use transistors to amplify ac signals.

EXAMPLE 27.4   How would you change the circuit in Fig. 27.11 for an *npn* **bipolar junction transistor?**

Solution In an *npn* transistor, the emitter and collector would be *n* portions, while the base would be a *p* portion, of the transistor. So that the emitter–base *p–n* junction would be forward biased and the base–collector *p–n* junction would be reverse biased, you'd merely reverse the polarity of the two batteries shown in Fig. 27.11.   ●

27.4
MOSFETs

When a manufacturer announces that it is producing an integrated circuit less than a centimeter square containing tens of thousands of transistors, the transistors are probably not bipolar junction transistors. More likely, each is some type of ***metal-oxide-semiconductor field-effect transistor* (MOSFET).** A common type of MOSFET is the ***N*-channel depletion type.** Figure 27.12 shows a cross-sectional view of one version of this type of MOSFET.

The source, gate, and drain are made of a metal. The **gate** is completely insulated from the source, drain, and semiconductor by the insulating oxide layer. Figure 27.12(a) shows a channel of *n*-type semiconductor, with *p*-type material below and around it, from the **source** to the **drain.** With the gate neutral, the current can easily flow from the source to the drain through this channel.

Now suppose that we give the gate a negative charge. The electric field of this charge will have the following effect: It will attract holes in the semiconductor toward the gate and repel electrons from the gate. This

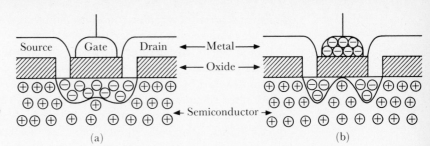

FIGURE 27.12
An N-channel depletion-type MOSFET in cross-section: (a) with the gate neutral; (b) with the gate well-charged.

process begins to make holes the majority carriers in the region below the gate. In fact, enough negative charge on the gate can even close off the channel, as shown in Fig. 27.12(b). Because one of the resulting p–n junctions will be reversed biased, the current will decrease considerably. The actual amount of charge placed on the gate is small, so that an extremely small current to and from the gate can control a much larger current through the channel.

EXAMPLE 27.5 Redraw Fig. 27.12 to show a **P-channel enhancement-type** MOSFET. In this type, charging the gate opens the channel.

Solution This time, we start with a p channel through n-type material. However, the purpose of the charge on the gate isn't to deplete the channel, but rather to enhance it. Therefore the channel initially could be closed when the gate is neutral, as in Fig. 27.13(a). A positive charge on the gate would close off the channel even more by attracting more electrons and repelling more holes below the gate. Therefore we need to give the gate a negative charge, attracting holes and repelling electrons to open the channel, as in Fig. 27.13(b). ●

Many of the dimensions in a MOSFET in an integrated circuit are measured in micrometers. That's why MOSFETs can have such a high

FIGURE 27.13
A P-channel enhancement-type MOSFET in cross-section: (a) with the gate neutral; (b) with the gate well-charged.

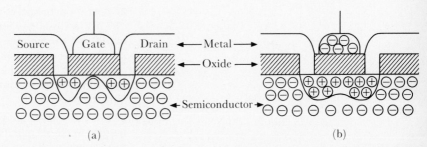

density on an integrated circuit chip. And that is also why it will be quite difficult to make them smaller than they are now.

We began this chapter discussing the $p-n$ junction and deriving an idealized current–voltage relation. We looked at deviations from that relation caused by avalanche and Zener breakdowns. We then used that relation for calculations to show the rectifying properties of the $p-n$ junction diode. We continued with the junction by describing LEDs, semiconductor lasers, solar cells, photodiodes, and CCDs. We then finished the chapter with discussions of types of junction transistors and MOSFETs, completing our introduction to solid state physics.

QUESTIONS AND PROBLEMS

27.1 Figure 27.14 shows a metal and an n-type semiconductor (a) separated from one another and (b) formed in contact. When they're in contact, the metal surface receives a negative charge.

a) Explain the negative charge.

b) Further explain why the surface region of the semiconductor becomes depleted of charge carriers and gains a positive charge.

c) Then explain the origin of the built-in electric field.

d) The potential energy barrier formed is called a Schottky barrier. Why is there a potential barrier?

e) Used as a two-terminal device, this metal-semiconductor junction rectifies and is called a Schottky diode. Why does it rectify?

f) Explain how a Schottky diode can be used as a detector of light.

27.2 The V in Eq. (27.1) is $V_{\text{battery}} - iR$. What is the R?

27.3 For small voltages, a $p-n$ junction is ohmic. Use $e^x \approx 1 + x$ for small x to show that the junction is ohmic and find the expression for the resistance at small voltages.

27.4 For small voltages, the resistance of a junction diode can be approximated by $R = kT/i_se$. a) Calculate the small-voltage resistance of the junction diode of Example 27.1. b) However, R is not directly proportional to T. Why not?

27.5 Why does Zener breakdown occur only for abrupt junctions?

27.6 A junction diode is placed in parallel with a galvanometer so that if the dc voltage across the galvanometer becomes dangerously high, the current will flow through the diode rather than the galvanometer. Why should this diode be reverse biased and by what mechanism will this junction diode break down and conduct? Explain.

27.7 At room temperature, a forward voltage of 157 mV across a $p-n$ junction results in a current of 209 mA through the junction. a) Calculate the maximum reverse current that is theoretically possible. b) How is this saturation current related to the generation current and the recombination current?

27.8 If a $p-n$ junction is heated, a) what will happen to the recombination current? b) what will happen to the generation current?

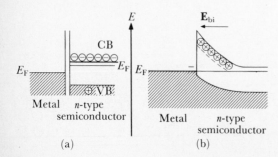

FIGURE 27.14

27.9 What voltage will give a reverse current through a junction diode that is a) 10%, b) 50%, and c) 90% of the saturation current at room temperature?

27.10 Graph power versus voltage at a p–n junction for forward and reverse biases.

27.11 The differential, or dynamic, resistance is defined by $R_d = dV/di$. Calculate the expression for R_d versus V for forward and reverse bias. Also show that R_d approaches kT/i_se for small V.

27.12 A diode is reverse biased. How can it be used as a detector of light? What wavelengths can it detect?

27.13 The elements Si and Ge aren't used for LEDs because very few photons are produced in them by recombination at the junction. What could be happening to the energy released by the recombinations in Si and Ge?

27.14 What wavelengths will make a crystalline Si solar cell produce energy? (See the paragraph preceding Example 25.1.)

27.15 If a crystalline CdTe solar cell will begin to deliver current when the light frequency is increased to above 3.51×10^{14} Hz, what does this tell you about crystalline CdTe?

27.16 The recombination length is a measure of how far electrons and holes will travel before recombining. Why should the recombination length be large for efficient solar-cell operation? (Short recombination lengths have been a problem in polycrystalline and amorphous materials.)

27.17 Rewrite Section 27.3 up to Example 27.3 for an npn bipolar junction transistor.

27.18 In a particular npn bipolar junction transistor, a current of 500 mA can flow from the collector, if a current of 8.0 mA is maintained to the base. What percentage of the electrons are diffusing all the way across the base? Assume that the base remains neutral.

27.19 In a particular pnp bipolar junction transistor, 1.3% of the holes recombine with the electrons in the base as the remaining 98.7% diffuse from the emitter to the collector. Calculate the current from the emitter if a current of 3.8 mA is drawn from the base and the base remains neutral.

27.20 Explain what you could do to get a current that varied with time in the "square" fashion shown in Fig. 27.15 through the collector of a bipolar junction transistor.

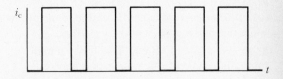

FIGURE 27.15

27.21 Does the common-base circuit of Fig. 27.11 give a) a current gain greater than 1? b) a voltage gain greater than 1? c) a power gain greater than 1?

27.22 The device shown schematically in Fig. 27.16(a) is an SCR (semiconductor, or silicon, controlled rectifier). You can understand its operation most easily by mentally converting it into two partially parallel transistors, as shown in Fig. 27.16(b). With the gate unconnected, both transistors almost immediately block the current flow. When it flows, the current's direction is to the right through this SCR.

a) If the gate is connected to withdraw electrons from the p part of the npn transistor, why does the npn transistor conduct?

b) Why does this action then allow the pnp transistor to conduct?

c) Why will the SCR stay "turned on" even if the gate is disconnected?

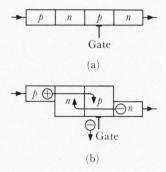

FIGURE 27.16

27.23 What does a negative charge on the gate do to the current from the source to the drain in
a) an N-channel enhancement-type MOSFET?
b) a P-channel depletion-type MOSFET? Explain your answers.

27.24 What type or types of MOSFET will decrease the current from the source to the drain when the gate is given a positive charge?

27.25 What type of MOSFET will have a minimum drain current when the gate is neutral rather than positively charged?

27.26 Redo Fig. 27.12 for an N-channel enhancement-type MOSFET.

27.27 Redo Fig. 27.12 for a P-channel depletion-type MOSFET.

27.28 Why does a MOSFET have such a high input impedance (to the gate)?

27.29 Could tunneling limit the miniaturization of MOSFETs?

27.30 A tunnel diode uses tunneling, but in not quite the same way as does a Zener diode.
a) Through outside reading, find out how a tunnel diode works.   b) Draw a sketch of the energy band structure for a tunnel diode.
c) Explain the forward bias i–V or V–i characteristics.

PART Two

General Relativity and Cosmology

CHAPTER **28**

The Nucleus

I always begin this part of a modern physics course with mixed feelings. The material is exciting, interesting, and fundamental to an understanding of the world around us. But its application has also given us devices that can destroy our world. I believe that most people agree that moral decisions must be made concerning the application of knowledge. For example, a person can learn to be a technically competent artist. But if that person uses his or her artistic ability to further some type of hate and prejudice against another group, that ability is being used for immoral purposes. Certainly, then, you face a moral decision in applying the knowledge you will gain about nuclear physics. Even if your career doesn't deal directly with nuclear applications, you're still involved in them as a citizen of the world. My hope is that you will make moral decisions about the uses of *all* the knowledge you acquire in this course and throughout your life. To quote G. K. Chesterton, "Morality, like art, consists of drawing a line somewhere."

28.1
NUCLIDES

The smaller something is, the harder it is to see. And the harder it is to see, the more difficult it is to describe. If we're to see something clearly using waves, as in Fig. 28.1, the wavelength scattered must be less than the size of the object we're observing. Otherwise, diffraction effects will begin to blur the image of the object.

In Rutherford's alpha particle scattering experiments (which we discussed in Section 11.1), he found the nucleus to have a radius of somewhat less than 10^{-14} m. Rutherford used 7.68-MeV alphas, and since 7.68 MeV is much less than 3.73 GeV, the rest energy of an alpha particle, we can use accurately the classical value $p = \sqrt{2m_0 KE}$ for the momentum. The

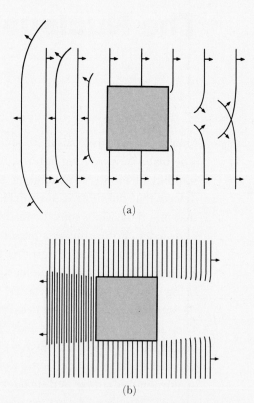

(a)

(b)

FIGURE 28.1
Waves moving to the right hit a square object. (a) Long wavelengths diffract through larger angles, giving a fuzzier image or shadow. (b) Short wavelengths diffract through smaller angles, giving a sharper image or shadow.

FIGURE 28.2
A two-dimensional model of a $_{4}^{9}$Be nucleus, for which $Z = 4$, $N = 5$, and $A = 9$.

de Broglie wavelength of Rutherford's alpha particles was

$$\lambda = \frac{h}{p} = \frac{h}{\sqrt{2m_0 KE}} = \frac{hc}{\sqrt{2m_0 c^2\, KE}} = \frac{1.240 \times 10^{-6}\ \text{eV} \cdot \text{m}}{\sqrt{2(3.73 \times 10^9\ \text{eV})(7.68 \times 10^6\ \text{eV})}}$$

$$= 5.18 \times 10^{-15}\ \text{m} = 5.18\ \text{fm},$$

where 1 fm = 1 femtometer = 10^{-15} m, sometimes called a *fermi*.

The wavelength $\lambda = 5.18$ fm is less than the diameter of a large nucleus, such as that of gold, but not a lot less. Therefore Rutherford couldn't "see" the gold nucleus too clearly in his scattering experiments. Higher energy α particles would have shorter wavelengths. However, their more violent collisions would make a nuclear reaction more likely to occur. This reaction might well destroy both the α particle and the nucleus that the particle was supposed to scatter from.

Because of the difficulty in "seeing" the nucleus, several different models have been developed to explain nuclear properties. We'll first relate some experimental facts about nuclei. Then we'll discuss two of the nuclear models.

The nucleus is made up of positive protons ($q = +e$) and neutral neutrons ($q = 0$). Because they make up nuclei, the term **nucleon** is used to mean either a proton or a neutron. The **atomic number,** Z, gives the number of protons in the nucleus. The net charge of the nucleus is then $+Ze$. A neutral atom has a net charge of zero: $+Ze$ from the nucleus and $-Ze$ from the atomic electrons. Therefore Z is also the number of electrons in the neutral atom. Each chemical element has a different Z.

The **neutron number,** N, is the number of neutrons in the nucleus. The **mass number,** A, is the number of nucleons in the nucleus. Therefore A is the sum of the number of neutrons and the number of protons, or

$$A = N + Z. \tag{28.1}$$

A nucleus with a certain A, N, and Z is called a **nuclide** and is represented, as in Fig. 28.2, by

$$_{Z}^{A}\text{E}, \tag{28.2}$$

where E is the element's chemical symbol. For example, $^{238}_{92}U$ is a famous nuclide. The U is the chemical symbol for the element uranium, and the nuclide contains 92 protons and 238 nucleons. It therefore has $N = A - Z = 238 - 92 = 146$ neutrons. Note that the 92 and the U are redundant; *all* U nuclides have 92 protons. Therefore we may also express $^{238}_{92}U$ as ^{238}U, U^{238}, or U-238. In fact, "U-238" is how we usually say it.

Isotopes of an element are nuclides with the same number of protons but different numbers of neutrons. Thus all the isotopes of a given element have the same chemical symbol and Z but different N's and A's. For example, uranium in nature is usually found to be 0.0055 percent $^{234}_{92}U$, 0.72 percent $^{235}_{92}U$, and 99.27 percent $^{238}_{92}U$. Each of these isotopes is uranium, with 92 protons and 92 electrons in the neutral atom. Each isotope has mostly the same chemical properties because they are mainly determined by the electrons of the atom. The main difference is that the three uranium isotopes have 234, 235, and 238 nucleons, respectively, giving them 142, 143, and 146 neutrons, respectively. Figure 28.3 shows the three isotopes of hydrogen.

Because of their different masses, isotopes can be separated by their different rates of diffusion (for example, U in the gas UF_6), by mass spectrometers, or by ultracentrifuges. Since the electrons interact slightly differently with the different nuclei of the isotopes, separation methods based

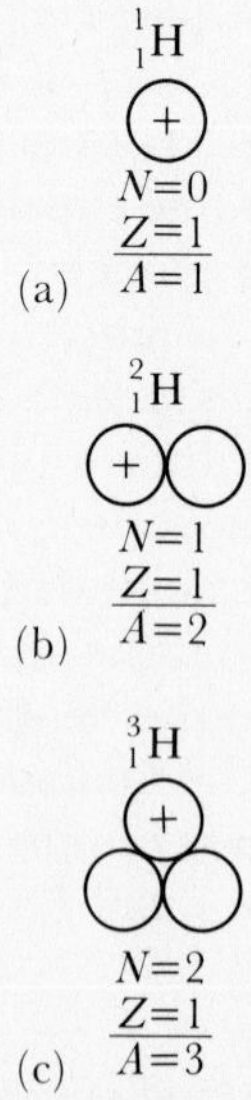

$^{1}_{1}H$

$+$

$N=0$
$Z=1$
(a) $A=1$

$^{2}_{1}H$

$+$◯

$N=1$
$Z=1$
(b) $A=2$

$^{3}_{1}H$

$+$

$N=2$
$Z=1$
(c) $A=3$

FIGURE 28.3
Three isotopes of hydrogen: (a) H-1 has one proton, no neutron, and one nucleon; (b) H-2 (deuterium) has one proton, one neutron, and two nucleons; (c) H-3 (tritium) has one proton, two neutrons, and three nucleons.

on the slightly different atomic energy levels of the isotopes are also possible. (The nuclear charge is the same, but the nuclear magnetic dipole moments and masses differ.) Some very few isotopes have extraordinarily different behaviors. For example, at 2.2 K, liquid $^4_2$He exhibits superfluid (zero viscosity) behavior, but $^3_2$He is a normal liquid at that temperature.

EXAMPLE 28.1   The isotope $^{239}_{94}$Pu is a dangerous product of nuclear reactors.

a) How many protons, neutrons, and nucleons are there in $^{239}_{94}$Pu?

b) How many electrons does a Pu^{2+} ion have?

c) How would you represent the plutonium isotope that has three more nucleons in its nucleus?

Solution

a) Comparing $^{239}_{94}$Pu with ^A_ZE, we see that $Z = 94$, so there are 94 protons in the nucleus; $A = 239$, so there are 239 nucleons in the nucleus. Since $A = N + Z$ gives $A - Z = N = 239 - 94 = 145$, there are 145 neutrons in $^{239}_{94}$Pu.

b) With 94 protons, a neutral Pu atom would have 94 electrons. Doubly ionized Pu^{2+} has 2 electrons missing. Therefore a Pu^{2+} ion has $94 - 2 = 92$ electrons.

c) Adding, we have 239 nucleons + 3 nucleons = 242 nucleons, so $A = 242$, and the isotope is $^{242}_{94}$Pu. By the way, all three of the extra nucleons must be neutrons, because all Pu isotopes have the same number of protons (94). Pu-242 has $145 + 3$ (or $242 - 94$) $= 148$ neutrons.   ●

28.2
THE STRONG NUCLEAR FORCE

Because of the $1/r^2$ dependence of Coulomb's law, the positively charged protons in all nuclei (except hydrogen) repel one another with a very strong force. Despite this strong repulsive force, nucleons are tightly bound to one another. Thus there must be another type of force—an attractive force—responsible for nuclear binding. This force is called the **strong nuclear force.**

Coulomb's law satisfactorily explains the scattering of protons from nuclei as long as the bombarding protons don't get too close to the nuclei. Deviations from Coulomb's law scattering occur only if the protons closely approach the nucleons. This condition shows that the strong nuclear force is a **short-range force.** It acts only at distances of less than a few fm.

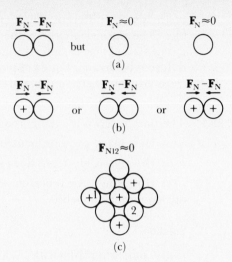

FIGURE 28.4
The strong nuclear force is (a) short range, (b) relatively charge-independent, and
(c) saturable (nucleons 1 and 2 have negligible attraction).

The strong nuclear force doesn't just attract nucleons to one another. Experiments have shown that it has a very short-range repulsive
term, operating at even smaller distances than its attractive term. Because
it is so very short-range, this term adds only slightly to the nuclear potential
energy.

The results of a number of experiments indicate that the proton–
proton, proton–neutron, and neutron–neutron nuclear forces are either
exactly or almost exactly the same. Therefore we say that the strong
nuclear force is relatively **charge-independent.** The strong nuclear force
also seems to be **saturable.** That is, each nucleon appears to interact with
only a limited number of its neighbors. In summary, the strong nuclear
force is a strong, short-range, relatively charge-independent, saturable
force between nucleons, as illustrated in Fig. 28.4.

EXAMPLE 28.2   In the hydrogen molecule, the two protons are separated by 74.1 pm.
What can we say about their strong nuclear force of attraction?

Solution Converting, $74.1 \text{ pm} = 74.1 \times 10^{-12} \text{ m} = 7.41 \times 10^4 \text{ fm}$. Since
this separation is at least 10^4 times the few fm range of the nuclear force, we
can safely say that the strong nuclear force of attraction is zero.   ●

EXAMPLE 28.3   Recall from your previous physics courses that forces which are always
attractive give a negative potential energy (with respect to infinite separa-

QUESTIONS AND PROBLEMS

28.1 Calculate the energy of electrons having a wavelength of 0.10 fm. Explain why these electrons would be useful in studying the nucleus.

28.2 Why can't we use visible light and some kind of "supermicroscope" to examine the nucleus?

28.3 How many electrons are in a neutral atom with a nucleus that contains 181 nucleons and 108 neutrons? What atom is it?

28.4 How many protons, electrons, and neutrons are there in a singly ionized carbon-14 atom?

28.5 Potassium has isotopes with mass numbers ranging from about 35 to 50. What can you say about the atomic number and neutron number of potassium?

28.6 Write the symbol for the nuclide that has 38 protons and 52 neutrons in it.

28.7 Write the symbol for the nuclide that has a mass number of 60 and a neutron number of 33.

28.8 Consider $^{70}_{32}\text{Ge}$. Write the symbols for the Ge isotopes containing 2, 3, 4, and 6 more nucleons.

28.9 Sketch a rough graph of force versus distance for p–p, p–n, and n–n interactions.

28.10 Sketch a rough graph of potential energy versus distance for p–p, p–n, and n–n interactions.

28.11 Compare the wavelengths of Rutherford's 7.68-MeV α particles to the diameter of the $^{197}_{79}\text{Au}$ nucleus.

28.12 Find a nuclide that has a radius of 5.0 fm.

28.13 Calculate the maximum cross-sectional area of a $^{235}_{92}\text{U}$ nucleus.

28.14 a) Relate the ratio of the surface areas of two different nuclides to the ratio of their mass numbers. b) By what percent is the surface area increased when the mass number is doubled?

28.15 What energy and wavelength gamma ray could break a deuteron up into a proton and a neutron? (The process is called *photodisintegration*.)

28.16 Calculate the binding energy of $^{12}_{6}\text{C}$ and the binding energy per nucleon.

28.17 For an alpha particle, calculate a) the binding energy, b) the binding energy per nucleon, c) the separation energy of a neutron, and d) the separation energy of a proton.

28.18 Calculate the energy released when $^{7}_{3}\text{Li}$ absorbs a neutron (of negligible kinetic energy) to become $^{8}_{3}\text{Li}$.

28.19 Calculate the binding energy and binding energy per nucleon of $^{232}_{90}\text{Th}$.

28.20 Calculate a) the separation energy of a proton, b) the separation energy of a neutron, and c) the binding energy per nucleon for $^{51}_{23}\text{V}$. Then compare and discuss the results.

28.21 How inaccurate is it to ignore the binding energies of the electrons in the neutral atom? Is the inaccuracy compensated for or worsened by using the mass of a neutral H-1 atom?

28.22 Calculate the binding energy per nucleon of $^{235}_{92}\text{U}$.

28.23 How well does "the whole equals the sum of its parts" apply to nuclear binding?

28.24 Calculate the separation energy of a neutron from $^{252}_{98}\text{Cf}$.

28.25 Calculate the separation energy of a proton from $^{5}_{3}\text{Li}$. Explain the result.

28.26 Calculate the separation energy of a deuteron from $^{12}_{6}\text{C}$.

28.27 Calculate the separation energy of an alpha particle from $^{226}_{88}\text{Ra}$. Explain the result.

CHAPTER 29

Nuclear Models

As we've previously mentioned, a "good" mental model of a system is one that agrees with experimental data. Unfortunately, at present there's no single model that can be used to explain everything about nuclear behavior. In this chapter we'll discuss how some important experimentally determined nuclear properties are explained by the liquid drop model and the shell model.

29.1
THE LIQUID DROP MODEL

One of the uses of the liquid drop model is to explain the variation of nuclear binding energy per nucleon. Figure 29.1 shows the graph of the binding energy per nucleon versus the mass number as a smooth curve, with only a few exceptions. The most notable exception is He-4; others include C-12 and O-16.

Most light nuclei have a small binding energy per nucleon. The curve increases rapidly with A, then peaks at a maximum value of 8.8 MeV/nucleon at Fe-56. For heavier nuclei, the curve drops off slowly to somewhat

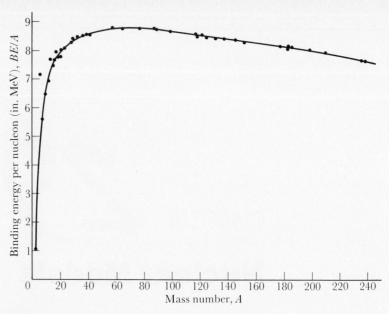

FIGURE 29.1

The binding energy-per-nucleon curve.

over 7.5 MeV/nucleon at the highest A. In the 1930s, physicists explained this curve by using the **liquid drop model.** Just as the molecules in a drop of liquid are attracted by neighboring molecules to form the drop, so do nucleons attract neighboring nucleons to form a nucleus.

However, those molecules that are instantaneously on the surface of the drop have no neighbors to the outside in the same manner as those nucleons that are instantaneously on the surface of the nucleus. This condition gives a larger (less negative) potential energy than if all particles were surrounded. A sphere has the smallest surface area for a given volume, and therefore the smallest percentage of particles on its surface, as illustrated in Fig. 29.2. So that potential energy can have a minimum

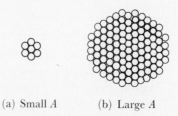

(a) Small A (b) Large A

FIGURE 29.2

(a) Surface nucleons have a smaller binding energy because they aren't surrounded by other nucleons. (b) As the total number of nucleons increases, a smaller fraction of the nucleons are on the surface.

value, liquid drops in "zero-gravity" space are spheres, and nuclei in their ground states are also mostly spherical.

The liquid drop model, then, indicates that small-mass-number nuclei have a smaller binding energy per nucleon because a larger percentage of their nucleons are on the surface. Why? Because the volume of a sphere and therefore the number of particles in the sphere are both proportional to the cube of the radius. In contrast, the surface area and therefore the number of particles on the surface are proportional only to the square of the radius.

Therefore as the number of nucleons increases, a smaller percentage of the nucleons appear on the surface. Thus the average binding energy increases for each nucleon. This average would asymptotically approach a value of about 16 MeV/nucleon (corresponding to zero percent nucleons on the surface), except for the Coulomb repulsion of the protons. This 16-MeV/nucleon quantity corresponds to the saturation of the strong nuclear force; it also is the absolute value of the sum of a potential energy term of about -50 MeV/nucleon and a kinetic energy term averaging over 30 MeV/nucleon.

The Coulomb repulsion of the positive protons for one another gives a positive potential energy, or a negative binding energy, proportional to $Z(Z-1)/A^{1/3}$. Each of the Z protons is repelled by $Z-1$ other protons, as shown in Fig. 29.3. Also the average separation is proportional to the nuclear radius, which in turn, is proportional to $A^{1/3}$. If you divide the expression again by A, you find that the Coulomb repulsion energy per nucleon is proportional to $Z(Z-1)/A^{4/3}$.

The Coulomb repulsion energy term causes the binding energy per nucleon to decrease after about $A = 56$. For example, the ratio of the Coulomb energy per nucleon for $^{206}_{82}$Pb to that for $^{56}_{26}$Fe is

$$\frac{(82)(81)/(206)^{4/3}}{(26)(25)/(56)^{4/3}} = 1.80.$$

That is, the Coulomb repulsion energy per nucleon is 80 percent greater for Pb-206 than for Fe-56.

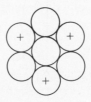

FIGURE 29.3
Each of the $Z = 3$ protons is repelled by the $Z - 1 = 2$ other protons. If A were larger, the protons would be farther apart on average.

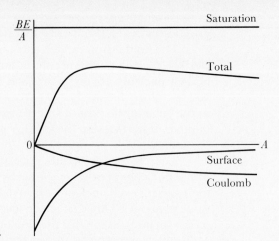

FIGURE 29.4
The saturation, surface, and Coulomb terms combine to give the total binding energy per nucleon.

EXAMPLE 29.1   The nuclear force gives a maximum binding energy per nucleon of about 16 MeV/nucleon. State the main reasons that the BE/A curve is much less for

   a) small A and   b) large A nuclides.

Solution

   a) For small A nuclides, the nuclear force isn't saturated for a large percentage of the nucleons (those nucleons instantaneously on the surface of the nucleus have no neighbors to their outside). Therefore the nuclear force can't give its maximum binding energy for those nucleons, dropping the average value well below 16 MeV/nucleon.

   b) For large A nuclides, the repulsive Coulomb force pushes apart the numerous protons, weakening the total nuclear binding and therefore decreasing the average binding energy per nucleon.   ●

   The contributions of these three terms (the saturation, surface, and Coulomb) to the total binding energy per nucleon are sketched in Fig. 29.4. Further terms can be added to obtain an even better fit to experimentally determined binding energies.

29.2
NUCLEAR ANGULAR MOMENTUM

As we have said before, protons and neutrons are fermions. That is, they have a spin angular momentum quantum number, s, of $\frac{1}{2}$. Like electrons, they have a spin angular momentum $\mathbf{S}$ of magnitude $\sqrt{\frac{1}{2}(\frac{1}{2} + 1)}\hbar = \sqrt{3}/2\,\hbar$.

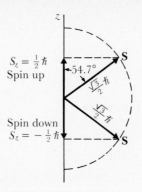

FIGURE 29.5
Space quantization of the spin angular momentum of a nucleon.

And their spin magnetic quantum number, m_s, equals $\pm\frac{1}{2}$, which gives z components of the spin angular momentum, S_z, of $-\hbar/2$ or $+\hbar/2$. As shown in Fig. 29.5, **S** is space quantized, with

$$\cos\theta = (\pm\hbar/2)/[(\sqrt{3}/2)\hbar] = \pm 1/\sqrt{3}$$

giving $\theta = 54.7356°$ or $125.2644°$, the same as for an electron.

    Recall that the z component of the spin magnetic dipole moment of an electron is $\mu_{sz} \approx \mp\mu_B$. The term μ_B is the Bohr magneton, which equals $e\hbar/2m$, where m is the electron mass. For nucleons, we define the **nuclear magneton**, μ_N, as

$$\mu_N = \frac{e\hbar}{2m_p}. \qquad (29.1)$$

Since m_p, the mass of a proton, is 1836 times larger than m, the mass of an electron, the nuclear magneton is 1836 times smaller than the Bohr magneton. Numerically, $\mu_N = 5.051 \times 10^{-27}$ J/T $= 3.152 \times 10^{-8}$ eV/T.

    Although $\mu_{sz} \approx \mp\mu_B$ for an electron, similar approximations don't hold for nucleons. For the proton, experiments give

$$\mu_{sz}(\text{proton}) = \pm 2.793\mu_N. \qquad (29.2)$$

The $\pm$ sign emphasizes the finding that the spin magnetic dipole moment of the positive proton is in the same direction as its spin angular momentum. (See Fig. 29.6.)

FIGURE 29.6
The spin magnetic dipole moment is parallel to the spin angular momentum for the proton.

FIGURE 29.7
Despite being neutral, the neutron has a spin magnetic dipole moment that is antiparallel to its spin angular momentum.

In classical physics, a magnetic dipole moment is caused by the motion of charge around a loop. So it would seem that the uncharged spinning neutron should have no magnetic dipole moment. However, experiments show that not only does the neutron have a magnetic dipole moment but that it is directed opposite the spin angular momentum. This result indicates that the neutral neutron is made up of a special distribution of equal amounts of plus and minus charge, as found in the quark model (which we'll introduce in Chapter 35). For the neutron, experimental results give

$$\mu_{sz}(\text{neutron}) = \mp 1.913\mu_{N}. \tag{29.3}$$

The $\mp$ emphasizes the opposite directions of μ_s and $\mathbf{S}$, as shown in Fig. 29.7.

EXAMPLE 29.2   Compare the magnetic potential energy for a free electron, a free proton, and a free neutron in a large, 2.0-T magnetic field. (Recall that the magnetic potential energy is $-\mu \cdot \mathbf{B} = -\mu_z B$, taking this potential energy as equal to zero when μ is perpendicular to $\mathbf{B}$.)

Solution For an electron,

$$-\mu_z B \approx \pm\mu_B B = (\pm 5.79 \times 10^{-5} \text{ eV/T})(2.0 \text{ T})$$
$$= \pm 1.16 \times 10^{-4} \text{eV}.$$

For a proton,

$$-\mu_z B = (-)(\pm 2.793\mu_N)B = \mp 2.793(3.152 \times 10^{-8} \text{ eV/T})(2.0 \text{ T})$$
$$= \mp 1.76 \times 10^{-7} \text{ eV}.$$

For a neutron,

$$-\mu_z B = (-)(\mp 1.913\mu_N)B = \pm 1.913(3.152 \times 10^{-8} \text{ eV/T})(2.0 \text{ T})$$
$$= \pm 1.21 \times 10^{-7} \text{ eV}. \qquad \bullet$$

Example 29.2 shows that the magnetic potential energies for the two nucleons are of the same order of magnitude. However, both are about three orders of magnitude smaller than those of the electron.

Transitions between the different energy states of nucleons or nuclei may be made when they're split apart by a magnetic field, as in Example 29.2 and Fig. 29.8. This process is the basis for the phenomenon called **nuclear magnetic resonance (NMR).** Magnetic resonance imaging (**MRI**) uses NMR, usually with protons (our bodies are 75 percent H_2O) and superconducting magnets. This use of NMR has become an important low electromagnetic radiation level medical diagnostic tool. A related phenomenon is the source of the 21-cm-wavelength hydrogen line used so often by radio astronomers. In this case, the magnetic field from the proton's magnetic dipole moment interacts with the electron's magnetic dipole moment to split the 1s hydrogen ground level into spin-up and spin-down states. These states are separated by $hc/(21 \text{ cm}) = 5.9 \times 10^{-6}$ eV, which is so small compared to the -13.6 eV$/n^2$ hydrogen energies that it's called **hyperfine splitting.**

The total angular momentum of a nucleus will be the vector sum of the spin and orbital angular momentums of all its nucleons. The total nuclear angular momentum, **I**, is quantized by the relation

$$I = \sqrt{i(i+1)}\hbar, \qquad (29.4)$$

where i is the total nuclear angular momentum quantum number. Unfortunately, i is usually called the **nuclear spin,** a very misleading name. Actually, i is a quantum number involving both spin and orbital angular momentums.

Defined by the same types of equations as the other angular momentums that you've studied, **I** is space quantized. That is, $I_z = m_i\hbar$, with m_i varying from $-i$ to $+i$ in steps of 1, and i itself either a positive integer, zero, or a positive odd half-integer ($\frac{1}{2}, \frac{3}{2}, \frac{5}{2}, \ldots$).

In the nuclear ground state, **even–even nuclei** (with an even number of protons, Z, and an even number of neutrons, N) have $i = 0$ only; **even–odd nuclei** or **odd–even nuclei** (with an odd number of nucleons, A) have $i = \frac{1}{2}, \frac{3}{2}, \frac{5}{2}, \ldots$; and **odd–odd nuclei** (odd Z, odd N, but an even A) have $i = 0, 1, 2, 3, \ldots$. These experimental results are summarized in Table 29.1.

FIGURE 29.8
An applied magnetic field splits the spin-down and spin-up proton energy levels.

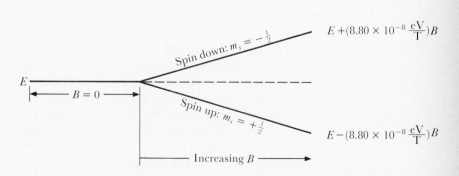

E

$B = 0$

Spin down: $m_s = -\frac{1}{2}$

Spin up: $m_s = +\frac{1}{2}$

$E + (8.80 \times 10^{-8} \frac{\text{eV}}{\text{T}})B$

$E - (8.80 \times 10^{-8} \frac{\text{eV}}{\text{T}})B$

Increasing B

TABLE 29.1
Ground State "Nuclear Spins"

Z	N	i
Even	Even	0
Even	Odd	$\frac{1}{2}, \frac{3}{2}, \frac{5}{2}, \ldots$
Odd	Even	$\frac{1}{2}, \frac{3}{2}, \frac{5}{2}, \ldots$
Odd	Odd	0, 1, 2, $\ldots$

EXAMPLE 29.3   Calculate the total nuclear angular momentum of the nuclear ground state of

    a) $^{73}_{32}$Ge, which has a nuclear spin of $\frac{9}{2}$, and   b) $^{28}_{14}$Si.

Solution

    a) With $i = \frac{9}{2}$, the magnitude of **I**, from Eq. (29.4), is

$$I = \sqrt{\tfrac{9}{2}(\tfrac{9}{2} + 1)}\hbar = \sqrt{99/4}\hbar = (3\sqrt{11}/2)\hbar.$$

The vector **I** can have $2(\frac{9}{2}) + 1 = 10$ different directions in space, with z components ranging from $-\frac{9}{2}\hbar$ to $+\frac{9}{2}\hbar$ in steps of $\hbar$.

    b) The nuclide $^{28}_{14}$Si has 14 protons and 14 neutrons. Since 14 is an even number, $^{28}_{14}$Si has an even–even nucleus. Therefore $i = 0$ and $I = \sqrt{i(i+1)}\hbar = 0.$ ●

    Example 29.3(b) shows that the total angular momentum of all even–even nuclei in the nuclear ground state is zero.

29.3
NUCLEAR STABILITY

There are currently 261 known stable nuclides, 60 percent are even–even nuclides; fewer are even–odd or odd–even, and only four ($^2_1$H, $^6_3$Li, $^{10}_5$B, and $^{14}_7$N) are odd–odd. The distribution is shown in Table 29.2.

    If we plot the stable nuclides on a graph of neutron number N versus proton number Z, as in Fig. 29.9, they initially fall on or close to the $N = Z$ line for the lighter elements. However, the number of neutrons exceeds the number of protons for all stable nuclides with $Z \geq 21$. The last completely stable nuclide is $^{208}_{82}$Pb, which contains 82 protons and 126 neutrons. For Pb-208, N exceeds Z by more than 50 percent! The **nuclear**

TABLE 29.2

The 261 Stable Nuclides

Z	N	Number
Even	Even	157
Even	Odd	51
Odd	Even	49
Odd	Odd	4

FIGURE 29.9

Stable nuclides. The tendency is for the number of neutrons to exceed the number of protons by an amount that increases with atomic number.

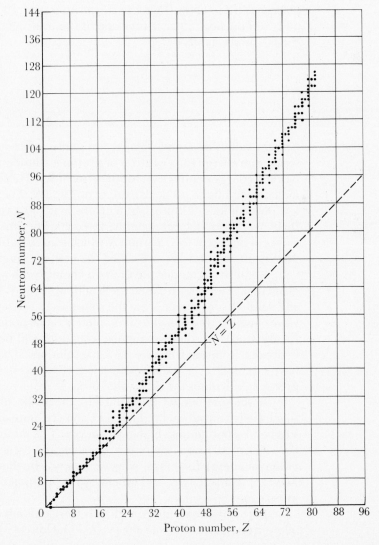

stability curve is a smooth curve drawn approximately through the stable nuclides.

The increasing excess of neutrons can be explained by the Coulomb repulsion term in the liquid drop model. Holding the nucleus together against a rapidly increasing repulsive force requires more neutrons with their attractive strong nuclear force. The added neutrons also spread the protons farther apart and thereby decrease the Coulomb repulsive force.

EXAMPLE 29.4   In terms of "even" and/or "odd," what kind of a nuclide is $^{134}_{55}$Cs? Is it stable?

Solution The nuclide $^{134}_{55}$Cs has 55 protons and 79 neutrons, so it's an odd–odd nuclide. Since it isn't one of the four light stable odd–odd nuclides, it's *not* stable.   ●

29.4
THE SHELL MODEL

Certain values of Z and N yield unusually stable nuclides. That is, the separation energy for a neutron or proton is considerably greater than the binding energy per nucleon for these nuclides. There are also more stable nuclides with these particular values of Z and N. These nuclides have very low probabilities of capturing neutrons in nuclear reactions compared to their neighbors. Experimentally, 2, 8, 20, 28, 50, 82, and 126 are the values of Z (except for 126) and N which lead to stability. The values 2, 8, 20, 28, 50, 82, and 126 are called the **nuclear magic numbers.**

Recall from atomic physics and chemistry that the especially stable elements are those with closed shells and subshells. Recall also that the shell and subshell ideas come from the solutions to the separated Schrödinger wave equation. If we assume a completely radial potential energy function for the nucleons, the two angular equations will be the same as for the hydrogen atom and will have the same solutions. That is, we can expect to have s, p, d, f, . . . states with $j = |l \pm \frac{1}{2}|$ and $m_j = -j$ to $+j$ in steps of 1 giving $2j + 1$ states for each value of j.

In the **shell model** of the nucleus, the magic numbers correspond to filled shells and subshells. The next allowed state above a filled shell or subshell is considerably higher in energy. The most successful version of the shell model uses a mixture of harmonic oscillator and square-well potential energy functions. This version was further improved by Maria Mayer, J. Hans Jensen, and others by the addition of a strong spin–orbit interaction.

In the hydrogen atom, with only the Coulomb force considered, the allowed energies were $-13.60 \text{ eV}/n^2$. The lowest n values had the lowest energies in a strict order. For example, all $n = 3$ states were lower than all $n = 4$ states. But more complicated atoms have more complicated interactions and, as a result, their 4s states are often lower in energy than their 3d states. In the nuclear case, with non-Coulomb forces and extremely strong spin–orbit interactions, the n values are even less related to the energies. For instance, one $n = 6$ level has a lower energy than one $n = 3$ level. Also, contrary to the atomic case, the $n - 1$ value of l has the lowest energy, not the highest; and the $j = l + \frac{1}{2}$ level is lower than the $j = l - \frac{1}{2}$ level.

The proton levels will differ somewhat from the neutron levels in the nuclear shell model, especially for large Z, because of the Coulomb repulsion energy of the protons. Nucleons are fermions, following the Pauli exclusion principle. For any level, this principle limits the maximum number of protons or neutrons to $2j + 1$. For instance, a $j = \frac{3}{2}$ proton level can contain up to $2(\frac{3}{2}) + 1 = 4$ protons (corresponding to $m_j = -\frac{3}{2}, -\frac{1}{2}, +\frac{1}{2}$, and $+\frac{3}{2}$). Figure 29.10 is an energy-level diagram of the predictions of the shell model.

In any level having an even number of protons or neutrons, the protons or neutrons will pair up and cancel each other's angular momentum in the ground state. Therefore $i = 0$ for even–even nuclei. If there is an odd number of nucleons, either one proton or one neutron will be unpaired, and the j value for that unpaired last odd nucleon is the i value for the entire nucleus. This shell model prediction usually, but not always, agrees with the value experimentally obtained from hyperfine interactions. Odd–odd nuclei have a total angular momentum that results from a complicated vector sum of the angular momentums of the unpaired last odd proton and last odd neutron.

We express the **parity** of a nuclear state as $(-1)^l$, where l is the orbital angular momentum quantum number of the unpaired last odd nucleon. (A reminder: s, p, d, f, g, h, i, . . . correspond to $l = 0, 1, 2, 3, 4, 5, 6, . . .$) Even–even nuclei have $+$ parity for their ground states.

A symmetry concept, parity is useful in determining allowed and forbidden transitions. Plus or even parity occurs when exchanging $-x, -y$, and $-z$ for $x, y,$ and z results in no change in the wave function, ψ, of a particular state. Minus or odd parity does result in a change in the sign of ψ.

EXAMPLE 29.5   Use one down arrow beside one up arrow to show paired nucleons and one lone down arrow for an unpaired nucleon in drawing a shell model energy-level diagram for a) $^{16}_{8}O$ and   b) $^{43}_{20}Ca$. Determine the spin and parity of the nuclear ground state according to the shell model.

Solution

a) Use only the part of the shell model energy-level diagram (Fig. 29.10) necessary. Remember that no more than $2j + 1$ protons or neu-

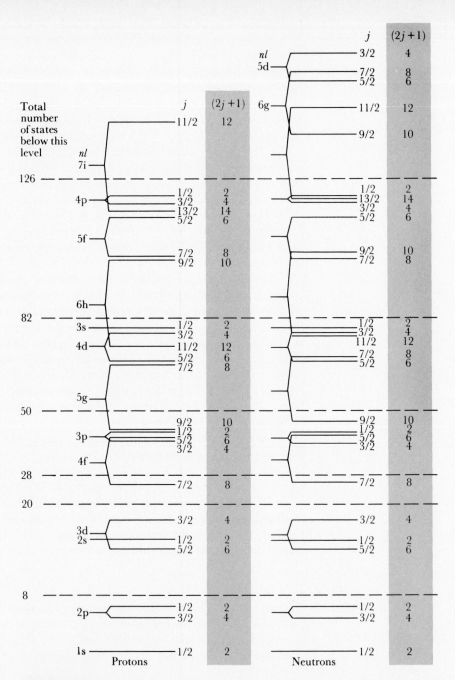

FIGURE 29.10

The shell-model energy levels for protons and neutrons. Any level may contain from zero to $2j + 1$ protons or neutrons.

Source: P. F. A. Klinkenberg, *Rev. Mod. Phys.*, **24**, 63 (1952).

FIGURE 29.11

Shell-model energy-level diagram for O-16.

trons can be in any level. The nuclide $^{16}_{8}$O has 8 protons and 8 neutrons, which gives Fig. 29.11. All the nucleons are paired in this even–even, doubly magic nuclide giving $i = 0$ and $+$ parity.

b) The nuclide $^{43}_{20}$Ca has 20 protons and 23 neutrons, and filling the levels gives Fig. 29.12. The unpaired nucleon is a neutron in a 4f state with $j = \frac{7}{2}$. Therefore the nuclear spin is $\frac{7}{2}$. For an f state, $l = 3$. Since $(-1)^l = (-1)^3 = -1$, the parity is $-$. The prediction of the shell model, $\frac{7}{2}-$, agrees with experimental results. ●

EXAMPLE 29.6 What does the shell model predict for the spin and parity of the ground state of

  a) $^{208}_{82}$Pb,   b) $^{97}_{42}$Mo, and   c) $^{133}_{55}$Cs?

Solution

a) Because the nuclide $^{208}_{82}$Pb has an even–even nucleus, its spin is zero, and its parity is $+$. The answer is $0+$.

b) The nuclide $^{97}_{42}$Mo has 42 protons and 55 neutrons, so the unpaired last odd nucleon is the 55th neutron. From Fig. 29.10, the "Total number of states below this level" column shows us that the 50th neutron fills

FIGURE 29.12

Shell-model energy-level diagram for Ca-43.

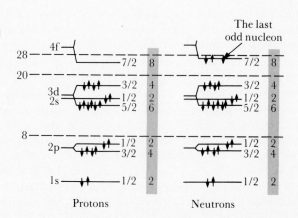

the $5g_\frac{9}{2}$ level. For neutrons (but not for protons), the $4d_\frac{5}{2}$ level is the next highest level and can hold up to $2(\frac{5}{2}) + 1 = 6$ neutrons. Thus the 5 remaining neutrons are in the $4d_\frac{5}{2}$ level, so $i = \frac{5}{2}$. The d state gives $l = 2$ and a parity of $(-1)^2 = +1$. The answer is $\frac{5}{2}+$.

c) The nuclide $^{133}_{55}$Cs has 55 protons and 78 neutrons. The 50th proton fills the $5g_\frac{9}{2}$ level. For protons, the $5g_\frac{7}{2}$ level is the next highest level and can hold up to $2(\frac{7}{2}) + 1 = 8$ protons. Thus the 5 remaining protons are in the $5g_\frac{7}{2}$ level, so $i = \frac{7}{2}$. The g state gives $l = 4$ and a parity of $(-1)^4 = +1$. The answer is $\frac{7}{2}+$.

All three answers agree with experimental results.   ●

In this chapter we used the liquid drop model to explain the general shape of the binding energy per nucleon curve and the tendency of N to be increasingly greater than Z for stable nuclides. In that model, small A nuclei have a larger fraction of their nucleons on the surface than do large A nuclei. Conversely, large A nuclei have a much larger Coulomb repulsion than do small A nuclei. Both these effects decrease the average binding energy per nucleon below that for the saturated nuclear force. Also, increasing the ratio of neutrons to protons dilutes the Coulomb repulsion. We showed that both the proton and the neutron have spin magnetic dipole moments. We found experimental nuclear magic numbers to be explained by the closed nuclear shells of the shell model. This model also enabled us to determine the so-called nuclear spin, as well as the parity, for nearly all nuclei, although we didn't go into the complicated methods used for odd–odd nuclei.

QUESTIONS AND PROBLEMS

29.1 The BE/A of H-2 is only 1.1 MeV/nucleon. Is the surface effect or the Coulomb effect responsible for this small value?

29.2 An equation for the binding energy per nucleon is $BE/A = a - b/A^{1/3} - cZ(Z-1)/A^{4/3}$. Explain the origin of each term.

29.3 Further corrections to the equation in Question 29.2 include a $-d(N-Z)^2/A^2$ term and another term, which is $+$ for even–even nuclides and $-$ for odd–odd nuclides. Give a reason for each of these terms.

29.4 Nucleus A has three times the radius of nucleus B. Calculate the ratios of their volumes,

their surface areas, and the fraction of nucleons on their surfaces.

29.5 Calculate the ratio of the Coulomb energy term for $^{54}_{26}$Fe to that for $^{27}_{13}$Al.

29.6 The magnetic potential energy of two dipoles is of the order $(\mu_0/4\pi)(\mu_1\mu_2/r^3)$. Calculate the order of magnitude spin magnetic potential energy for a) two protons touching, b) two neutrons touching, c) a proton and a neutron touching, and d) the proton and the electron in the Bohr atom.

29.7 When a nucleon is placed in a magnetic field, its energy is split into two levels. Calculate the photon frequency as a function of B that will result in a transition from the lower level to the upper level for   a) a proton and b) a neutron.

29.8 The total nuclear angular momentum of a nucleus has a magnitude of $(\sqrt{35}/2)\hbar$. Calculate the nuclear spin.

29.9 Why is *nuclear spin* a poor choice of name for i?

29.10 For a nucleus, $I = 0$. What can you say about that nucleus?

29.11 He-3 is the only stable nuclide with $Z > N \neq 0$. Why is $Z > N$ so uncommon?

29.12 Is $^{232}_{90}$Th stable?

29.13 Give examples of even–even, odd–odd, even–odd, and odd–even stable nuclides. (See the Appendix.)

29.14 All nuclides found above $Z = 82$ so far have been unstable. Some experimenters are trying to create or find a $Z = 126$ element, which they hope will be stable. Why might it be stable? Why is it very unlikely that sticking two $^{153}_{63}$Eu nuclei together to form $^{306}_{126}$X would produce a stable nuclide? (*Hint:* Consider the nuclear stability curve.)

29.15 Why are the quantum numbers l, j, and m_j used for nuclei as well as atoms?

29.16 What splits apart the $2p_{\frac{3}{2}}$ and $2p_{\frac{1}{2}}$ levels in a shell model?

29.17 The parity operator converts x, y, and z in Cartesian coordinates to $-x$, $-y$, and $-z$. Recall that $x = r \sin \theta \cos \phi$, $y = r \sin \theta \sin \phi$, and $z = r \cos \theta$. The distance, r, remains plus under the parity operation. What does the parity operator do to θ and ϕ in spherical coordinates?

29.18 Draw a shell model energy diagram and determine the spin, total angular momentum, and

parity of the ground state of the nuclides a) $^4_2$He, b) $^{27}_{13}$Al, and c) $^{61}_{28}$Ni.

29.19 What does the shell model predict for the spin, total angular momentum, and parity of the ground state of the following nuclides? a) $^{47}_{21}$Sc, b) $^{91}_{40}$Zr, c) $^{131}_{53}$I, and d) $^{203}_{81}$Tl and $^{205}_{81}$Tl.

29.20 Draw a shell model energy diagram and determine the spin, total angular momentum, and parity of the first excited level of $^{17}_8$O.

29.21 a) Prove that the integral over all space of a function with $-$ or odd parity equals zero.

b) Will all volume integrals of $-$ parity functions therefore equal zero?

c) Must the integral over all space of a function with $+$ or even parity be nonzero?

29.22 In many texts, the levels in Fig. 29.10 are labeled 1s, 1p, 2s, 1d, 1f, 2p, 1g, and so on. The first number then isn't the principal quantum number n, but the radial quantum number r.

a) What does r equal in terms of n and l?

b) Is a 1i state then possible?

c) Are any rl states impossible?

d) Is there any reason that the principal quantum number n isn't especially appropriate for the shell model?

29.23 Some nuclei, especially in their excited states, deviate considerably from a spherical shape, resulting in a potential energy function that isn't spherically symmetric. Will this have any effect on the angular portion of the wave functions used in the shell model?

29.24 Use parity to explain why $\int_{\text{all space}} \psi_i^* z \psi_j \, dv$ is zero when Δl is even for the states i and j. (This integral is useful in providing selection rules for transitions.)

CHAPTER **30**

Radioactivity I

Radioactivity has many useful applications in medicine, research labs, and industry. And radioactivity is literally a part of your everyday life. As a result, questions about radioactivity safety are a continuing concern, not only for the scientists and engineers involved with its use, but also for active citizens, their elected lawmakers, and appointed regulators. In this chapter we will introduce you to such fundamental concepts as activity, half-life, and biological effect. We'll also discuss how radioactivity can be used to determine the age of some objects.

30.1
RADIOACTIVE DECAY

There may be 261 known stable isotopes, but there are many more unstable nuclides. These unstable nuclides decay into different nuclides with the emission of some type of radiation. The most common types of radiation, as illustrated in Fig. 30.1 and listed in Table 30.1, are alpha (α) particles, beta (β) particles, and gamma (γ) rays. In the case of γ rays, the nucleus is changing from an excited state to a less excited state. Therefore

FIGURE 30.1

Paths of the most common types of radiation in a perpendicular magnetic field. The initial velocity is toward the top of the page; the magnetic field is directed out of the page. Positive, massive alpha particles are deviated somewhat to the right. Positive, low-mass beta-plus particles are deviated to the right in a tight circle with negative, low-mass beta-minus particles of equal energy following a mirror-image path to the left. Neutral gamma rays are undeviated.

the atomic and mass numbers of the nuclide don't change for γ radiation; the difference is in the initial and final energies of the nucleus.

In radioactive decay, the original nucleus is called the **parent,** and the resulting nucleus is called the **daughter.** For example, $^{239}_{94}\text{Pu}$ decays by the emission of an alpha particle to $^{235}_{92}\text{U}$. The parent is Pu-239 and the daughter is U-235 in this particular radioactive decay.

Another indication that quantum mechanics governs the behavior of nucleons is that the laws of probability determine whether an unstable nucleus will decay. The probability per time that the nucleus will decay is given the symbol λ. Called the **decay constant** or the **disintegration constant,** λ has units of (time)$^{-1}$, usually s^{-1}. For example, λ is about 2×10^{-9} s^{-1} for ^3_1H. So, if we choose one particular ^3_1H nucleus, the probability is about 2×10^{-9} (one chance in five hundred million) that it will decay in any given second.

Suppose that at some time t we have a large number N of parent nuclei. The probability per time that a single nucleus will decay is λ. Therefore, in a time dt, we'll have $-dN$ parents decaying, where $dN = -\lambda N\, dt$. The minus sign indicates that dN is negative: dN is a *decrease* in N, the number of parent nuclei. (We are assuming that no parent nuclei are being created by other processes.)

TABLE 30.1

The Most Common Types of Radiation from Nuclei

Name	Identification
α particle	^4_2He nucleus
β particle	Electron (β^-) or positron (β^+)
γ ray	High-energy photon

The first thing we can do with $dN = -\lambda N\, dt$ is to divide both sides by dt. This gives us dN/dt, the rate at which parent nuclei are decaying, or disintegrating, into the daughter nuclei. The absolute value of the rate is called the **activity,** A, or

$$A = -\frac{dN}{dt} = \lambda N. \qquad (30.1)$$

We measure the activity in disintegrations per second (dis/s). Currently, the becquerel (Bq) is an uncommon unit that equals 1 dis/s. A more common unit is the curie (Ci), named after Marie Curie: 1 Ci is defined as exactly 3.7×10^{10} dis/s. Therefore the millicurie (mCi) and the microcurie (μCi) equal exactly 3.7×10^7 dis/s and 3.7×10^4 dis/s, respectively. Figure 30.2 represents the relationship among activity, decay constant, and number of parent nuclei. Here the letters A and N stand for activity and number of parent nuclei; please don't confuse them with the A and N that stand for the mass number and neutron number.

EXAMPLE 30.1   There are an estimated 10^{17} kg of uranium in the top 20 km of the earth's crust. Of the total amount, 5.5×10^{-3} percent of the U atoms are U-234, 0.72 percent are U-235, and the remaining 99.27 percent are U-238. The decay constant for U-234 is 8.97×10^{-14} s^{-1}, for U-235, it is 3.12×10^{-17} s^{-1}; and for U-238, it is 4.92×10^{-18} s^{-1}. Calculate the activity of the total amount of uranium in curies.

Solution First, we need to find N by using the atomic mass of uranium, 238.029 g/mol, and Avogadro's number, 6.02×10^{23} nuclei/mol. (Avogadro's number gives the number of molecules/mole, but for uranium there is 1 atom/molecule and one nucleus/atom.) Arranging the quanti-

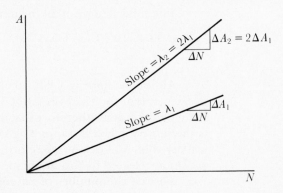

FIGURE 30.2
The activity is directly proportional to both the number of parent nuclei present and to the decay constant.

ties to cancel all the units but "nuclei" we have

$$N = \frac{(10^{17} \times 10^3 \text{ g})(6.02 \times 10^{23} \text{ nuclei/mol})}{238.029 \text{ g/mol}} = 2.53 \times 10^{41} \text{ nuclei}.$$

Now 5.5×10^{-3} percent of these nuclei are U-234, so

$$N_{234} = (5.5 \times 10^{-5})(2.53 \times 10^{41} \text{ nuclei}) = 1.39 \times 10^{37} \text{ nuclei}.$$

Then $A = \lambda N$ gives

$$A_{234} = (8.97 \times 10^{-14} \text{ s}^{-1})(1.39 \times 10^{37} \text{ nuclei})$$

$$= (1.25 \times 10^{24} \text{ dis/s})\left(\frac{1 \text{ Ci}}{3.7 \times 10^{10} \text{ dis/s}}\right) = 3.4 \times 10^{13} \text{ Ci}.$$

The "nuclei" here are disintegrating, so nuclei/s becomes dis/s. Similarly,

$$N_{235} = (7.2 \times 10^{-3})(2.53 \times 10^{41} \text{ nuclei}) = 1.82 \times 10^{39} \text{ nuclei},$$

and

$$A_{235} = (3.12 \times 10^{-17} \text{ s}^{-1})(1.82 \times 10^{39} \text{ nuclei})\left(\frac{1 \text{ Ci}}{3.7 \times 10^{10} \text{ dis/s}}\right)$$

$$= 1.5 \times 10^{12} \text{ Ci}.$$

Also,

$$N_{238} = 0.9927(2.53 \times 10^{41} \text{ nuclei}) = 2.51 \times 10^{41} \text{ nuclei},$$

and

$$A_{238} = (4.92 \times 10^{-18} \text{ s}^{-1})(2.51 \times 10^{41} \text{ nuclei})\left(\frac{1 \text{ Ci}}{3.7 \times 10^{10} \text{ dis/s}}\right)$$

$$= 3.3 \times 10^{13} \text{ Ci}.$$

The total activity of all the uranium is

$$A_T = 3.4 \times 10^{13} \text{ Ci} + 1.5 \times 10^{12} \text{ Ci} + 3.3 \times 10^{13} \text{ Ci} = 6.9 \times 10^{13} \text{ Ci}.$$

Although there is 1.8×10^4 times as much U-238 as U-234, the decay constant for U-234 is 1.8×10^4 times larger than that for U-238. Example 30.1 shows therefore that the activity (to the accuracy of the data) is the same for U-238 as for U-234. The 69 million million curie total activity is huge. Luckily, we're ordinarily shielded from virtually all the radiation released by this uranium.

To find the number of parent nuclei remaining at any time t, we rearrange $dN = -\lambda N \, dt$ as $dN/N = -\lambda \, dt$. With the variables N and t each on its own side of the equal sign, we can now integrate. There will be N_0 parent nuclei at $t = 0$ and N parent nuclei remaining at $t = t$. The equation then is $\int_{N_0}^{N} dN/N = -\lambda \int_0^t dt$. Integrating and substituting the limits gives $\ln N/N_0 = -\lambda t$. (Remember that $\ln N - \ln N_0 = \ln N/N_0$.) Making both sides a power of e is the same as taking the natural antilog of both sides, so

$$N/N_0 = e^{-\lambda t} \quad \text{or} \quad N = N_0 e^{-\lambda t}.$$

Equation (30.1) becomes $A_0 = \lambda N_0$ at $t = 0$. Multiplying $N = N_0 e^{-\lambda t}$ by λ gives $A = A_0 e^{-\lambda t}$.

A much-used property of radioactive nuclides is their **half-life**, $T_{\frac{1}{2}}$. The half-life of a radioactive nuclide is the time required for one-half the original parent nuclei to decay. Since the activity is directly proportional to the number of nuclei ($A = \lambda N$), the half-life is also the time it takes for the activity to drop to one-half its original value. The source for the half-lives and much of the other nuclear data in this book is the *Chart of the Nuclides*, Copyright R 1984 by General Electric Company.

The greater the probability of decay, the smaller is the time needed for half the nuclei or activity to decay. To find the relation between $T_{\frac{1}{2}}$ and λ, we simply set $t = T_{\frac{1}{2}}$. At that time, N will equal $N_0/2$ (or A will equal $A_0/2$) by the definition of $T_{\frac{1}{2}}$. Solving $N = N_0/2 = N_0 e^{-\lambda T_{\frac{1}{2}}}$, we get $\ln(1/2) = -\lambda T_{\frac{1}{2}}$, or $-\ln(1/2)/\lambda = T_{\frac{1}{2}}$. With $-\ln(1/2) = +0.69315$, we have

$$T_{\frac{1}{2}} = \frac{0.69315}{\lambda}. \tag{30.2}$$

EXAMPLE 30.2 The half-life of Co-57 is 271.8 days. Calculate its decay constant in units of s^{-1}.

Solution For $T_{\frac{1}{2}} = 271.8$ days, $T_{\frac{1}{2}} = 0.69315/\lambda$ becomes

$$\lambda = \frac{0.69315}{T_{\frac{1}{2}}}$$

$$= \frac{0.69315}{(271.8 \text{ days})(24 \text{ hr}/1 \text{ day})(3600 \text{ s}/1 \text{ hr})}$$

$$= 2.952 \times 10^{-8} \text{ s}^{-1}. \qquad \bullet$$

If $N = N_0 e^{-\lambda t}$ and $\lambda = -\ln(1/2)/T_{\frac{1}{2}}$, then $N = N_0 e^{[+\ln(1/2)]t/T_{\frac{1}{2}}} = N_0(\frac{1}{2})^{t/T_{\frac{1}{2}}}$. Then $A = \lambda N$ gives $A = A_0(\frac{1}{2})^{t/T_{\frac{1}{2}}}$. Therefore

$$N = N_0 e^{-\lambda t} = N_0(\tfrac{1}{2})^{t/T_{\frac{1}{2}}} \tag{30.3}$$

and

$$A = A_0 e^{-\lambda t} = A_0(\tfrac{1}{2})^{t/T_{\frac{1}{2}}}. \tag{30.4}$$

Which version is most useful depends on the problem being worked.

EXAMPLE 30.3   An experimenter notes that the activity of a sample of Ba-137 drops from 2.85 mCi to 0.73 mCi in 5.00 min. Calculate the decay constant of Ba-137 from these data.

Solution We start with $A_0 = 2.85$ mCi, $A = 0.73$ mCi, and $t = (5.00 \text{ min})(60 \text{ s}/1 \text{ min}) = 300$ s. We want λ. Solving $A = A_0 e^{-\lambda t}$ by dividing by A_0, taking the ln of both sides, and then dividing by $-t$, we have $-[\ln(A/A_0)]/t = +[\ln(A_0/A)]/t = \lambda$. Therefore

$$\lambda = \frac{+\ln(2.85 \text{ mCi}/0.73 \text{ mCi})}{300 \text{ s}} = 4.5 \times 10^{-3} \text{ s}^{-1}. \qquad \bullet$$

EXAMPLE 30.4   a) If half the parent nuclei have decayed after 1 half-life, does this mean that all will have decayed after 2 half-lives?

b) If not, how many will have decayed then?

c) When will all have decayed?

Solution

a) No. One-half decay during the first half-life, then one-half of the remaining half decay during the second half-life, and so on.

b) Half of one-half is one-fourth, so one-fourth are left and three-fourths must have decayed. In calculation form,

$$N = N_0(\tfrac{1}{2})^{t/T_{\frac{1}{2}}} = N_0(\tfrac{1}{2})^{2T_{\frac{1}{2}}/T_{\frac{1}{2}}} = N_0(\tfrac{1}{2})^2 = N_0/4.$$

Of the original N_0 nuclei, $N_0/4$ remain. Therefore $N_0 - N_0/4 = (3/4)N_0$ have decayed.

c) All will have decayed ($N \rightarrow 0$ in Eq. 30.3) only as t approaches infinity. $\bullet$

You sometimes hear arguments that seem to say that radioactive isotopes will become safe after their half-lives have passed. However, Example 30.4 and Fig. 30.3 show that in 1 half-life the activity drops by only a factor of 2. If there is 10 times too much radioactivity from some radioactive waste, after 1 half-life there will still be 5 times too much, after 2 half-lives 2.5 times too much, and after 3 half-lives still 1.25 times too much. Only as time approaches infinity will the activity approach zero. Thus the average, or mean, lifetime of a radioactive nuclide, $\overline{T}$, is longer than its half-life. Problem 30.8 will show that $\overline{T}$ is 44.3 percent longer than $T_{\frac{1}{2}}$ and that $\overline{T}$ simply equals the reciprocal of the decay constant.

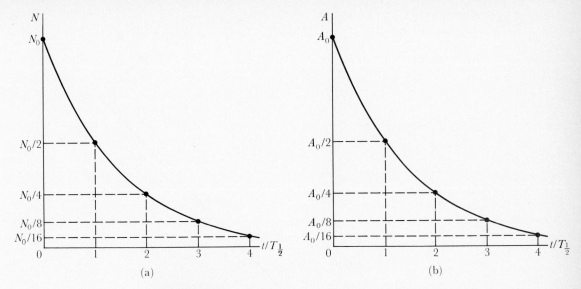

FIGURE 30.3

(a) The number of parent nuclei and (b) the corresponding activity decrease by one half their previous values each half-life (assuming that no parent nuclei are being created).

30.2
NATURAL RADIOACTIVITY

Natural radioactivity is the decay of those unstable nuclides found in nature. It's easy to understand why there's still natural radioactivity from U-238. It has a half-life of 4.468×10^9 years, and the farthest back most models put the creation of the universe is at about 20×10^9 years.

But why has Po-214 been detected in nature with its half-life of 163.7 μs? The answer is that Po-214 is part of a natural radioactive series. A **radioactive series** is a group of nuclides, each of which decays into another. For example, Fig. 30.4 shows that U-238 decays into Th-234, which decays into Pa-234, and so on (including a branch to Po-214) until the series ends with the stable nuclide Pb-206. Therefore the natural radioactive series beginning with U-238 will also contain many other radioactive nuclides. An equilibrium property of such a series gives the same activity to U-238 and U-234, which we calculated in Example 30.1. In addition to this uranium series, there are two other natural radioactive series. Many other radioactive series have been created artificially. Irène Curie and Frederick Joliot discovered **artificially induced radioactivity** in 1934.

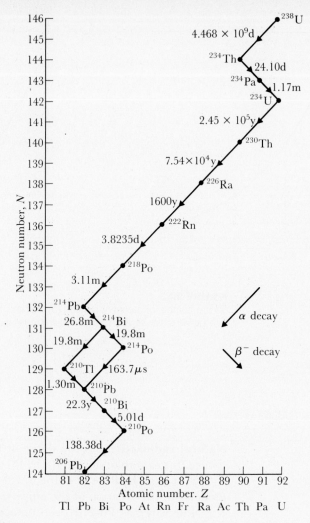

FIGURE 30.4
A natural radioactive series. This one is called the uranium series.

EXAMPLE 30.5   What is an artificial radioactive series?

Solution Such a series is a group of nuclides, each of which decays into another but not all occur naturally on earth. Rather, the entire series would be found after the artificial production of the parent.   ●

30.3
RADIOACTIVE DATING

Going out with your sweetie for an evening at the nearest nuclear power plant is *not* an example of radioactive dating. **Radioactive dating** is a method of determining the age of material by measuring its natural radio-activity. One example of this phenomenon occurs because cosmic ray interactions in the earth's upper atmosphere continually produce $^{14}_{6}C$. This C-14 then diffuses down to the earth's surface, where it mingles with stable C-12 and C-13. Since all living matter grows from carbon compounds, all living matter is composed of a certain fraction of C-14. This fraction currently gives an activity of about 15.3 dis/min for each gram of carbon in the living matter.

When an organism dies (for instance, when a tree is cut down) its intake of carbon stops. From that moment the amount and activity of the C-14 begin to decrease with a half-life of 5730 years, as shown in Fig. 30.5. For example, 7.65 is one-half of 15.3. Therefore an activity of 7.65 dis/min for each gram of carbon contained in a mummy would tell us that the mummy was 5730 years old (assuming that C-14 was produced in the upper atmosphere at the same rate then as it is now).

EXAMPLE 30.6 An archeological sample contains 400 mg of carbon. During one hour you measure the occurrence of 139 disintegrations. Calculate the age of the sample.

Solution You first determine A_0. With (15.3 dis/min)/g of carbon and 400 mg = 0.400 g of carbon, you have

$$0.400 \text{ g} \times (15.3 \text{ dis/min})/\text{g} = 6.12 \text{ dis/min} = A_0.$$

FIGURE 30.5
A semilog plot of the decay of the activity of a 1 g C-14 sample. Ages can't be determined exactly from this graph, because the rate of C-14 production has varied somewhat in the past.

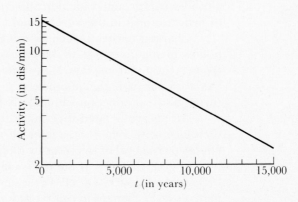

Activity (in dis/min)

t (in years)

Then $A = 139$ dis/hr $= 139$ dis/60 min $= 2.317$ dis/min. Solving $A = A_0(\frac{1}{2})^{t/T_{\frac{1}{2}}}$ gives $A/A_0 = (\frac{1}{2})^{t/T_{\frac{1}{2}}}$. Then $\ln(A/A_0) = (t/T_{\frac{1}{2}})\ln(\frac{1}{2})$, and finally $T_{\frac{1}{2}} \ln(A/A_0)/\ln(\frac{1}{2}) = t$. From the text, $T_{\frac{1}{2}} = 5730$ years, so

$$5730 \text{ yr } \ln(2.317/6.12)/\ln(\tfrac{1}{2}) = 8030 \text{ yr}$$

is the sample's age.   ●

You can also solve Example 30.6 graphically. Calculate the activity of a gram of carbon, which would be

$$(1 \text{ g})(2.317 \text{ dis/min})/0.400 \text{ g} = 5.79 \text{ dis/min}.$$

Then use Fig. 30.5 to find that this activity corresponds to $t = 8000$ yr.

Other isotopes with different half-lives are also used for radioactive dating. It's now possible to use a particle accelerator as a mass spectrometer to count carbon atoms one at a time, and determine the ratio of C-14 atoms to a stable C isotope. This ratio is cut in half each half-life. Thus we can calculate the half-life by knowing the original ratio and measuring the current ratio. The advantage of the accelerator method is that a thousandth as much carbon is needed for the same accuracy.

30.4
BIOLOGICAL EFFECTS OF RADIATION

The question that may worry you most about radiation is, "What is it doing to me?" Many of the answers to this question would take us over into a controversial area of biology, but we can say something about the physics of radiation exposure. Much of the radiation that we receive is natural **background radiation,** of which cosmic rays are one source. The cosmic ray flux is about 120 per m^2 per s at sea level, doubling every 2000 m of altitude. Background radiation also comes from the natural radioactivity of many isotopes in the world around us (and even within our bodies).

Charged particles, such as α and β particles, ionize atoms in our bodies, giving up energy in the process. The β^+ particles annihilate electrons, creating gamma rays. X-rays and gamma rays give up their energy to electrons through the photoelectric and Compton effects, as well as through pair production within body tissue.

Atoms are bonded together into molecules by the interactions of their electrons. When these interactions are changed by the addition of energy or removal of electrons by radiation, the molecular structure may be changed. Changes in the molecular structure in turn may affect the biological processes of a cell. For instance, damaging the DNA molecules may result in changes in cell reproduction. Some of this damage can be repaired by normal biological processes. Some of the damage is irreversible.

The cell damage is not necessarily a negative result. As early as 1902 a positive use for radiation—the killing of unwanted cells, especially cancer cells—was recognized. The problem is how to deliver the energy to just the cancerous tissue while sparing all the healthy cells around it. Currently, radiation treatment can't do so; its aim is to do the most damage to the cancer cells and the least damage to the healthy cells.

An **absorbed dose,** AD, is the energy from radiation actually absorbed, E_A, per mass of absorbing matter, m. That is,

$$AD = \frac{E_A}{m}. \tag{30.5}$$

The SI unit for the absorbed dose is the gray (Gy). One gray is equal to one joule of energy from radiation absorbed per kilogram of absorber. A more commonly used unit is the rad (not the abbreviation for radian): 1 rad = 10^{-2} Gy.

EXAMPLE 30.7 Sr-90 is found in the fallout from atmospheric tests of nuclear weapons. It replaces calcium in places such as bones and milk. It has a half-life of 29 years, and the net energy absorbed from each decay will average about 1.1 MeV. Calculate the absorbed dose from 1.0 μg of Sr-90 in 1 year if the energy is absorbed in 50 kg of body tissue.

Solution First, we calculate N_0 for 1.0 μg of Sr-90:

$$N_0 = \frac{(1.0 \times 10^{-6} \text{ g})(6.02 \times 10^{23} \text{ nuclei/mol})}{90 \text{ g/mol}} = 6.7 \times 10^{15} \text{ nuclei.}$$

Then using Eq. (30.3), the number of nuclei that decay in a time t is $N_0 - N = N_0[1 - (\frac{1}{2})^{t/T_{\frac{1}{2}}}]$. So in 1 year with a half-life of 29 years we have 6.7×10^{15} nuclei$[1 - (\frac{1}{2})^{1/29}] = 1.6 \times 10^{14}$ dis. Finally, we calculate the absorbed energy in J and divide it by the mass to get the absorbed dose:

$$AD = \frac{E_A}{m} = \frac{(1.6 \times 10^{14} \text{ dis})(1.1 \times 10^6 \text{ eV/dis})(1.6 \times 10^{-19} \text{ J/eV})}{50 \text{ kg}}$$

$$= 0.56 \text{ Gy}(1 \text{ rad}/10^{-2} \text{ Gy}) = 56 \text{ rad.} \qquad \bullet$$

The biological effects of the absorbed dose depend on a large number of variables. These variables include the size of the dose; the type of radiation; the rate at which it is absorbed; the species, age, and physical condition of the organism; the type of tissue; and others. The quantity that measures the biological effect of the radiation is the **dose equivalent,** DE (also known as just the **dose**). The SI unit for the dose equivalent is the

TABLE 30.2
Some *RBEs*

Type of Radiation	*RBE* (Sv/Gy or rem/rad)
X-rays, γ rays, β particles > 30 keV	1.0
β particles < 30 keV	1.7
Neutrons and protons < 10 MeV, alphas	10
Low-speed, heavy nuclei	20

sievert (Sv). The more commonly used unit is the rem, which equals 10^{-2} sievert.

Different types and energies of radiation cause different amounts of biological damage. X-rays, gamma rays, and β particles (with energy above 30 keV) cause the least damage for a given absorbed dose. The relative biological effectiveness, *RBE*, is defined as the dose equivalent, *DE*, per absorbed dose, *AD*. Therefore

$$DE = RBE \times AD. \tag{30.6}$$

Table 30.2 lists some *RBEs*. The *RBE* for x-rays and gamma rays, as well as for β particles above 30 keV, is 1.0 Sv/Gy = 1.0 rem/rad. Lower energy β particles, as well as heavier particles, will stop in shorter distances, causing more localized and greater damage. The *RBE* for them is therefore greater than 1 Sv/Gy. For example, one of the reasons that inhaling less than one microgram of Pu-239 will almost certainly cause death is that Pu-239 is an alpha emitter, and alpha particles have an *RBE* of 10 Sv/Gy. Another is that the massive, doubly-charged α particles will be stopped in a short distance, depositing their kinetic energy in a small mass of tissue.

EXAMPLE 30.8 Radiation from Sr-90 involves β^- particles and gamma rays, all having energies greater than 0.1 MeV. Calculate the dose from Example 30.7.

Solution We want the *DE*, knowing that $AD = 0.56$ Gy = 56 rad. The β^- particles involved have energies of > 30 keV, so the *RBE* for them and the gamma rays is 1.0 (Sv/Gy or rem/rad). Therefore

$$DE = RBE \times AD = (1.0 \text{ Sv/Gy})(0.56 \text{ Gy}) = 0.56 \text{ Sv}$$

or

$$(1.0 \text{ rem/rad})(56 \text{ rad}) = 56 \text{ rem}$$

Check: $(0.56 \text{ Sv})(1 \text{ rem}/10^{-2} \text{ Sv}) = 56 \text{ rem}$. ●

If all the radiation is absorbed in about a week or less and is spread over the entire human body, external symptoms will occur with a dose of about 1.0 Sv = 100 rem. The chance of short-term recovery is excellent. However, the long-term effects may be serious. Scientists generally (but not unanimously) agree that there is no minimum threshold dose that can be absorbed with perfect safety. The chance of short-term recovery is about 80 percent for 3 Sv = 300 rem and declines to near 50 percent at about 5 Sv = 500 rems. Death is virtually certain above 6 Sv = 600 rems, usually with symptoms showing after 2 hours and death occurring after 2 weeks.

These numbers were obtained from reports of accidental exposures, data from the Japanese victims of the nuclear bombs (neither source being well-controlled experiments), and the extrapolation of animal tests to humans. Since the data aren't of good quality, correct values may vary considerably from the numbers given. Also remember that probabilities are involved. One person may receive a short-term, whole-body dose of 6 Sv and eventually recover. Another person may receive 1 Sv, show recovery, but then eventually die from long-term effects. Marie Curie and her daughter Irène Joliot-Curie both died from leukemia. We can't state with certainty that their research killed them. We can only say that their leukemia probably resulted from all the radiation they were exposed to during their experiments.

Natural background radiation contributes a dose equivalent of about 1 mSv, or 100 mrem, per year at sea level. A large fraction of this dose, possibly over one-half on average but varying widely, is from radon gas. Radon is produced by the decay of radium in the soil. Being a gas, radon then diffuses up into homes and other buildings. About one-third of background radiation comes from radioactive nuclides within our bodies.

Medical x-rays average nearly 700 μSv = 70 mrem per year per person, but x-ray exposures for diagnosis on the order of tens of mSv or several rems are quite common. We're still receiving about 40 μSv = 4 mrem per year from fallout from nuclear bomb tests in the atmosphere. Nuclear power plants contribute only about 0.03 μSv = 0.003 mrem per year per person. This low value, however, is averaged over the entire U.S. population. Some people get more because of where they live or work and/or because of accidents. For example, more than fifteen thousand people who lived near the Chernobyl reactor received an average dose of 0.5 Sv = 50 rem as a result of the April 1986 accident. The prediction is that, overall, about 9000 people will die sooner than they would have otherwise because of the Chernobyl accident; 1000 of them live outside the Soviet Union. To put this prediction in a perspective, however, an estimated 9000 people will die early in a single year in the United States alone as a result of pollution from coal-fired electric power plants. And approximately another 9000 will die early each year in the United States from exposure to radon.

In this chapter you learned how to calculate the activity, the rate at which parent nuclei decay into daughter nuclei, and how to relate the decay constant to the half-life. We derived the equations that describe how the activity and the number of parent nuclei decrease with time and how this knowledge can be used to date certain materials. Finally, we discussed some of the measurements of radiation received by biological systems, such as absorbed dose and dose equivalent, and the approximate effects of certain doses on human beings.

QUESTIONS AND PROBLEMS

30.1 What is the decay constant of a stable nuclide in its ground state? If the stable nuclide has an excited state, is your answer the same for the excited state?

30.2 The decay constant of Co-60 is 4.17×10^{-9} s^{-1}. Calculate the activity of 1.00 g of Co-60.

30.3 If you are of average mass, about 360 million nuclei in your body undergo radioactive decay each day. Express your activity in nCi.

30.4 A parent nuclide with a decay constant λ_p decays into a daughter nuclide with a decay constant λ_d.
   a) Prove that $dN_d/dt = \lambda_p N_p - \lambda_d N_d$.
   b) Explain why dN_d/dt isn't the activity of the daughter.
   c) In "secular equilibrium," $dN_d/dt = 0$. Relate the number of parent and daughter nuclei to their half-lives in secular equilibrium.

30.5 The nuclide Ra-226 has a decay constant of 1.37×10^{-11} s^{-1}.
   a) Calculate the activity of 1.0 g of Ra-226 (the original basis of the curie).
   b) Calculate the time by which 0.5 g of that Ra-226 will have decayed.

30.6 Some 0.0117% of natural potassium is K-40, a radioactive isotope with a half-life of 1.25×10^9 years. Calculate the activity of 100 lb (4.54×10^4 g is its mass) of natural potassium. The atomic mass of K is 39.1 g/mol.

30.7 Tritium (H-3) is formed in nuclear reactors and has a half-life of 12.3 years. How much tritium will have an activity of 1.0 Ci?

30.8 The average (or mean) life, $\overline{T}$, of a radioactive nuclide is longer than the half-life.

a) Explain why. b) $\overline{T} = \int_0^{N_0} t \, dN/N_0$; prove that $\overline{T} = 1/\lambda = 1.443 T_{\frac{1}{2}}$.

30.9 A parent nucleus decays into a stable daughter. Derive the relationship for the number of daughter nuclei at time t, if there are N_0 parent nuclei and zero daughter nuclei at $t = 0$.

30.10 The activity of a radioactive source is 25 mCi after 3 half-lives. What was its original activity?

30.11 Of all Rb atoms, 27.83% have the radioactive Rb-87 nucleus; the rest have the stable Rb-85 nucleus. The half-life of Rb-87 is 4.89×10^{10} years. Assuming that no Rb has been formed since then, what was the percentage of Rb-87 4.6×10^9 years ago when our solar system was formed?

30.12 A theory concludes that Th-232 and U-238 were originally produced at the same time with a ratio of 19 Th-232 nuclei to every 10 U-238 nuclei, and none has been produced since. On earth, there are currently 39 Th-232 nuclei to every 10 U-238 nuclei. The half-life of Th-232 is 1.40×10^{10} yr and that of U-238 is 4.468×10^9 yr. When were earth's Th-232 and U-238 produced?

30.13 Radioactive decay is a random process, so the actual number of decays, $-\Delta N$, in a time Δt will fluctuate mostly within the range $-\Delta N \pm \sqrt{-\Delta N}$. Suppose that you start with a milligram of parent nuclei. What will be the range of decays in the first minute if the substance is a) Rb-88, with a half-life of 17.7 min? b) Rb-87, with a half-life of 4.89×10^{10} yr?

30.14 Calculate the activity of the U-238 in the earth's crust 2.0×10^9 years ago. Use the data in Example 30.1 and assume that no U-238 has been formed since then.

30.15 The decay constant for Co-57 is 2.95×10^{-8} s^{-1}. A 25-mCi Co-57 source is prepared for a Mössbauer effect experiment. What will its activity be 5.00 years later?

30.16 A hospital purchases a 10.0-mg sample of Co-60, which has a half-life of 5.272 years. Calculate its initial activity and its activity after exactly 1, 10, and 50 years.

30.17 The amount of different nuclides in a radioactive series depends on their half-lives. Will more of the nuclides have short half-lives or long half-lives? Explain, using an analogy of moving a group of people standing in a series of lines from point A to point B; some are served by very quick clerks and some by slow clerks.

30.18 The atomic mass of all carbon is 12.011 g/mol; it is 14.003 g/mol for C-14. What fraction of all carbon is C-14 currently?

30.19 A scroll supposedly written by John the Evangelist contains 26 g of 1900-year-old carbon. How many disintegrations per second should occur from the carbon in the scroll?

30.20 Each pound of carbon (mass = 454 g) from timbers at an archeological site provides 5010 counts per minute. What date does this give the site?

30.21 A substantial fraction of the radioactive substances released in the Chernobyl accident were isotopes of the noble gases Xe and Kr. Even when inhaled, however, they give a small absorbed dose compared to another radioactive gas emitted, I-131. Consider the chemistry of these three elements and explain why.

30.22 About one-eighth of the Cs-137 present in the Chernobyl reactor was released. The nuclide Cs-137 has a half-life of 30.17 years and decays with the emission of a total of 1.17 MeV of energy released per decay; 0.51 MeV goes to an emitted electron and the remaining 0.66 MeV to a gamma ray. This nuclide is absorbed by plants that are eaten by livestock and humans. How many Cs-137 atoms would have to be present in each kg of body tissue if the dose equivalent is 3 Sv in a week? Assume that all the energy from the decays is deposited in that kg of tissue.

30.23 On average, 0.21% of the human body is potassium, of which 0.012% is radioactive K-40. The nuclide K-40 has a half-life of 1.25×10^9 years. From each decay, about 0.50 MeV of β^- (> 30 keV) and gamma-ray radiation is absorbed by the body. Calculate the absorbed dose in rad and Gy, as well as the dose equivalent in rem and Sv, for 70 years from natural body potassium.

30.24 Suppose that 2.0 μCi of tritium are accidentally ingested by a 50-kg person. Assume that all the tritium remains in and spreads throughout the body. Assume further that the average decay leads to the absorption of 5.5 keV of (< 30 keV) β^- energy. The half-life of tritium is 12.3 years. Calculate the absorbed dose and the dose equivalent for one year.

30.25 A person ingests 0.30 μCi of a very long lifetime α source. It lodges in the lungs, where all the 4.0-MeV alphas emitted are absorbed within a 0.50-kg mass of tissue. Calculate the absorbed dose and the dose equivalent for one year.

30.26 How many 700-keV electrons from P-32 must be absorbed by an 80-kg person throughout his body to receive a 600-rem dose from this source?

30.27 A 60-kg person absorbs over her entire body one million 5.0-MeV protons in an accelerator accident. Calculate the absorbed dose and the dose equivalent.

30.28 An 80-kg worker receives a whole-body 1.4 Sv dose, 60% due to alphas and 40% due to gammas. Calculate the energy absorbed for each type of radiation.

30.29 Multiplying the number of persons exposed to a given amount of radiation times the average dose equivalent gives the number of person-rems. A reasonable estimate is that 10^{-4} extra deaths are caused by each person-rem. (This estimate involves a somewhat controversial extrapolation.) For the 250 million people in the United States, what is the average number of extra deaths per year caused by a) medical x-rays, b) fallout, and c) nuclear power plants (at present level)?

30.30 In words, compare the cause–effect relationship between radioactive exposure and illness to that for smoking and illness. (By the way, researchers have suggested that natural radioactivity in tobacco-smoke particles is a major cause of lung cancer in cigarette smokers.)

CHAPTER 31

Radioactivity II

Why are stable isotopes stable? Why don't they undergo alpha, beta, or gamma decay? What determines the energies of the radiation emitted from decaying unstable isotopes? These are three of the questions we'll answer in this chapter by applying conservation laws to radioactive decay. In the process, we'll meet a new particle—the electron's neutrino—along with its antineutrino. We'll discuss a technique that can measure changes of less than one part in ten million million in nuclear energy levels. Finally, we'll take a look at two nuclear processes that cause atoms to emit, randomly and by themselves, characteristic x-rays.

31.1
ALPHA DECAY

In **alpha decay** (α decay), the parent nuclide decays to the daughter nuclide by emitting an alpha particle. If you recall that an alpha particle is a $^{4}_{2}\text{He}$ nucleus, you can see that two protons and two neutrons (for a total of four nucleons) leave the parent nucleus. Experimental results show that

417

the equation

$$\ce{_Z^A P} \longrightarrow \ce{_{Z-2}^{A-4} D} + \ce{_2^4 He} \qquad (\alpha \text{ decay}) \qquad (31.1)$$

describes α decay.

Both conservation of charge and conservation of nucleons occur in α decay. **Conservation of charge** follows from the experimental finding that the total nuclear charge is $+Ze$ before the decay and $+(Z-2)e + 2e = +Ze$ after the decay. **Conservation of nucleons** follows from the experimental finding that the total number of nucleons is A before the decay and $(A-4) + 4 = A$ nucleons after the decay. For example, $\ce{_{94}^{239}Pu}$ undergoes an α decay described by $\ce{_{94}^{239}Pu} \rightarrow \ce{_{92}^{235}U} + \ce{_2^4He}$. The total nuclear charge is conserved at $94e = 92e + 2e$ and the total number of nucleons at $239 = 235 + 4$.

EXAMPLE 31.1   All naturally occurring bismuth is $\ce{_{83}^{209}Bi}$, which undergoes α decay with a half-life $> 10^{19}$ years. Apply conservation of charge and nucleons to find the daughter nuclide.

Solution We begin with $\ce{_{83}^{209}Bi}$. The parent nucleus has a charge of $+83e$. The alpha particle takes a charge $+2e$ away during the decay. To conserve charge, the daughter nucleus must have a charge of $83e - 2e = 81e$. Then the subscripts in the equation $\ce{_Z^A P} \rightarrow \ce{_{Z-2}^{A-4} D} + \ce{_2^4 He}$ will balance ($83 = 81 + 2$).

The parent nucleus contains 209 nucleons. The alpha particle takes 4 nucleons away during the decay. To conserve nucleons, the daughter nucleus must contain $209 - 4 = 205$ nucleons. Then the superscripts in the equation will balance ($209 = 4 + 205$). A Z of 81 gives the element Tl: The reaction is $\ce{_{83}^{209}Bi} \rightarrow \ce{_{81}^{205}Tl} + \ce{_2^4He}$. The daughter is the stable nuclide $\ce{_{81}^{205}Tl}$.   ●

The kinetic energies of atoms at or below normal temperatures are much less than the energies of typical emitted alpha particles and recoiling daughter nuclides. So for simplicity, let's consider a reference frame in which the parent nucleus is at rest, as in Fig. 31.1(a). This condition gives us zero initial kinetic energy and zero initial linear momentum. When an α particle is emitted in one direction with a kinetic energy KE_α, the daughter nucleus must recoil with an equal and opposite momentum and a recoil kinetic energy KE_D, as in Fig. 31.1(b). We'll assume that linear momentum and mass–energy are conserved in the process.

Experiments have shown that the kinetic energies in α decay are much less than the rest energies. Therefore the classical relation $KE = p^2/2M$, or $p^2 = 2M\,KE$, holds. Conservation of mass–energy gives

$$m_P c^2 = (m_D + m_\alpha)c^2 + KE_D + KE_\alpha. \qquad (A)$$

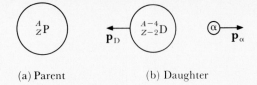

(a) Parent   (b) Daughter

FIGURE 31.1
In the reference frame of the parent nucleus, the alpha particle and the daughter nucleus have equal magnitude, oppositely directed linear momentums after alpha decay.

The problem with Eq. (A) is that m_P and m_D are the masses of parent and daughter nuclei only. The Appendix gives the masses of neutral *atoms*.

Rearranging Eq. (A) and adding electrons of mass m gives

$$KE_D + KE_\alpha = [(m_P + Zm) - (m_D + (Z-2)m) - (m_\alpha + 2m)]c^2, \quad \textbf{(B)}$$

or

$$Q = KE_D + KE_\alpha = (M_P - M_D - M_{He\text{-}4})c^2, \quad (31.2)$$

where Q is called the **disintegration energy,** or the energy released in a reaction. In α decay Q is the sum of the kinetic energies of the daughter nucleus and the alpha particle. The terms M_P, M_D, and $M_{He\text{-}4}$ are the masses of the neutral parent, daughter, and He-4 atoms respectively. Because of the subtraction, the error introduced by ignoring the atomic binding energies of the Z, $Z-2$, and 2 electrons is negligible. Since KE must be positive, you can see from Eq. (31.2) that spontaneous α decay is possible only if $Q > 0$; that is, if $M_P > (M_D + M_{He\text{-}4})$. If $Q < 0$ for any radioactive decay process, the nuclide is stable against that process.

EXAMPLE 31.2   Of all naturally occurring platinum, 0.01 percent is $^{190}_{78}\text{Pt}$. Can it spontaneously decay by α emission? If so, what is the disintegration energy?

Solution The reaction is $^{190}_{78}\text{Pt} \rightarrow {}^{186}_{76}\text{Os} + {}^{4}_{2}\text{He}$. Appendix A gives

$$Q = (M_P - M_D - M_{He\text{-}4})c^2$$

$$= (189.959917 \text{ u} - 185.953830 \text{ u} - 4.002603 \text{ u})931.5 \text{ MeV/u}$$

$$= 3.25 \text{ MeV}.$$

Since $Q > 0$, or $M_P > (M_D + M_{He\text{-}4})$, $^{190}_{78}\text{Pt}$ can spontaneously decay, with a disintegration energy of 3.25 MeV.   ●

We can use conservation laws to find how the disintegration energy is shared by the α particle and the recoiling daughter nucleus. Conservation of linear momentum gives $0 = \mathbf{p}_\alpha + \mathbf{p}_D$ or $p_\alpha^2 = p_D^2$. Using $p^2 = 2M\,KE$, we

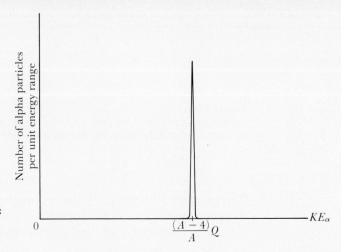

FIGURE 31.2
The alpha particle kinetic energy spectrum for a single Q.

have $2M_\alpha KE_\alpha = 2M_D KE_D$. To an accurate enough approximation for $A \gg 4$ nuclides (which comprise the bulk of the α emitters), $M_\alpha = 4$ u and $M_D = (A - 4)$ u. If we substitute these masses into $M_\alpha KE_\alpha = M_D KE_D$ and solve for KE_D, we find that $KE_D = 4KE_\alpha/(A - 4)$. Then, that KE_D in $Q = KE_D + KE_\alpha$ gives

$$Q = KE_\alpha[4/(A - 4) + 1] = KE_\alpha[4 + (A - 4)]/(A - 4) = KE_\alpha A/(A - 4).$$

Therefore, in α decay, the kinetic energy of the alpha particle is

$$KE_\alpha = Q \frac{A - 4}{A}, \tag{31.3}$$

where A is the mass number of the parent nuclide.

   Note that in this two-particle decay only one energy is allowed for a given Q, as shown in Fig. 31.2. Thus we say that the alpha particles are **monoenergetic.**

EXAMPLE 31.3   Calculate the kinetic energies of

   a) the α particle and

   b) the recoil nucleus in the alpha decay of Pt-190.

Solution

   a) In Example 31.2, we calculated $Q = 3.25$ MeV for the α decay of Pt-190. Also, $A = 190$ since the parent nuclide is Pt-190. Therefore

$$KE_\alpha = Q \frac{A - 4}{A} = 3.25 \text{ MeV} \frac{190 - 4}{190} = 3.18 \text{ MeV}.$$

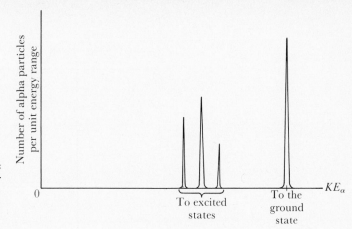

FIGURE 31.3
The alpha particle kinetic energy spectrum for four different Q's. Decays to excited states make this spectrum possible.

b) Using $Q = KE_D + KE_\alpha$, we get $KE_D = Q - KE_\alpha = 3.25$ MeV $-$ 3.18 MeV $= 0.07$ MeV. Or a bit more accurately,

$$KE_D = Q - Q\frac{A-4}{A} = Q\frac{4}{A} = 3.25 \text{ MeV} \frac{4}{190} = 0.068 \text{ MeV.} \quad \bullet$$

To find the recoil speed of the daughter nucleus, we use $KE_D = \frac{1}{2}M_D v^2$ (converting MeV to J and u to kg to obtain v in m/s).

The Pt-190 nuclei decay with just one group of monoenergetic α's at 3.18 MeV. Many other alpha emitters emit groups of α's at several different well-defined energies, as shown in Fig. 31.3. Different energies are possible because radioactive decay doesn't always go to the ground state of the daughter. Some decay goes to excited states instead. The daughter nucleus in an excited state has more energy and therefore more mass than in the ground state. Increasing M_D decreases both Q and KE_α, as you can see from Eqs. (31.2) and (31.3).

The fact that a nuclide *can* decay doesn't mean that it *will* do so immediately. For example, Pt-190 has the extremely long half-life of 7×10^{11} years. Experimental results show a trend: Nuclides that emit higher energy α's tend to have shorter half-lives. Gamow, Gurney, and Condon used tunneling to explain this trend in 1928.

The potential energy for an alpha particle in a nucleus looks something like Fig. 31.4. At $r = R$ a potential energy barrier keeps the α particle inside the nucleus. Recall that quantum mechanics tells us that there is some probability that an α particle can tunnel through the barrier with no energy loss. An α particle with the higher kinetic energy (corresponding to E_2) encounters a narrower and relatively shorter barrier. Both of those properties give the higher energy α particle a much greater probability of

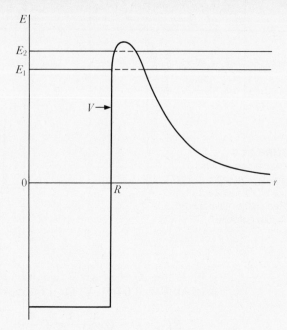

FIGURE 31.4
Tunneling through the
alpha particle potential energy barrier for two
different kinetic energies.
Remember: $KE = E - V$.

tunneling out of the nucleus each time it hits the barrier than an α particle
has with a lower kinetic energy corresponding to E_1. Also the α particle
having the higher kinetic energy will hit the barrier more often during a
given time. Both the higher probability per collision and more collisions
per unit of time lead to a shorter $T_{\frac{1}{2}}$ for a higher KE_α.

31.2
BETA DECAY

There are three different types of **beta decay** (β decay): β^- decay, β^+
decay, and electron capture. In β^- **decay,** a negative electron is emitted.
Since the electron has a charge of $-1e$ and is not a nucleon, the reaction
$_Z^AP \rightarrow {}_{Z+1}^AD + _{-1}^0e$ conserves charge and the number of nucleons.

Beta-minus decay seems to be another two-particle decay like α
decay. However, all β^- decay experiments show that the electrons aren't
monoenergetic. You can see this result by comparing Fig. 31.5 with Fig.
31.2. A possible explanation of the data is that mass–energy and linear
momentum aren't conserved in β^- decay, violating two of the most fundamental conservation laws of physics.

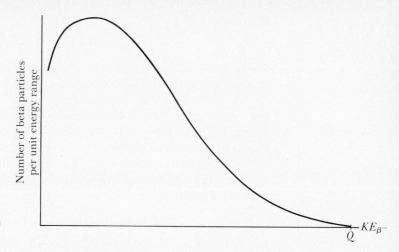

FIGURE 31.5
A typical beta-minus kinetic energy spectrum for a single Q. Compare this curve to the curve in Fig. 31.2.

If we examine what occurs within the parent nucleus in β^- decay, we see that a neutron appears to decay to a proton plus an electron. However, the neutron, proton, and electron all have spin angular momentum quantum numbers of $\frac{1}{2}$. This gives z components of their spin angular momentum of $\pm\frac{1}{2}\hbar$. Since $\pm\frac{1}{2}\hbar \neq \pm\frac{1}{2}\hbar + \pm\frac{1}{2}\hbar$ in any combination, angular momentum apparently isn't conserved either.

To preserve these well-established conservations laws, W. Pauli suggested in 1931 that a third particle was produced in the decay. This particle would have a rest mass of zero (or close to zero), no charge, and a spin angular momentum quantum number of $\frac{1}{2}$. Enrico Fermi named the particle **neutrino** (little neutral one), its antiparticle is an **antineutrino.** The symbols for these particles are ν and $\bar{\nu}$, respectively. For a neutrino, **S** and **p** (the spin angular momentum and linear momentum) have opposite components. For an antineutrino, **S** and **p** have parallel components. Fermi's 1934 paper on the theory of beta decay is considered to be one of the most important in the history of nuclear physics, but was published only after it had first been rejected by the editor of *Nature*.

Including the antineutrino, the β^- decay reaction then is

$$\begin{matrix}A\\Z\end{matrix}P \longrightarrow \begin{matrix}A\\Z+1\end{matrix}D + \begin{matrix}0\\-1\end{matrix}e + \bar{\nu} \qquad (\beta^- \text{ decay}), \qquad (31.4)$$

with $\begin{smallmatrix}0\\0\end{smallmatrix}\bar{\nu}$ shortened to $\bar{\nu}$. Figure 31.6 illustrates beta decay. In the nucleus,

FIGURE 31.6
The parent decays to three particles in beta emission.

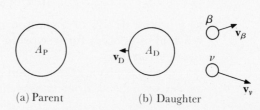

(a) Parent   (b) Daughter

the reaction is

$$\ _0^1n \longrightarrow \ _1^1p + \ _{-1}^0e + \bar{\nu},$$

as a neutron decays to a proton plus a negative electron plus an antineutrino. A free neutron also decays this way, with a half-life of 10.5 minutes.

Conservation of mass–energy gives $Q = KE_D + KE_{\beta^-} + KE_{\bar{\nu}} = (m_P - m_D - m)c^2$. Any rest mass–energy the antineutrino may have is negligible. We can replace m_P by $M_P - Zm$ and m_D by $M_D - (Z+1)m$ to use the masses of neutral atoms shown in the Appendix. Also, $Q = (KE_{\beta^-})_{max}$ with sufficient accuracy, because $KE_{\bar{\nu}}$ can approach zero and KE_D is a small recoil energy. (The recoil energy is small because m_D is so much greater than m.) Therefore

$$Q = (KE_{\beta^-})_{max} = (M_P - M_D)c^2 \qquad (\beta^- \text{ decay}). \qquad (31.5)$$

Equation (31.5) tells us that spontaneous β^- decay is possible if $Q > 0$ or if the mass of the neutral parent atom exceeds that of the neutral daughter atom.

EXAMPLE 31.4   Can $_6^{14}C$ decay by β^- emission? If so, what kinetic energies can the electrons have?

Solution   The reaction would be $_6^{14}C \rightarrow \ _7^{14}N + \ _{-1}^0e + \bar{\nu}$. From the Appendix,

$$Q = (M_P - M_D)c^2 = (14.003242 \text{ u} - 14.003074 \text{ u})(931.5 \text{ MeV/u})$$
$$= 0.156 \text{ MeV}.$$

Since $Q > 0$, the reaction can occur. And $Q = (KE_{\beta^-})_{max}$ tells us that the electrons can have any kinetic energy up to 0.156 MeV.   ●

Example 31.4 describes how C-14, used in radioactive dating, decays with $T_{\frac{1}{2}} = 5730$ years.

In the second type of beta decay, β^+ **decay,** a positive electron (positron) and a neutrino are emitted when the parent nucleus decays to the daughter. The β^+ decay reaction then is

$$_Z^AP \longrightarrow \ _{Z-1}^AD + \ _1^0e + \nu \qquad (\beta^+ \text{ decay}). \qquad (31.6)$$

In the nucleus, the reaction is

$$_1^1p \longrightarrow \ _0^1n + \ _1^0e + \nu,$$

as a proton decays to a neutron plus a positron plus a neutrino. This reaction can't occur for free protons because the proton mass is less than the neutron mass.

Again applying conservation of mass–energy and replacing nuclear masses by neutral atom masses minus the correct number of electron masses, we obtain

$$Q = (KE_{\beta^+})_{max} = (M_P - M_D - 2m)c^2 \qquad (\beta^+ \text{ decay}). \qquad (31.7)$$

You can see that the $Q > 0$ mass requirement for β^+ decay is stronger than for β^- decay. It isn't enough that $M_P > M_D$. The parent neutral atom mass must exceed the daughter neutral atom mass by two electron masses. For calculations with Eq. (31.7), you should use $2mc^2 = 1.022$ MeV, or more conveniently, $2m = 0.001097$ u.

EXAMPLE 31.5 Can ^{57}Co decay by β^+ emission?

Solution The reaction would be $^{57}_{27}\text{Co} \rightarrow {}^{57}_{26}\text{Fe} + {}^{0}_{1}\text{e} + \nu$. Then

$$Q = (56.936294 \text{ u} - 56.935396 \text{ u} - 0.001097 \text{ u})(931.5 \text{ MeV/u}) < 0.$$

The answer is no. Since $Q < 0$, the reaction can't occur spontaneously, showing that Co-57 is stable against β^+ decay.   ●

EXAMPLE 31.6 The nuclide $^{12}_{7}$N decays by β^+ emission with a maximum positron energy of 16.316 MeV. Calculate the neutral atomic mass of the parent atom.

Solution The reaction is $^{12}_{7}\text{N} \rightarrow {}^{12}_{6}\text{C} + {}^{0}_{1}\text{e} + \nu$. Since $Q = (KE_{\beta^+})_{max}$, $Q = 16.316$ MeV $= (M_P - M_D - 2m)c^2$. The quantity M_D, the mass of C-12, is exactly 12 u by definition of the unified atomic mass unit. We want M_P, the mass of N-12, so

$$\frac{16.316 \text{ MeV}}{c^2} + M_D + 2m = M_P$$

or

$$\frac{16.316 \text{ MeV}}{931.5 \text{ MeV/u}} + 12 \text{ u} + 0.001097 \text{ u} = M_P = 12.018613 \text{ u}. \qquad ●$$

The probability density of the atomic electrons does not equal zero when r approaches nuclear dimensions. Therefore some probability of the inner electrons interacting strongly with the nucleus exists. In **electron capture (EC)**, a K-shell electron or even an L-shell electron is captured by a proton in the nucleus in the reaction

$$_{-1}^{0}\text{e} + {}^{1}_{1}\text{p} \longrightarrow {}^{1}_{0}\text{n} + \nu.$$

You may think of this reaction as a result of adding $_{-1}^{0}\text{e}$ to both sides of $^{1}_{1}\text{p} \rightarrow {}^{1}_{0}\text{n} + {}^{0}_{1}\text{e} + \nu$, with the electron and positron canceling each other. Electron capture is the third type of beta decay, and thus the EC decay reaction is

$$-_1^0e + {}_Z^A P \longrightarrow {}_{Z-1}^A D + \nu \qquad \text{(EC decay)}. \qquad (31.8)$$

Note that the mass number, A, remains constant in all three types of beta decay. The atomic number, Z, *decreases* by 1 in both EC and β^+ decay. The atomic number *increases* by 1 in β^- decay. The number of nucleons and the charge are thereby conserved.

Applying conservation of mass–energy and replacing the nuclear masses by the neutral atom masses minus the correct number of electron masses, we obtain

$$Q = (M_P - M_D)c^2 \qquad \text{(EC)}. \qquad (31.9)$$

Immediately after EC, the daughter atom is in an excited atomic state. We are ignoring the small increase this excitation gives to M_D.

In electron capture, almost all the Q goes to the energy of the low-mass neutrino. The high-mass daughter nucleus recoils with a kinetic energy only in the eV range. The kinetic energies of the neutrino and the daughter are both monoenergetic because only two particles remain after the decay (as in alpha decay). The $Q > 0$ requirement for EC is the same as for β^- decay: The neutral parent atom must have a greater mass than the neutral daughter atom.

EXAMPLE 31.7   Can ${}_{27}^{57}$Co decay by EC? If so, calculate the kinetic energies of the neutrino and daughter nucleus.

Solution The reaction is $-_1^0e + {}_{27}^{57}$Co $\rightarrow {}_{26}^{57}$Fe $+ \nu$. Then

$$Q = (56.936294 \text{ u} - 56.935396 \text{ u})(931.5 \text{ MeV/u}) = 0.836 \text{ MeV}.$$

Since $Q > 0$, yes, EC decay can occur. Virtually all the disintegration energy goes to the neutrino, so $KE_\nu = 0.836$ MeV. The neutrino has zero, or close to zero, rest mass. For zero rest mass, $E = KE$ and $p = E/c = 0.836$ MeV/c. The momentum of the recoiling Fe-57 daughter will be equal and opposite to the neutrino's. Therefore, using $c^2 = 931.5$ MeV/u, the daughter's kinetic energy will be

$$\frac{p^2}{2M_D} = \frac{(0.836 \text{ MeV}/c)^2}{2(56.935 \ldots \text{ u})} = \frac{(0.836 \text{ MeV})^2}{2(56.935 \ldots \text{ u} \times c^2)}$$

$$= \frac{(0.836 \text{ MeV})^2}{2(56.935 \ldots \text{ u} \times 931.5 \text{ MeV/u})} = 6.60 \text{ eV}. \qquad \bullet$$

Examples 31.5 and 31.7 show that Co-57 can't decay by β^+ emission but can decay by electron capture. Example 31.7 also demonstrates numerically that the few eV of daughter recoil energy is negligible compared to the neutrino's energy.

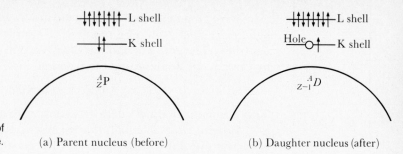

FIGURE 31.7
A representation of
K-electron capture.

(a) Parent nucleus (before)   (b) Daughter nucleus (after)

Neutrinos aren't the only radiation emitted in electron capture. As diagramed in Fig. 31.7, capturing inner electrons leaves holes in the inner electron shells of the daughter atoms. As other electrons drop down to fill the holes, characteristic x-rays can be emitted. It is these x-rays that show immediately that EC has occurred. Electron capture is a random process, like other radioactive decay processes. As Fig. 31.8 shows, EC, beta, and alpha radioactive decay processes all tend to move nuclides toward the stability curve.

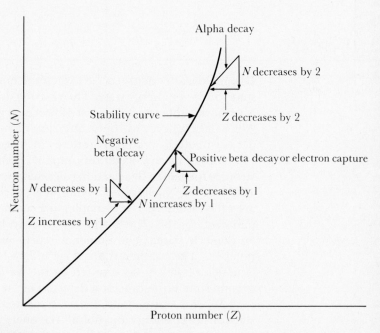

FIGURE 31.8
Alpha and beta decays tend to proceed toward the nuclear stability curve as drawn approximately through the stable nuclides (see Fig. 29.9).

Source: A. Beiser, *Modern Technical Physics,* 4th ed. Benjamin/Cummings, Menlo Park, Calif. © 1983, p. 764.

Detecting neutrinos or antineutrinos is very difficult. A slab of lead light years thick would be required to appreciably attenuate a beam of neutrinos! The antineutrino was discovered first — but not until 1956 — by Cowan and Reines using the reaction

$$\bar{\nu} + {}_1^1\text{p} \longrightarrow {}_0^1\text{n} + {}_1^0\text{e}.$$

This is essentially the β^+ decay reaction, with a $\bar{\nu}$ added to both sides.

Proving experimentally that the rest mass of a neutrino is or isn't zero is even more difficult. If neutrinos have a rest mass, they can't move exactly at c. An interpretation of the data from Supernova 1987A indicates that the light and the neutrinos from the supernova took the same time (to within an accuracy of a few hours) to travel 163,000 light years. This interpretation places an upper limit to the neutrino mass of about $20 \text{ eV}/c^2$. Since neutrinos are produced so often in nature and are so difficult to absorb, huge numbers of them are present in the universe. Any extra mass they might possess because of a possible rest mass could be important in models of the expanding universe, as well as in theories that attempt to link all or most of the forces in nature.

31.3
GAMMA DECAY

Gamma decay (γ decay) occurs when the nucleus decays from an excited state to a lower energy state. The reaction does *not* change Z or A. Therefore the daughter nuclide is just a lower energy state of the parent nuclide. Alpha or beta decay will quite often leave a nucleus in an excited state, so γ decay will often accompany α decay or β decay.

Most γ decays occur very rapidly. Some γ decays, however, are so strongly forbidden by selection rules that they have half-lives greater than 10^{-6} second. Nuclides remaining in an excited state longer than 10^{-6} s are called **isomers.** For example, in a common lab or demonstration experiment, Cs-137 ($T_{\frac{1}{2}} = 30.17$ years) decays by β^- emission to Ba-137m in an isomeric state. (The isomer is represented by the m after the mass number.) The half-life of Ba-137m is 2.552 min, far longer than 10^{-6} s. The 661.65-keV γ ray emitted by the Ba-137m in its transition to the ground state is the radiation detected in the experiment.

When a γ ray is emitted in the transition from a state of energy E_2 to a state of lower energy E_1, not all the energy ($E_2 - E_1$) goes to the γ-ray photon. As shown in Fig. 31.9, a small amount of recoil energy KE_r goes to the nucleus. As small as KE_r is, however, it's enough to ordinarily keep the emitted γ ray from being absorbed by a similar nucleus. If absorption were to occur, the absorbing nucleus would also have to recoil to conserve

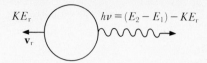

FIGURE 31.9
The γ-ray energy $h\nu$ is less than $(E_2 - E_1)$ because of the recoil kinetic energy given to the nucleus.

momentum. Therefore the absorbing nucleus requires $h\nu' = (E_2 - E_1) + KE_r$. To summarize, the emitted gamma has $h\nu = (E_2 - E_1) - KE_r$, and an absorbed gamma must have $h\nu' = (E_2 - E_1) + KE_r$. And $h\nu$ and $h\nu'$ differ by $2KE_r$, as you can see in Fig. 31.10.

In the **Mössbauer effect,** the recoil energy, $KE_r = p^2/2M_D$ is taken up by all the atoms in a solid instead of by a single nucleus. Thus M_D effectively increases from the mass of a single atom to the mass of the whole solid. Dividing by this hugely increased mass leaves KE_r essentially zero. Therefore the γ ray emitted by one nucleus can be absorbed by another, similar nucleus.

Using the Mössbauer effect, we can detect changes in nuclear energy levels of one part in 10^{13}, or better, when these changes make $(E_2 - E_1)$ for the source different from $(E_2 - E_1)$ for the absorber. We can eliminate the resulting small differences in the frequency of the emitted γ ray and the resonant frequency for absorption, making use of the Doppler effect by giving the source and absorber some relative motion. A result is shown in the data graphed in Fig. 31.11. The only ultimate limit on the sensitivity is the uncertainty principle. These small changes in nuclear energy levels can occur because of the interaction of the nuclei with different electronic states or because of electric or magnetic field interactions. We can also

FIGURE 31.10
The ΔE is the spread in a peak. Its minimum value is obtained from the Heisenberg uncertainty principle.

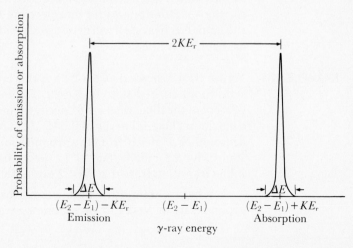

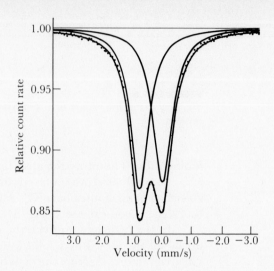

FIGURE 31.11

The dots are the Mössbauer-effect data. In this experiment, we used a 1 percent Fe in rutile absorber and measured nuclear energy level changes as small as 10^{-9} eV. There are two resonances because of a deviation from sphericity of the nuclear charge distribution, which interacted with the crystal's nonuniform electric field.

Source: T. R. Sandin, D. Schroeer, and C. D. Spencer, *Phys. Rev. B* **13**, 4784 (1976).

detect γ-ray photon frequency changes, which are caused by relativistic time dilation or gravitational shifts.

A γ ray is not always emitted as a result of the transition to a less-excited nuclear state. In the **internal conversion** process, the nucleus gives the energy $(E_2 - E_1)$ directly to one of the electrons in one of the atom's inner shells. The energy is great enough to kick that electron out of the atom. The electron will leave with a kinetic energy of $(E_2 - E_1)$ minus the ionization energy for that electron. As in electron capture, a hole in the inner electron shell will result, and characteristic x-rays can be emitted as that hole is filled.

EXAMPLE 31.8   Figure 31.12 shows that Na-22 has two types of decays from its ground state to an excited state of Ne-22. Electron capture decay occurs 10.2 percent of the time, whereas β^+ decay occurs 89.8 percent of the time, with a maximum kinetic energy of 0.545 MeV. Also, 0.06 percent of the decays are another β^+ decay from the ground state of Na-22 directly to a ground state of Ne-22. Finally, Fig. 31.12 shows that a 1.275-MeV γ ray is emitted in the transition from that Ne-22 excited state to the Ne-22 ground state.

a) What percent of the decays from the ground state of Na-22 to the ground state of Ne-22 involve gamma rays?

b) What is the maximum kinetic energy of the positrons emitted in any decay shown?

c) What is the ground state neutral atomic mass of Na-22?

d) What is the neutral atomic mass of Ne-22 in that excited state?

Solution

a) Both the EC and β_1^+ decays first go to the excited state shown. Then the transition from that excited state to the ground state yields a 1.275-MeV γ ray. Adding 10.2% and 89.8% appears to give 100%, but only because of round-off error. Actually, since 0.06% of the decays are directly to the ground state, the remaining $100.00\% - 0.06\% = 99.94\%$ of the decays involve a 1.275-MeV γ ray.

b) The energy released in the β_1^+ transition plus the γ-ray energy is 0.545 MeV $+ 1.275$ MeV $= 1.820$ MeV. However, the β_2^+ transition is directly to the ground state, so 1.820 MeV is the maximum kinetic energy of its emitted positron (β^+ particle).

c) We use $Q = (KE_{\beta+})_{max} = (M_P - M_D - 2m)c^2$, with $(KE_{\beta+})_{max} = 1.82$ MeV, $M_D = 21.991383$ u (the ground state neutral atomic mass of Ne-22 from the Appendix), and $2m = 0.001097$ u to obtain

$$M_P = 1.82 \text{ MeV}/(931.5 \text{ MeV/u}) + 21.991383 \text{ u} + 0.001097 \text{ u}$$
$$= 21.994434 \text{ u}.$$

d) In the excited state, Ne-22's energy is 1.275 MeV greater, so by $E = mc^2$, its mass is 1.275 MeV/c^2 greater. That is, its

$$\text{mass} = 21.991383 \text{ u} + 1.275 \text{ MeV}/(931.5 \text{ MeV/u}) = 21.992752 \text{ u}.$$

●

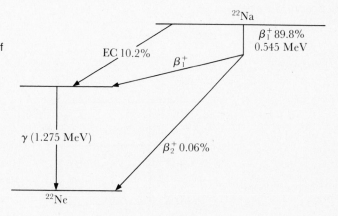

FIGURE 31.12

Nuclear energy level diagram for the decay of ^{22}Na.

^{22}Na

EC 10.2%

β_1^+

β_1^+ 89.8%
0.545 MeV

γ (1.275 MeV)

β_2^+ 0.06%

^{22}Ne

The 21.992752 u in Example 31.8 is the M_D that you would have to use to find Q for the β_1^+ process or the EC process, if Q weren't given. You would have to use it, rather than the value from the Appendix, because both processes go to the excited state, not the ground state, of Ne-22. The Appendix contains only ground state neutral atomic masses.

EXAMPLE 31.9   Three of the α's and three of the γ rays emitted when Am-241 decays to Np-237 are shown in Fig. 31.13. The particle α_6 has a *KE* of 5.443 MeV and α_7 has a *KE* of 5.486 MeV. Calculate the *KE* of α_9 and the energies of γ_1, γ_2, and γ_3.

Solution Because Q_9 is the parent ground state to daughter ground state Q, Eq. (31.2) gives

$$Q_9 = (M_P - M_D - M_{He-4})c^2$$
$$= (241.056824 \text{ u} - 237.048168 \text{ u} - 4.002603 \text{ u})(931.5 \text{ MeV/u})$$
$$= 5.638 \text{ MeV},$$

using the masses from the Appendix. Therefore Eq. (31.3) yields

$$KE_{\alpha_9} = Q_9 \frac{A-4}{A} = 5.638 \text{ MeV} \frac{237}{241} = 5.545 \text{ MeV}.$$

We now need to find the other Q's. We know that $KE_{\alpha_7} = 5.486$ MeV, so

$$Q_7 = KE_{\alpha_7} \frac{A}{A-4} = 5.486 \text{ MeV} \frac{241}{237} = 5.579 \text{ MeV}.$$

FIGURE 31.13
Nuclear energy level diagram for the decay of ^{241}Am.

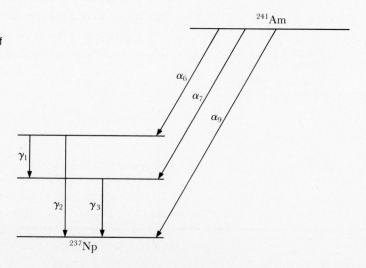

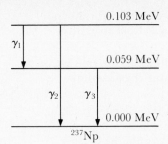

FIGURE 31.14
Nuclear energy level diagram of the daughter nucleus.

Similarly, we know that $KE_{\alpha_6} = 5.443$ MeV, so

$$Q_6 = 5.443 \text{ MeV } \frac{241}{237} = 5.535 \text{ MeV}.$$

The excited state for α_7 has $(5.638 - 5.579)\text{MeV} = 0.059$ MeV less disintegration energy than that for α_9 and therefore must be 0.059 MeV above the ground state. Similarly, the excited state for α_6 is $(5.638 - 5.535)$ MeV $= 0.103$ MeV above the ground state. Therefore Fig. 31.14 shows γ_3 has an energy of $(0.059 - 0.000)\text{MeV} = 0.059$ MeV, γ_2 has an energy of $(0.103 - 0.000)\text{MeV} = 0.103$ MeV, and γ_1 has an energy of $(0.103 - 0.059)\text{MeV} = 0.044$ MeV.   ●

We began this chapter discussing α decay, describing how the parent and daughter atomic numbers and mass numbers are related by conservation of charge and nucleons. We showed how to determine the disintegration energy and the single alpha particle kinetic energy for each Q. You saw that decays to excited states would give different Q's and KE_α's. Then we explained a half-life trend in terms of tunneling. We moved to beta decay and neutrinos, showing how to describe the reactions and calculate the Q values for β^-, β^+, and EC decay. We explained the characteristic x-rays emitted along with electron capture and internal conversion. Finally, we concluded the chapter by discussing gamma decay and the Mössbauer effect.

QUESTIONS AND PROBLEMS

31.1 Find the missing nuclide in these α decay reactions:

a) $^{251}_{98}\text{Cf} \rightarrow \text{D} + \alpha$ c) $\text{P} \rightarrow \alpha + \alpha$

b) $\text{P} \rightarrow ^{234}_{90}\text{Th} + \alpha$ d) $^5_3\text{Li} \rightarrow \text{D} + \alpha$

31.2 Do conservation of charge and number of nucleons hold for α, β, and γ decay?

31.3 If the α's are monoenergetic, why are α energies listed for U-238 as 4.20, 4.15, . . . MeV?

31.4 Calculate the disintegration energy for the 4.15-MeV alpha particle from U-238.

31.5 Why isn't the spontaneous reaction $^{16}_8\text{O} \rightarrow ^{12}_6\text{C} + ^4_2\text{He}$ possible?

31.6 Some decays not possible from the ground state of a parent are possible from an excited state. Explain.

31.7 The nuclide $^{186}_{76}$Os decays by the emission of a 2.76-MeV alpha. Calculate the mass of the daughter nuclide.

31.8 The nuclide $^{257}_{103}$Lr is the daughter resulting from an α decay with an 8.93-MeV α particle. a) Calculate the recoil energy of the Lr-257 nucleus. b) Calculate the neutral atomic mass of Lr-257. c) What assumptions about ground states did you make?

31.9 a) Prove that the recoil speed of the daughter nucleus equals $(3.0 \times 10^8 \text{ m/s})\sqrt{8Q/(M_D c^2 A)}$ in α decay. b) Calculate the recoil speed for the Pt-190 decay of Example 31.2. Check your answer against the results of Example 31.3.

31.10 Isotopes from Fr-218 to Fr-221 are α emitters with energies of 7.87 MeV, 7.31 MeV, 6.68 MeV, and 6.34 MeV, respectively, and half-lives of 0.7 ms, 20 ms, 27.4 s, and 4.8 min. Can you explain this trend?

31.11 Does the disintegration constant increase or decrease with the α-particle energy for α decay? Why (from basic concepts)?

31.12 For every 10^9 decays of Ra-223, 1 decay emits a C-14 nucleus. Show that Q for this process is 31.85 MeV.

31.13 Show that the kinetic energy of the C-14 nucleus of Problem 31.12 is 29.85 MeV.

31.14 Of the three kinds of decay, β^-, β^+, and EC, determine which is possible for
   a) $^{115}_{49}$In   c) $^{97}_{43}$Tc
   b) $^{64}_{29}$Cu   d) $^{236}_{93}$Np

31.15 The nuclide In-112 has 2.59 MeV/c^2 more mass than Cd-112 and 0.66 MeV/c^2 more than Sn-112. Is EC possible? If β^- and/or β^+ decay are possible, what are their values of KE_{max}?

31.16 A parent nucleus decays to Mg-25 by β^- emission, with a maximum kinetic energy of 3.83 MeV. What is the parent nuclide and its mass? Only ground states are involved.

31.17 An Na-24 atom in its ground state has 5.514 MeV/c^2 more mass than an Mg-24 atom in its ground state. In each β^- decay the β^- is followed by two γ rays of energy 2.754 MeV and

1.369 MeV. What is the maximum kinetic energy of the electron emitted?

31.18 The nuclide Ga-66 decays by β^+ emission to a nuclide 0.005556 u smaller. What is the maximum kinetic energy of the positron emitted? What are the kinetic energies of the neutrino and daughter for this positron kinetic energy?

31.19 The nuclide Tc-94 decays by β^+ emission with a maximum positron kinetic energy of 0.82 MeV. Each positron is followed by three γ rays of energies 0.8710 MeV, 0.7026 MeV, and 0.8497 MeV. What is the mass of a Tc-94 atom?

31.20 Derive Eq. (31.7).

31.21 Derive Eq. (31.9).

31.22 Why is it more probable that a 2s electron will be involved in EC than a 2p electron?

31.23 What reaction with a nucleon could be used to detect a neutrino?

31.24 When Cu-64 decays by β^+ emission to the ground state of Ni-64, the positron has a maximum kinetic energy of 0.653 MeV. But Cu-64 also has two EC decay modes: one to the ground state, the other to an excited state followed by a 1.346-MeV gamma transition to the ground state.

   a) Calculate the energies of the neutrinos emitted in the two EC decay modes.

   b) The nuclide Cu-64 also decays to the ground state of Zn-64 by β^- emission with a maximum KE of 0.578 MeV. Calculate the neutral atomic mass of Ni-64.

31.25 a) Calculate the recoil energy of an Fe-57 atom when a 14.4-keV γ-ray photon is emitted. b) Repeat that calculation for the recoil energy of a mole of Fe-57 atoms. (This photon is commonly used in Mössbauer experiments.)

31.26 Is internal conversion a kind of photoelectric effect?

31.27 Some 98.6% of the time, Tc-99m ($T_{\frac{1}{2}} = 6.01$ hr) decays to its ground state with the emission of a 2.17-keV gamma followed by a 0.1405-MeV gamma. The other 1.4% of the time, Tc-99m decays to its ground state with the emission of a single gamma. Calculate the energy of that gamma.

31.28 Changes that alter the electron probability density near the nucleus may produce changes in some types of half-lives. The largest currently known change is 3.6% for Nb-90m in the metal compound with the pentafluoride complex. Consider the various ways that decay can occur and state which might have a half-life change.

31.29 Refer to Fig. 31.15. Calculate the energies of the four gammas and the four alphas.

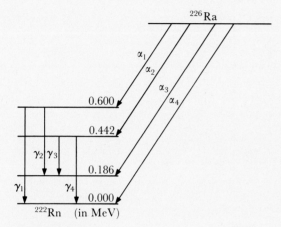

FIGURE 31.15

31.30 Is the following statement true? "The Mössbauer effect occurs when no phonons are created or absorbed in the solids in both the emission and the absorption of the gamma ray." Explain.

31.31 In one form of the very rare double-beta decay, the parent nuclide simultaneously emits two electrons and two antineutrinos.

a) Show that this double-beta decay reaction is $_Z^A\text{P} \rightarrow \,_{Z+2}^A\text{D} + 2_{-1}^0\text{e} + 2\bar{\nu}$.

b) Show that $Q = (M_\text{P} - M_\text{D})c^2$ for this decay.

31.32 The nuclide Se-82 undergoes the double-beta decay discussed in Problem 31.31 with a record-setting half-life of 1.1×10^{20} yr. This phenomenon was discovered by S. Elliot, A. Hahn, and M. Moe in 1987 using a sample containing about 10^{23} Se-82 atoms.

a) How many decays occurred in an average hour?

b) The half-life was once estimated by finding traces of the daughter in ancient selenium ores. What was the daughter nuclide?

c) Why doesn't Se-82 undergo one of the simpler beta decays discussed in this chapter?

d) What was the disintegration energy for the reaction?

CHAPTER **52**

Nuclear Reactions

How is an artificial radioactive isotope, such as a β^+ emitter for medical diagnosis using positron emission tomagraphy, produced? These and similar isotopes are made in a conceptually simple collision procedure that you'll study in this chapter. You'll see how much energy is released or required to convert one isotope to another. We'll then discuss the short-lived intermediate compound nucleus formed in such a conversion. Finally, you'll learn why the probability that the conversion will occur is proportional to an area measured in units of barns.

32.1
LOW-ENERGY NUCLEAR REACTIONS

The nuclear reactions you've studied so far are random processes. They involve the α, β, and/or γ decay of nuclei. They can occur spontaneously only if $Q > 0$. However, other nuclear reactions can be made to occur even though $Q < 0$. These non-random processes can be used to create isotopes, to produce energy by fission and fusion, to do neutron activation analysis, and for other purposes.

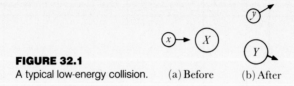

FIGURE 32.1
A typical low-energy collision.   (a) Before   (b) After

To initiate the type of low-energy nuclear reactions that we'll discuss, a low-mass incident particle with a kinetic energy of less than 100 MeV strikes a nuclide at rest. Typical low-mass particles are protons, neutrons, deuterons, and α particles. A γ-ray photon will also behave like a particle in these reactions. The reactions produce another nuclide and another low-mass particle, as illustrated in Fig. 32.1. The incident low-mass particle is represented by x, the target nuclide by X, the resulting nuclide by Y and the other low-mass particle by y. The nuclear reaction then is

$$x + X \rightarrow Y + y. \tag{32.1}$$

A simpler version of this reaction is

$$X(x, y)Y. \tag{32.1a}$$

Rutherford was the first to detect a nuclear reaction of this type in 1919. He bombarded $^{14}_{7}N$ with α particles and obtained $^{17}_{8}O$ and protons. That is, Rutherford *transmuted* N-14 into O-17.

EXAMPLE 32.1   Write Rutherford's reaction.

Solution With x being the α particle, X is $^{14}_{7}N$, Y is $^{17}_{8}O$, and y is the proton. The reaction is

$$^{4}_{2}He + ^{14}_{7}N \rightarrow ^{17}_{8}O + ^{1}_{1}H$$

or

$$^{14}_{7}N(\alpha,p)^{17}_{8}O. \qquad \bullet$$

Note in the first answer in Example 32.1 that the charge and nucleon number are conserved. The subscripts $2 + 7 = 8 + 1$ show charge conservation. The superscripts $4 + 14 = 17 + 1$ show nucleon number conservation.

Neutrons are difficult to detect directly because they have no charge. However, bombarding boron with neutrons gives the reaction $^{10}_{5}B(n,\alpha)^{7}_{3}Li$, with $Q = +2.79$ MeV. The α particles resulting from this reaction can then be easily detected. Chadwick discovered the neutron in

1932 in the reaction $^9_4$Be$(\alpha,n)^{12}_6$C. The neutrons from his reaction knocked protons out of paraffin, and he detected the protons.

32.2
ENERGY IN NUCLEAR REACTIONS

The dream of the ancient alchemist was to turn base (plentiful and cheap) metals into gold. The closest thing to a base metal near the $Z = 79$ of gold is mercury with $Z = 80$. The gold desired would be $^{197}_{79}$Au, because all other gold isotopes are unstable. Two possible reactions using mercury are $^{200}_{80}$Hg$(p,\alpha)^{197}_{79}$Au and $^{197}_{80}$Hg$(n,p)^{197}_{79}$Au. Since Hg-197 is unstable and 23.1 percent of natural Hg is Hg-200, the reaction $^{200}_{80}$Hg$(p,\alpha)^{197}_{79}$Au is the one to use.

The question, then, is how much energy must the proton have to initiate the reaction $^{200}_{80}$Hg$(p,\alpha)^{197}_{79}$Au? Let's calculate this required energy in general for the reaction $X(x, y)Y$. For simplicity, we assume that the target nucleus X is at rest, so $KE_X = 0$. Conservation of mass–energy then gives

$$m_x c^2 + KE_x + m_X c^2 + 0 = m_Y c^2 + KE_Y + m_y c^2 + KE_y. \qquad \text{(A)}$$

The net energy released in the reaction is $Q = \Sigma KE_{\text{after}} - \Sigma KE_{\text{before}}$. From Eq. (A), we have

$$Q = (KE_Y + KE_y - KE_x) = (m_x + m_X - m_Y - m_y)c^2. \qquad \text{(B)}$$

Again, these masses are the nuclear masses. To use the ground state neutral atom masses, M_X and M_Y, we have to replace m_X and m_Y by $(M_X - Z_X m)$ and $(M_Y - Z_Y m)$ in Eq. (B). The electron masses cancel if the x and y particles either are neutral or are made neutral by the addition of electrons. Then the net gain in kinetic energy is

$$Q = (KE_Y + KE_y - KE_x) = [(M_x + M_X) - (M_Y + M_y)]c^2, \qquad \text{(32.2)}$$

where M_x or M_y is the mass of a neutral $^4_2$He atom for α particles, a neutral $^2_1$H atom for deuterons, a neutral $^1_1$H atom for protons, a neutron for neutrons, and is zero for γ-ray photons.

Exoergic reactions have $Q > 0$. **Endoergic reactions** have $Q < 0$. Like exothermic reactions in chemistry, exoergic reactions release energy. Figure 32.2 shows that more KE exists after the reaction than before in exoergic reactions (and correspondingly less rest mass–energy). Endoergic reactions, like endothermic chemical reactions, can occur only if given energy. Figure 32.3 shows that less KE exists after the reaction than before in endoergic reactions (and correspondingly more rest mass–energy).

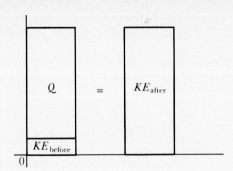

FIGURE 32.2
Exoergic reactions have a positive Q, so KE_{after} is greater than KE_{before}. Exoergic reactions are possible even if KE_{before} approaches zero.

EXAMPLE 32.2   Is the $^{200}_{80}Hg(p,\alpha)^{197}_{79}Au$ reaction endoergic or exoergic?

Solution We need to determine whether $Q < 0$ or $Q > 0$. From the Appendix,

$$Q = [(M_x + M_X) - (M_Y + M_y)]c^2$$
$$= [(1.007825 + 199.968300)\text{ u}$$
$$\quad - (196.966543 + 4.002603)\text{ u}](931.5\text{ MeV/u})$$
$$= +6.501\text{ MeV}.$$

Since $Q > 0$, the reaction is exoergic.   ●

In Example 32.2, the alpha particle and the Au-197 atom share the +6.501 MeV, plus any kinetic energy brought in by the proton. Conservation of momentum shows that the lower mass alpha particle will have much more of this total kinetic energy than the higher mass Au-197 atom. A positively charged incident particle will be repelled by the positively charged target nucleus. Therefore the proton in this example will usually need sufficient kinetic energy to overcome the Coulomb potential energy

FIGURE 32.3
Endoergic reactions have a negative Q, so KE_{after} is less than KE_{before}. Endoergic reactions aren't possible below a certain threshold value of KE_{before}.

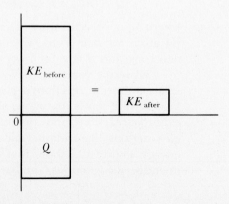

barrier and get close enough to the Hg-200 nucleus to interact through the strong nuclear force. The probability that the charged incident particle will tunnel through the barrier into the nucleus is much smaller.

32.3
THRESHOLD ENERGY

Let's consider the reaction $^{56}_{26}\text{Fe}(n,p)^{56}_{25}\text{Mn}$. It has

$$Q = [(1.008665 + 55.934939)\text{ u}$$
$$- (55.938907 + 1.007825)\text{ u}](931.5\text{ MeV/u})$$
$$= -2.914\text{ MeV}.$$

Since $KE_Y + KE_y$ can't be negative, will $KE_x = +2.914$ MeV make this endoergic reaction possible? The answer is no.

The bombarding particle carries momentum, as well as kinetic energy, into the collision. For momentum to be conserved, the Y atom and the y particle must move after the collision. Therefore $KE_Y + KE_y$ cannot equal zero, and so $Q = (KE_Y + KE_y - KE_x)$ shows that KE_x cannot equal $-Q$. In fact, KE_x must be larger than $-Q$ for an endoergic reaction.

The minimum value of KE_x needed to initiate an endoergic reaction is called the **threshold energy**, KE_{th}. If $KE_x < KE_{th}$, the reaction can't occur. Figure 32.4 illustrates the minimum-energy endoergic collision: Particle x runs into X at rest; then x and X stick together forming Y and y, which do not have enough energy to fly apart.

We can let primes refer to values after this minimum-energy collision. Then the total rest masses are $\Sigma m = m_x + m_X$ before and $\Sigma m' = m_Y + m_y$ after. Conservation of mass–energy with both sides squared gives $E^2 = E'^2$. Then

$$E^2 = (\Sigma mc^2 + KE_{th})^2 = E'^2 = p'^2c^2 + (\Sigma m'c^2)^2. \tag{C}$$

Conservation of momentum gives $p_x = p'$, which along with the combina-

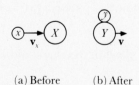

(a) Before   (b) After

FIGURE 32.4
A threshold-energy collision.

tion of $E_x^2 = p_x^2 c^2 + (m_x c^2)^2$ and $E_x = KE_{th} + m_x c^2$, yields $p'^2 c^2 = KE_{th}^2 + 2 m_x c^2 KE_{th}$. We then substitute this expression into Eq. (C) to get $(\Sigma m c^2 + KE_{th})^2 = KE_{th}^2 + 2 m_x c^2 KE_{th} + (\Sigma m' c^2)^2$. Solving for the threshold energy, we obtain

$$KE_{th} = -\frac{(\Sigma m + \Sigma m')}{2 m_x} (\Sigma m - \Sigma m') c^2. \tag{D}$$

Using the masses of neutral atoms for the nuclei in Eq. (D), the term $(\Sigma m - \Sigma m') c^2$ is another version of Eq. (32.2) and gives us Q. The term $(\Sigma m + \Sigma m')$ is the total mass of all the particles involved in the reaction. Therefore for our $X(x, y)Y$ reaction, Eq. (D) becomes

$$KE_{th} = -Q \left(\frac{M_x + M_X + M_Y + M_y}{2 M_X} \right). \tag{E}$$

But Q is much less than the total rest energy, allowing us to replace $M_Y + M_y$ by $M_x + M_X$ in Eq. (E) and to finally arrive at

$$KE_{th} = -Q \frac{M_x + M_X}{M_X} \tag{32.3}$$

for low energy endoergic reactions.

EXAMPLE 32.3 Calculate the kinetic energy of the neutron required for the reaction $^{56}_{26}\text{Fe}(n,p)^{56}_{25}\text{Mn}$.

Solution We've already calculated $Q = -2.914$ MeV. Therefore

$$KE_{th} = -(-2.914 \text{ MeV}) \frac{1.009 + 55.935}{55.935} = +2.967 \text{ MeV},$$

using Eq. (32.3). Since KE_{th} is the minimum KE_x, the neutron can have any kinetic energy ≥ 2.967 MeV. ●

The answer in Example 32.3 is the same as you'd calculate using 1.008665 u for M_x and 55.934939 u for M_X. We need the extra decimal places only when subtracting masses, as in the calculation of Q. At the threshold, the proton and the Mn-56 atom (the y and Y particles) will share $(2.967 - 2.914)$ MeV $= 0.053$ MeV of kinetic energy.

EXAMPLE 32.4 Calculate the threshold energy for the reaction $^{200}_{80}\text{Hg}(p, \alpha)^{197}_{79}\text{Au}$.

Solution Example 32.2 showed this reaction to be exoergic, not endoergic. Therefore no KE_{th} is needed to conserve momentum. ●

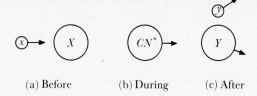

FIGURE 32.5
A compound nucleus exists for a short time during a low-energy nuclear reaction.

(a) Before (b) During (c) After

32.4
THE COMPOUND NUCLEUS

There is considerable experimental evidence that low-energy nuclear reactions proceed in two steps, as shown in Fig. 32.5:

1. Particle x and nuclide X combine to form a single nucleus in an excited state called the **compound nucleus.**

2. This compound nucleus then decays into nuclide Y and particle y.

The compound nucleus has a long lifetime compared to the 10^{-21} s needed for a new nucleus to come to equilibrium. (A nucleon moving at about 10^7 m/s can move across the typical nucleus more than once in 10^{-21} s.) But this lifetime, of the order of 10^{-14} s, is too short to be measured directly at present.

One of the main reasons for using the compound nucleus model is that the same products of different reactions are formed in the same ratios, even though different initial x and X combinations were used (provided that the same compound nucleus is formed in the same excited state). Suppose that different $x + X$ reactions form the same compound nucleus excited state, as in Fig. 32.6. The compound nucleus then exists long enough that it "forgets how it was formed," and the same ratio of Y and y products is found for the different $x + X$'s.

FIGURE 32.6
Different $x + X$ forming the compound nucleus in the same excited state result in the same ratio of $Y + y$ products.

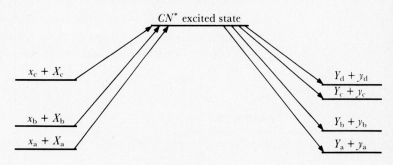

We represent the compound nucleus in an excited state by an asterisk, as in the reactions

$$^4_2\text{He} + ^{14}_7\text{N} \rightarrow ^{18}_9\text{F}^* \rightarrow ^{17}_8\text{O} + ^1_1\text{H} \quad \text{and} \quad ^1_0\text{n} + ^{10}_5\text{B} \rightarrow ^{11}_5\text{B}^* \rightarrow ^7_3\text{Li} + ^4_2\text{He}.$$

EXAMPLE 32.5   Find the compound nucleus for the reactions:

a) $^{200}_{80}\text{Hg}(\text{p}, \alpha)^{197}_{79}\text{Au}$ and   b) $^{56}_{26}\text{Fe}(\text{n}, \alpha)^{53}_{24}\text{Cr}.$

Solution   We use conservation of charge and nucleon number to determine the compound nucleus, along with a periodic table of the elements for part (a) to find that the element with $Z = 81$ is Tl.

a) $^1_1\text{H} + ^{200}_{80}\text{Hg} \rightarrow ^{1+200}_{1+80}CN^* = ^{201}_{81}\text{Tl}^*.$

b) $^1_0\text{n} + ^{56}_{26}\text{Fe} \rightarrow ^{1+56}_{0+26}CN^* = ^{57}_{26}\text{Fe}^*.$ ●

32.5
CROSS SECTION

When particle x approaches nucleus X, there is a certain probability that a reaction will occur. Suppose that you're on a farm, tossing a ball at the side of a barn. You have a much greater probability of hitting the broad side of the barn than of hitting a much smaller tool shed the same distance away. The reason is that the side of a barn has much more area than the side of a tool shed. This relationship between probability and area has been brought from the farm to nuclear physics by introducing the notion of cross section.

The **cross section,** σ, is an area that is proportional to the probability that a given reaction will occur. (See Fig. 32.7.) To determine σ, we assume that:

1. This cross-sectional area has its plane perpendicular to the velocity of the incoming particle.

2. The reaction will occur if the particle strikes the area.

3. The sample is thin enough or the cross section small enough that no nuclear cross section blocks out the cross sections of the other nuclei.

FIGURE 32.7
Different cross section sizes.

(a) Smaller
     cross section

(b) Larger
     cross section

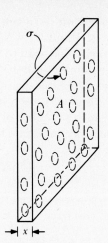

FIGURE 32.8
A sample of material having area A and thickness x and containing n nuclei per volume. Each nucleus has a cross section σ.

We can let n be the number of nuclei per volume and x be the thickness of the sample. Figure 32.8 shows that A is the area of the sample and Ax is its volume. Therefore the total cross section is (cross section/nucleus)(nuclei/volume)(volume), or $\sigma n A x$.

The probability that a reaction will occur is then the total cross section of the sample (its "active area") divided by its total area; that is, $\sigma n A x / A = \sigma n x$. Therefore if N_0 particles are incident normally on a sample, $N_0 \sigma n x$ will have a reaction. If we let $N_r =$ the number of particles reacting, then

$$N_r = N_0 \sigma n x. \qquad (32.4)$$

Equation (32.4) is accurate only for $\sigma n x \ll 1$. Otherwise, some nuclear cross sections will begin to block out others, violating assumption 3.

Each different process for the same nucleus will have its own cross section. For example, there will be one σ for elastic scattering, another σ for inelastic scattering, a third for neutron capture, and so on. A nucleus with a large cross section for a particular reaction would be "as easy to hit as the broad side of a barn." This old saying gave the name to the unit normally used for nuclear cross sections: 1 barn $= 10^{-24}\,\text{cm}^2 = 10^{-28}\,\text{m}^2$. You need to remember that the cross section has no relation to the actual cross-sectional area of the nucleus. The cross section is just a measure of a reaction's probability.

EXAMPLE 32.6 The cross section for the reaction $^{10}_{5}\text{B}(n,\alpha)^{7}_{3}\text{Li}$ is a very large 3838 barns for slow neutrons (called thermal neutrons). For fast neutrons, the cross section for the reaction $^{10}_{5}\text{B}(n,\gamma)^{11}_{5}\text{B}$ is 0.5 barn. Boron has an atomic mass of 10.81 g/mol, is 20 percent B-10, and has a density of 2.31 g/cm³. Calculate the boron thickness that will cause 1 percent of normally incident neutrons to undergo the given reactions.

Solution You're given $N_r/N_0 = 1$ percent $= 0.01$. Therefore $\sigma n x = 0.01 \ll 1$ and $N_r = N_0 \sigma n x$ is accurate. You're also given the σ's for two reactions. You want to find the corresponding x's, so you need to calculate n, the number of B-10 nuclei per volume. You can do this by treating the units as algebraic quantities. Starting with the density (which gives you the "per volume"), then dividing by the atomic mass (actually the mass of Avogadro's number of atoms), and multiplying by Avogadro's number (1 nucleus per molecule here), for boron you get

$$\left(\frac{2.31 \text{ g/cm}^3}{10.81 \text{ g/mol}}\right)(6.02 \times 10^{23} \text{ nuclei/mol}) = 1.286 \times 10^{23} \text{ nuclei/cm}^3.$$

Since you were given the information that 20 percent of these B nuclei are B-10, you have

$$n = 0.20(1.286 \times 10^{23} \text{ nuclei/cm}^3) = 2.57 \times 10^{22} \text{ nuclei/cm}^3.$$

Then $N_r = N_0 \sigma n x$ gives

$$\frac{N_r/N_0}{\sigma n} = x = \frac{0.01}{\sigma(2.57 \times 10^{22} \text{ nuclei/cm}^3)} = \frac{3.89 \times 10^{-25} \text{ cm}^3/\text{nucleus}}{\sigma}.$$

For the reaction $^{10}_{5}\text{B}(n, \alpha)^{7}_{3}\text{Li}$,

$$\sigma = 3838 \text{ barns} = 3838 \times 10^{-24} \text{ cm}^2/\text{nucleus}$$

and

$$x = \frac{3.89 \times 10^{-25} \text{ cm}^3/\text{nucleus}}{3838 \times 10^{-24} \text{ cm}^2/\text{nucleus}} = 1.01 \times 10^{-4} \text{ cm} = 1.01 \ \mu\text{m}.$$

A mere $1.01 \ \mu$m of B will cause 1 percent of all thermal neutrons to be absorbed, producing α's and Li-7.

The reaction $^{10}_{5}\text{B}(n, \gamma)^{11}_{5}\text{B}$ has the much smaller cross section of 0.5 barn. Being only about $0.5/3838 = 1/7700$ as probable, the same 1 percent reaction will require 7700 times the thickness of B, or $7700(1.01 \times 10^{-4} \text{ cm}) = 0.8$ cm. Alternatively,

$$x = \frac{3.89 \times 10^{-25} \text{ cm}^3/\text{nucleus}}{0.5 \times 10^{-24} \text{ cm}^2/\text{nucleus}} = 0.8 \text{ cm.} \qquad \bullet$$

Cross sections depend on energy as well as on reaction. As a simple example, $\sigma = 0$ for all endoergic reactions below KE_{th}, and $\sigma > 0$ above KE_{th}. Cross sections often show peaks when graphed versus energy. These peaks, called **resonances,** correspond to the excited states of the compound nucleus.

Being neutral, neutrons have no Coulomb potential energy barriers to cross (or tunnel through) to get close enough to the nucleus that the nuclear force can take over. Aside from any resonances, the faster a neu-

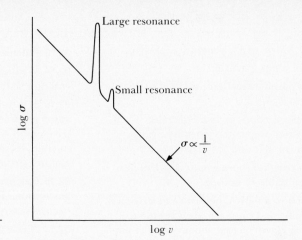

FIGURE 32.9

A 1/v cross section dependence, plus resonances.

tron is moving, the less likely it is to interact with the nucleus of most nuclides. The time that a neutron spends in the vicinity of a nucleus is inversely proportional to the speed, v, of the neutron. This time is also proportional to the probability that an allowed reaction will occur. That is, for exoergic reactions, $\sigma \propto 1/v$. Therefore a log-log graph like that in Fig. 32.9 has a slope of -1 (aside from resonances) for these nuclei.

EXAMPLE 32.7   The cross section for a particular reaction is 150 barns for 1.4×10^4 m/s neutrons. If there are no resonances, what cross section for that reaction would you expect for 5.0×10^3 m/s neutrons?

Solution Assuming that $\sigma \propto 1/v$, we can write

$$\frac{\sigma_2}{\sigma_1} = \frac{v_1}{v_2},$$

or

$$\sigma_2 = \sigma_1 \frac{v_1}{v_2} = 150 \text{ barns} \frac{1.4 \times 10^4 \text{ m/s}}{5.0 \times 10^3 \text{ m/s}} = 420 \text{ barns.} \qquad \bullet$$

In this chapter you studied $X(x, y)Y$ reactions, where x and y are small particles and KE_x is less than 100 MeV. We saw that the reactions conserved the number of nucleons, charge, linear momentum, and mass–energy just as in radioactive decay. We discussed exoergic reactions and the threshold energy for endoergic reactions, with the energies being obtained from neutral atomic masses and $E = mc^2$. We introduced the compound nucleus, then concluded with an explanation of cross sections, the nuclear physics way of specifying the probability that a particular reaction will occur.

QUESTIONS AND PROBLEMS

32.1 What particle striking $^{209}_{83}Bi$ could yield $^{206}_{82}Pb$ and an alpha particle?

32.2 Write a possible reaction for $^{112}_{48}Cd$ to be transmuted to $^{109}_{47}Ag$.

32.3 Write the reactions giving detectable charged particles or gammas when neutrons interact with a) ^3_2He and b) $^{180}_{72}Hf$.

32.4 Write a reaction to transmute Hg-199 into an even–even stable Pt isotope.

32.5 Explain how a 6-MeV neutron can strike a block of paraffin and cause a 6-MeV proton to emerge. (*Hint:* Have you ever shot marbles or pool?)

32.6 A 3.2-MeV proton initiates the reaction $^{11}_5B(p, \gamma)^{12}_6C$. Calculate the sum of the KE's of the γ and the C-12.

32.7 In the reaction $^{60}_{28}Ni(\alpha, n)Y$, the neutron and the Y nucleus together have 7.907 MeV less kinetic energy than the α particle. a) What is the Y nucleus? b) Calculate its neutral atom mass. c) Is the reaction exoergic or endoergic? d) Calculate the minimum kinetic energy the α particle must have to initiate the reaction.

32.8 Calculate the Q value of the following reactions, stating whether they are endoergic or exoergic:

a) $^2_1H(p, \gamma)Y$ d) $^{120}_{50}Sn(\alpha, p)Y$

b) $^{15}_7N(p, \alpha)Y$ e) $^2_1H(\gamma, y)^1_1H$

c) $^2_1H(d, n)Y$

32.9 Calculate the kinetic energy (in MeV) needed for a bombarding proton to just make it to the surface of a $^{200}_{80}Hg$ nucleus. [Use $PE = (9.0 \times 10^9 \text{ N} \cdot \text{m}^2/\text{C}^2)q_1q_2/r$.]

32.10 Redo Example 32.3 a) using 1.008665 u and 55.934939 u in place of the values used there and b) using the more exact relation, $KE_{th} = -Q(M_x + M_X + M_Y + M_y)/2M_X$. Compare your answers to the answer in Example 32.3. What do you conclude?

32.11 Calculate KE_{th} for the reactions in Problem 32.8.

32.12 Consider the reaction $^{12}_6C(n, y)^{11}_5B$. a) What is the particle y? b) What minimum neutron kinetic energy is needed? c) What is the total KE after the reaction?

32.13 Calculate KE_{th} for the reaction $^4_2He(p, d)Y$.

32.14 What is the compound nucleus for every reaction in Problem 32.8?

32.15 a) List the nuclides that must be targets for gammas, protons, neutrons, deuterons, and α particles to give $^{27}_{13}Al^*$. b) What are the resulting nuclides if $^{27}_{13}Al^*$ then decays by the emission of a gamma, proton, neutron, deuteron, or α particle?

32.16 Calculate KE_{th} for a reaction in which the absorption of a proton gives $^{47}_{21}Sc^*$, which then decays by emission of an alpha particle.

32.17 A reaction involves $^{116}_{50}Sn$, $^{117}_{50}Sn^*$, and $^{115}_{49}In$. Calculate both the minimum kinetic energy of the bombarding particle needed to initiate the reaction and the corresponding total KE after the reaction.

32.18 a) What is the total momentum at all times in the frame of reference of the compound nucleus?

b) Describe in words the motions of the particles in the reaction $X(x, y)Y$ with respect to that frame of reference.

c) What is the ratio KE_y/KE_Y in that frame of reference?

32.19 a) Particle x doesn't have quite enough kinetic energy to "climb over" the x–X Coulomb potential energy barrier, but if $Q > 0$ the probability is still greater than zero that the compound nucleus will be formed. Explain how this is possible.

b) For the situation in part (a), will σ be large?

32.20 In what region is the cross section for a reaction with a negative Q equal to zero? Why?

32.21 Since $r \propto A^{1/3}$ and area $\propto r^2$, is $\sigma \propto A^{2/3}$? Explain your answer.

32.22 If σ refers to an absorption process, the absorption coefficient, α, is defined by $\alpha = n\sigma$. What is the physical meaning of α?

32.23 Let Φ_0 be the flux of particles incident on a sample of volume V. Let the reaction rate, R, be the number of reactions occurring per unit time. Prove that $R = n\sigma\Phi_0 V$. What are your assumptions?

32.24 A reactor has a thermal neutron flux of 2.5×10^{14} neutrons/cm$^2 \cdot$ s. Natural Co is 100 percent Co-59, has a density of 8.9 g/cm^3 and an atomic mass of 58.9 g/mol. A 0.15-g sample of natural Co is placed in the reactor for 60 s, producing some Co-60, with an (n, γ) cross section of 37 barns. The half-life of Co-60 is 5.27 years. Calculate the activity of the Co-60 source produced (see Problem 32.23).

32.25 Some 12.22% of natural Cd is Cd-113, with a thermal neutron capture cross section of 2.00×10^4 barns. The cross sections for the other Cd isotopes are negligible in comparison. Cadmium has an atomic mass of 112.41 g/mol and a density of 8.65 g/cm^3. If 8.3×10^{15} thermal neutrons are incident normally on a 2.0 μm thick Cd film, how many will be absorbed?

32.26 Thermal neutrons absorbed by U-235 cause immediate fission 85.6% of the time. The remaining 14.4% of the time the reaction $^{235}_{92}U(n, \gamma)^{236}_{92}U$ occurs with a radiative capture cross section of 98 barns. Calculate the fission capture cross section.

32.27 Explain the origin of the capture cross section resonances.

32.28 If KE_x results in a resonance, give two reasons why the compound nuclear excited state is not KE_x above the ground state.

32.29 If $\sigma \propto 1/v$, what is the slope of the more common log-log graph of σ versus KE_x for nonrelativistic neutrons?

32.30 For the reaction $^1_1H(n, \gamma)^2_1H$, $\sigma = 0.333$ barn for 0.025-eV neutrons, and σ follows the $1/v$ rule for this reaction in this region of energy.

a) Calculate the radiative capture cross section for 0.050-eV neutrons incident on H-1.

b) What fraction of 0.025-eV neutrons are absorbed in passing through 1.00 m of H_2 gas at STP where 1 mol occupies 22.4×10^{-3} m^3? Hydrogen is 99.985 percent H-1; the remaining fraction is H-2, with a much smaller capture cross section.

32.31 For 1-keV neutrons interacting with natural B in an (n, α) reaction, $\sigma = 4$ barns. The $1/v$ rule is obeyed, the mixture begins with 20% B-10 and 80% B-11, and $\sigma_{11} \ll \sigma_{10}$. Find σ for 0.1-eV neutrons for this reaction with pure B-10.

CHAPTER

Fission and Fusion

In this chapter we'll relate the tremendous energy releases in fission reactions and fusion reactions to the average binding energy per nucleon curve. We'll discuss fission in terms of the liquid-drop model, explain generally the operation and control of a nuclear reactor, and introduce the concept of a breeder reactor (one that produces more fuel than it uses). Then we'll move to fusion, describing some of the energies released and the fusion reactions that apparently power the stars. We'll finally present some general strategies for obtaining controlled thermonuclear reactions.

33.1
FISSION

If U-235 absorbs a low-energy neutron, the compound nucleus U-236* is formed. Rather than decaying by emitting a low-mass particle, U-236* usually undergoes a **fission** reaction. That is, it breaks into two medium-mass nuclides plus some low-mass particles. The nuclide U-236 has almost 1.6 neutrons in its nucleus for every proton, but the stable medium-mass

FIGURE 33.1

Hiroshima, August 1945, the result of supercritical fissioning of a few moles of U-235. "The unleashed power of the atom has changed everything save our modes of thinking and thus we drift toward unparalleled catastrophe." (Albert Einstein, 1946.)

Source: Photograph courtesy Shigeo Hayashi. Print provided by *Newsweek*.

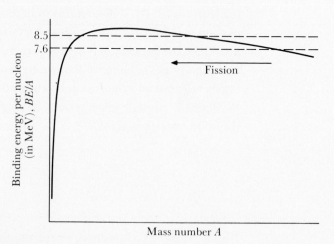

FIGURE 33.2

Fission reactions release energy through an increase in the average binding energy per nucleon.

nuclides have closer to 1.3 neutrons for every proton. Therefore the final products of fission reactions must decrease the neutron-to-proton ratio as they approach stability. This decrease results mainly from the emission of neutrons and β^- decay.

Two typical fission reactions are

$$\,^1_0n + \,^{235}_{92}U \longrightarrow \,^{236}_{92}U^* \longrightarrow \,^{144}_{56}Ba + \,^{89}_{36}Kr + 3\,^1_0n$$

and

$$\,^1_0n + \,^{235}_{92}U \longrightarrow \,^{236}_{92}U^* \longrightarrow \,^{140}_{54}Xe + \,^{94}_{38}Sr + 2\,^1_0n.$$

The fission fragments undergo as many as six β^- decays to reach a stable nuclide. For example,

$$\,^{140}_{54}Xe \xrightarrow{\beta^-} \,^{140}_{55}Cs \xrightarrow{\beta^-} \,^{140}_{56}Ba \xrightarrow{\beta^-} \,^{140}_{57}La \xrightarrow{\beta^-} \,^{140}_{58}Ce.$$

Enrico Fermi, Emilio Segre, and others first bombarded uranium with neutrons and detected the β^-'s in 1934. They thought, however, that they had produced a $Z > 92$ element. In 1938, Otto Hahn and Fritz Strassmann detected barium after the reaction. In 1939, Lise Meitner and Otto Frisch further explained the process, soon named *fission.*

Other nuclides besides U-235, such as U-233 and Pu-239, can be induced to fission. Also, other particles besides the neutron can cause fission to occur. Some very heavy nuclides (such as several isotopes of $_{98}Cf$) even decay by random **spontaneous fission,** with no particle bombardment needed to trigger the breakup.

Many different fission products can be produced. In fact, the even splitting of $\,^{236}_{92}U^*$ into two $\,^{117}_{46}Pd$ and two $\,^1_0n$ occurs only about 0.01 percent of the time. The resulting mass numbers vary from about $A = 72$ to $A = 163$, but the most probable yields occur in the vicinity of $A = 95$ and $A = 138$. Occasionally the breakup is into three nuclides. For instance, another 0.01 percent of the time U-236* fissions in nuclear reactors to produce two medium-mass nuclides plus H-3 (tritium).

Figure 33.1 gives the stark and horrifying evidence that every fission reaction releases a great amount of energy. In Chapter 29 you learned how to calculate the binding energy per nucleon of a nuclide. Examination of the binding energy per nucleon curve in Fig. 33.2 shows that BE/A is about 7.6 MeV/nucleon for the heavy elements such as uranium, but increases to about 8.5 MeV/nucleon for the medium-mass fission fragments. Therefore with 236 nucleons involved, the binding energy increases by about (236 nucleons)(8.5 − 7.6) MeV/nucleon = 200 MeV in a fission reaction.

Recall that an increase in the binding energy means that the total energy of the system has decreased. You can see, then, that the internal energy of the nuclides has decreased by about 200 MeV per fission, thereby releasing about 200 MeV of energy for each fission. About 85 percent of this energy goes to the *KE* of the fission fragments. The remaining 15 percent is divided among the neutrons, β^-'s, antineutrinos, and γ rays produced.

A chemical reaction (for example, a hydrocarbon $+ O_2 \rightarrow CO_2 + H_2O$) involves the valence electrons of the atoms. Chemical reactions can't release more energy than the energy changes of the valence electrons, so a release of 20 eV per chemical reaction would be considerable. However, about 200×10^6 eV per reaction is released in fission, giving over $200 \times 10^6/20 = 10^7$, or ten million, times the energy output per reaction! This explains the immense energy output of fission processes from a relatively small amount of material.

EXAMPLE 33.1   A possible fission reaction is

$$_{0}^{1}n + _{92}^{235}U \longrightarrow _{92}^{236}U^* \longrightarrow _{47}^{121}Ag + _{45}^{113}Rh + 2_{0}^{1}n.$$

a) Is this reaction very probable?

b) Will the reaction stop there?

c) In principle, how could you calculate the energy released in this part of the reaction?

Solution

a) No. Since the reaction is close to an even split with $A_{Ag} = 121$ and $A_{Rh} = 113$, the probability of fission is very low.

b) No. The unstable nuclides Ag-121 and Rh-113 will decay toward the stability curve.

c) Use $Q = $ (total rest mass before $-$ total rest mass after)c^2.   ●

The fission fragments Ag-121 and Rh-113 in Example 33.1 emit β^-'s to finally reach Sb-121 and Cd-113. Of course, these β^- decays also release energy.

The liquid-drop model helps us to understand the fission process. The nucleus in its ground state is nearly spherical. The compound nucleus is in an excited state, which means that the compound nucleus must have less binding energy than the ground state. The liquid-drop model states that this decrease in binding energy can occur when the excited state has more of the nucleons on the surface of the nucleus than the ground state.

The larger number of nucleons on the surface of a particular nucleus means the surface area increases. For a fixed volume, the surface area can increase only if the nucleus is distorted from its nearly spherical ground state shape. As illustrated in Fig. 33.3, the nucleus (a) can go to an ellipsoidal shape (b), then to a dumbbell shape (c). With a large repulsive Coulomb

FIGURE 33.3
Steps in the fission
process in the liquid drop
model.

(a)   (b)   (c)   (d)

force between the two halves of the dumbbell and only a small attractive nuclear force in the narrow neck, the dumbbell shape breaks into parts (d) and the nucleus fissions.

EXAMPLE 33.2 Let $^{144}_{56}$Ba and $^{89}_{36}$Kr be two spherical nuclei just touching one another. Calculate their approximate repulsive Coulomb potential energy, $(1/4\pi\epsilon_0)(q_1q_2/r)$, where $1/(4\pi\epsilon_0) = 9.0 \times 10^9$ N $\cdot$ m^2/C^2.

Solution For the Ba nucleus, $q_1 = 56(1.60 \times 10^{-19}$ C). For the Kr nucleus, $q_2 = 36(1.60 \times 10^{-19}$ C). Since r is the center-to-center separation of the nuclei, r equals the sum of the radii, $r_{144} + r_{89}$. Equation 28.3 gives $r = (1.2 \times 10^{-15}$ m$)A^{1/3}$ as the nuclear radius, so

$$r_{144} = (1.2 \times 10^{-15} \text{ m})144^{1/3} = 6.29 \times 10^{-15} \text{ m}$$

and

$$r_{89} = (1.2 \times 10^{-15} \text{ m})89^{1/3} = 5.36 \times 10^{-15} \text{ m}.$$

Adding, we get $r_{144} + r_{89} = 11.65 \times 10^{-15}$ m, as shown in Fig. 33.4. The approximate potential energy then equals

$$9.0 \times 10^9 \, \frac{\text{N} \cdot \text{m}^2}{\text{C}^2} \, \frac{(56)(36)(1.60 \times 10^{-19} \text{ C})^2}{11.65 \times 10^{-15} \text{ m}}$$

$$= 4.0 \times 10^{-11} \text{ J} \, \frac{1 \text{ MeV}}{1.60 \times 10^{-13} \text{ J}} = 250 \text{ MeV}.$$

●

FIGURE 33.4
^{144}Ba and ^{89}Kr nuclei as
spheres of positive charge.

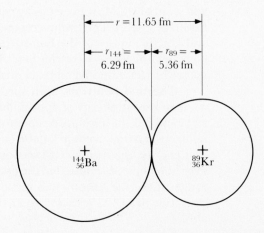

The 250 MeV estimated in Example 33.2 isn't the total potential energy for the two fission fragments. The negative (attractive) potential energy from the strong nuclear force between the fragments is another large term that decreases the total potential energy. This smaller potential energy can then be converted to the approximately 170 MeV of kinetic energy that the fission fragments carry away in the fission reaction.

33.2
NUCLEAR REACTORS

On average, 2.5 neutrons are produced in every slow neutron-induced fission of U-235. The neutrons in one fission may go on to be absorbed, causing more fissions in what is called a **chain reaction.** The average number of neutrons from one fission that actually cause another fission is called the **multiplication factor.** For example, if on average, less than 1 of the 2.5 neutrons produces another fission, the chain reaction will eventually die out. If so, we call the system **subcritical.**

If the multiplication factor is exactly 1, the reaction will continue at a constant rate. We then call the system **critical.** Finally, if the multiplication factor is greater than 1, as in Fig. 33.5, the reaction rate increases and we call the system **supercritical.** A nuclear fission bomb (the "atom bomb") is a terrifying example of a deliberately caused supercritical reaction.

A self-sustaining critical reaction requires that 1.5 of the 2.5 neutrons be removed from the reactions for U-235 fission. Some neutrons will simply leak out of the walls of the system. Because of this loss, a certain minimum mass, called the **critical mass,** must be present for a self-sustaining critical reaction. The critical mass depends on the material and its geometry. Other neutrons will be absorbed within the system in nonfission reactions.

A **nuclear reactor** is a system in which chain reactions can be initiated and controlled. Most present-day reactors in the United States utilize the fission of U-235. The fission cross section of U-235 increases as the speed and kinetic energy of the bombarding neutrons decrease. Slow-moving neutrons therefore have the best chance of causing fission before they escape or cause nonfission reactions.

Thermal neutrons, with speeds of a few thousand m/s, have an average kinetic energy of approximately kT, where k is Boltzmann's constant and T is the absolute temperature of the reactor core. Thus thermal neutrons have an average kinetic energy of a few hundreths of an eV. At that energy, the fission cross section of U-235 is about 500 barns. However, the neutrons emitted in the fission process have energies in the MeV range where the fission cross section is only about 1 barn. For efficient operation of a U-235 reactor, then, the neutrons must be slowed considerably.

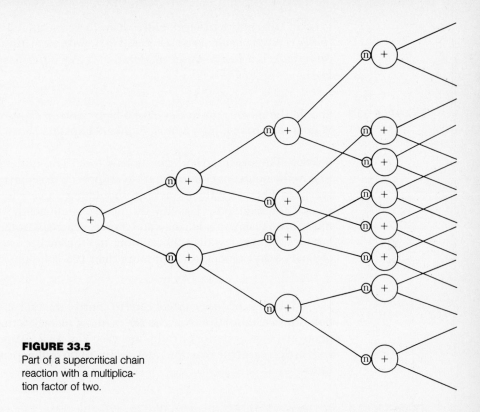

FIGURE 33.5
Part of a supercritical chain
reaction with a multiplication factor of two.

The material that slows the neutrons in a reactor is called the **moder-ator.** (See Fig. 33.6.) The most effective slowing is done by means of elastic collisions with nuclides having a mass close to that of the neutron. (A marble can stop dead after hitting a marble of equal mass but will bounce off a boulder with little energy loss.) In 1942, Fermi directed construction of the first man-made reactor on the southside of Chicago. Carbon in the form of graphite was used as the moderator in that reactor, as it was in the

FIGURE 33.6
A moderator slows
neutrons having (a) large
initial velocities, v_i, to (b)
small final velocities, v_f.

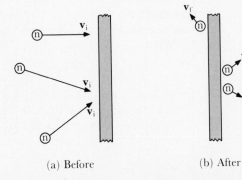

(a) Before   (b) After

Russian Chernobyl reactors more than 40 years later. In most modern power reactors, ordinary water is used to slow the neutrons. In Canada, the preference is a **heavy water** moderator (heavy water has H-2 instead of H-1).

EXAMPLE 33.3   Should the moderator nuclei have a high neutron capture cross section? Should they have a high atomic number? Explain.

Solution The answer to both questions is "no." The moderator is supposed to slow the neutrons efficiently. It isn't supposed to remove them from the reaction. Therefore the moderator should have a very low capture cross section for neutrons. To slow the neutrons efficiently, the moderator nuclei need to have a small mass and therefore a small atomic number. (For instance, it takes less than 20 collisions to "thermalize" neutrons with an ordinary water moderator, but more than 100 collisions with graphite.)

•

Most reactors use **control rods** to control the rate at which the reaction is proceeding. Control rods are made of materials that easily absorb neutrons, such as boron or cadmium. (See Fig. 33.7.) Moving the control rods in decreases the reaction rate; pulling them out increases the reaction rate.

EXAMPLE 33.4   A reactor is to maintain a constant reaction rate. On average, 1.0003 neutrons per fission are initiating new fissions. What should the reactor operator do?

Solution To keep a constant rate, 1.0000 neutrons should be initiating new fissions. Therefore the operator must move the control rods in far enough to absorb 0.0003 more neutrons per fission, on average. The reaction is

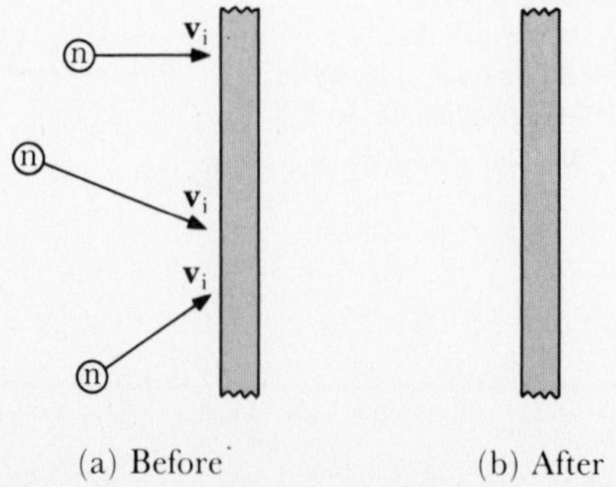

FIGURE 33.7
A control rod absorbs
neutrons.

(a) Before   (b) After

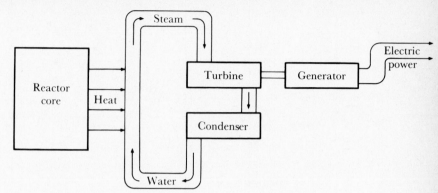

FIGURE 33.8
Heat is removed from this reactor core to boil water. The resulting steam spins the turbine, the turbine drives a generator, and the useful output is electric power.

supercritical with a multiplication factor of 1.0003. It needs to be made critical, with a multiplication factor of exactly 1. ●

We call the neutrons emitted during fission **prompt neutrons.** They cause a new reaction in about 30 μs. An operator simply cannot move control rods in and out that quickly. Control of a U-235 thermal neutron reactor is possible, however, because 0.6 percent of the neutrons are emitted only after the beta decay of a fission fragment. These **delayed neutrons** do not appear for an average time of more than a second after fission. Even so, 1 s is not a long time in terms of human response, and 0.6 percent is not a very large safety margin. Therefore careful start-ups and automatic controls with safety backups are essential. In the Chernobyl reactor accident, some automatic controls were turned off so that an electrical engineer could run an experiment. Then the control rods were pulled completely out and could not be put back in rapidly enough to stop the disaster.

Most of the energy produced by the fission process is converted into thermal energy in the reactor by the slowing of fission fragments and neutrons. In reactors used to produce electric power, the thermal energy is then utilized to drive some type of heat engine, typically a steam turbine, to provide the output work, as shown in Fig. 33.8. Since heat engines are used, the efficiency is limited by the second law of thermodynamics. Therefore at current operating temperatures, the majority of the energy produced isn't available as output work, but must be disposed of as waste heat.

Operation of a nuclear reactor produces many radioactive isotopes. The disposal or use of these radioactive wastes have created severe problems for nuclear reactors. Some of the radioactive nuclides produced have practical uses. However, the separation of highly radioactive elements is no easy task.

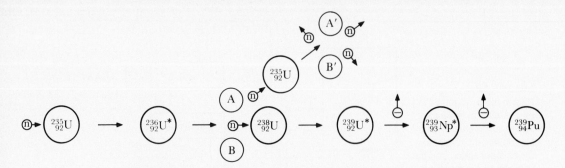

FIGURE 33.9
In a breeding reaction, U-235 captures a neutron, then fissions, emitting two neutrons. One of the two neutrons causes another U-235 fission. The other neutron is captured by U-238, which forms (through two β^- decays) the fissionable nucleus Pu-239.

Natural uranium contains 0.72 percent U-235. For efficient reactor operation with a natural water moderator, the uranium must be enriched to a few percent U-235. Enrichment isn't necessary for heavy-water moderators, because the neutron capture cross section of H-2 is 650 times smaller than that of H-1. (For nuclear weapons, 90 percent enrichment is necessary, making the enrichment process very expensive.) Therefore in a uranium fueled power reactor, the greatest percentage of the uranium by far is U-238, which does absorb some neutrons. Occasionally, the U-239* produced by reactor neutrons will fission, but most of the time the U-239* undergoes two β^- decays to produce Pu-239. Easily fissionable, Pu-239 produces an average of 3 neutrons per fission by fast neutrons; Pu-239 reactors can be built. (The first nuclear explosive device and the Nagasaki bomb both used Pu-239.)

In the U-235 reactor, 1.0 of the 2.5 neutrons produced per fission is used to cause further fissions. More than 1.0 of the remaining 1.5 neutrons per fission can be used to convert nonfissionable U-238 to fissionable Pu-239, as illustrated in Fig. 33.9. In principle, more fissionable material can be produced than is being used. A **breeder reactor** is a reactor that produces more fissionable nuclei than it uses. For example, a breeder reactor might produce 1.1 Pu-239 nuclei for every 1.0 U-235 nucleus involved in a fission reaction. Some current nuclear research is aimed at solving the major technical and safety problems associated with breeder reactors.

EXAMPLE 33.5   Another breeder reaction occurs when fissionable U-233 is produced from Th-232 (after Th-232 absorbs a neutron and undergoes two β^- decays). Explain how a Pu reactor could be a breeder, using Th-232 and producing up to two fissionable nuclei for each Pu-239 fission.

Solution On average, fast Pu-239 fissions produce three neutrons. One of the three neutrons from the Pu-239 fission would be used to continue the critical chain reaction in the reactor. Each of the other two could be absorbed by Th-232, eventually yielding fissionable U-233. Each average Pu-239 fission could therefore produce as many as two U-233 nuclei. The reactor is then a breeder, producing more fissionable nuclei than it uses.

●

33.3
FUSION

In a **fusion** reaction, small nuclei come together to form a larger nucleus. Energy is released in fusion for the same reason that it is released in fission: The binding energy per nucleon is greater after the reaction than before, as shown in Fig. 33.10. Some examples of fusion involving hydrogen isotopes are

$$^2_1H + ^2_1H \longrightarrow ^3_2He + ^1_0n \qquad (Q = 3.3 \text{ MeV})$$
$$^2_1H + ^2_1H \longrightarrow ^3_1H + ^1_1H \qquad (Q = 4.0 \text{ MeV})$$
$$^2_1H + ^3_1H \longrightarrow ^4_2He + ^1_0n \qquad (Q = 17.6 \text{ MeV}).$$

EXAMPLE 33.6 Verify the $Q = 3.3$ MeV value of the first hydrogen isotope fusion reaction.

Solution Using Eq. (32.2), we obtain

$$Q = [(M_x + M_X) - (M_Y + M_y)]c^2$$
$$= [2(2.014102) \text{ u} - (3.016029 + 1.008665) \text{ u}](931.5 \text{ MeV/u})$$
$$= 3.27 \text{ MeV}.$$

●

FIGURE 33.10
Fusion reactions also release energy through an increase in the average binding energy per nucleon. Compare this curve to the one in Fig. 33.2.

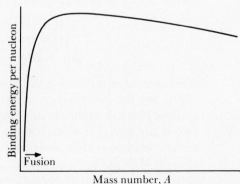

Binding energy per nucleon

Fusion

Mass number, A

Obtaining the deuterium (H-2) for a fusion reaction isn't very difficult. One hydrogen atom in every 6600 is deuterium, and hydrogen is the sixth most abundant element in the earth's crust (think of all that H_2O, for instance). Getting two deuterium nuclei together (or in current research, H-2 and H-3) is much more difficult. Their centers would have to be about 5 fm apart before the short-range nuclear force could take over. The Coulomb potential energy at that distance is

$$9.0 \times 10^9 \frac{N \cdot m^2}{C^2} \frac{(1.60 \times 10^{-19} \, C)^2}{5 \times 10^{-15} \, m} = 4.6 \times 10^{-14} \, J \frac{1 \, eV}{1.60 \times 10^{-19} \, J}$$
$$= 3 \times 10^5 \, eV.$$

Setting this energy equal to the average kinetic energy of an ideal gas molecule, $\frac{3}{2}kT$, we get

$$T = \frac{2(3 \times 10^5 \, eV)}{3(8.6 \times 10^{-5} \, eV/K)} = 2 \times 10^9 \, K.$$

Actually, temperatures lower than 10^8 K are sufficient to produce fusion through collisions. The reasons are that many nuclei are moving much faster than the rms speed, extra energy is available in a head-on collision, and some tunneling through the Coulomb potential energy barrier is possible.

At such high temperatures, the energies are much greater than the binding energies of the outer electrons. Therefore the gas becomes highly ionized. Remember that a gas of charged particles is called a plasma. In a plasma of hydrogen isotopes and electrons at a sufficiently high temperature, fusion reactions can occur. Because of the high temperature, these fusion reactions are called **thermonuclear** reactions.

Thermonuclear reactions are the source of energy from the stars (one of which is our sun). Two types of cyclic thermonuclear reactions thought to occur in stars are the **proton–proton cycle** and the carbon cycle. The proton–proton cycle "burns up" protons and evidently is the main cycle in our sun. About 10^{56} protons have been "burned up" so far in its lifetime. The proton–proton cycle involves the following reactions:

$$^1_1H + ^1_1H \longrightarrow ^2_1H + ^0_1\beta^+ + ^0_0\nu$$
$$^1_1H + ^2_1H \longrightarrow ^3_2He$$
$$^3_2He + ^3_2He \longrightarrow ^4_2He + ^1_1H + ^1_1H.$$

Note that we must double the first two reactions to obtain the two He-3 nuclei for the last reaction. Overall, six protons are converted to two protons plus one alpha particle, two positrons, two neutrinos, and a number of gammas that aren't shown. Thus $6 - 2 = 4$ protons are "burned up" in this fusion cycle with the release of almost 25 MeV of energy, or about 6 MeV per proton burned. This 6 MeV per proton is far more than the 0.9 MeV of energy per nucleon released in a fission reaction.

The **carbon cycle** is more involved:

$$
\begin{aligned}
{}^{1}_{1}\text{H} + {}^{12}_{6}\text{C} &\longrightarrow {}^{13}_{7}\text{N} \\
{}^{13}_{7}\text{N} &\longrightarrow {}^{13}_{6}\text{C} + {}^{0}_{1}\beta^{+} + \nu \\
{}^{1}_{1}\text{H} + {}^{13}_{6}\text{C} &\longrightarrow {}^{14}_{7}\text{N} \\
{}^{1}_{1}\text{H} + {}^{14}_{7}\text{N} &\longrightarrow {}^{15}_{8}\text{O} \\
{}^{15}_{8}\text{O} &\longrightarrow {}^{15}_{7}\text{N} + {}^{0}_{1}\beta^{+} + \nu \\
{}^{1}_{1}\text{H} + {}^{15}_{7}\text{N} &\longrightarrow {}^{12}_{6}\text{C} + {}^{4}_{2}\text{He}.
\end{aligned}
$$

Overall, four protons are converted to one alpha particle, two positrons, two neutrinos, and a number of gammas that aren't shown. We say that the C-12 acts as a catalyst because it is present at both the beginning and end of the cycle, and the cycle could not occur without it. The overall effect of the carbon cycle is virtually the same as that of the proton–proton cycle. Again, two neutrinos and almost 25 MeV are released per cycle.

Although the physics of solar fusion energy production seems well-understood, experiments have detected neutrinos being emitted at only one-third the expected rate. However, these experiments just measure the number of the electron neutrino reactions. Evidently, there are two other types of neutrinos besides the electron neutrino. Therefore if neutrinos were to oscillate evenly between the three types, the experiments would detect them only one-third of the time. Other experiments underway use nuclear reactors (which are controlled antineutrino sources) to attempt to measure any possible antineutrino oscillations.

Thermonuclear fusion reactions have been produced on earth. Unfortunately, their first use was in the hydrogen bomb. This device uses a nuclear fission reaction to provide the extremely high temperatures required for a fusion reaction to occur. Currently, the nuclear arsenals of the world are stocked with tens of thousands of these weapons with a total explosive yield of more than one ton of TNT for every person on earth. For more humane uses, the problem with controlled thermonuclear reactions is that of producing and containing a plasma having a temperature well over ten million degrees. So far, no controlled thermonuclear reaction has managed to produce more energy than that needed to make it occur.

Methods of controlling the plasma include magnetic confinement and inertial confinement. In **magnetic confinement** schemes, various geometries of magnetic fields are used to hold in the charged plasma. Magnetic confinement basically uses the $q\mathbf{v} \times \mathbf{B}$ magnetic force on moving charged particles. This force is zero for particles moving parallel (or antiparallel) to the magnetic field and is perpendicular to both $\mathbf{v}$ and $\mathbf{B}$ otherwise. In any future magnetic confinement fusion power reactor, superconducting magnets will have to be used for most of the large volumes of strong magnetic fields required. Without superconductivity, the

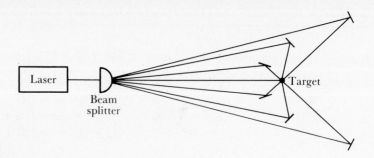

FIGURE 33.11
A diagram of a laser inertial confinement fusion experiment.

power needed to drive the magnets would exceed the power produced by the reactor.

Inertial confinement schemes include imploding a pellet of fuel by bombarding it from all sides with high-energy beams. These may be laser beams, as in Fig. 33.11, electron beams, or ion beams. Bombardment is done quickly, so that the combination of the inertia of the incoming energy and the inertia of the created plasma keeps the plasma together long enough to react.

There is no question that tremendous amounts of energy can be produced by fusion processes. Fortunately, the radioactive waste problem for fusion reactors should be minor. However, the technical problems involved in actually producing and utilizing this energy in a controlled manner have, so far, proved extraordinarily difficult.

EXAMPLE 33.7   Confinement time (the time a plasma can be held together) and plasma density (the number of ions per volume) are two important parameters in producing controlled thermonuclear reactions. Why?

Solution There are probabilities (therefore, cross sections) that the desired fusion reactions will occur. The longer the plasma holds together, the greater are the number of collisions and the greater is the total probability that fusion will occur. A larger plasma density means that a larger number of nuclei are present in a given volume, a greater number of collisions can occur, and the total probability of fusion is greater.   ●

Since both confinement time and plasma density are important, their product is a criterion of the possible performance of any fusion device. The *Lawson criterion* suggests that this product should be greater than 3×10^{20} s · ions/m³.

A different method of inducing nuclear fusion doesn't require high temperatures or energies. This area of current research is called **muon-catalyzed fusion,** or **cold fusion.** In an ordinary H^{2+} ion, the two protons of the H-1 nuclei are bound together by the sharing of an electron. The equilibrium separation of the protons is about 100 pm or 10^5 fm (about twice the radius of the ground state of the Bohr hydrogen atom).

However, in muon-catalyzed fusion, the H^{2+} ion is made up of a deuteron (H-2 nucleus) and a triton (H-3 nucleus), with the electron replaced by a muon. An unstable muon has the same charge, spin, and so on as an electron, but it has about 200 times more mass. As a result, the equilibrium separation of the deuteron and triton is about 200 times less than 100 pm, or about 500 fm.

At a separation of 500 fm, the deuteron and triton are still much too far apart for the strong nuclear force to fuse them, but we cannot correctly describe them as point particles kept apart by a Coulomb barrier. They can tunnel through the barrier to one another and fuse with the release of 17.6 MeV of energy. The muon remains after the fusion reaction and ordinarily can go on to replace other electrons, perhaps catalyzing hundreds of fusions before decaying. Considerable research, including finding methods to produce muons cheaply, will be needed if muon-catalyzed fusion ever is to become a practical energy source.

This chapter was mostly descriptive. We introduced concepts such as fission and chain reactions; subcritical, critical, and supercritical systems; and nuclear reactors. We described the functions of moderator, control rods, prompt and delayed neutrons, and breeder reactors. We showed the energy release for several fusion reactions, presented the proton–proton cycle and the carbon cycle, and discussed the need for extremely high temperatures and some type of plasma confinement in controlled thermonuclear fusion. We ended with a discussion of muon-catalyzed fusion, which is another application of tunneling.

QUESTIONS AND PROBLEMS

33.1 Why are so many antineutrinos associated with fission reactions?

33.2 The nuclide Cf-252 undergoes spontaneous fission. Explain what this means. Why then is Cf-252 utilized as a neutron source?

33.3 What are the approximate A values for the maximum and minimum fission yields from Fig. 33.12 for the fission of U-236*?

33.4 What is the approximate amount of energy released when a neutron induces the fission of a) U-233? b) Pu-239?

33.5 Does U-236* always fission? When it does fission, does it always break into two nuclides?

33.6 Why are fission fragments invariably radioactive?

33.7 In the liquid-drop model of the fission

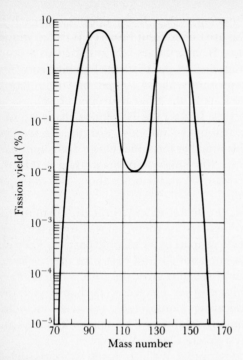

FIGURE 33.12

process, a) why isn't the compound nucleus spherical? b) must the compound nucleus fission? c) what force holds the compound nucleus together, what force pushes it apart, and why does one "win" over the other? d) why doesn't U-236* break into 236 nucleons?

33.8 Redo Example 33.2 for $^{140}_{54}$Xe and $^{94}_{38}$Sr.

33.9 Is a multiplication factor of 2.5 actually possible for U-235 fission? Explain your reasoning.

33.10 Would the critical mass be the same for a sphere of a material as for a flat plate of the material? Explain.

33.11 For a sphere, how do both the leakage and the number of fissionable nuclei depend on the radius? From your answer, explain why there is a critical mass.

33.12 Show that 2.2 km/s gives a kinetic energy of kT for neutrons, where T is room temperature.

33.13 At Oklo in the Gabon Republic of Africa some 2×10^9 years ago, a local concentration of uranium and water apparently allowed a low-power, natural reactor to operate for about 500,000 years. The average power output evidently was above 20 kW.

a) Why was there a much larger percentage of U-235 present in natural uranium 2×10^9 years ago than now?

b) What was the function of the water?

c) Why was the uranium necessary for the operation of the reactor?

d) As a result, what nuclides in the area would now have a larger-than-usual concentration?

e) One-third of the U-235 fissioned in the reaction came from U-238, which captured a neutron and then underwent two β^- decays and an alpha decay. Write the reactions, showing the intermediate nuclides.

33.14 Most (98 percent) of the reactor energy is originally deposited in the fuel. Explain why.

33.15 The cladding is the material used to contain reactor fuel.

a) Should it have a high or low neutron capture cross section?

b) Should it have a high or low thermal conductivity?

c) If a cladding nucleus does undergo a capture reaction, why might a radiative capture reaction, (n, γ), be advantageous to the cladding's strength?

33.16 Should the reactor fuel and the moderator material be in separate parts of the reactor? Explain.

33.17 a) Should the control rod material have a low or high neutron capture cross section? Explain. b) Why is a control rod material called a "poison"? (What does it kill?) c) Is the atomic mass of the control rod element important to its operation?

33.18 In Pu-239 fast neutron fission, more neutrons are produced and a much smaller percentage are delayed neutrons than for U-235 thermal neutron fission. What are the ramifications for the safe operation of a fast Pu-239 reactor?

33.19 For every 1000 U-235 fissions, 1007 neutrons cause new fissions. Of the 1007, six are delayed neutrons that cause fissions after several seconds. The average prompt neutron causes a fission in 30 μs.

a) Starting with 1000 fissions at $t = 0$, how many fissions will be occurring at $t = 0.10$ s?

b) What type of chain reaction is described in part (a)?

c) What is the significance of the answers to parts (a) and (b) of this problem?

33.20 Even an hour after the chain reaction is stopped, more than 1 percent of the reactor operating power is still being produced. It's *not* from continuing fission reactions. What is it from?

33.21 One fission fragment in a reactor is I-135, which decays with a half-life of 6.585 hr to Xe-135. The nuclide Xe-135 has the extraordinary thermal neutron capture cross section of 2.6 million barns and a half-life of 9.10 hr. However, when a reactor is operating, the Xe-135 concentration is low, with almost all Xe-135 nuclei absorbing a neutron (becoming stable Xe-136) long before they have a chance to decay.

a) If a reactor is shut down quickly, why does the Xe-135 concentration in the reactor increase for several hours?

b) Why does this condition make it quite difficult or even impossible to restart the reactor for many hours?

33.22 Approximately how many grams of U-235 are fissioned each month in the operation of a 3000 MW(T) (megawatts thermal) power reactor?

33.23 If the power demand goes up, the moderator water circulating through a pressurized water reactor returns from the heat exchanger cooler. The cooler temperature makes the water denser. Why does this condition then cause more fissions per time, thereby raising the thermal power output of the reactor?

33.24 In the operation of a nuclear reactor, when do the operators want the multiplication factor to be a) <1? b) $=1$? c) >1?

33.25 If the neutrons are much higher in energy than thermal neutrons, more neutrons are produced per average fission and the probability of producing Pu-239 from U-238 is increased. In the liquid-metal fast-breeder reactor (LMFBR) liquid sodium is used as the core coolant, and fast neutrons are employed for breeding and fissioning.

a) Why isn't water used as the core coolant?

b) As time goes by, the Na will become contaminated with Mg. Why?

c) An Na atom is not an especially good moderator. Why is that desirable?

d) Explain the words abbreviated as LMFBR.

33.26 Should a power reactor ever be a) supercritical? b) critical? c) subcritical?

33.27 a) How would you "scram" a reactor (shut it down as fast as possible)?

b) What would happen to a reactor if it lost its coolant?

c) If a LOCA is a loss of coolant accident and the ECCS is the emergency core cooling system, answer the following jargon-filled question. If the reactor is scrammed in response to a LOCA, why is an ECCS still needed to avoid meltdown?

33.28 a) Verify that 4.0 MeV is the Q value of the fusion reaction $^2_1H + ^2_1H \rightarrow ^3_1H + ^1_1H$.

b) If each of the deuterons had 0.25 MeV of KE before a head-on collision, how much KE do the triton (H-3 nucleus) and the proton each have after the reaction?

33.29 a) Verify that 17.6 MeV is the Q value of the fusion reaction $^2_1H + ^3_1H \rightarrow ^4_2He + ^1_0n$.

b) Why would the Coulomb barrier be lower for this reaction than for the reaction involving two deuterons?

c) If the total kinetic energy in the center of mass system is 400 keV before the reaction, how much KE do the alpha particle and the neutron each have after the reaction in the center of mass system?

33.30 A plasma is often called the fourth state of matter. Why does a plasma behave differently than a gas?

33.31 About 5 percent of the energy in fission reactions is lost in antineutrino production. Is this type of loss process present in fusion reactions?

33.32 Each fission reaction releases more than 10 times the energy of a fusion reaction. a) Why then are fusion reactions considered more powerful? b) What are the advantages of fusion reactions as an energy source?

33.33 Explain why the simple magnetic field represented in Fig. 33.13 "reflects" the charged particle.

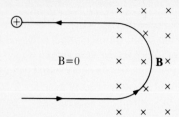

FIGURE 33.13

33.34 Explain the two main classes of plasma confinement now being used. Why are they needed? Why is a plasma needed?

33.35 What kind of plasma confinement is used in our sun? Could it be used on earth?

33.36 Would $^6_3\text{Li} + ^6_3\text{Li} \rightarrow ^{12}_6\text{C}$ be a fusion reaction? Why would a plasma of deuterons require a lower temperature for fusion than a plasma of Li-6?

33.37 How would you get electric power from the fusion reaction of deuterium (H-2) and tritium (H-3)?

33.38 What force opposes the establishment of a denser plasma?

33.39 Compare the radioactive waste problems of fission with those of fusion.

33.40 Can you envision some kind of a combination fission–fusion reactor using deuterium and U-238? Describe it, including the reactions.

33.41 A mean muon has a lifetime of 2.2 μs. Suppose that after production, a mean muon reaches the reaction chamber in 0.2 μs. There it is to catalyze 200 fusions. On average, how much time does a mean muon have to catalyze each fusion?

33.42 a) First the good news: Many negative particles in nature have much more rest mass than a muon. How would this larger mass increase the probability of a cold fusion reaction?

b) Now the bad news: These particles have much, much shorter mean lifetimes than does a muon. What effect would these very short lifetimes have on cold fusion?

CHAPTER

Detectors and Accelerators

If you get caught in a rainstorm, you know it immediately. You can see the drops and feel them hitting you. But you can't ordinarily see or feel individual alphas, betas, gammas or x-rays. How can you detect them? In the first part of this chapter we'll discuss some of the more basic types of detectors of radioactivity. Much of the progress in nuclear physics has been gained through the data from these detectors in conjunction with controlled sources of radiation from particle accelerators. In the last part of the chapter, we'll describe various types of particle accelerators, explaining why some of today's accelerators have ten thousand times the radius of the first cyclotron and why some physicists want a machine with a circumference of more than 80 kilometers.

34.1
DETECTORS OF RADIOACTIVITY

Many different types of devices are used to detect charged particles and uncharged particles (including photons). Modern detectors now rely heavily on electronics. In many types of these devices the detection process begins when the particle first creates directly or indirectly some kind of an electrical signal. This electrical signal is then detected and may tell the energy of the particle, as well as indicating its presence. Other types of detectors utilize a track created by the particle to gain information.

Scintillation detectors use the light emitted when particles strike a sensitive material called a **phosphor.** The electrons of the phosphor atoms are knocked into excited states by the collision. As they drop back toward their ground states, they emit photons, as shown in Fig. 34.1. You're familiar with this effect when the particles are electrons and the phosphor coats the screen of a cathode-ray tube in a television set, an oscilloscope, or a computer terminal. In fluorescent lights, the incident particles include ultraviolet photons.

A **photomultiplier tube** is used to increase the sensitivity of a scintillation detector well past that of the human eye. In a photomultiplier tube, as Fig. 34.2 shows, an incident photon is absorbed at the cathode, causing the emission of a photoelectron. This photoelectron is then accelerated through a potential difference of about 100 V to strike the first dynode. At the first dynode, the accelerated photoelectron knocks out several electrons. Each of these electrons is then accelerated through about 100 V to strike a second dynode. Each again knocks out several more electrons and the process continues. With ten dynodes, a greatly increased pulse of current is produced.

This current pulse is nearly proportional to the energy of the particle detected. Therefore the particle energy can be determined with proper calibration. Other advantages of the scintillation detector include the ability to detect at high counting rates and a high efficiency. If the light from the scintillator is adequate, photodiodes (see Section 27.2) may be used in place of the photomultiplier tube.

FIGURE 34.1
The scintillation process.

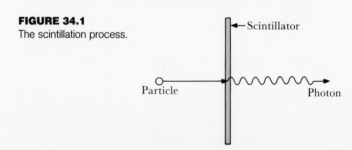

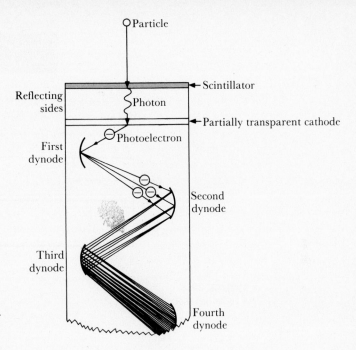

FIGURE 34.2
Part of a photomultiplier
tube.

EXAMPLE 34.1   What two methods of removing electrons from a material are used in the photomultiplier tube?

Solution The photoelectric effect at the cathode and secondary emission at the dynodes.   ●

Gas-filled detectors include ionization chambers, proportional counters, and Geiger–Müller (G–M) counters. In these devices, charged particles or high energy photons pass through a thin window into a gas-filled chamber. The particles or photons then knock electrons off gas atoms, creating ion pairs made up of a positive ion and a negative electron. Typically, one ion pair is created for each 30 or so eV of energy deposited in the chamber. The actual average ionization energy depends on the gas in the chamber.

The electrons of the ion pairs are then attracted to a positive electrode and the ions are attracted to a negative electrode. This motion of charges gives a current pulse through a resistor, R, external to the detector, as shown in Fig. 34.3. The resulting voltage pulse across R is detected by the electronic circuit.

In the low-voltage region 1 in Fig. 34.4, many of the ion pairs simply recombine, decreasing the charge collected. In region 2 (typically less than 300 V), the electric field is sufficient to prevent recombination of the ion

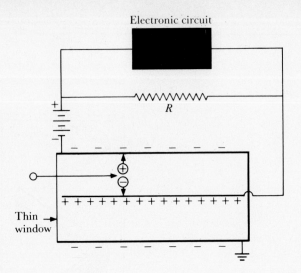

FIGURE 34.3
A basic gas-filled detector.

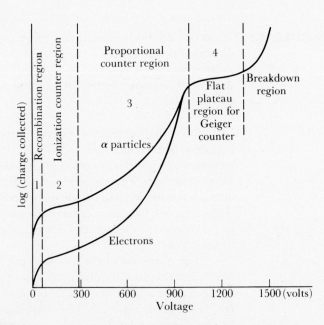

FIGURE 34.4
The charge collected at the electrodes (on a log scale) for alpha particles and electrons as a function of the voltage applied to a gas-filled detector.

pairs. Recall that one ion pair is created on the average for every 30 or so eV of energy deposited in the chamber. Therefore the charge collected in an **ionization chamber** is a measure of the energy deposited in it.

EXAMPLE 34.2 An average of 35 eV is required to create an ion pair in air. How much charge will be collected in an air-filled ionization chamber if a 0.60-MeV electron loses all its energy in the chamber and negligible recombination occurs?

Solution Treating the units as algebraic quantities, we get

$$(0.60 \times 10^6 \text{ eV})/(35 \text{ eV/ion pair}) = 1.71 \times 10^4 \text{ ion pairs.}$$

Since negligible recombination occurs, all 1.71×10^4 ion pairs are collected at the electrodes. Each ion and electron has $|q| = e = 1.60 \times 10^{-19}$ C, so

$$(1.71 \times 10^4 \text{ ion pairs})(1.60 \times 10^{-19} \text{ C/ion pair}) = 2.7 \times 10^{-15} \text{ C}$$

is collected. ●

In region 3 (typically between 300 V and 900 V) the stronger electric field will accelerate the electrons of the ion pairs to the energy that enables them to create more ion pairs. Then the electrons from these ion pairs can create even more ion pairs, and so on. This multiplication of ion pairs in region 3 may give up to a million times the charge collected for the same incident particle in region 2.

A gas-filled detector operating in region 3 is called a **proportional counter.** Example 34.2 gave about 10^{-15} C for an ionization chamber, but the charge collected could be more like 10^{-9} C for a proportional counter. The word *proportional* is part of the name because the charge collected is directly proportional to the energy deposited in the detector in much of region 3. However, the voltage applied must be held quite constant. Otherwise, the charge collected will vary widely for the same energy deposited because of the steepness of the log curve of Fig. 34.4.

EXAMPLE 34.3 A 1.33-MeV gamma ray gives 2.04×10^{-9} C of charge collected in a proportional counter. Another gamma ray gives 1.79×10^{-9} C. What is the energy of the second gamma ray?

Solution The charge collected is directly proportional to the energy deposited, so $q_1/q_2 = E_1/E_2$, or

$$E_2 = E_1 \frac{q_2}{q_1} = 1.33 \text{ MeV} \frac{1.79 \times 10^{-9} \text{ C}}{2.04 \times 10^{-9} \text{ C}} = 1.17 \text{ MeV.}$$ ●

The assumptions we made in Example 34.3 are that the potential difference remains constant, that all the gamma-ray energy is deposited in the chamber, and that the counter is being operated in the strictly proportional part of region 3.

At higher voltages (typically in the region of 1000 V), there is a **plateau region** (region 4 in Fig. 34.4). If the chamber is being operated in the plateau region, any ionizing radiation will trigger a large momentary breakdown in the gas that starts near the central electrode (where the electric field is the greatest). The gas-filled detector acts as a **Geiger–Müller (G–M) counter** in this voltage region. For a G–M counter, the charge collected does not depend on the energy of the particle. Therefore this type of counter is used as a relatively simple method of counting photons or the particles of certain types of ionizing radiation.

EXAMPLE 34.4   A Geiger–Müller counter doesn't require a well-regulated power supply. Why not?

Solution Figure 34.4 shows that changing V along the plateau region results in little change of the charge collected. The charge collected doesn't give the energy of the particle, anyway.   ●

If the voltage is increased even farther into region 5 in Fig. 34.4, the electric field at the central electrode will exceed the dielectric strength of the gas. Therefore the gas in the tube will break down and conduct even without ionizing radiation. The result is a continuous discharge in the detector.

Semiconductor detectors act much like solar cells. Ionizing radiation creates electron–hole pairs in the semiconductor. If these pairs are sufficiently energetic, they may lose energy in collisions that create other electron–hole pairs. The process continues until the average energy deposited per pair is a few times the width of the energy gap of the semiconductor.

In the junction region, the built-in electric field and any applied field sweep these pairs apart so that the charge collected is proportional to the particle energy deposited. Diffusing Li into the junction increases its width. The Si(Li) and Ge(Li) detectors are lithium-drifted Si and Ge detectors. (They are humorously called "silly" and "jelly" detectors.) Their resolution is much better than that of proportional counters. Better resolution means that they might distinguish two different particle energies that are very close to one another, while the proportional counter would see just one comparatively smeared-out energy. The Si(Li) and Ge(Li) detectors also have a short response time.

An airplane moving through air faster than the speed of sound creates a shock wave of air, which we hear as a "sonic boom." Similarly, a charged particle moving through a transparent medium faster than the speed of light in that medium creates an electromagnetic shock wave in that medium. The energy in this electromagnetic shock wave is called **Cerenkov radiation.** A detector having a photomultiplier tube or other device to detect a particle by means of this radiation is called a **Cerenkov detector.**

EXAMPLE 34.5 The speed of light in water is 2.25×10^8 m/s. Within what speed range will electrons create Cerenkov radiation while moving through water?

Solution The electron, a charged particle, must move faster than the speed of light in the water to create Cerenkov radiation. However, the electron cannot move at or greater than c, the speed of light in a vacuum. Therefore the range of the speed v is 3.00×10^8 m/s $> v > 2.25 \times 10^8$ m/s. ●

34.2
PARTICLE TRACK RECORDERS

The particles of nuclear and subnuclear physics are much too small for us to see, even through the finest microscope. However, we can use a number of different devices to display the tracks of submicroscopic charged particles. These tracks often pass through a known applied magnetic field, **B**. We can then apply $qv_{\perp}B = mv_{\perp}^2/r$ to the measured track, obtaining information about the particle's charge and momentum.

A **nuclear emulsion** is simply thicker and more sensitive than the photographic emulsion used in ordinary camera film. The passage of a charged particle exposes the grains of the emulsion along its path. The emulsion is later developed and examined by microscope. Because of its light weight and simplicity, a nuclear emulsion is often used for high altitude rocket or balloon studies of cosmic rays.

Conceptually, **plastic track detectors** are similar to nuclear emulsions. Some plastics are easily damaged by the passage of energetic charged particles, especially heavy ions. Chemical etches can then be used to enlarge the damaged areas, making them clearly visible. The size of each resulting etch pit depends on the charge and the energy, so stacked sheets of these plastics can be used as track detectors.

The **cloud chamber,** first developed by C. T. R. Wilson in 1911, is basically a container filled with a supersaturated vapor. When a charged particle moves through the vapor, it creates ions along its path. The ions then act as condensation centers for droplets of liquid. We see this trail of droplets in Fig. 34.5, not the particle itself. (This effect is much like seeing a high-flying jet plane's contrail but not being able to see the plane itself.)

FIGURE 34.5
Carl Anderson's 1932 cloud chamber photograph, the first recognized experimental
evidence of the positron's existence. Dirac had predicted its existence the previous year
(as a hole in a sea of otherwise filled electron states). The magnetic field was directed into
the page and the positron entered the chamber from the bottom of the photo at high
speed. The dark band across the center was a 6-mm thick slab of lead that slowed the positron.
Source: Carl Anderson.

Wilson originally began work on the chamber in 1895 to reproduce in his
lab some beautiful effects of sunlight on mist that he'd noticed.

Donald Glaser developed the **bubble chamber** in 1952 for studying
high-energy interactions. It is basically a container filled with a superheated liquid. The liquid is three orders of magnitude denser than the
vapor in the cloud chamber. High-energy particles are thus much more
likely to interact within the chamber. When a charged particle moves
through the superheated liquid, it creates ions along its path. These local
additions of energy cause small volumes of boiling. That is, a track of vapor
bubbles appears in the superheated liquid, as shown in Fig. 34.6. Despite
the well-known — but apocryphal — story, Glaser didn't get the idea for
the device while watching the bubbles rise from the bottom of a glass of
beer. However, he did study the properties of that beverage as a superheated liquid during his development of the bubble chamber.

EXAMPLE 34.6   The bubble chamber has been called the "inverse of the cloud chamber."
Explain why.

Solution In both types of chambers, ionizing particles leave a track through
a metastable medium. In the cloud chamber, the track is of liquid in a
vapor. In a bubble chamber, the track is of vapor in a liquid.   ●

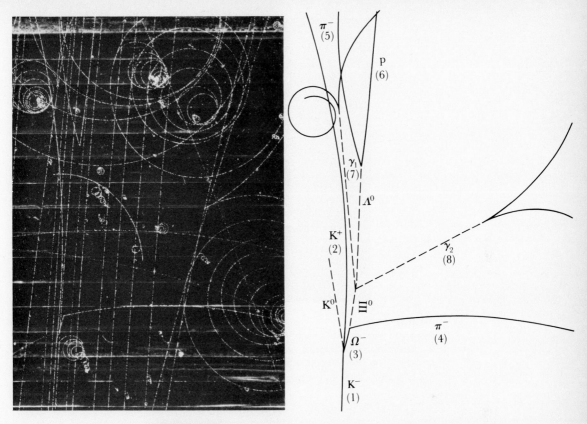

FIGURE 34.6

A bubble chamber photograph, which established the Ω^- particle's existence. An energetic K^- particle (1) entered the chamber from the bottom of the photo and struck a proton at rest in the liquid of the bubble chamber, producing a K^+ particle (2), an invisible neutral K^0 particle, and the Ω^- particle (3), having the properties predicted from quark theory by Gell-Mann. Debris farther down the line include two invisible gamma rays (7 and 8) from the almost immediate decay of an invisible Ξ^0 particle. Both gamma rays disappear with the production of electron–positron pairs.

Source: N. P. Samios, Brookhaven National Laboratory.

In a **spark chamber** and a **streamer chamber** a high electric field is set up in a gas so that the gas is almost ready to break down and conduct. Passing a charged particle through creates ion pairs in the gas. These ion pairs then multiply, creating a spark along the track of the particle. The positions of the sparks can be detected photographically (as in Fig. 34.7) or electronically and used to study the track.

A **drift chamber,** represented in Fig. 34.8, contains an array of closely spaced (a few mm apart) wires and a low-pressure gas. Charged

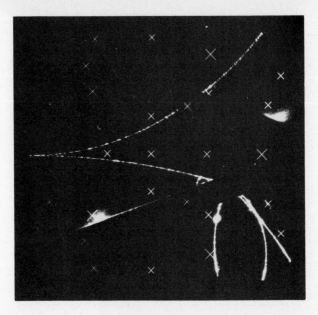

FIGURE 34.7

Streamer chamber photograph: An invisible neutral kaon enters from the left. It decays into a charged muon (curving up), an oppositely charge pion (curving down), and a neutrino (invisible). The pion interacts with a lead plate, and other particles are then produced.

Source: SLAC, funded by the U.S. Department of Energy.

particles ionize the gas and ions drift to the wires, depositing their charge. The charge collected and time information from each wire is then processed on a computer to reconstruct the trajectories.

Besides those we've discussed, there are various other types of particle track recorders. The type used for a particular experiment depends on such variables as energy, charge, time between events, mass, lifetime, and — of course — budget.

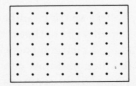

FIGURE 34.8

A cross-section of a cylindrical drift chamber with vertical planes of closely spaced wires. The events studied would typically occur near the central axis of the chamber.

34.3
LINEAR ACCELERATORS

Particle accelerators are used in nuclear physics to obtain the threshold or resonant energy needed for some nuclear reactions and to obtain the high momentum (and corresponding small de Broglie wavelength) needed to "see" small structure. Accelerators are also used for ion implantation for selective doping of semiconductors, for alloying with minute quantities of rare metals, for surveying for hydrocarbons surrounding well shafts, for the production of medical isotopes and for selective radiation, for changing the properties of plastics, for radioactive dating, and for many other purposes. In almost all particle accelerators, the charged particles move in an excellent vacuum, as low as 10^{-13} atmosphere. Otherwise, they would lose energy, be scattered in different directions, or even be absorbed in collisions before reaching their intended target.

Linear accelerators speed up charged particles in a straight line. The acceleration results from the force applied by an electric field. In a **Cockroft–Walton accelerator,** the electric field is provided by a large potential difference. A particle of charge q moving through a potential drop V can gain a kinetic energy:

$$KE = qV. \tag{34.1}$$

The high voltage (up to about 2 MV) may be obtained by charging capacitors with a voltage multiplier circuit.

EXAMPLE 34.7   From their first machine, invented in 1932, Cockroft and Walton obtained 800-keV protons. What was the potential drop?

Solution The kinetic energy, KE, is 800 keV. For a proton, $q = e$. Then $KE/q = V = 800$ keV$/e = 800$ kV.   ●

Using $KE = qV$ is almost trivial in eV units, but recall that $q = +2e$ for alpha particles.

In 1929, Robert J. Van de Graaff invented an electrostatic generator, which is now called the **Van de Graaff accelerator.** It uses a moving belt to carry charge to the inside of a hollow conductor. Recall that all excess charge resides on the surface of an electrostatic conductor (unless that charge is insulated from the surface). Therefore the charge carried to the inside by the belt is rapidly conducted to the outside surface of the conductor. Then the excess charge, external electric field, and potential difference build up and are limited only by the insulating properties of the surroundings. Potential differences of 30 MV are even possible.

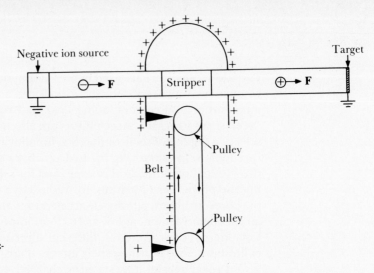

FIGURE 34.9
Schematic drawing of a
tandem Van de Graaff ac-
celerator.

In the ingenious **tandem Van de Graaff accelerator,** ions are first
negatively charged and accelerated toward a positively charged conductor, as shown in Fig. 34.9. Inside the conductor electrons are stripped off
to make the moving ions positive. Acceleration of the ions then continues,
but away from the positively charged conductor.

EXAMPLE 34.8   Describe how 30-MeV protons could be obtained from a tandem Van de
Graaff accelerator operating at 15 MV above ground. Start with H-1
atoms.

Solution First, an extra electron is added to each hydrogen atom, giving H^-
ions. The H^- ions are then accelerated from ground to $+15$ MV. Therefore $+15$ MV is the potential *rise,* so the potential *drop* is -15 MV. For H^-,
$q = -e$. Equation 34.1 then tells us that the H^- ions gain a kinetic energy of
$qV = (-e)(-15 \text{ MV}) = +15$ MeV.

Then both electrons are stripped off each moving H^- ion, leaving
bare protons. The charge is now $+e$. The protons then continue from
$+15$ MV to ground, dropping 15 MV. Therefore they gain an additional
$qV = (+e)(+15 \text{ MV}) = 15$ MeV of kinetic energy for a total *KE* gain of
30 MeV.   ●

A **drift-tube linear accelerator** (or **linac**) is composed of a series of
coaxial conducting cylinders (drift tubes), as illustrated in Fig. 34.10. The
cylinders are alternately connected to an oscillating voltage source. The
electric field inside the conductors will be almost zero, but between
the conductors it will always be in the direction needed to accelerate the

FIGURE 34.10

A drift-tube linear accelerator.

Source: A. Arya, *Elementary Modern Physics.* © 1974, Addison-Wesley Publishing Co., Inc., Reading, Massachusetts. P. 445, Fig. 13.20. Reprinted with permission.

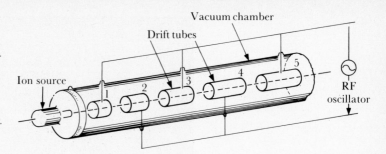

charged particle. For example, suppose that a proton leaves the ion source. Drift tube 1 is made negative, attracting the proton into the cylinder. While the proton moves (drifts) through the tube, the charge on 1 is made positive and drift tube 2 becomes negative, so that the proton is further accelerated between 1 and 2. Then 2 is made positive and 3 negative, again accelerating the proton as it passes between 2 and 3. This process continues for the length of the linac. The drift tubes are designed with increasing lengths to compensate for the speed increase, but their lengths soon approach a constant value as v approaches c.

Low-mass charged particles, such as electrons and positrons, are usually accelerated to high energies in a **traveling-wave linear accelerator.** This type of linear accelerator is basically a long, hollow, evacuated conductor, or waveguide, designed so that electromagnetic waves traveling down the tube have an electric-field component parallel to its axis. This electric field accelerates the charged particles. The waveguide is also designed so that the speed of the electromagnetic wave (which is less than the free vacuum value of c) matches the increasing speed of the particles. The Stanford Linear Accelerator Center (SLAC) has a traveling-wave linear accelerator over three kilometers (two miles) long, as shown in Fig. 34.11. Essentially a 2856 MHz waveguide, it accelerates electrons and positrons to an energy that has exceeded 50 GeV. Like surfboarders riding a wave, the electrons in this accelerator ride an electromagnetic wave down the waveguide.

EXAMPLE 34.9 Can electrons and positrons be accelerated in the same bunch in a linear accelerator? (*Hint:* $\mathbf{F} = q\mathbf{E}$.)

Solution Even if you could figure out a way to keep them from annihilating one another, the answer is still "no." In all types of accelerators, an electric field provides the accelerating force. The relation $\mathbf{F} = q\mathbf{E}$ tells you that the positive positrons are accelerated in the direction of the electric field but that the negative electrons are accelerated in the opposite direction. ●

FIGURE 34.11
The two-mile-long linear accelerator at the Stanford Linear Accelerator Center (SLAC).
Alternate bunches of positrons and electrons are accelerated to about 50 GeV, then
separated into the dashed arcs to provide head-on e^+e^- collisions. The particles radiate out
about one GeV in the half-turn of the collider arcs. The electron and positron beams must
have cross-sectional radii of about 1 μm to provide a sufficient number of reactions.

Source: SLAC, funded by the U.S. Department of Energy.

In Example 34.9, the phrase "in the same bunch" provides the main
restriction. In drift-tube and traveling-wave linear accelerators, the electric field reverses direction every half cycle. Therefore a bunch of positrons can be accelerated and then, a half cycle later, a bunch of electrons
can follow.

34.4
CYCLIC ACCELERATORS

In **cyclic accelerators,** magnetic fields bend charged particles around
closed paths and electric fields speed them up. When a charged particle of
mass m and charge q moves in a circular path of radius r perpendicular to a
magnetic field **B**, $\mathbf{F} = m\mathbf{a}$ becomes $qvB = mv^2/r$. Solving for the momen-

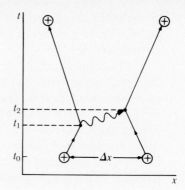

FIGURE 35.1

The repulsion of like charges. Two protons are moving toward one another (Δx decreases as t increases). At t_1 one proton emits a virtual photon and recoils. At t_2 the other proton absorbs the virtual photon and also recoils. (This is an example of a *Feynman diagram*.)

disappear a short time later without being directly observed. Conservation of momentum and energy may even be temporarily violated in the production of a virtual particle. Its existence is allowed by the Heisenberg uncertainty principle. In the form

$$\Delta E \, \Delta t \geq h, \qquad (35.1)$$

this principle says that in order for us to measure the energy of a state to within a spread ΔE, then that state must exist for a time Δt which is $h/\Delta E$ or *longer*. A virtual particle is created by an energy change in a system of ΔE, where ΔE is the energy of the virtual particle. We can't directly observe the exchange of the virtual particle because it exists for a time *shorter* than $\Delta t = h/\Delta E$.

This phenomenon has always reminded me of a shifty bookkeeper, who can "borrow" money from his company so long as he keeps it for a short enough time that the auditor doesn't find it missing during her audits. Although the "borrowed" money can't be directly observed in the company books, its effects on the bookkeeper can be. If caught, he may be repelled from the company and attracted to jail.

For example, if the strong nuclear force is to act over a spread in distance of 6 fm by the exchange of a virtual particle moving at an appreciable fraction of the speed of light, the particle could be exchanged in a time, Δt, of $(6 \times 10^{-15}$ m$)/(3 \times 10^8$ m/s$) = 2 \times 10^{-23}$ s, or so. The energy fluctuation, ΔE, could therefore be as large as

$$\frac{h}{\Delta t} \approx \frac{4 \times 10^{-15} \text{ eV} \cdot \text{s}}{2 \times 10^{-23} \text{ s}} = 2 \times 10^8 \text{ eV}.$$

Giving most of this amount to the rest energy of the virtual particle would give a rest energy over 100 MeV. Despite his geographical isolation in Japan, Hideki Yukawa in 1935 predicted the rest energy and spin of this virtual particle.

Physicists generally agree that most of these virtual particles have their real counterparts in nature. Indeed, there is a subfamily of particles called **mesons** with rest energies greater than 100 MeV. The first mesons were discovered in 1947. The part of the strong nuclear force having the longest range was believed to result from the exchange of mesons between nucleons. However, more recent theories and experiments suggest that both the nucleons and the mesons are made up of even more elementary particles that we call **quarks.** The fundamental virtual particle for the strong force between quarks evidently is something that we call a **gluon.**

We explain the covalent bonding of molecules by the exchange of electrons between atoms. However, the more basic force that binds atoms together is the electrostatic attraction of positive and negative charges. In a somewhat similar manner, we can explain many aspects of nuclear binding by the exchange of virtual mesons. However, the more basic force evidently is the strong nuclear force between quarks.

Long-range forces, such as the electromagnetic force, require that Δt be much larger and that ΔE be much smaller than for short-range forces. In fact, quantum field theory holds that long-range forces are due to zero rest-mass particles. The electromagnetic force is caused by the exchange of virtual photons, whereas the gravitational force is caused by the exchange of (undiscovered) virtual **gravitons.**

The weak nuclear force (responsible for beta decay and other reactions involving neutrinos) is a short-range force caused by the exchange of massive virtual particles called W^+, W^-, and Z^0 particles. The $W^\pm$ and Z^0 particles were discovered in 1983 at CERN. (See Fig. 35.2.) In this discovery, evidence was first obtained for four W^- particles and one W^+ particle from the examination of 10^9 $p\bar{p}$ collisions.

If we assign the strong nuclear force between protons a strength of 1 at typical nuclear separations, Table 35.1 then gives the strengths of the electromagnetic force, the weak nuclear force, and the gravitational force at that distance.

EXAMPLE 35.2   Estimate roughly the range of the weak nuclear force if the $W^\pm$ and Z^0 particles have a rest energy of about 90 GeV.

Solution The solution is the opposite of the estimate we made of the strong force earlier in this section. We begin with a rest energy of about 90 GeV. Let's add several more GeV for kinetic energy to give a total energy of about 100 GeV, so that

$$\Delta E = 100 \text{ GeV} = 100 \times 10^9 \text{ eV}.$$

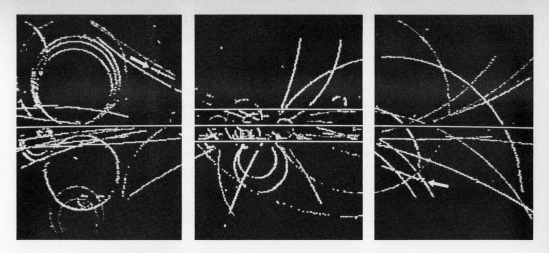

FIGURE 35.2

Computer reconstruction of the tracks of particles resulting from a May 1983 proton–antiproton collision. The tracks of the electron and positron resulting from the decay of a Z^0 particle are indicated by the arrows.

Source: CERN, the European Laboratory for Particle Physics.

Then $\Delta E\, \Delta t = h$ gives

$$\Delta t = h/\Delta E = \frac{4 \times 10^{-15}\ \text{eV} \cdot \text{s}}{100 \times 10^9\ \text{eV}} = 4 \times 10^{-26}\ \text{s}.$$

With v being somewhat less than 3×10^8 m/s, the range is roughly $(3 \times 10^8\ \text{m/s})(4 \times 10^{-26}\ \text{s}) = 10^{-17}$ m (or only a hundredth of a femtometer). ●

EXAMPLE 35.3   Explain why the gravitational force exerted on us by the entire earth is so strong if the gravitational force shown in Table 35.1 is so weak.

TABLE 35.1

The Four Forces of Nature

Force	Relative strength	Virtual particles exchanged			
		Name	Rest mass	Charge	Spin
Strong nuclear	1	Gluon	0	0	1
Electromagnetic	10^{-2}	Photon	0	0	1
Weak nuclear	10^{-13}	$W^{\pm}$, Z^0	82, 93 GeV/c^2	$\pm e$, 0	1
Gravitational	10^{-38}	Graviton	0	0	2

Solution The strong and weak nuclear forces are extremely short-range forces. Therefore we are much too far from any of the earth's matter to be attracted or repelled by the two nuclear forces. For electromagnetic forces, virtually every positive charge in the earth is balanced by a negative charge and the currents caused by the net motion of charge are small. Thus the potentially very strong electromagnetic field is canceled to almost zero. The net charge and current of our own bodies also are almost zero, so that the total electric and magnetic fields of the earth exert little force on us.

However, the force of gravitation exerted on us by the particles of the earth has no opposing sources in the earth to cancel it out, and it is a long-range force. Therefore we feel the vector sum of the gravitational forces between us and every bit of mass in the entire planet. This sum gives a considerable force. In fact, many people find this force to be so large that they diet to reduce it.   ●

35.2
UNIFIED THEORIES

One of the main goals of Einstein's later years was to begin to derive a **unified field theory,** that is, a single theory encompassing the four forces of nature. He attempted to unify the electromagnetic and gravitational forces. He didn't succeed, and neither has anyone else to date.

The first unification of forces is generally attributed to Maxwell, who unified the electric and magnetic forces through what we call Maxwell's equations. The next successful unification was developed during the 1960s by Steven Weinberg, Abdus Salam, and Sheldon Glashow. Called the **electroweak theory,** it ties together the weak nuclear and electromagnetic forces.

You may have trouble imagining what the weak nuclear force and the electromagnetic force have in common, but that probably is because of the regions in which they are ordinarily encountered. Similarly, for electromagnetism, if you consider only charges at rest and stationary permanent magnets, the electrical forces between the charges and the magnetic forces between the magnets might seem completely unrelated. But if you move the charges to give currents and changing electric fields, you would find that magnetic fields and forces are produced. Or if you move the magnets to give changing magnetic fields, you would find that electric fields and forces are produced. That is, you can see more unification as you increase the kinetic energy from zero.

Also, if you were to examine the permanent magnets on a small enough scale — at the atomic level — you would see that their magnetism results from the motions of electrical charges. That is, you see more unification as you move toward smaller scale to about 10^{-10} m. A similar situation occurs with respect to the electroweak theory, except that the unification becomes readily apparent only at energies above 10^3 GeV or distances of less than 10^{-18} m.

The electroweak theory and other modern unified theories are based heavily on symmetry principles. (Symmetry principles have applications in most areas of physics, as in the study of crystalline solids.) Most physicists believe that the next step is to find the symmetry principles that will unify the weak nuclear, the electromagnetic, and the strong nuclear forces. These theories are called **grand unified theories** (GUTs). The electroweak theory is considered successful because it explains known experimental results and correctly predicts the masses and other properties of particles such as the W and Z. The GUTs predict that a proton has a finite probability of decaying, but with an average lifetime of more than 10^{31} years. These theories also predict the existence of magnetic monopoles (isolated north or south poles), which should be very massive particles. And the GUTs predict that the neutrinos should have a small rest mass. Experiments are underway to attempt to verify all these predictions. However, performing these experiments is very difficult. Laboratory limitations on energies and distances don't allow us to observe easily or accurately the symmetries and therefore the effects of the GUTs.

The energies and distances at which unification of the weak, electromagnetic, and strong forces occurs are more than 10^{15} GeV and less than 10^{-31} m. These conditions apply in the Big Bang model of the creation of the universe. The GUTs are being used in an effort to understand why there seems to be so much more matter than antimatter in the universe, as well as other questions.

The next step in unified field theories would be to include the fourth force, gravity, in a single theory. This may well involve more revolutionary concepts. Some physicists are suggesting that other forces exist in nature besides the four we conventionally list and/or that other dimensions exist than the four (x, y, z, t) we can easily detect. For instance, **superstring theories** describe the universe in terms of several, maybe ten, dimensions. In superstring theories, elementary particles aren't pointlike, but one-dimensional strings only about 10^{-35} m long. There are certainly more questions than answers in the whole area of unified field theories.

EXAMPLE 35.4   At greater distances or lower energies than those given in the preceding discussion "symmetry breaking" occurs for the electroweak theory and for the GUTs. What could the phrase "symmetry breaking" mean?

Solution   The unified theories are based on symmetry principles, such that unification based on these symmetries is readily apparent only below the distances or above the energies given. Therefore at greater distances or lower energies, the symmetry isn't easy to detect. The symmetry appears to be lost, or broken, as the unified force breaks into two or three distinct forces.   ●

35.3
FAMILIES OF PARTICLES

The hundreds of particles in the subnuclear zoo do have some similarities. These similarities allow us to group them into three main families. The first family includes the **gauge bosons,** which are responsible for the electroweak force. The gauge bosons appear to be truly fundamental, that is, not to be composed of any even smaller particles. The first member of the family is the photon. The recently discovered W^+, W^-, and Z^0 complete the gauge boson family. Other predicted, related bosons include the graviton ($q = 0$ and $s = 2$), the gluons ($q = 0$ and $s = 1$), and the Higgs particle or particles ($q = 0$, or 0 and $\pm e$, and $s = 0$).

The next family of particles is yet to be completely discovered. Its members aren't affected by the strong nuclear force and appear to be truly fundamental. They are some of the lightest particles and so are called **leptons,** from the Greek word for light (mass). The leptons are the electron and its neutrino, the muon and its neutrino, and the tau particle and its neutrino.

The largest family, with well over 100 members, is called the **hadron family.** All hadrons interact by the strong nuclear, electroweak, and gravitational forces. For most of the hadrons, we give a number $+1$ to particles and a -1 to antiparticles. This number evidently has been conserved in all nuclear processes observed so far. (But possibly this number was not conserved close to the big bang, and it is not completely conserved if protons decay.) The conserved number is called the **baryon number,** because those hadrons that follow this conservation law are called **baryons.** For instance, in the reaction $\gamma \rightarrow p + \bar{p}$ (pair production of a proton and an antiproton), the baryon number before the reaction is zero (a photon is not a baryon, it's in the gauge boson family). The total baryon number after the reaction is $+1 + (-1)$, which also equals zero. With zero before and zero after, the total baryon number has been conserved. Those hadrons having no baryon number are the mesons. **Lepton numbers** also are conserved. We assign $+1$ to particles and -1 to antiparticles in the lepton family. Not only is the total lepton number conserved, but evidently the lepton numbers of each of the three particle–neutrino pairs in the lepton family also are conserved. This conservation explains why radioactive beta emission always involves either the electron–electron antineutrino pair or the antielectron (positron)–electron neutrino pair.

EXAMPLE 35.5 Show that conservation of both baryon and lepton numbers occurs in the radioactive decay of a free neutron, $n \rightarrow p + e^- + \bar{\nu}_e$.

Solution The baryon number for both the neutron and the proton is $+1$ (they are both baryon particles, not baryon antiparticles). The baryon number for the electron and the electron antineutrino is zero (they are not baryons). Therefore $1 = 1 + 0 + 0$, or $1 = 1$, and the total baryon number is conserved.

Lepton numbers for both the neutron and the proton are zero (they are not leptons). The electron–electron neutrino pair lepton number is $+1$ for the electron (a lepton particle) and is -1 for the electron antineutrino (a lepton antiparticle). Therefore $0 = 0 + 1 + (-1)$, or $0 = 0$, and the total lepton number is also conserved. ●

Another quantum number that can be assigned to these particles is called **strangeness.** A number of particles were found experimentally to be created only in certain combinations in some reactions, not to be created in other reactions where they were expected, and to have lifetimes much longer than expected because certain decay modes seemed to be forbidden. These were strange and unexpected results, so these particles were called strange particles. M. Gell-Mann, T. Nakano, and K. Nishijina found that they could explain these phenomena if strangeness quantum numbers were assigned and the total strangeness was conserved in strong nuclear reactions. In weak nuclear reactions, the change in the strangeness could be zero or ± 1.

EXAMPLE 35.6 The reaction $\Omega^- \rightarrow \Xi^- + \pi^0$ is found to occur in 8 percent of all Ω^- decays. Show that this reaction conserves baryon number, and that it is forbidden as a strong nuclear reaction but is allowed as a weak nuclear reaction. The baryon number and the strangeness of each particle are $(1, -3)$, $(1, -2)$, and $(0, 0)$, respectively.

Solution First, let's consider the baryon numbers: $1 = 1 + 0$. Therefore the baryon number is conserved (1 before and 1 after). Second, let's consider the strangeness: $-3 \neq -2 + 0$. Therefore strangeness is not conserved, which means that this reaction is forbidden as a strong nuclear reaction. (Strong nuclear reactions require conservation of strangeness.) However, the reaction is allowed as a weak nuclear reaction because the strangeness does change by $+1$. (Weak nuclear reactions allow a change in strangeness of -1, 0, or $+1$.) ●

In Table 35.2, we summarize some properties of a few small particles. The inside headings give the family names, with the meson and baryon subfamilies of the hadron family in parentheses. The first column

TABLE 35.2
Some Properties of Selected Subnuclear Particles

Name	Symbol	Charge	E_0 (MeV)	Mean life
				Gauge bosons
Photon	γ	0	0	Stable
W	W^+	$+e$	82,000	$>10^{-25}$ s
Z	Z^0	0	93,000	$>10^{-25}$ s
				Leptons
Electron	e^-	$-e$	0.511	Stable
Neutrino$_e$	ν_e	0	0	Stable
Muon	μ^-	$-e$	105.66	2.197 μs
Neutrino$_\mu$	ν_μ	0	<0.25	Stable
Tau	τ^-	$-e$	1784	0.3 ps
Neutrino$_\tau$	ν_τ	0	<70	Stable
				Hadrons (mesons)
Pions	π^+	$+e$	139.57	26.0 ns
	π^0	0	134.96	87 as
Kaons	K^+	$+e$	493.67	12.4 ns
	K^0_S	0	497.7	89.2 ps
	K^0_L	0	497.7	51.8 ns
Eta	η^0	0	549	0.6 as
Charmed D's	D^+	$+e$	1869	0.9 ps
	D^0	0	1865	0.4 ps
	D^+_S	$+e$	1970	0.3 ps
J/Psi	J/ψ	0	3097	0.01 as
Bottom B's	B^+	$+e$	5271	1.4 ps
	B^0	0	5275	1.4 ps
Upsilon	Υ	0	9460	0.02 as
				Hadrons (baryons)
Proton	p	$+e$	938.280	Stable
Neutron	n	0	939.573	898 s
Lambda	Λ^0	0	1115.60	263 ps
Sigmas	Σ^+	$+e$	1189.37	80.0 ps
	Σ^0	0	1192.46	0.06 as
	Σ^-	$-e$	1197.34	148 ps
Delta Resonances	Δ	0, $\pm e$, and $+2e$	1232	6×10^{-24} s
Xi's	Ξ^0	0	1314.9	0.29 ns
	Ξ^-	$-e$	1321.3	0.164 ns
Omega	Ω^-	$-e$	1672.5	82 ps
Charmed lambda	Λ^+_C	$+e$	2281	0.2 ps

Spin	Strangeness	Some main decay modes
1	0	Stable
1	0	$e^+\nu_e$, $\mu^+\nu_\mu$
1	0	e^+e^-, $\mu^+\mu^-$
$\frac{1}{2}$	0	Stable
$\frac{1}{2}$	0	Stable
$\frac{1}{2}$	0	$e^-\bar{\nu}_e\nu_\mu$
$\frac{1}{2}$	0	Stable
$\frac{1}{2}$	0	$\mu^-\bar{\nu}_\mu\nu_\tau$, $e^-\bar{\nu}_e\nu_\tau$, $\rho^-\nu_\tau$
$\frac{1}{2}$	0	Stable
0	0	$\mu^+\nu_\mu$
0	0	$\gamma\gamma$, γe^+e^-
0	+1	$\mu^+\nu_\mu$, $\pi^+\pi^0$
0	+1	$\pi^+\pi^-$, $\pi^0\pi^0$
0	+1	$\pi^0\pi^0\pi^0$, $\pi^\pm e^\mp\bar{\nu}_e$ or ν_e, $\pi^\pm\mu^\mp\bar{\nu}_\mu$ or ν_μ
0	0	$\gamma\gamma$, $\pi^0\pi^0\pi^0$, $\pi^+\pi^-\pi^0$
0	0	$K^0\pi^+\pi^+\pi^-$, $\overline{K}^0\pi^+\pi^0$
0	0	$K^-\pi^+\pi^0$, $K^-\pi^+\pi^+\pi^-$
0	+1	$\phi\pi^+$, $\phi\pi^+\pi^+\pi^-$
1	0	e^+e^-, $\mu^+\mu^-$
0	0	$\overline{D}^0\pi^+$, $D^{*-}\pi^+\pi^-$
0	0	$\overline{D}^0\pi^+\pi^-$, $D^{*-}\rho^+$
1	0	$\tau^+\tau^-$, $\mu^+\mu^-$, e^+e^-
$\frac{1}{2}$	0	Stable
$\frac{1}{2}$	0	$pe^-\bar{\nu}_e$
$\frac{1}{2}$	−1	$p\pi^-$, $n\pi^0$
$\frac{1}{2}$	−1	$p\pi^0$, $n\pi^+$
$\frac{1}{2}$	−1	$\Lambda^0\gamma$
$\frac{1}{2}$	−1	$n\pi^-$
$\frac{3}{2}$	0	$N\pi$ (N = nucleon)
$\frac{1}{2}$	−2	$\Lambda^0\pi^0$
$\frac{1}{2}$	−2	$\Lambda^0\pi^-$
$\frac{3}{2}$	−3	$\Lambda^0 K^-$, $\Xi^0\pi^-$, $\Xi^-\pi^0$
$\frac{1}{2}$	0	$pK^-\pi^+$, $p\overline{K}^0$

shows the name of the particle, the second column the symbol of the particle, and the third column the charge of the particle in terms of e (1.60×10^{-19} C).

The fourth column lists the rest energy in MeV and the fifth column the mean lifetime, $\overline{T}$. Two columns of quantum numbers follow: The sixth gives the spin angular momentum quantum number and the seventh the strangeness quantum number, labeled "spin" and "strangeness" for short. The last column gives some main decay modes in concise notation. For example, $\pi^{\pm}e^{\mp}\overline{\nu}_e$ or ν_e for K_L^0 refers to the decays

$$K_L^0 \longrightarrow \pi^+ + e^- + \overline{\nu}_e \quad \text{and} \quad \overline{K}_L^0 \longrightarrow \pi^- + e^+ + \nu_e.$$

Every particle listed in the table except the photon, the Z^0, the neutral pion, the eta, the J/psi, and the upsilon has an antiparticle. For them, we don't know a method to distinguish particle from antiparticle, so we say that each particle is its own antiparticle. Because the photon is its own antiparticle, we can't determine by examining their light whether distant galaxies are made of matter or antimatter. The photons from an atom made of protons, neutrons, and electrons are indistinguishable from those from antiatoms made of antiprotons, antineutrons, and positrons. (Other evidence indicates antimatter galaxies do not exist; the universe seems to contain much more matter than antimatter.)

We identify an antiparticle as before by placing a bar over the symbol for the particle. Thus, $\overline{K}^+$ is the antiparticle of K^+. The rest energy, mean lifetime, and spin angular momentum quantum number are the same for particle and antiparticle, but the charge and strangeness have opposite signs. (Therefore $\overline{K}^+$ is also K^-.) To obtain the antiparticle decay modes from the particle decay modes listed in Table 35.2, every particle is replaced by its antiparticle (if any) and vice versa.

EXAMPLE 35.7 What are the charge, strangeness, and main decay modes of negative kaons?

Solution A negative kaon is a K^- or $\overline{K}^+$, the antiparticle of K^+. Its charge and strangeness have signs opposite from those of K^+, and therefore are $-e$ and -1, respectively, from Table 35.2. The decay mode $K^+ \rightarrow \mu^+ + \nu_\mu$ becomes $\overline{K}^+ \rightarrow \mu^- + \overline{\nu}_\mu$ (with μ^+ being the antiparticle of μ^-), or $K^- \rightarrow \mu^- + \overline{\nu}_\mu$. The decay mode $K^+ \rightarrow \pi^+ + \pi^0$ becomes $\overline{K}^+ \rightarrow \overline{\pi}^+ + \overline{\pi}^0$, or $K^- \rightarrow \pi^- + \pi^0$. ●

One of the oddest particles in the table is K^0. The wave function describing this particle can be written as the sum of wave functions of two different particles. The particles have masses that differ by less than 1 part in 10^{14} but have mean lifetimes, because of conservation principles involving charge and parity, that are much different. As a result, one-half of all K^0 decay in a short time (K_S^0 has $\overline{T} = 89 \times 10^{-12}$ s), and the other one-half

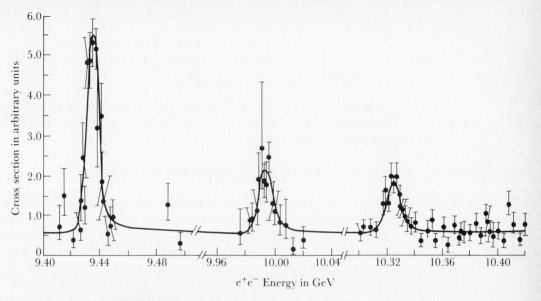

FIGURE 35.3

Resonances used to determine the rest mass and approximate lifetime of three members of the upsilon family.

Source: T. Böhringer, et al., *Phys. Rev. Letters* **44**, 1111 (1980). By permission of the American Physical Society and the CUSB collaboration.

of all K^0 decay in a longer time (K_L^0 has $\overline{T} = 52 \times 10^{-9}$ s, or 580 times longer).

Some particles exist for such a short time that they may be properly called **resonances**, rather than particles. Many of these resonances are considered to be excited states of particles or other resonances that have lower mass. Even when it moves at nearly the speed of light, the delta resonance (Table 35.2) will travel a mean distance of only about the diameter of a nucleus. (If $v = 0.99c$, $\gamma = 7.1$, giving a dilated mean lifetime of $\gamma \overline{T}_0 = 7.1(6 \times 10^{-24}$ s$) = 4 \times 10^{-23}$ s. Speed times lifetime then is $0.99(3 \times 10^8$ m/s$)(4 \times 10^{-23}$ s$) = 12$ fm.)

No detector can measure a track only 12 fm long, so we can infer the existence of such a resonance from a peak in the graph of the measured reaction cross section versus the energy of the outgoing particles, as in Fig. 35.3. The energy position of the peak yields the rest mass of the resonance. If the detector has sufficient resolution, the width of the peak combined with the uncertainty principle gives the mean lifetime.

EXAMPLE 35.8 Consider the possibility of the reaction $\pi^- + p \rightarrow \Lambda^0 + K^0$ from the standpoints of conservation of mass–energy, charge, and spin angular momentum. The proton is initially at rest.

Solution We use the data in Table 35.2. For mass–energy,

$$Q = (\text{Total rest mass before} - \text{Total rest mass after})c^2$$
$$= (139.57 + 938.28) \text{ MeV} - (1115.6 + 497.7)\text{MeV}$$
$$= (1077.85 - 1613.3)\text{MeV} = -535.45 \text{ MeV}.$$

This endoergic reaction will require a KE_{th}. Substituting the total mass, $(1077.85 + 1613.3)\text{MeV}/c^2 = 2691.15 \text{ MeV}/c^2$, and the mass of the target proton, $938.28 \text{ MeV}/c^2$ into Eq. (34.5), we obtain

$$KE_{th} = -\frac{2691.15}{2(938.28)}(-535.45 \text{ MeV}) = 768 \text{ MeV}.$$

(A π^- with almost 0.77 GeV of kinetic energy is required to initiate the reaction if the proton is at rest.) For charge, $(-e) + (+e) = 0 + 0$, so $0 = 0$ in the reaction, and charge is conserved. For spin angular momentum, the particles' spin angular momentum quantum numbers give $0 + \frac{1}{2} = 0 + \frac{1}{2}$, so spin angular momentum can be conserved.   ●

You can also tell from Table 35.2 that the baryon number, lepton number, and strangeness are conserved in the reaction in Example 35.8. Many particles/resonances are not included in Table 35.2, a large fraction of which are excited states of the listed particles. Current theories predict even more particles: The Higgs particle or particles are predicted as an explanation of spontaneous symmetry breaking. **Supersymmetry,** which is part of the superstring theories, gives every boson a fermion superpartner and every fermion a boson superpartner. This prediction of supersymmetry has led for searches for sleptons, photinos, squarks bound together by gluinos, and even winos.

35.4
QUARKS

In 1963, Gell-Mann and Zweig independently attempted to bring some order to the particle zoo by postulating the existence of truly fundamental particles, which we now know as quarks. The gauge bosons and leptons appear to be elementary particles, but the hadrons have structure and appear to be made up of quarks. Originally three quarks (each with its antiquark) were postulated. (Quarks were named by Gell-Mann from the call of a bartender in James Joyce's *Finnegans Wake:* "three quarks for Muster Mark.") The number of quarks now stands at six.

Quark physics is replete with whimsical terminology. The six quarks are named **up, down, strange, charmed, top,** and **bottom** (some would rather call the last two truth and beauty). Therefore we often hear it said that there are six **flavors of quarks.** Quarks have some extraordinary

TABLE 35.3

Some Properties of Quarks

Name	Symbol	Charge	Strangeness	Charm
Up	u	$+\frac{2}{3}e$	0	0
Down	d	$-\frac{1}{3}e$	0	0
Charmed	c	$+\frac{2}{3}e$	0	+1
Strange	s	$-\frac{1}{3}e$	−1	0
Top	t	$+\frac{2}{3}e$	0	0
Bottom	b	$-\frac{1}{3}e$	0	0

Note: An antiquark has the opposite signs for charge, strangeness, and charm. For example $\bar{c}$ has a q of $-\frac{2}{3}e$, a strangeness of 0, and a charm of −1.

properties. Table 35.3 shows that their charges, for instance, are $\pm\frac{1}{3}e$ and $\pm\frac{2}{3}e$ (called fractional charges). All quarks are fermions, with $s = \frac{1}{2}$.

In a theory called **quantum chromodynamics (QCD),** each set of quarks has a dynamic property called **color.** This term has nothing to do with visual color, but is merely a convenient way of postulating a property to explain quark behavior. The three colors are *red, green,* and *blue* for quarks and *antired, antigreen,* and *antiblue* for antiquarks. In the theory, all hadrons are *white.* White can be obtained by combining red, green, and blue (as with light) or by combining antired, antigreen, and antiblue. Alternatively, white can be made by adding a color to its anticolor, for instance blue plus antiblue.

Mesons are combinations of a quark and an antiquark. The positive pion, for example, is composed of an up quark and a down antiquark ($u\bar{d}$). This combination gives the pion a charge of $+\frac{2}{3}e + [-(-\frac{1}{3}e)] = +1e$. If the up quark is red, the down antiquark is antired, with red plus antired giving white for the pion (since all hadrons are white). If the quark and the antiquark in a meson have the same flavor, they can easily annihilate each other. This rapid annihilation explains why the neutral pion has so much shorter a lifetime than the charged pions.

Taking another example, the upsilon meson is made of a bottom quark and a bottom antiquark (or, if you prefer, a beauty quark and a beauty antiquark) and therefore has a net charge of $-\frac{1}{3}e + [-(-\frac{1}{3}e)] = 0$ and a short lifetime. Its bottom quark and bottom antiquark "canceled its bottomness," giving it "hidden bottom" (or "hidden beauty"). Therefore the search for a meson composed of a b quark and a $\bar{u}$ or $\bar{d}$ antiquark was called a search for "bare bottom" or "naked beauty," resulting in a lot of off-color jokes. The search eventually led to the discovery of the B mesons.

The baryons are made up of some combination of three quarks (or three antiquarks for antibaryons) in quark theory. The proton has two ups

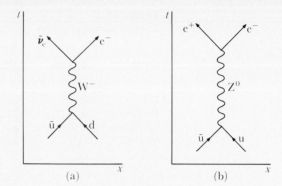

FIGURE 35.4
Two possible reactions from the collision of a proton (uud) and an antiproton ($\overline{u}\overline{u}\overline{d}$). (a) A d quark from the proton and a $\overline{u}$ quark from the antiproton form a W⁻ particle. The W⁻ particle decays into an electron and an electron antineutrino. (b) A u quark from the proton and a $\overline{u}$ from the antiproton form a Z⁰ particle. The Z⁰ decays to an electron–positron pair (or a muon–antimuon pair). The four other quarks in each collision are involved in the formation of hadrons.

and a down (uud) for a charge of $+\frac{2}{3}e + \frac{2}{3}e + (-\frac{1}{3}e) = +e$. The antiproton, however, consists of two up and one down antiquarks ($\overline{u}\overline{u}\overline{d}$) with a resulting charge of $-e$. Figure 35.4 follows some of the quarks in a p$\overline{p}$ collision. The neutron has two downs and an up (ddu), which add up to zero net charge. In the proton and neutron, two of the spin angular momentum vectors are antialigned, giving a total spin angular momentum quantum number of $\frac{1}{2} + (-\frac{1}{2}) + \frac{1}{2} = \frac{1}{2}$.

The omega minus particle is made up of three strange quarks. The result is a strangeness of -3 and a charge of $3(-\frac{1}{3}e) = -1e$, as well as a spin angular momentum quantum number of $3(\frac{1}{2}) = \frac{3}{2}$. Without color, the omega minus particle would have three identical quarks in the same state, violating the Pauli exclusion principle. Therefore in the omega minus particle (as in all baryons), there are one red quark, one green quark, and one blue quark.

Moreover, the delta resonance has a spin angular momentum quantum number of $\frac{3}{2}$ and four charge states. The four correspond to spin-aligned ddd ($q = -e$), ddu ($q = 0$), duu ($q = +e$), and uuu ($q = +2e$) quark configurations. In terms of quarks, the $\Delta^{2+} \rightarrow p + \pi^+$ decay is uuu $\rightarrow$ uud + u$\overline{d}$. In this decay, the three u quarks remain and a d$\overline{d}$ is produced.

EXAMPLE 35.9   Discuss the neutron decay reaction in terms of the quarks involved.

Solution Example 35.5 and Table 35.2 remind us that the reaction is n $\rightarrow$ p + e⁻ + $\overline{\nu}_e$. From our discussion, the neutron is ddu and the proton is uud. The electron and its antineutrino are fundamental particles, not made up of quarks. Therefore we have ddu → uud, telling us that 1 down quark was converted into an up quark in the decay. ●

In quantum electrodynamics, the electromagnetic force comes from the exchange of virtual photons. Photons are neutral and therefore don't interact with the electromagnetic field that they're setting up. The property called color is analogous to charge (and is sometimes called *color charge*) in quantum chromodynamics. In QCD, the strong force results from the interchange of virtual gluons. However, gluons carry a net color and do interact with the quarks emitting them and with each other. One result of these extra interactions is that the forces between quarks behave quite differently than do the forces between other charged particles.

As an example of a QCD interaction, let's consider a meson initially made up of a blue quark and an antiblue antiquark. The blue quark could emit a blue–antigreen virtual gluon (with the blue quark changing to a green quark in the process to conserve color charge). The blue–antigreen virtual gluon would then be absorbed by the antiblue antiquark, converting it to an antigreen antiquark. Besides binding the quark–antiquark pair together, the emission and absorption of the virtual gluon has converted a blue–antiblue pair to a green–antigreen pair.

Quarks seem to move about rather freely within hadrons. However, despite the use of higher and higher energies in accelerators, no sighting of a free quark has occurred. Rather, as Fig. 35.5 shows, quark–antiquark

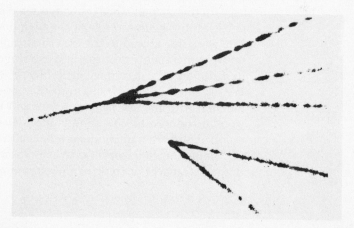

FIGURE 35.5

A 20-GeV photon interacts with a proton in a bubble chamber to produce two particles, one containing a charmed quark and the other a charmed antiquark. The positively charged charmed particle travels 0.86 mm from the left production point, then decays into three charged particles, while the invisible neutral charmed particle travels 1.8 mm before decaying into two charged particles. Both decays also contain invisible neutral particles.

Source: SLAC, funded by the U.S. Department of Energy.

FIGURE 35.6

(a) Pointlike quarks "rattle around" freely inside a baryon. (b) If a quark tries to escape, it encounters a force that increases with distance.

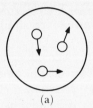

(a)

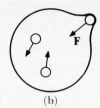

(b)

production occurs. The **bag model** explains this short-range free movement and long-range binding by allowing the quarks to move freely as if inside a hadron-sized elastic bag, as shown in Fig. 35.6. Moving the quarks farther apart would stretch the bag and require more and more energy. Only the application of infinite energy could separate them completely. The exchange particle for the strong force between quarks must therefore be extraordinarily "sticky"—a good reason to call it a gluon. Actually, there are evidently eight types of gluons, as well as combinations of gluons called *glueballs*.

The discussion in this chapter has probably seemed less exact and the ideas less patterned than in the other chapters of this book. This area of physics is a new research frontier, with many more questions than answers so far. The new theories look promising, but certainly none is *the* definitive theory. Researchers in this field continually try to obtain better data and develop better theories to explain and predict the phenomena they are dealing with. These researchers find their work to be exciting, and from the sound of the terms they use, they find it fun, too.

In this chapter we investigated some properties of many of the small particles studied as a result of the high energies of particle accelerators. You were introduced to the idea of virtual particles, allowed by the Heisenberg uncertainty principle and exchanged to create forces. Attempts to unify forces were described, including the electroweak theory and GUTs. We classified small particles into the gauge boson, lepton, and hadron families, with the last having meson and baryon subfamilies. Then you were introduced to new conservation laws involving baryon and lepton numbers and a new quantum number called strangeness. We concluded by explaining the six flavors of quarks and some of their properties. All in all, this is a strange, but colorful and charming, area of physics.

QUESTIONS AND PROBLEMS

35.1 How is the de Broglie wavelength of a particle related to its kinetic energy in the extreme relativistic region $E \gg E_0$?

35.2 What advantages are there in building even

higher energy particle accelerators? What disadvantages? Why not just use cosmic rays (energies $> 10^{20}$ eV)?

35.3 Discuss how two children on roller skates can

repel or attract each other by exchanging
  a) balls or
  b) boomerangs.

35.4 What virtual particles are exchanged between two electrons?

35.5 What virtual particles are exchanged between two quarks?

35.6 In view of the concept of virtual particles, is mass–energy conserved exactly?

35.7 Approximate the rest energy of a virtual particle that acts over a range of up to $1/3$ fm at a speed somewhat less than c.

35.8 If an entire star is made of nothing but neutrons with a density exceeding that of a nucleus, the gravitational force holding it together is important. Explain why, in comparison with the other three forces of nature.

35.9 A gamma ray moving through a vacuum disappears. Its place is taken by an electron–positron pair. They soon annihilate each other to form a single gamma ray with the same energy and momentum as the original gamma ray. Explain why this is possible with no nearby heavy nucleus if it happens over a small enough range, Δx.

35.10 If the average lifetime of a proton were 2.5×10^{31} years, what fraction of those protons created 10^{10} years ago in the Big Bang will have decayed?

35.11 Assume that about half your body mass is protons and that protons have an average lifetime of 2.5×10^{31} years. At what rate are your body's protons decaying?

35.12 Suppose that you were only 10^{-16} m tall and could see only for nuclear distances. What would you see as the symmetry of your surroundings if you were at the center of a Si nucleus? Now let your size and range of vision increase until you can see many atomic spacings. What happened to the symmetry of your surroundings as the scale increased?

35.13 Which family or families of particles appear to a) have no structure? b) to be made up of more fundamental particles?

35.14 If the baryon and lepton numbers of the universe were zero, what would that mean?

35.15 "The baryon number of the universe is about 10^{78}." What does that mean?

35.16 What does baryon number have to do with the conservation of mass number law for medium-energy nuclear reactions and radioactive decay?

35.17 Show that lepton number is conserved in electron–positron pair production.

35.18 For the tau particle decay modes specified in Table 35.2, show that the total lepton number and the lepton number for each particle–neutrino pair are conserved.

35.19 Use some of the main decay modes in Table 35.2 to show that if a meson number were defined, it wouldn't be conserved.

35.20 What conservation laws would be broken in the decays (allowed by some GUTs): a) $p \rightarrow e^+ + \pi^0$? b) $p \rightarrow \mu^+ + K^0$? c) $p \rightarrow \bar{\nu} + \pi^+$?

35.21 Give two decay modes of the D^-.

35.22 Determine whether mass–energy, charge, spin angular momentum, baryon number, lepton number, and strangeness can be conserved in the following reactions.
  a) $p + p \rightarrow p + \pi^+ + \Lambda^0 + K^0$
  b) $K^- + p \rightarrow K^0 + K^+ + \Omega^-$
  c) $p + p \rightarrow p + n + K^+$
  d) $p + p \rightarrow p + \pi^+ + \Lambda^0 + \overline{K}^0$
  e) $\pi^- \rightarrow \mu^- + \bar{\nu}_\mu$
  f) $\overline{\Omega}^- \rightarrow \overline{\Xi}^- + \pi^0$
  g) $\mu^- \rightarrow e^- + \nu_\mu + \bar{\nu}_e$
  h) $\gamma \rightarrow \pi^+ + \pi^+$

35.23 K^0 is not its own antiparticle because $K^0 + p$ doesn't give $\Sigma^+ + \pi^0$ ($\overline{K}^0 + p$ does). Using strangeness, show that this behavior is to be expected.

35.24 Find the mean length of a track in a detector of a D^+ with a kinetic energy of 4.652 GeV. Ignore energy losses in forming the track.

35.25 Show that the following constructions of particles by quarks agree with their charge, spin angular momentum quantum number, and strangeness.
  a) $\Lambda^0 = uds$
  b) $\pi^0 = u\bar{u} + d\bar{d}$
  c) $\overline{K}^+ = s\bar{u}$
  d) $\bar{n} = \overline{ddu}$
  e) $\pi^- = d\bar{u}$
  f) $J/\psi = c\bar{c}$

35.26 What colors could the quarks have in Problem 35.25?

35.27 In terms of the quark model, why do all mesons have $s = 0$ or $s = 1$?

35.28 Explain this statement: "The $-e$ and $+2e$ delta resonances would have been forbidden by the Pauli exclusion principle were it not for the property of color."

35.29 In the decay of the W particle, it appears that a $t\bar{b}$ particle is formed. Identify the family of the $t\bar{b}$ particle and give its charge, strangeness, charm and possible spin angular momentum quantum numbers.

35.30 Explain why it makes sense to give each quark a baryon number of $\frac{1}{3}$. What then is the baryon number of an antiquark?

35.31 If a π^0 is $u\bar{u} + d\bar{d}$, what happens to the quarks in π^0 decay?

35.32 Discuss what happens to the quarks in the decay of the $J/\psi = c\bar{c}$ and $Y = b\bar{b}$.

35.33 Discuss two Λ^0 decay modes in terms of the quarks involved; $\Lambda^0 = $ uds.

35.34 Consider the behavior of quarks as described by the bag model and try to figure out what the terms *asymptotic freedom* and *infrared slavery* refer to.

35.35 Explain why gluons appear to attract one another to form glueballs, but photons don't attract one another (to form photballs?).

PART Six

Nuclear and Particle Physics

CHAPTER 36

General Relativity

You began this course studying relativity and will return to it in this chapter. We'll begin with a simple thought experiment and then look at some of its far-reaching consequences. These include the bending of light rays near a massive body, gravity's effect on the measurement of time intervals, deviations from Newton's law of gravitation, and the possible existence and properties of black holes.

36.1
THE PRINCIPLE OF EQUIVALENCE

In Chapters 2–5, we considered Einstein's special theory of relativity. The word "special" refers to the fact that it was developed especially for inertial reference frames. These are frames of reference that are not accelerating: Objects at rest in these frames remain at rest when the net external force on them is zero. Einstein's **general theory of relativity (GTR),** then, would seem to apply to frames of reference that are accelerating with respect to inertial reference frames. However, the GTR is even more

515

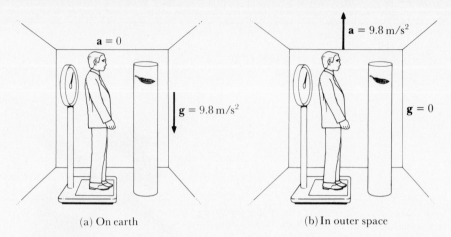

(a) On earth   (b) In outer space

FIGURE 36.1
Both on earth (with $a = 0$ and $g = 9.8$ m/s²) and in outer space (with $a = 9.8$ m/s² and $g = 0$), the feather accelerates toward the floor in a vacuum at 9.8 m/s² with respect to the observer, and the scale reads the same force.

general than that, as Einstein's "thought experiment" shows:

> Suppose that you are in a small closed room. You have any measuring equipment with you that you might desire (such as meter sticks, clocks, etc.). You drop an object in a vacuum from rest and you find that its acceleration is 9.8 m/s² toward the floor. Where is your room?

One possible answer is that your room is at rest on the surface of the earth, where all freely falling bodies accelerate downward at $g = 9.8$ m/s². (If your room is at rest, its acceleration with respect to the surface of the earth is zero.) But Fig. 36.1 also shows another possible answer: Your room is in distant space far from any large masses (in a region where $g = 0$), but your room is accelerating "upward" at a constant 9.8 m/s². The point is that you have no way of deciding from your experiment whether your closed room is at rest in the earth's gravitational field or accelerating in distant space. This is the basis of Einstein's **principle of equivalence:**

No experiment can distinguish between a uniform gravitational field and an equivalent uniform acceleration.

EXAMPLE 36.1   Suppose that the closed room is in orbit about the earth. Ignoring the small nonuniformity of the earth's gravitational field, could you distinguish being in orbit from zero acceleration in distant space?

Solution In orbit, everything is in free fall. That is, **g** and **a** are equal in both magnitude and direction so that a released object doesn't fall with respect to the room (giving what is called weightlessness). But that's the same behavior measured when both **a** and **g** are zero, so the answer is "no."

●

One result of this principle is that gravitational mass and inertial mass are equivalent. In Fig. 36.1, the gravitational mass of the dropped object gives the force to accelerate it toward the floor on earth at 9.8 m/s². The inertial mass of the released object keeps it moving at constant velocity in outer space, while the floor accelerates toward it at 9.8 m/s².

36.2
SPACETIME IN THE GTR

Recall that in special relativity space and time are not absolute quantities but are related to each other. In the general theory of relativity, the relationships between space and time are even further removed from classical ideas. We use a new conceptual term, speaking of **spacetime,** a four-dimensional reference frame. The equations of special relativity give a **flat spacetime.** That is, the spacetime of special relativity follows the rules of Euclidean geometry.

You know from Euclidean geometry that a straight line is the shortest distance between two points, lines that are parallel in one region will never meet in any other region, the angles of a triangle add up to 180°, and the ratio of the circumference to the diameter of any circle is π. Figure 36.2 shows examples of Euclidean geometry on a flat surface.

FIGURE 36.2
Examples of Euclidean geometry on a flat surface.

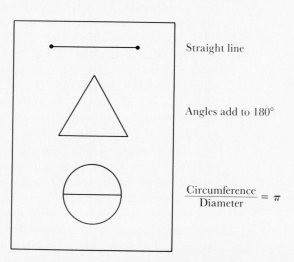

Straight line

Angles add to 180°

$\dfrac{\text{Circumference}}{\text{Diameter}} = \pi$

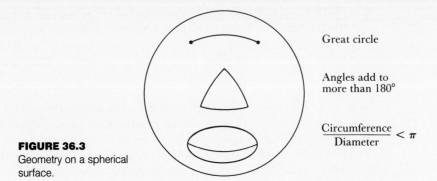

Great circle

Angles add to
more than 180°

$$\frac{\text{Circumference}}{\text{Diameter}} < \pi$$

FIGURE 36.3
Geometry on a spherical
surface.

However, the equations of the GTR give possible **curved space-times.** Flat spacetimes can be represented in three dimensions (one each for x, y, and z) at a particular instant of time. To represent the curvature of spacetime, we need a fourth dimension. Since most of us aren't familiar with four-dimensional solids, let's mentally discard a dimension, for instance the z dimension, and use a three-dimensional solid to represent x, y, and the curvature at some instant of time. For example, consider lines that are restricted to lie on the surface of a sphere, as in Fig. 36.3. On such a surface, as in curved spacetime, Euclidean geometry no longer holds. A great circle is the shortest distance between two points, lines parallel at the equator meet at the poles, the angles of a triangle add to more than 180°, and the ratio of the circumference to the diameter is less than π.

Now suppose that a small hole is bored in the side of your previously closed room in distant space. At the instant your relative speed is zero, an external inertial observer sends a pulse of light through the hole and parallel to the floor. The pulse travels in a straight line with respect to the inertial observer, but your floor is accelerating upward, moving toward the pulse at a faster and faster rate as the pulse moves across the room. That means you would measure the light to be moving toward the floor in a curved path, as depicted in Fig. 36.4. However, the principle of equivalence says that you should measure the same curved path if you are in an equivalent gravitational field. Therefore this particular light pulse doesn't move in a straight line but in a curved path in a gravitational field. Only light moving directly toward the ceiling or floor (opposite or parallel to the gravitational field) would travel in a straight line.

Let's consider a light pulse emitted at the floor from a source moving at zero relative speed in distant space. By the time the light gets to the ceiling, a detector there would be moving away from the source and would therefore register a decrease in frequency. By the equivalence principle, this phenomenon is the so-called **gravitational redshift:** The frequency (and therefore photon energy) decreases as an electromagnetic wave moves against an apparent gravitational field. The gravitational redshift

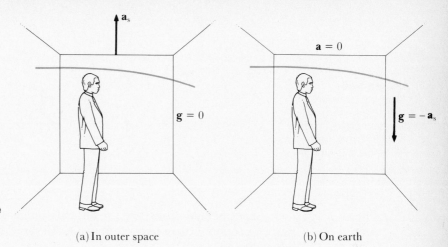

FIGURE 36.4
Curvature of the light pulse (greatly exaggerated) in two equivalent situations.

(a) In outer space   (b) On earth

for the equivalent observer at rest in a gravitational field isn't the result of any relative motion that would cause wave crests to spread. Rather, it is because of an increase in period. This result leads to the prediction that observers above the surface of a massive object will measure a time dilation for events occurring on the surface.

The path that light follows is the shortest distance between two points. Such a path is called a **geodesic.** The gravitational fields and forces of classical physics are set up by masses. But since geodesics near masses in the GTR generally aren't straight lines, masses give us curved spacetimes. Like centrifugal and Coriolis forces in classical mechanics, gravity is a fictitious force in the GTR. Centrifugal and Coriolis forces were invented in order to apply inertial reference frame concepts to noninertial (rotating) systems. Gravity is invoked to apply flat spacetime concepts to curved spacetime. In the GTR, masses move toward other masses because those masses have caused spacetime near them to be curved. The larger, more dense, and closer the object, the greater is the curvature. Figures 36.5 and 36.6 show a model of spacetime curved by a large mass.

FIGURE 36.5
The rubber-sheet model of a two-dimensional spacetime. With no masses, spacetime is flat and special relativity is correct. Coordinate lines are represented by straight lines inked onto the sheet.

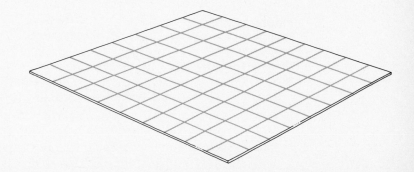

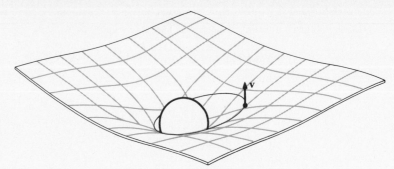

FIGURE 36.6
A lead ball represents a heavy, dense mass that curves spacetime. Ignoring rolling friction, a marble can be made to orbit the heavy mass.

Experimental results are the ultimate judge of any theory. Einstein himself proposed three tests of his GTR. All three, and two others, verify its predictions.

One of the first problems that Einstein attacked using his GTR was the **precession of the perihelion** of the planet Mercury. If Mercury were the only planet in orbit about a spherically symmetric sun, its classical orbit would be a closed ellipse. The perihelion is the point of closest approach of the ellipse to the sun. With respect to the sun, each orbit would have the perihelion at exactly the same distance and angle. However, observations show that Mercury reaches its perihelion at an angle that increases slightly with each orbit. As shown in Fig. 36.7, *precession* is the term used for this motion of the perihelion. Since the increase in angle occurs each orbit, it is cumulative.

The precession of Mercury's perihelion occurs mostly because of the effects of the other planets in the solar system. However, the most careful calculations using Newton's law of gravitation left a precession of 43 seconds of arc per century unaccounted for. When Einstein made his calculations, he came up with 43.0 seconds as the effect of the curvature of spacetime near the sun. In his words, "for a few days, I was beside myself with joyous excitement" after the calculations' agreement. In fact, he was so excited that he got heart palpitations.

FIGURE 36.7
The precession of the perihelion (greatly exaggerated) of a planet about the sun.

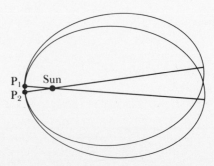

EXAMPLE 36.2 The measured relativistic precessions of the perihelions of Venus, Earth, and the asteroid Icarus also agree with general relativistic predictions (to within the limits of experimental error). Are these precessions of Venus and Earth smaller or larger than Mercury's?

Solution They are smaller, with the Earth's smaller than Venus's. The main reason for relativistic precession is the curvature of space, which is greatest near the sun. Therefore Venus and Earth, being farther than Mercury from the sun, will have smaller precessions. ●

TABLE 36.1
General Relativistic Precession of Perihelions in Seconds of Arc per Century

Orbiting body	Observed	Predicted
Mercury	43.1 ± 0.5	43.0
Icarus	9.8 ± 0.8	10.3
Venus	8.4 ± 0.5	8.6
Earth	5.0 ± 1.2	3.8

Table 36.1 gives the actual numbers for the general relativistic precessions of four bodies in our solar system. A much larger (4.23° per year) general relativistic precession is that of the binary (double) star system containing the pulsar PSR 1913 + 16.

A second test of the GTR is the deflection of electromagnetic waves as they pass close to the sun (see Fig. 36.8). This successful test made headlines around the world in 1919. More recent tests have used radio waves from quasars, obtaining agreement within their 1 percent measurement error.

EXAMPLE 36.3 Why don't photons move in straight lines as they skim by the sun?

Solution Our light-pulse illustration (see Fig. 36.4) showed that light (or any electromagnetic wave) moving past a massive object will be deflected toward that object. The closer the photons of the electromagnetic wave are to the sun, the more their paths are curved. ●

The third test proposed by Einstein was the gravitational redshift. This frequency- and time-interval change was first verified to within 10 percent by Pound and Rebka in 1959 using the Mössbauer effect and then to within 0.007 percent by Vessot and others in 1976 using a hydrogen

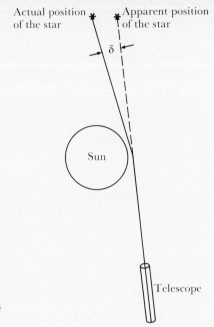

FIGURE 36.8
The deflection of starlight (greatly exaggerated) as it moves past the sun. The GTR predicts that δ equals only 1.74 seconds of arc.

maser clock on a rocket. In both cases, the v^2/c^2 special relativistic effects of relative motion also had to be considered.

A fourth test, time delay, is related to the deflection of electromagnetic waves. Electromagnetic waves take longer to move past the sun following a curved geodesic than they would by taking a Newtonian straight-line path. Signals from Viking spacecraft and landers on Mars gave agreement with GTR predictions to within 0.1 percent experimental error.

A fifth test of the GTR involves Einstein's prediction that certain changes in mass distributions would produce changes in their curved spacetime that would propagate out at the speed of light. Since this curvature of spacetime gives rise to what we call gravity, these changes moving out from the mass would appear as **gravitational pulses** or **gravitational waves.** To date no gravitational waves have been directly detected. However, the rotation of the binary pulsar PSR 1913 + 16 (evidently two neutron stars rotating about their common center of mass) is slowing down in a manner consistent with their expected masses and general relativistic predictions of energy radiated away by gravitational waves.

At the beginning of Chapter 3, we considered a pulse of light spreading from the origin at $t = 0$ in S_1 and S_2. In either system, we described it in space and time by

$$x^2 + y^2 + z^2 - c^2 t^2 = 0.$$

At a time dt after $t = 0$, we can write its equation in S_1 or S_2 as

$$0 = c^2 (dt)^2 - (dx)^2 - (dy)^2 - (dz)^2.$$

In the flat spacetime of special relativity, we define the **interval,** ds, in Cartesian coordinates by

$$(ds)^2 = (c\ dt)^2 - (dx)^2 - (dy)^2 - (dz)^2, \tag{36.1}$$

or in spherical coordinates by

$$(ds)^2 = (c\ dt)^2 - (dr)^2 - (r\ d\theta)^2 - (r \sin \theta\ d\phi)^2. \tag{36.2}$$

For light, the interval is zero.

EXAMPLE 36.4   What is the interval for two events that

a) occur simultaneously?   b) occur at the same position?

Solution

a) Simultaneous occurrence means that there is no difference in time between the two events (in the frame of reference being used). Therefore $dt = 0$. Equation (36.1) then gives $ds = i[(dx)^2 + (dy)^2 + (dz)^2]^{1/2}$, which is the square root of -1 times the spatial distance between the two events.

b) At the same position means that there is no difference in r, θ, or ϕ, so $dr = d\theta = d\phi = 0$, and Eq. (36.2) gives $ds = c\ dt$, which is the distance that light traveled between the times of the two events.   ●

We could have used either Eq. (36.1) or (36.2) to answer either question in Example 36.4.

Karl Schwarzschild used Einstein's general relativistic equations in 1916 to arrive at the following relation for the interval at a distance r from the center and outside of a nonrotating, spherically symmetric mass M:

$$(ds)^2 = g(c\ dt)^2 - \frac{(dr)^2}{g} - (r\ d\theta)^2 - (r \sin \theta\ d\phi)^2,$$

where   (36.3)

$$g = 1 - \frac{2GM}{c^2 r}.$$

The dimensionless coefficient g is *not* the acceleration due to gravity. Equation (36.3) gives curved spacetime unless $g = 1$.

EXAMPLE 36.5   Show that spacetime is almost flat at the surface of the earth.

Solution We use $M = 5.98 \times 10^{24}$ kg and $r = 6.37 \times 10^6$ m for the mass and radius of the earth. The constants $G = 6.673 \times 10^{-11}$ N $\cdot$ m^2/kg^2 and $c = 2.998 \times 10^8$ m/s yield a value for $2GM/c^2r$ of only 1.39×10^{-9}. Therefore g in Eq. (36.3) is equal to 1 to within almost 1 part in 10^9. Placing $g = 1$ in Eq. (36.3) gives Eq. (36.2), which is the expression for flat spacetime.   ●

In other words, Example 36.5 shows us that general relativistic effects are small at the earth's surface. Thus classical physics or special relativity accurately describe most of what happens on earth.

However, spacetime becomes more and more curved as $2GM/c^2r$ approaches 1. If $2GM/c^2r = 1$, the $(dr)^2$ term in Eq. (36.3) is divided by zero, giving rise to an infinity (called a *singularity*). The value of r at which a singularity occurs is called the **Schwarzschild radius,** r_s. Therefore $2GM/c^2r_s = 1$ gives

$$r_s = \frac{2GM}{c^2} = (1.485 \times 10^{-27} \text{ m/kg})M. \qquad (36.4)$$

Calculations show that the escape velocity exceeds the speed of light for anything inside this radius. That is, it seems that nothing, not even light, can escape. Therefore we can't tell what's happening inside r_s, so the sphere of radius r_s surrounding a mass M of smaller radius is called the **event horizon** (see Fig. 36.9). However, although seemingly not even light can escape, the huge curvature of space (huge gravitational field) just outside the event horizon means that matter and energy can readily be sucked in. Thus a mass with a radius of less than r_s is called a **black hole.**

EXAMPLE 36.6   To what radii would

a) earth and

b) our sun have to be compressed to become black holes? Their masses are 5.98×10^{24} kg and 1.99×10^{30} kg, respectively, and their present radii are 6.37×10^6 m and 6.96×10^8 m, respectively.

Solution They would have to be compressed to less than the Schwarzschild radius, r_s, given by Eq. (36.4).

a) For earth,

$$r_s = (1.485 \times 10^{-27} \text{ m/kg})(5.98 \times 10^{24} \text{ kg}) = 8.88 \text{ mm.}$$

b) For our sun,

$$r_s = (1.485 \times 10^{-27} \text{ m/kg})(1.99 \times 10^{30} \text{ kg}) = 2.96 \text{ km.} \qquad ●$$

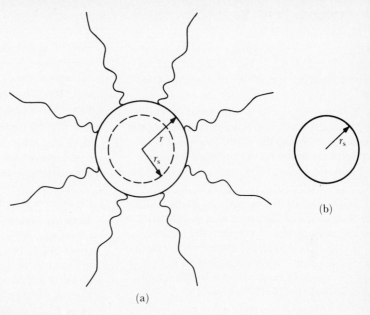

FIGURE 36.9
(a) For $r > r_s$, redshifted light can escape the surface. (b) For $r < r_s$, the escape velocity is greater than the speed of light, and very little information can be determined about whatever is within the event horizon.

If no electromagnetic radiation can escape, we can tell very little about a black hole. We can't tell what it's made of, only its mass, charge, and angular momentum. The baldness of this information is sometimes expressed by the statement "black holes have no hair." (Look at Fig. 36.9 (b) again.) However, theorist Stephen Hawking has devised a method involving virtual pair production and tunneling that removes energy (and, therefore, mass) from black holes. This allows black holes to "evaporate" away, with an evaporation time proportional to the cube of the mass. For a one-solar-mass black hole, the evaporation time would be 10^{66} years.

Einstein's GTR is not the only general theory of relativity, but it seems to rest on the firmest experimental grounds and therefore has a much higher level of acceptance than any other theory. Black holes, however, are a different story. At this time, the experimental evidence for black holes is only circumstantial but has convinced many. On the theoretical side, Einstein did not believe that his GTR equations could be applied to regions of extreme curvature, much less to a singularity. For black holes, both the interpretation of experimental evidence and theory may have been pushed too far. But then again, they may not. After all, as you have seen again and again in your study of modern physics, new insights have often been possible only by leaping far beyond the boundaries of accepted science.

QUESTIONS AND PROBLEMS

36.1 Consider the accelerating room in space illustrated in Fig. 36.1. How does an inertial observer describe the "fall" of the "dropped" object?

36.2 Describe the motion of the closed room in Fig. 36.1 if dropped objects accelerate toward the floor at 9.8 m/s^2 and the closed room is at 0.75 earth radii above the earth's surface where $g = 3.2$ m/s^2.

36.3 A meter stick floating in an orbiting space vehicle will tend to line up with the earth's field. (The bottom half of the meter stick will experience a slightly stronger force than the top half, giving a net torque at all but three exact angles.) There is no tendency toward alignment in a gravity-free inertial room. Does this difference invalidate the principle of equivalence?

36.4 If you dropped two balls simultaneously from the same height, but some horizontal distance apart in the closed room at rest on the earth's surface, the earth's gravitational field would move them closer together as they fell. Why? Would this same effect be present in the accelerating room in distant space?

36.5 What restriction does Question 36.4 place on Einstein's thought experiment?

36.6 From outside reading, discuss the current status of the Eötvös experiment.

36.7 Consider lines restricted to the surface of a sphere. Describe a triangle having angles that add up to 270°.

36.8 Find the ratio of the circumference to the diameter of a great circle, considering only lines restricted to the surface of the sphere.

36.9 Atomic clocks flown around the world in commercial jetliners showed different time intervals than those kept in one place on the earth's surface. The time difference reflected both special relativistic time dilation and a GTR effect. Why was there a GTR effect?

36.10 The difference in times between a clock at rest on the surface of a massive object and one at rest above the surface depends on the strength of the gravitational field, the distance above the surface, and therefore the difference in gravitational potential energy. Explain these reasons by reference to the energy loss of a photon moving away from the surface.

36.11 Reconsider Problem 36.10 in the equivalent accelerating room in distant space.

36.12 Consider the possibility of a particle that could travel between two points at a speed approaching c along a shorter path than a photon would take between those points. Relate this to the definition of a geodesic.

36.13 Suppose that you are visiting a hospitalized friend who fell while rock climbing. You decide to cheer her up by explaining that gravity is a fictitious force. Do so by examining fictitious forces in a closed room centered on a rapidly moving merry-go-round.

36.14 From outside reading, describe how a gravitational lens can make a distant astronomical object appear beside itself and relate that to the calculation that made Einstein feel beside himself.

36.15 Calculate the ratio of the general relativistic precession of PSR 1913 + 16 to that of Mercury.

36.16 a) Appreciable deflection of electromagnetic waves occurs only when they pass close to the sun. Why? b) Why did the first deflection measurements have to await a solar eclipse?

36.17 What were the relative positions of Earth and Mars when the greatest relativistic time delay occurred for signals received from the Viking spacecraft?

36.18 From outside reading, discuss the current state of gravitational wave detection.

36.19 Show that Eq. (36.1) is consistent with the motion of a pulse of light that spreads from the origin at $t = 0$.

36.20 Show that Eq. (36.2) is consistent with the motion of a pulse of light that spreads from the origin at $t = 0$.

36.21 Show that for certain limits of M or r, curved spacetime becomes flat spacetime.

36.22 Using the data in Example 36.6, discuss the flatness of spacetime at the surface of the sun.

36.23 The term g in Eq. (36.3) is not the acceleration due to gravity, but it is related to the force of gravity. a) How strong is the force of gravity if $g = 1$ exactly? b) What happens to g as the gravitational field gets stronger?

36.24 In Eqs. (36.1) and (36.2), the time part has a positive sign and the space parts have negative signs. a) Show that the same is true outside the event horizon. b) Give a reason for the statement, "space and time swap places in a black hole," considering your answer to (a).

36.25 Consider a photon moving directly away from the event horizon. Show that the time, dt, for the photon to move a distance dr approaches infinity as the initial position of the photon approaches the event horizon (the result of increased time dilation).

36.26 Is it true that "once a black hole gets you in its gravitational field, there's no escape"?

36.27 From outside reading, discuss the current experimental evidence for black holes.

36.28 Calculate the coefficient g 1 m outside your event horizon if you were crushed to sufficient density to become a black hole.

36.29 Calculate the coefficient g 3.00 m from the surface of a black hole that has a Schwarzschild radius of 2.00 m.

36.30 If black holes can evaporate, are they truly black?

36.31 Draw a log-log graph of evaporation time for black holes as a function of mass.

36.32 a) From the information in the text, write an equation for the evaporation time of a black hole. b) If the universe is about 10^{10} years old, all black holes of less than a certain mass and radius that were formed at its beginning must have evaporated. Calculate that mass and radius.

36.33 In 1956, Einstein stated: "For large densities of field and matter, the [GTR] field equations and even the field variables which enter into them have no real significance." What, therefore, do you believe that he would say about the prediction of black holes?

36.34 Considering the views of Einstein (see Question 36.33 and the end of Section 13.1), what do you think his response would be to this statement of Stephen Hawking: "Consideration of particle emission from black holes seems to suggest that God not only plays dice, but also sometimes throws them where they cannot be seen."

CHAPTER **57**

Cosmology

In this chapter you'll see results of applying the general theory of relativity to models of the universe. Some models reflect its apparent current expansion and can be projected forward in time to the eventual fate of the universe and back in time to the Big Bang. Starting from shortly after the Big Bang, we'll describe how the physics of the smallest particles in nature is used to explain the history of the largest thing we can measure, our universe.

37.1
COSMOLOGICAL MODELS AND EXPANSION

Cosmology is the study of the universe, which is done by means of models. Most cosmological models utilize the **cosmological principle,** the assumption that the universe is homogeneous and isotropic. That is, the assumption is made that the properties of the universe are the same at all places and in all directions. For example, the cosmological principle requires the density, ρ, of the universe to be constant in space (but not in time). This assumption, of course, isn't true at small scale. Matter exists in clumps, ranging from leptons and quarks to you and me to galaxies and to super-clusters of galaxies. In fact, the farther and the finer that astronomers can see, the more nonuniformity they seem to find in the distribution of mass. At the universal scale, however, the cosmological principle seems to be a reasonable first assumption.

Many early cosmological models made the earth the unmoving center of the universe. More modern cosmologies consider the earth to have no exceptional location in a vast universe made up of about 10^{11} galaxies, each containing an average of 10^{11} stars. However, modern cosmologists do still assume one earth-centered concept: The laws of physics that we discover here on earth are the same as the laws everywhere else in the universe.

EXAMPLE 37.1   Does the assumption about earth's laws applying throughout the universe follow from the cosmological principle? Explain.

Solution The answer is "yes." The cosmological principle states that the properties of the universe are the same everywhere. Therefore the laws that cause these properties must be the same everywhere. Regional changes in the laws would give regional differences in measured properties.   ●

Thus if an astronomer finds that all the spectral lines in the light from a star or group of stars have shifted to different frequencies, he assumes that this shift is the result of known phenomena, such as the Doppler effect. Only as a last resort would the astronomer assume that the laws of nature are different at the source of the light.

Vesto Slipher measured the first frequency shift of spectral lines from a galaxy at the Lowell Observatory in Flagstaff, Arizona, in 1912. It was a blueshift (a shift to higher frequencies) and it indicated that the Andromeda galaxy has a relative velocity component of about 300 km/s toward us. He then continued his measurements on other galaxies.

Meanwhile, Einstein applied his GTR to the whole universe, publishing the results in 1917. He utilized the cosmological principle, assuming a constant density and a constant positive curvature of spacetime. He also decided that he wanted a static universe. Newton had realized that a finite static universe was impossible. A finite universe would eventually collapse because of the gravitational attraction of each mass for all the others. However, Newton thought (incorrectly) that an infinite static universe was possible.

Einstein similarly found that his field equations didn't yield a static-universe solution, even for an infinite universe. To arrive at a static equilibrium, he introduced a term called the **cosmological constant**, Λ, into his equations. (This step was equivalent to inventing a repulsive fifth force of nature to oppose gravity.) However, because of later astronomical data and proof that the static equilibrium was unstable, Einstein's model fell from favor.

Slipher's continued measurements showed that the Andromeda blueshift was unusual. By 1923, of 41 galaxies studied, 36 had redshifts and only five had blueshifts. This result seemed to relate to another cosmological model published in 1917, the decidedly odd model of the Dutch astronomer Willem de Sitter.

Attempting to arrive at a theory for a static universe, de Sitter used the cosmological constant. He also used a constant density, but did so by letting ρ equal zero. The de Sitter universe contained no matter at all! However, if particles of mass were added to this model, they acted as if they were moving away from one another, just as most of the galaxies seemed to be moving away from us.

Then, in 1924 at Mt. Wilson in California, Edwin Hubble began to improve techniques to measure distances to galaxies. With the help of Milton Humason, he discovered what appeared to be a linear relationship between the distances of the galaxies from us and their speeds. Hubble and Humason determined that the few galaxies moving toward us are nearby. More-distant galaxies are all moving away from us. The universe seemed to be expanding. This linear relationship is called the **Hubble law:**

$$v_r = Hs, \tag{37.1}$$

where v_r is the **recession velocity,** or the velocity at which a galaxy is being expanded away from us; s is the distance the galaxy is away from us; and H is a quantity best labeled the **Hubble parameter.** Apparently constant for all galaxies now (and therefore commonly called the Hubble constant), H may change with time. It has no precisely determined value because of the great uncertainties in measuring distances for galaxies. The current value of the Hubble parameter is $H = (23 \pm 8)$ km/s·Mc·yr. For most of our purposes, we'll use the average value of 23 km/s·Mc·yr. Recall that Mc·yr is our abbreviation for 10^6 light years and that one light year is the distance traveled in one year at the speed of light.

EXAMPLE 37.2   The recession velocity of a galaxy in Ursa Major is 1.5×10^4 km/s. How far away from us is this galaxy?

Solution Using $v_r = 1.5 \times 10^4$ km/s and the average value of the Hubble parameter, $H = 23$ km/s·Mc·yr in $v_r = Hs$, we get

$$\frac{v_r}{H} = s = \frac{1.5 \times 10^4 \text{ km/s}}{23 \text{ km/s·Mc·yr}} = 6.5 \times 10^2 \text{ Mc·yr},$$

or 650 million light years away from us.   ●

In the 1920s, cosmological theory also indicated that the universe was not static, but was expanding. The chief theoreticians were Alexander Friedmann, a Russian, and Georges Lemaitre, a Belgian. Their models of the universe used the cosmological principle with a nonzero density. Lemaitre included a cosmological constant (but not Einstein's value), but Friedmann used no cosmological constant. Einstein realized that his rather arbitrary addition of Λ kept him from discovering the concept of the expanding universe and called it "the biggest blunder of my life."

When you first hear of the **expanding universe,** your mental picture may well be of a bunch of stars and stuff all moving outward through space from some central point. That is *not* what is happening according to the dynamic general relativistic cosmologies. Rather, the "stars and stuff" are relatively fixed in space, and *it is space itself that is expanding.*

Our problem in trying to visualize what is happening is that we need five dimensions, four for spacetime and one for curvature, to properly display expansion. However, suppose that we again ignore one dimension and discuss the evolution of a three-dimensional representation in time to try to get an understanding of the expanding universe. Figure 37.1 shows one model—the raisin-bread-dough model. However, we'll use the common spherical rubber-balloon model.

First we mark longitude and latitude lines on the balloon, just like those on a globe of the earth. Then randomly here and there we glue pieces of confetti on the surface with a flexible glue. (See Fig. 37.2a.) Each piece of confetti then represents a galaxy (or even some larger clustering of galaxies). All displacements are restricted to the surface of the balloon. The flexible glue enables us to move each piece of confetti a bit from the latitude and longitude coordinates peculiar to that piece. This type of motion gives rise to what is called the **peculiar velocity** of a galaxy, that is, its motion with respect to its coordinates in space.

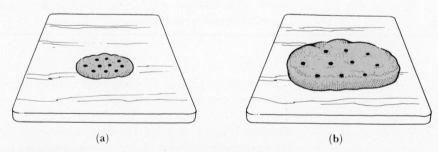

(a)   (b)

FIGURE 37.1
The raisin-bread-dough model of the expanding universe. (a) The newly mixed dough is the universe. The raisins randomly scattered throughout the dough are the galaxies. (b) As the dough rises uniformly in a warm room, each raisin moves away from every other raisin at a rate directly proportional to their distances apart.

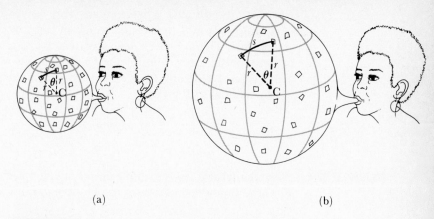

(a)   (b)

FIGURE 37.2
The spherical rubber-balloon model of the expanding universe.

Now let's blow up the balloon slowly. What happens to the pieces of confetti? Ignoring their peculiar velocities, we see that their coordinates don't change, but the distances between them (as measured along great circles on the surface) do. (See Fig. 37.2b.) Every piece of confetti moves away from every other piece of confetti. The distance between pieces is given by $s = r\theta$, where r is the radius of the balloon and θ is the angle between two radii from the center to two pieces of confetti. The latitude and longitude angles don't change, so θ is also constant. The recession velocity of the pieces is $v_r = ds/dt = (dr/dt)\theta$. Therefore v_r at any time is directly proportional to how far apart the pieces are at that time.

EXAMPLE 37.3   Prove that last statement in the text.

Solution First, $s = r\theta$ gives $\theta = s/r$. We then substitute $\theta = s/r$ into $v_r = (dr/dt)\theta$ to arrive at $v_r = [(dr/dt)/r]s$. At any time, $(dr/dt)/r$ is the same for all points on the balloon, showing that v_r is directly proportional to s at that time.   ●

The Andromeda galaxy is about two million light years away from us and has a velocity component of 300 km/s toward us. We might be tempted to try to analyze its motion using $v_r = Hs$, but we would be wrong to do so. The cosmological models undergirding the Hubble law assume a uniform density of matter at any given time and therefore a uniform curvature of spacetime. On the scale of only a few million light years, matter is manifestly not uniformly distributed. The 300 km/s is mainly a component of a peculiar velocity, not a recession velocity.

You should remember that the expansion of the universe involves the expansion of space itself in most cosmological models. Therefore we can write the distance s between any two distant objects as the product of a **scale factor** R times the change in the space coordinates between the objects. This approach is similar to converting a distance on a map to an actual distance by multiplying by the map's scale factor; for example, 2.6 (map cm) $\times$ 50 km/(map cm) = 130 km. Solving $v_r = ds/dt = Hs$ with s proportional to R gives

$$H = \frac{1}{R}\frac{dR}{dt}. \tag{37.2}$$

EXAMPLE 37.4 Assume that R follows a simple power law, $R = Ct^n$, where C is a constant. Find the time-dependence of the Hubble parameter.

Solution Differentiating, we have

$$\frac{dR}{dt} = \frac{d}{dt}(Ct^n) = nCt^{n-1} = \frac{nR}{t}.$$

Therefore

$$H = \frac{1}{R}\frac{dR}{dt} = \frac{1}{R}\frac{nR}{t} = \frac{n}{t}. \qquad \bullet$$

Because the scale factor is changing with time, distances between points are also changing with time, including the distance between adjacent crests in waves. This distance is the wavelength of the wave. Consider a wave emitted at time t_0 with wavelength λ_0 and frequency v_0 when the scale factor was R_0. We detect the wave at time t with wavelength λ and frequency v when the scale factor is R. Since λ is directly proportional to R and v is inversely proportional to λ, we can write

$$\frac{v}{v_0} = \frac{\lambda_0}{\lambda} = \frac{R_0}{R}. \tag{37.3}$$

EXAMPLE 37.5 Is the redshift of distant galaxies a special relativistic Doppler shift as given by Eq. (3.5)?

Solution The answer is "no." Equation (37.3) shows that this redshift (lower frequencies and longer wavelengths) is a direct result of the expansion of the universe, because expansion means that R (now) is greater than R_0 (when the galaxies emitted the light). $\qquad \bullet$

Most cosmological models predict that the expansion of the universe is slowing down or decelerating because of the attraction of the masses of the universe to one another. If there is a deceleration, dR/dt decreases with time, making d^2R/dt^2 negative. The deceleration parameter, q, is a measure of this decrease:

$$q = -\frac{R(d^2R/dt^2)}{(dR/dt)^2}. \qquad (37.4)$$

The deceleration parameter has no units and is positive for a decelerating universe.

EXAMPLE 37.6   Assume that R follows a simple power law, $R = Ct^n$, where C is a constant. Calculate the deceleration parameter.

Solution As in Example 37.4, $dR/dt = nCt^{n-1} = nR/t$. Then

$$\frac{d^2R}{dt^2} = \frac{d}{dt}(nCt^{n-1}) = n(n-1)Ct^{n-2} = \frac{n(n-1)R}{t^2}.$$

Therefore

$$q = \frac{-R(d^2R/dt^2)}{(dR/dt)^2} = \frac{-R[n(n-1)R/t^2]}{(n^2R^2/t^2)} = \frac{1}{n} - 1. \qquad \bullet$$

An area of great current interest is the determination of the average density of the universe. Below a certain **critical density** of about 10^{-26} kg/m^3, the universe will continue to expand, and cool, forever. Above that critical value, the universe will eventually stop expanding and begin to collapse, as shown in Fig. 37.3. That collapse is sometimes called the "Big

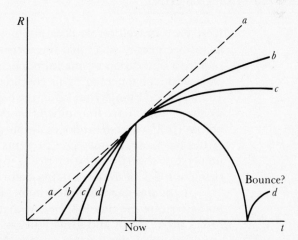

FIGURE 37.3
The scale factor of the universe as a function of time for certain cosmological models, where a represents zero density, b represents a density below the critical density, c represents the critical density (the slope approaches zero as t approaches infinity), and d represents a density above the critical density with a possible "Big Bounce."

Crunch." In some models, the universe will then begin to expand again, with a new oscillation, turning the Big Crunch into a "Big Bounce." The best current estimates of the average density of the universe range from 0.03 to 10 times the critical density.

How old is the universe? We can get a rough idea from the relation $R = Ct^n$, which gives $H = n/t$. Letting the "beginning" be $t = 0$, our time now is $t = T$, the age of the universe. Therefore $T = n/H = n(15 \pm 5) \times 10^9$ yr. Different cosmological models give different values for n, with typical values of $\frac{1}{2}$ to $\frac{2}{3}$, giving $T = (9 \pm 4) \times 10^9$ yr. As you'll see, however, we can't really answer the question with our current level of understanding.

If we mentally travel backward in time, our expanding universe becomes a contracting universe, with the scale factor approaching zero as t approaches zero in most cosmological models. That means that the universe started out with a very small scale factor and corresponding very small distances between everything — and then expanded. This expansion is called the **Big Bang.** Georges Lemaitre is called "the father of the Big Bang," not because he was a Catholic priest, but because he first investigated the idea (but gave it a different name). Despite its name, the Big Bang is not an explosion *in* space, but a rapid expansion *of* space.

37.2
THE HISTORY OF THE UNIVERSE

If we try to travel too far back in time, we run into a problem. The scale factor becomes so small that the corresponding separations and densities require a unification of quantum mechanics and general relativity to explain what is happening. This condition corresponds to a time of about 10^{-43} s (called **Planck time**) after the beginning of the Big Bang. However, we simply don't have the unified theory needed to define just what time *is* before 10^{-43} s, and so cannot truly talk about the "when" of the beginning of the Big Bang. Figure 37.4 presents an overview of the history of the universe, with all the question marks in the upper left of the figure emphasizing our lack of understanding before the time 10^{-43} s.

In the **standard model** of the Big Bang, all four forces of nature were unified before the time 10^{-43} s. At that time, the equivalence of the four forces broke down and gravity began to act separately. At 10^{-43} s, the temperature of the universe in the standard model was about 10^{32} K. Since Boltzmann's constant, 8.62×10^{-5} eV/K, is approximately 10^{-13} GeV/K, the average energy per particle, approximately kT, was about 10^{19} GeV. By 10^{-35} s, the decrease in density and temperature (to about 10^{27} K) of the expanding universe brought about the separation of

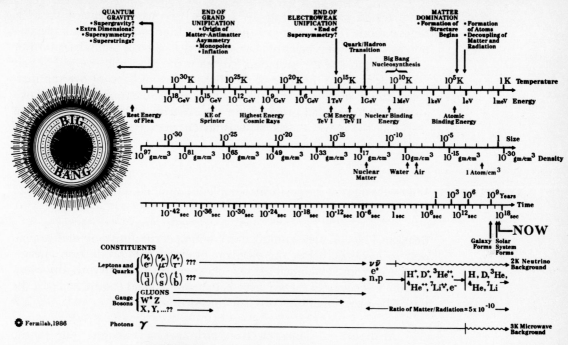

FIGURE 37.4

A schematic history of the universe.

Source: Used by permission of Fermi National Accelerator Laboratory. Available from Friends of Fermilab, Mail Stop 105, Fermi National Accelerator Laboratory, P.O. Box 500, Batavia, IL 60510.

the strong nuclear force from the electroweak force. Therefore the 10^{-43} s to 10^{-35} s interval is the GUTs era, or the time for which the grand unified theories explain what is happening. One prediction of GUTs is that the baryon number need not be exactly conserved, so it is in this era that the GUTs predict a very slight excess of quarks over antiquarks.

In **inflationary models,** this separation of the strong nuclear force corresponded to a phase change (like the phase change between liquid and gas phases of a substance, with its corresponding latent heat of vaporization). As a result, one inflationary model predicts that the scale factor, R, increased by 10^{50} in 10^{-32} s. This model, with a relatively small number of assumptions, claims to explain the overall homogeneity and isotropy of the universe, the great age of the universe, the reason that the total density of the universe is apparently so close to the critical density, the apparent rarity of magnetic monopoles, and the small scale fluctuations that eventually led to the formation of galaxies and other features of the universe.

The universe — a mixture of quarks, leptons, and gauge bosons — continued to expand and cool from the inflationary period to 10^{-6} s when the quarks began to bind together to form baryons and their antiparticles. Before about 10^{-2} s, most radiation had enough energy that pair production by two photons balanced baryon–antibaryon pair annihilation. After that time, most of the radiation fell below the threshold energy for pair production. Then, because of the slight excess of baryons over antibaryons, almost all the antibaryons and most of the baryons annihilated one another. After similar processes for the electron–positron pairs (at about $t = 14$ s), the universe was left with its current great preponderance of matter over antimatter.

EXAMPLE 37.7   Why were most of the positrons completely annihilated so much later than most of the antibaryons?

Solution This complete annihilation couldn't occur until most of the radiation had energy below the electron–positron pair production energy. Since electrons have much less rest mass–energy than baryons, this event wouldn't occur until the universe had expanded and cooled well past the time when baryon-antibaryon pair production ceased.   ●

At about $t = 1$ s, most electrons no longer had enough energy for the reaction $e + p \rightarrow n + \nu_e$. The neutrinos also had a lower average energy and the matter of the universe was more spread out, so equilibrium reactions involving the absorption of neutrinos began to decrease rapidly. Thus at this time the flux of neutrinos and antineutrinos throughout the universe "decoupled" from the rest of the universe. Because of the extraordinarily low cross section for neutrino absorption, this flux is still present, although cooled by expansion to an average temperature of about 2 K (according to the standard model, but no one has been able to figure out an experiment to test this prediction).

Also at about $t = 1$ s, the ratio of protons to neutrons was $e^{-\Delta E/kT}$, where ΔE is the rest energy difference between the two nucleons, $(m_n - m_p)c^2$. With a temperature of roughly 10^{10} K, this relation gives about nine protons for every two neutrons. However, neutrons are radioactive with a half-life of 630 s, so the proton–neutron ratio continued to increase until $t = 225$ s. At that time, T was below 10^9 K and very few photons had energies greater than the 2.22-MeV binding energy of the deuteron. Therefore a proton and a neutron could combine and remain combined as a deuteron, H-2. This combination halted the decay of the free neutrons at a ratio of about seven protons to every one neutron. The nuclide H-2 can absorb a neutron to form H-3 or a proton to form He-3, and the H-3 and He-3 can absorb a proton and a neutron respectively to both form He-4, but there the building of nuclei almost completely stops. The reason is that no stable or long-lived nuclides have a mass number of five, and the cross section for further reactions is very small.

EXAMPLE 37.8   Almost all the protons and neutrons in the seven-to-one ratio either made He-4 or remained as H-1. Therefore, by mass, what percent of H and He should there have been?

Solution   The nuclide He-4 contains 2 protons and 2 neutrons, so we need at least 2 neutrons in the $7:1$ (proton-to-neutron) ratio. The ratio then is 14 protons to 2 neutrons. The 2 neutrons and 2 of the protons make up 1 He-4 nucleus. The remaining 12 protons have no neutrons to combine with and so become H-1 nuclei. Since the masses of H-1 and He-4 are about 1 u and 4 u, respectively, there were 12 u of H-1 for every 4 u of He-4, and a total of $12\,u + 4\,u = 16\,u$. Therefore H was $12/16 = 75$ percent and He was $4/16 = 25$ percent by mass.   ●

The average energy per particle at that time was still much too high to allow electrons to start binding to nuclei to form atoms. Atomic binding didn't occur until after 10^{13} s, or nearly 700,000 years later, when the temperature had dropped to about 3000 K. After that very few free electrons were left to scatter electromagnetic radiation and so the flux of electromagnetic radiation throughout the universe decoupled from the matter of the universe. The radiation has continued to expand its wavelengths to form the **2.7 K blackbody radiation** that we receive (spatially uniform to 0.01 percent) from space. This radiation is one of the main experimental pillars of the standard model of the Big Bang.

EXAMPLE 37.9   By approximately what ratio has the universe expanded since $t = 700,000$ years?

Solution   For blackbody radiation, Eq. (6.1) gives $\lambda_p = \text{constant}/T$. Therefore we can write Eq. (37.3) as

$$\frac{\text{Constant}/T_0}{\text{Constant}/T} = \frac{R_0}{R} \quad \text{or} \quad \frac{R}{R_0} = \frac{T_0}{T}.$$

At $t = 700,000$ yr, T was about 3000 K and now it is 2.7 K (or about 3 K), giving $R/R_0 = 3000\,\text{K}/3\,\text{K} = 1000$. Universal distances are proportional to the scale factor R, so the R/R_0 ratio is also the distance ratio. Thus the universe has expanded by approximately a factor of 1000.   ●

The energy radiated from stars is produced in fusion reactions in which H-1 becomes He-4 (two possible cycles are shown in Section 33.3). When the hydrogen is about used up, the inward gravitational pressure exceeds the outward radiation and gas pressure, and the star's core begins to contract. As it does so, its gravitational potential energy decreases, and the kinetic energy of its atoms increases. For stars of sufficient mass, there is both enough energy and the necessary density to begin helium fusion.

Two He-4's first fuse to form Be-8*. The Be-8* has an extremely short half-life (10^{-16} s) but is compensated for by an unusually high resonance cross section for He-4 absorption at these energies. Therefore a reasonable fraction of the Be-8* fuses with more He-4 to give stable C-12. (Three He-4's have fused to form one C-12 in what is called the **triple-alpha process.**) Then successive fusions with He-4 give O-16, Ne-20, and Mg-24. All these reactions release energy to heat the star so that C-12 and O-16 can fuse to form elements having higher and higher atomic numbers.

The binding energy per nucleon curve peaks out at Fe, so exoergic fusion reactions stop with Fe, but successive neutron captures can continue the synthesis of heavy elements. A step in the evolution of some massive stars is an explosion of the star (a supernova). Such explosions release into space the heavy elements that were produced by the earlier processes. In space, the debris and other interstellar matter gravitationally bunch together to form stars and planets. Thus it is more than poetic to say that the earth is formed of stardust.

In this chapter we explained how physicists and astronomers have tried to come up with a comprehensive theory of the development of the universe by bringing together topics from many areas of modern physics. The concepts and models in this chapter have considerably different levels of acceptance among scientists. Many of the ideas, from the expanding universe on, have reputable critics, and other models and other interpretations of the limited data are available. At present we have no way of knowing which, if any, of the cosmological models is correct. The probability is good that in your lifetime some of the ideas in this chapter (or the rest of this book, for that matter) will be more generally accepted as true, some will be proven false, and some will remain in dispute. Stay tuned for further developments. Or even better, be part of the developments!

QUESTIONS AND PROBLEMS

37.1 The steady-state model of the universe assumes a "perfect cosmological principle," one that says the properties are the same at all times, at all places, and in all directions. Explain why this model then requires that matter be created constantly throughout an expanding universe.

37.2 Assume that the universe has an edge. Placing yourself at the edge, show that this assumption violates the cosmological principle.

37.3 Does the fact that all distant galaxies are moving away from us mean that we are at the center of the universe? Explain.

37.4 Explain why the universe has no edge and no center if it can be modeled by the expanding balloon.

37.5 Explain the five blueshifts of the galaxies measured by Slipher.

37.6 Convert a Mc·yr to meters.

37.7 Suppose that you decided to use the Hubble law for the Andromeda galaxy, $2\frac{1}{4}$ million light years away, despite the warning in the text. How would the recessional velocity calculated compare to its 300 km/s peculiar velocity of approach?

37.8 At what distances do recession velocities and peculiar velocities have about the same magnitude? (Get a rough estimate by using a value given in the text and in Problem 37.7 for the Andromeda galaxy.)

37.9 The light from quasar PKS 2000-330 may indicate a recession velocity of $0.916c$. If so, how far away from us is it according to the Hubble law, assuming that H is constant?

37.10 An astronomer determines that a galaxy in Virgo is receding from us at 12 million m/s and that it is 800 million light years away from us. Are these values reasonable?

37.11 Intensity measurements indicate that a galaxy in Bootes is $1.7 \times 10^9 \, c \cdot$yr from us. What would you expect its recession velocity to be?

37.12 The Hubble length is defined as the distance at which the recession velocity equals c. a) Find the limiting values of the Hubble length. b) What is the Hubble length for a galaxy $5 \, Gc \cdot$yr from us?

37.13 A galaxy can't move through space at a velocity greater than c, but both its recession velocity and the sum of its recession and peculiar velocities can be greater than c. Explain.

37.14 Show that $v/v_0 \approx 1 - v_r/c$ for a small $\Delta R = R - R_0$. (*Hint:* Use $\Delta R \approx (dR/dt)\Delta t$.)

37.15 If the Hubble parameter is a constant, a) find how the scale factor depends on time (assuming a value of R_0 at $t = 0$) and b) calculate the deceleration parameter and explain the sign.

37.16 Assume that R follows a simple power law, $R = Ct^n$, where C is a constant. a) What value of n gives no deceleration? b) For that n, how does the Hubble parameter depend on time? c) What range of n gives deceleration?

37.17 What is the approximate critical density of the universe, as measured in hydrogen atoms per cubic meter?

37.18 The intensity emitted from an ideal blackbody is $c/4$ times the energy density of its electromagnetic waves. Therefore what does the 2.7 K radiation contribute to the average density of the universe?

37.19 What exploded to make the Big Bang?

37.20 Define Planck time as $t_P = G^x\hbar^yc^z$. The SI units of G are $N \cdot m^2/kg^2 = m^3/kg \cdot s^2$, of $\hbar$ are $J \cdot s = kg \cdot m^2/s$, and of c are m/s. a) Use the units to find the expression for t_P (that is, find x, y, and z). b) Show that $t_P = 5.39 \times 10^{-44}$ s.

37.21 Define Planck length as $l_P = G^x\hbar^yc^z$. The SI units of G are $N \cdot m^2/kg^2 = m^3/kg \cdot s^2$, of $\hbar$ are $J \cdot s = kg \cdot m^2/s$, and of c are m/s. a) Use the units to find the expression for l_P (that is, find x, y, and z). b) Show that $l_P = 1.616 \times 10^{-35}$ m.

37.22 A possible relation in the GTR between the scale factor R, the density ρ, and the curvature of the universe (through the constant k) is $(dR/dt)^2 = 8\pi G\rho R^2/3 - kc^2$. Assume that $k = 0$ (or that kc^2 is negligible) and that $R = Ct^n$.

a) Find how the density of the universe depends upon time.

b) Assume that ρ is all due to matter and that ρR^3 is constant, solve for n, and show that $\rho = 1/(6\pi Gt^2)$.

c) Assume that ρ is all due to radiation and is therefore proportional to T^4, that $\lambda_p T$ is constant, and that λ_p is proportional to R. Solve for n and show that $\rho = 3/(32\pi Gt^2)$.

37.23 If the abundance of nuclides in the universe by mass is about 25 percent He-4 and 75 percent H-1, what percent of all nuclei are He-4?

37.24 a) Refer to Example 37.9. By approximately what ratio has the density of the universe changed since $t = 700,000$ yr?

b) In some cosmological models, the density of both matter and radiation is inversely proportional to the square of time since the Big Bang. What is the approximate age of the universe (since Planck time) according to these models?

37.25 Why must a star be heated "so that C-12 and O-16 can fuse"?

37.26 Why can it be said that you and I are partially made of stardust?

37.27 Consider a star of mass M and radius R made entirely of neutrons. You can use Eq.

(21.7), with m replaced by m_n, the mass of a neutron, for the Fermi energy because $T \ll T_F$. This gives a total kinetic energy of $3NE_{F0}/5$. The neutron star's gravitational potential energy is $-3GM^2/5R$. Other energy terms are constant or small.

a) Use $dE_t/dR = 0$ at equilibrium to show

that the radius of a neutron star is $R = (3/2\pi)^{4/3}h^2/(4Gm_n^3N^{1/3})$.

b) Calculate this radius for a neutron star having a mass twice that of the sun.

c) How close is this radius to the radius of an equal-mass black hole?

APPENDIX

Table of Neutral Atomic Masses

Atom	A	Mass	Atom	A	Mass	Atom	A	Mass
$_1$H	1	1.007825	($_4$Be)	8*	8.005305	($_6$C)	13	13.003355
	2	2.014102		9	9.012182		14*	14.003242
	3*	3.016049		10*	10.013534		15*	15.010599
				11*	11.021658			
$_2$He	3	3.016029				$_7$N	13*	13.005739
	4	4.002603	$_5$B	8*	8.024606		14	14.003074
	5*	5.01222		9*	9.013329		15	15.000109
	6*	6.018886		10	10.012937		16*	16.006100
				11	11.009305		17*	17.008450
$_3$Li	5*	5.01254		12*	12.014353			
	6	6.015121				$_8$O	14*	14.008595
	7	7.016003	$_6$C	10*	10.016856		15*	15.003065
	8*	8.022486		11*	11.011433		16	15.994915
$_4$Be	7*	7.016928		12	12.000000		17	16.999131

(continued)

---

* Radioactive isotope.

Note: Masses are in u: $1\ u = 931.5\ \text{MeV}/c^2$. Some other masses: Two electrons, 0.001097: $_0^1$n, 1.008665: $_1^1$p, 1.007276. The last figure shown isn't necessarily the last significant figure of the mass.

Sources: Data were extracted from the *1983 Atomic Mass Table,* A. H. Wapstra and G. Audi, Nuc. Phys. **A432,** 1 (1985); *Chart of the Nuclides.* Copyright © 1984 by the General Electric Company. All rights reserved.

Atom	A	Mass	Atom	A	Mass	Atom	A	Mass
($_8$O)	18	17.999160	$_{16}$S	30*	29.984903	($_{21}$Sc)	46*	45.955170
	19*	19.003577		31*	30.979554		47*	46.952409
	20*	20.004076		32	31.972071		48*	47.952235
				33	32.971458			
$_9$F	17*	17.002095		34	33.967867	$_{22}$Ti	44*	43.959690
	18*	18.000937		35*	34.969032		45*	44.958124
	19	18.998403		36	35.967081		46	45.952629
	20*	19.999981		37*	36.971126		47	46.951764
	21*	20.999948					48	47.947947
			$_{17}$Cl	33*	32.977452		49	48.947871
$_{10}$Ne	18*	18.005710		34*	33.973763		50	49.944792
	19*	19.001880		35	34.968853		51*	50.946616
	20	19.992436		36*	35.968307		52*	51.946898
	21	20.993843		37	36.965903			
	22	21.991383		38*	37.968011	$_{23}$V	48*	47.952257
	23*	22.994465		39*	38.968005		49*	48.948517
	24*	23.993613					50*	49.947161
			$_{18}$Ar	34*	33.980269		51	50.943962
$_{11}$Na	21*	20.997650		35*	34.975256		52*	51.944778
	22*	21.994434		36	35.967546		53*	52.944340
	23	22.989768		37*	36.966776			
	24*	23.990961		38	37.962732	$_{24}$Cr	48*	47.954033
	25*	24.989953		39*	38.964314		49*	48.951338
	26*	25.992586		40	39.962384		50	49.946046
				41*	40.964501		51*	50.944768
$_{12}$Mg	22*	21.999574		42*	41.963050		52	51.940510
	23*	22.994124					53	52.940651
	24	23.985042	$_{19}$K	37*	36.973377		54	53.938882
	25	24.985837		38*	37.969080		55*	54.940842
	26	25.982594		39	38.963707		56*	55.940643
	27*	26.984341		40*	39.963999			
	28*	27.983877		41	40.961825	$_{25}$Mn	52*	51.945568
				42*	41.962402		53*	52.941291
$_{13}$Al	25*	24.990429		43*	42.960717		54*	53.940361
	26*	25.986892					55	54.938047
	27	26.981539	$_{20}$Ca	38*	37.976318		56*	55.938907
	28*	27.981910		39*	38.970718		57*	56.938285
	29*	28.980446		40	39.962591			
				41*	40.962278	$_{26}$Fe	52*	51.948114
$_{14}$Si	26*	25.992330		42	41.958618		53*	52.945310
	27*	26.986704		43	42.958766		54	53.939613
	28	27.976927		44	43.955481		55*	54.938296
	29	28.976495		45*	44.956185		56	55.934939
	30	29.973770		46	45.953689		57	56.935396
	31*	30.975362		47*	46.954543		58	57.933277
	32*	31.974148		48	47.952533		59*	58.934877
				49*	48.955672		60*	59.934078
$_{15}$P	29*	28.981803						
	30*	29.978307				$_{27}$Co	57*	56.936294
	31	30.973762	$_{21}$Sc	43*	42.961150		58*	57.935755
	32*	31.973907		44*	43.959404		59	58.933198
	33*	32.971725		45	44.955910			

Atom	A	Mass	Atom	A	Mass	Atom	A	Mass
($_{27}$Co)	60*	59.933820	$_{33}$As	73*	72.923827	($_{38}$Sr)	89*	88.907450
	61*	60.932478		74*	73.923928		90*	89.907738
$_{28}$Ni	56*	55.942134		75	74.921594	$_{39}$Y	87*	86.910882
	57*	56.939799		76*	75.922393		88*	87.909508
	58	57.935346		77*	76.920646		89	88.905849
	59*	58.934349	$_{34}$Se	73*	72.926768		90*	89.907152
	60	59.930788		74	73.922475		91*	90.907303
	61	60.931058		75*	74.922522	$_{40}$Zr	88*	87.910225
	62	61.928346		76	75.919212		89*	88.908890
	63*	62.929670		77	76.919912		90	89.904703
	64	63.927968		78	77.917308		91	90.905644
	65*	64.930086		79*	78.918498		92	91.905039
$_{29}$Cu	61*	60.933461		80	79.916520		93*	92.906474
	62*	61.932586		81*	80.917991		94	93.906315
	63	62.929599		82	81.916698		95*	94.908042
	64*	63.929766		83*	82.919117		96	95.908275
	65	64.927793	$_{35}$Br	77*	76.921378		97*	96.910950
	66*	65.928872		78*	77.921144	$_{41}$Nb	91*	90.906991
	67*	66.927747		79	78.918336		92*	91.907192
$_{30}$Zn	62*	61.934332		80*	79.918528		93	92.906377
	63*	62.933214		81	80.916289		94*	93.907281
	64	63.929145		82*	81.916802		95*	94.906835
	65*	64.929243		83*	82.915179	$_{42}$Mo	90*	89.913933
	66	65.926035	$_{36}$Kr	77*	76.924610		91*	90.911755
	67	66.927129		78	77.920396		92	91.906808
	68	67.924846		79*	78.920084		93*	92.906813
	69*	68.926552		80	79.916380		94	93.905085
	70	69.925325		81*	80.916590		95	94.905841
	71*	70.927727		82	81.913482		96	95.904678
$_{31}$Ga	67*	66.928204		83	82.914135		97	96.906020
	68*	67.927982		84	83.911507		98	97.905407
	69	68.925580		85*	84.912531		99*	98.907711
	70*	69.926028		86	85.910616		100	99.907477
	71	70.924700		87*	86.913360		101*	100.910345
	72*	71.926365	$_{37}$Rb	83*	82.915144	$_{43}$Tc	95*	94.907657
	73*	72.925169		84*	83.914390		96*	95.907870
$_{32}$Ge	68*	67.928096		85	84.911794		97*	96.906364
	69*	68.927969		86*	85.911172		98*	97.907215
	70	69.924250		87*	86.909187		99*	98.906254
	71*	70.924954		88*	87.911326		100*	99.907657
	72	71.922079	$_{38}$Sr	83*	82.917566	$_{44}$Ru	94*	93.911361
	73	72.923463		84	83.913430		95*	94.910414
	74	73.921177		85*	84.912937		96	95.907599
	75*	74.922858		86	85.909267		97*	96.907556
	76	75.921402		87	86.908884		98	97.905287
	77*	76.923548		88	87.905619			

(continued)

Atom	A	Mass	Atom	A	Mass	Atom	A	Mass
($_{44}$Ru)	99	98.905939	$_{50}$Sn	111*	110.907741	($_{54}$Xe)	130	129.903509
	100	99.904219		112	111.904826		131	130.905072
	101	100.905582		113*	112.905176		132	131.904144
	102	101.904348		114	113.902784		133*	132.905888
	103*	102.906323		115	114.903348		134	133.905395
	104	103.905424		116	115.901747		135*	134.907130
	105*	104.907744		117	116.902956		136	135.907214
				118	117.901609		137*	136.911557
$_{45}$Rh	101*	100.906159		119	118.903311			
	102*	101.906814		120	119.902199	$_{55}$Cs	131*	130.905444
	103	102.905500		121*	120.904239		132*	131.906431
	104*	103.906651		122	121.903440		133	132.905429
	105*	104.905686		123*	122.905722		134*	133.906696
				124	123.905274		135*	134.905885
$_{46}$Pd	101*	100.908287		125*	124.907785			
	102	101.905634				$_{56}$Ba	129*	128.908642
	103*	102.906114	$_{51}$Sb	119*	118.903948		130	129.906282
	104	103.904029		120*	119.905077		131*	130.906902
	105	104.905079		121	120.903821		132	131.905042
	106	105.903478		122*	121.905179		133*	132.905988
	107*	106.905127		123	122.904216		134	133.904486
	108	107.903895		124*	123.905938		135	134.905665
	109*	108.905954		125*	124.905252		136	135.904553
	110	109.905167					137	136.905812
	111*	110.907660	$_{52}$Te	119*	118.906411		138	137.905232
				120	119.904048		139*	138.908826
$_{47}$Ag	105*	104.906520		121*	120.904947			
	106*	105.906662		122	121.903050	$_{57}$La	137*	136.906460
	107	106.905092		123*	122.904271		138*	137.907105
	108*	107.905952		124	123.902818		139	138.906347
	109	108.904756		125	124.904428		140*	139.909471
	110*	109.906111		126	125.903310		141*	140.910896
				127*	126.905221			
$_{48}$Cd	105*	104.909459		128	127.904463	$_{58}$Ce	134*	133.908890
	106	105.906461		129*	128.906594		135*	134.909117
	107*	106.906613		130	129.906229		136	135.907140
	108	107.904176		131*	130.908528		137*	136.907780
	109*	108.904953					138	137.905985
	110	109.903005	$_{53}$I	125*	124.904620		139*	138.906631
	111	110.904182		126*	125.905624		140	139.905433
	112	111.902757		127	126.904473		141*	140.908271
	113*	112.904400		128*	127.905810		142*	141.909241
	114	113.903357		129*	128.904986			
	115*	114.905430				$_{59}$Pr	139*	138.908917
	116	115.904755	$_{54}$Xe	123*	122.908469		140*	139.909071
	117*	116.907228		124	123.905894		141	140.907647
				125*	124.906397		142*	141.910039
$_{49}$In	111*	110.905109		126	125.904281		143*	142.910814
	112*	111.905536		127*	126.905182			
	113	112.904061		128	127.903531	$_{60}$Nd	141*	140.909594
	114*	113.904916		129	128.904780		142	141.907719
	115*	114.903882					143	142.909810

Atom	A	Mass	Atom	A	Mass	Atom	A	Mass
($_{60}$Nd)	144*	143.910083	($_{65}$Tb)	158*	157.925411	$_{71}$Lu	173*	172.938929
	145*	144.912570		159	158.925342		174*	173.940336
	146	145.913113		160*	159.927163		175	174.940770
	147*	146.916097		161*	160.927566		176*	175.942679
	148	147.916889					177*	176.943752
	149*	148.920145	$_{66}$Dy	156*	155.924277			
	150	149.920887		157*	156.925460	$_{72}$Hf	174*	173.940044
	151*	150.923825		158	157.924403		175*	174.941507
				159*	158.925735		176	175.941406
$_{61}$Pm	143*	142.910930		160	159.925193		177	176.943217
	144*	143.912588		161	160.926930		178	177.943696
	145*	144.912743		162	161.926795		179	178.945812
	146*	145.914708		163	162.928728		180	179.946546
	147*	146.915135		164	163.929171		181*	180.949096
	148*	147.917473		165*	164.931700			
	149*	148.918332				$_{73}$Ta	179*	178.945930
			$_{67}$Ho	163*	162.928731		180*	179.947462
$_{62}$Sm	143*	142.914626		164*	163.930285		181	180.947992
	144	143.911998		165	164.930319		182*	181.950149
	145*	144.913409		166*	165.932281		183*	182.951369
	146*	145.913053		167*	166.933127			
	147*	146.914894				$_{74}$W	179*	178.947067
	148*	147.914819	$_{68}$Er	161*	160.929996		180	179.946701
	149*	148.917180		162	161.928775		181*	180.948192
	150	149.917273		163*	162.930030		182	181.948202
	151*	150.919929		164	163.929198		183	182.950220
	152	151.919728		165*	164.930723		184	183.950928
	153*	152.922094		166	165.930290		185*	184.953416
	154	153.922205		167	166.932046		186	185.954357
	155*	154.924636		168	167.932368		187*	186.957153
				169*	168.934588			
$_{63}$Eu	149*	148.917926		170	169.935461	$_{75}$Re	183*	182.950817
	150*	149.919702		171*	170.938027		184*	183.952530
	151	150.919847					185	184.952951
	152*	151.921742	$_{69}$Tm	167*	166.932848		186*	185.954984
	153	152.921225		168*	167.934170		187*	186.955744
	154*	153.922975		169	168.934212			
	155*	154.922889		170*	169.935798	$_{76}$Os	183*	182.953290
				171*	170.936427		184	183.952488
$_{64}$Gd	152*	151.919786					185*	184.954041
	153*	152.921745	$_{70}$Yb	167*	166.934946		186*	185.953830
	154	153.920861		168	167.933894		187	186.955741
	155	154.922618		169*	168.935186		188	187.955830
	156	155.922118		170	169.934759		189	188.958137
	157	156.923956		171	170.936323		190	189.958436
	158	157.924099		172	171.936378		191*	190.960920
	159*	158.926384		173	172.938208		192	191.961467
	160	159.927049		174	173.938859		193*	192.964138
	161*	160.929664		175*	174.941273			
				176	175.942564	$_{77}$Ir	189*	188.958712
$_{65}$Tb	157*	156.924023		177*	176.945253		190*	189.960580

(continued)

Atom	A	Mass	Atom	A	Mass	Atom	A	Mass
($_{77}$Ir)	191	190.960584	($_{83}$Bi)	209*	208.980374	$_{93}$Np	235*	235.044056
	192*	191.962580		210*	209.984095		236*	236.046550
	193	192.962917		211*	210.987255		237*	237.048168
	194*	193.965069					238*	238.050941
			$_{84}$Po	208*	207.981222		239*	239.052933
$_{78}$Pt	190*	189.959917		209*	208.982404			
	191*	190.961665		210*	209.982848	$_{94}$Pu	236*	236.046032
	192	191.961019		211*	210.986627		238*	238.049555
	193*	192.962977		212*	211.988842		239*	239.052158
	194	193.962655					240*	240.053808
	195	194.964766	$_{85}$At	209*	208.986149		241*	241.056846
	196	195.964926		210*	209.987126		242*	242.058737
	197*	196.967315		211*	210.987469			
	198	197.967869		212*	211.990725	$_{95}$Am	240*	240.055278
	199*	198.970552					241*	241.056824
			$_{86}$Rn	211*	210.990576		242*	242.059542
$_{79}$Au	195*	194.965013		219*	219.009479		243*	243.061375
	196*	195.966544		220*	220.011368			
	197	196.966543		222*	222.017571	$_{96}$Cm	243*	243.061382
	198*	197.968217					244*	244.062747
	199*	198.968740	$_{87}$Fr	212*	211.996130		245*	245.065484
				221*	221.014230		246*	246.067218
$_{80}$Hg	195*	194.966640		222*	222.017560		247*	247.070347
	196	195.965807		223*	223.019733		248*	248.072343
	197*	196.967187					250*	250.078352
	198	197.966743	$_{88}$Ra	223*	223.018501			
	199	198.968254		224*	224.020186	$_{97}$Bk	245*	245.066357
	200	199.968300		225*	225.023604		247*	247.070300
	201	200.970277		226*	226.025403		249*	249.074980
	202	201.970617		228*	228.031064			
	203*	202.972848				$_{98}$Cf	248*	248.072183
	204	203.973467	$_{89}$Ac	225*	225.023205		249*	249.074845
	205*	204.976047		226*	226.026084		250*	250.076400
				227*	227.027750		251*	251.079580
$_{81}$Tl	201*	200.970794		228*	228.031015		252*	252.081621
	202*	201.972085						
	203	202.972320	$_{90}$Th	229*	229.031755	$_{99}$Es	251*	251.079986
	204*	203.973839		230*	230.033128		252*	252.082944
	205	204.974401		232*	232.038051		253*	253.084818
	206*	205.976084		234*	234.043593		254*	254.088019
			$_{91}$Pa	231*	231.035880	$_{100}$Fm	252*	252.082466
$_{82}$Pb	202*	201.972134		232*	232.038565		253*	253.085173
	203*	202.973365		233*	233.040242		254*	254.086846
	204	203.973020		234*	234.043303		255*	255.089948
	205*	204.974458					256*	256.091767
	206	205.974440	$_{92}$U	232*	232.037130		257*	257.095099
	207	206.975872		233*	233.039628			
	208	207.976627		234*	234.040947	$_{101}$Md	255*	255.091081
	209*	208.981065		235*	235.043924		256*	256.093960
				236*	236.045563			
$_{83}$Bi	207*	206.978446		237*	237.048725	$_{102}$No	252*	252.08895
	208*	207.979717		238*	238.050785		254*	254.09095

Atom	A	Mass	Atom	A	Mass	Atom	A	Mass
$(_{102}No)$	256*	256.09425	$_{104}Rf$	259*	259.1055	106	263*	263.1182
	257*	257.09685		261*	261.1087			
	259*	259.10093				107	262*	262.1229
			$_{105}Ha$	260*	260.1110			
$_{103}Lr$	259*	259.10290		261*	261.1118			
	260*	260.10532		262*	262.1138			

Odd-Numbered Answers

1.1 Approximately when it's at rest or moving in a straight line at constant speed with respect to (w.r.t.) the earth. When accelerating w.r.t. the earth (by changing its speed and/or its direction of motion).

1.3 For motions of the stars and planets, Ptolemy's frame of reference was fixed in the earth and Copernicus's in the sun. Copernicus's was more nearly inertial.

1.5 The Galilean coordinate transformations assume that S_2 moves at a constant velocity w.r.t. S_1. Any frame of reference that moves at constant velocity w.r.t. an inertial reference frame is also an inertial reference frame.

1.7 $x_2 = -5.0$ m, $y_2 = 1.0$ m, $z_2 = -0.5$ m, and $t_2 = 2.0$ s. In the sketch the origin of S_2 will be at $x_1 = 8.0$ m.

1.9 $x_1 = -9$ km (9 km north), $y_1 = 18$ km (18 km east), and $z_1 = 0.6$ km (0.6 km up).

1.11 $L_1 = L_2$.

1.13 As in Example 1.3, except that M and m move together at the same velocity after the collision.

1.15 No. (For example, an object could be at rest w.r.t. one observer but moving w.r.t the other.)

1.17 W.r.t. the river: $x_2 = -71$ ft, $y_2 = 71$ ft, $z_2 = 0$, $v_{2x} = -3.5$ ft/s, $v_{2y} = 3.5$ ft/s, and $v_{2z} = 0$. W.r.t. the earth: $x_1 = 229$ ft, $y_1 = 71$ ft, $z_1 = 0$, $v_{1x} = 11.5$ ft/s, $v_{1y} = 3.5$ ft/s, and $v_{1z} = 0$. $\mathbf{a}_1 = 0 = \mathbf{a}_2$.

1.19 a) 31.4° N of E. b) 360 s.

1.21 9.7 m/s, 21.1° N of E and 9.7 m/s, 21.1° S of W.

1.23 a) The matrix of Problem 1.22 with β replaced by $-\beta$. b) Use the laws of matrix multiplication to do so.

2.1 No.

2.3 Just replace all subscript 1's by subscript 2's.

2.5 Certainty could never be reached, only approached as the number of experiments approached infinity. One correct experiment could prove it false.

2.7 a) 5.0 m/s.   b) 3.0×10^8 m/s.

2.9 Just in case the earth in its orbit happened to be at rest w.r.t. the ether during the first measurement. In case the ether was dragged less higher up than at the surface.

2.11 Zero.

2.13 No. It has nothing to do with the first postulate, and the second postulate refers to the speed of light in a vacuum, not in a material.

2.15 Yes. It's not forbidden because the light is not moving in a vacuum.

2.17 No. The light itself is not moving at $v > c$.

3.1 Because the speed of light is independent of the source.

3.3 It doesn't give one point in S_1 for one point in S_2.

3.5 $v = c\sqrt{1 - 1/\gamma^2}$.   **3.7** Infinity. c.

3.9 $(-0.88, -3.0, 0)$ km and 3.7 μs. $(-0.18, -3.0, 0)$ km and 4.0 μs.

3.11 $x_1 = (ct/\beta)(1 - 1/\gamma)$, and $x_2 = -(ct/\beta)(1 - 1/\gamma)$.

3.13 $x_1 = 3 \times 10^{10}$ m, and $t_1 = 100$ s.

3.15 Equation (3.3) plus lots of algebra gives Eq. (3.4).

3.17 (a) The matrix of Problem 3.16 with both β's replaced by $-\beta$'s.

3.19 First add subscript 1's to all the components in the field matrix. Then do the matrix multiplication and equate your result to the field matrix with subscript 2's added to all the components.

3.21 $0.926c$.

3.23 Any distance from us at an angle of 95.8° with our $+x$ axis.

3.25 Equation (3.5) plus the binomial expansion.

3.27 a) $\cos \theta_1 = vt_1/\sqrt{v^2 t_1^2 + y_1^2}$.   b) Points include $(1.86, -5)$, $(1.68, -3)$, $(1.16, -1)$, $(0.80, 0)$, $(0.61, 1)$, $(0.52, 3)$, and $(0.51, 5)$.   c) 2.00 and 0.50 corresponding to motion directly toward and directly away.   d) $v = v_0/\{1 + (2.6 \times 10^{-5})[1 + (100/3.9 t_1)^2]^{-1/2}\}$.

3.29 No. It also occurs when $\gamma[1 + (v/c)\cos \theta_1] = 1$.

4.1 Use $V = \int dV = \int\int\int dL_x dL_y dL_z$, with $dL_y = dL_{y0}$, $dL_z = dL_{z0}$, and $dL_x = dL_{x0}/\gamma$.

4.3 0.20 m.

4.5 $0.4048c$, using 1 yd = 0.9144 m.

4.7 Show $t_2' \neq t_2$ when $x_1' \neq x_1$ using Eq. (3.3).

4.9 6000 m.

4.11 a) 400 m.   b) 1.36 μs w.r.t. particle, 6.8 μs w.r.t. the mountain.

4.13 Still saucer shaped, but rotated.

4.15 18 min.   **4.17** $0.995c$.

4.19 a) $0.866c$.   b) 4.7×10^{11} m. c) 2.3×10^{11} m.

4.21 3.1 ms w.r.t. rest system, 2.7 ms w.r.t. isotope.

4.23 4.2×10^{-16} s.

4.25 Let $v_{1x} = -c$ and $v = v$, and find $|v_{2x}| = c$.

4.27 $v_{1x} = 0.052c$ and $v_{1y} = 0.69c$.

4.29 $v_{1x} = -0.90c$ and $v_{1y} = v_{1z} = 0$.

4.31 a) $v_{2x} = -0.80c$, $v_{2y} = 0.60c$, and $v_{2z} = 0$.   b) 143°.

4.33 a) 87 m.   b) 14.7 m.

4.35 a) 30 MHz.   b) 6.9 MHz.

4.37 The frames themselves have no acceleration, but objects may accelerate w.r.t. the frames.

5.1 a) 9.14×10^{-31} kg.   b) 15.2×10^{-31} kg.

5.3 2.89×10^8 m/s.   **5.5** 2.2×10^7 kg.

5.7 $0.140c = 4.2 \times 10^7$ m/s.

5.9 1.65×10^{-14} N.

5.11 a) 1.0001.   b) 1.01.   c) 50.

5.13 44 pg w.r.t. rest observers, 0 w.r.t sprinter.

5.15 22 fg using energy $= \frac{1}{2}k'x^2$.

5.17 Your rest mass times c^2.

5.19 5.609×10^{29} MeV/kg.

5.21 a) When E_0 and $m_0 = 0$.   b) When $E \gg E_0$, $\gamma \gg 1$, or other equivalent answers.   c) When $E_0 \gg KE$, $v \ll c$, or other equivalent answers.

5.23 1001 GeV, 1067, 1001 GeV/c^2, $0.9999996c$, and 1001 GeV/c.

5.25 172 MeV, 172 MeV/c^2, 32 MeV, 1.23, and $0.58c$.

5.27 120 MeV/c.

5.29 a) 1 kg. b) A: 1.67 kg and B: 1.81 kg.

5.31 8.3×10^9 kg.

6.1 Much lower efficiency than rough black surfaces (assuming reflection at all wavelengths).

6.3 5.27×10^3 K.

6.5 No, the laws of physics will be the same in that inertial system as in any other.

6.7 Equation (6.3) can't be used because the filament isn't an ideal blackbody.

6.9 2.6 days.

6.11 Multiply T by $\sqrt{2}$.

6.13 a) $8\pi hc \, d\lambda/[\lambda^5(\exp(hc/\lambda kT) - 1)]$. b) Differentiate u_λ w.r.t. λ, set equal to zero, and solve for λ_p.

6.15 2.6×10^{14} Hz.

6.17 a) 3.4×10^{14} Hz. b) 880 nm. c) 500 nm. d) 6.0×10^{14} Hz. e) Wavelength is inversely proportional to frequency, so a distribution per unit wavelength is not proportional to a distribution per unit frequency. f) 7.0×10^8 m.

6.19 49 pW/m².

7.1 a) 0.114 eV. b) 0.114 V. c) Because $v \ll c$.

7.3 Pt/hv, where P = the radiated power of the station in W, t = the length of the song in s, h = Planck's constant in J · s, and v = the frequency of the station in Hz.

7.5 3.10 eV to 1.77 eV.

7.7 0.8 eV and 5×10^5 m/s.

7.9 Most electrons require more than the minimum work (the work function) to remove them from the surface, leaving less KE.

7.11 hv is the energy carried in by the photon and transferred to the electron. ϕ is the minimum amount of that energy used to remove an electron from the surface. KE_{max} is the maximum kinetic energy the electron can have when completely released from the surface.

7.13 Intensity is energy per area per time. The energy per photon is smaller for laser 1, but it must have a sufficiently larger number of photons emitted per area per time.

7.15 4.1×10^{-15} eV · s and 3.0 eV.

7.17 Slope: Planck's constant. Vertical intercept: negative of the work function. Horizontal intercept: threshold frequency.

7.19 No photoelectrons can be emitted in a single photon process.

7.21 2.3×10^5 m/s. (The laser's power doesn't matter here.)

7.23 $E = E_0\sqrt{1 - v^2/c^2}/[1 + (v/c)\cos \theta_1]$.

7.25 a) 6.23×10^{15} Hz. b) 674 nm.

8.1 In the equipment, electrons accelerated through the high voltages will be stopped or slowed, emitting x-rays.

8.3 Photoelectric effect: A photon gives up all its kinetic energy to produce a photoelectron. This process: An electron gives up all its kinetic energy to produce a photon.

8.5 1.7×10^7 m/s.

8.7 $0.833c = 2.5 \times 10^8$ m/s.

8.9 83 kV.

8.11 It contains frequencies up to 6.3×10^{18} Hz or wavelengths down to 48 pm.

8.13 Light would have to have $v < c$ or carry infinite mass and energy.

8.15 2.5×10^{-31} N.   **8.17** 32 keV/c.

8.19 $m = E/c^2 = pc/c^2 = p/c$.

8.21 $v_s = 1/[1/v_o + (h/m_0c^2)(1 - \cos \theta)]$.

8.23 a) 106°. b) all angles.

8.25 0.99 MeV. 0.01 MeV to the recoiling electron.

8.27 5.909×10^{-23} N · s. The vector difference is the momentum of the recoiling electron.

8.29 4.4×10^{-24} kg · m/s.

8.31 2.599×10^{-28} kg · m/s.

9.1 1.8766 GeV, 4.537×10^{23} Hz, and 6.608×10^{-16} m.

9.3 No. Charge must be conserved.

9.5 1.235×10^{20} Hz.

9.7 a) To conserve momentum. b) 1.022 MeV.

9.9 0.538 MeV, 0.538 MeV/c, 2.30×10^{-12} m, and 1.301×10^{20} Hz.

9.11 a) 2.43×10^{-12} m.   b) 1.322×10^{-15} m.

9.13 $P = F/A = (dp/dt)/A = [d(E/c)/dt]/A = [(dE/dt)/A]/c = I/c$ for complete absorption; twice as much momentum change gives $2I/c$ for complete reflection.

9.15 2.3 cm^{-1}.   **9.17** a) $0.693/\mu$.   b) 1/4.

9.19 Both give exponential decay curves, but hard x-rays have a smaller absorption coefficient and so the (b) curve approaches the x axis less rapidly.

9.21 $x = (1/\mu_1) \ln(v_2/v_1)$. (Therefore, $v_2 > v_1$.)

9.23 0.3 m.   **9.25** 64 MeV/m$^2 \cdot$ s.

9.27 126 MeV/m$^2 \cdot$ s.

10.1 $\lambda = hc/(E^2 - m_0^2 c^4)^{1/2}$.

10.3 1.1×10^{-38} m. This is immeasurably small, so there is no experimental evidence that $\lambda = h/p$ applies to such objects.

10.5 3.96 km/s.

10.7 If $m = 66$ kg, $v = 1.0 \times 10^{-35}$ m/s. 2.5×10^{27} yr. No.

10.9 Interfere. Bounce off one another. Sometimes act like particles, sometimes like waves.

10.11 p increases. Δp_x increases. The interaction changes the system—for example, the photons or particles hitting the object in this question.

10.13 $v_{ph} = c/[1 - (m_0^2 c^4/E^2)]^{1/2}$.

10.15 a) $(\omega_{max} a/2)\sin ak$.
b) $(\omega_{max}/2k)(1 - \cos ak)$.
c) $\omega_{max} a^2 k/2 = 2\omega_{max} a^2 k/4$.

10.17 (Constant)$/2\sqrt{k} = \frac{1}{2}$(constant)$/\sqrt{k}$.

10.19 At least 4.2×10^{-8} eV.

10.21 a) The spread in the x component of the momentum $\geq 1.33 \times 10^{-24}$ kg $\cdot$ m/s.
b) Nothing.

10.23 100%.

10.25 0.38 GeV. (This must be done relativistically.)

11.1 10^{29} m/s^2 away from the nucleus.

11.3 121.5 nm, 102.6 nm, and 91.2 nm.

11.5 $Z^2 me^4/(4n^3 \epsilon_0^2 h^3)$ for large n.

11.7 762 nm.

11.9 4, $+0.85$ eV, and 12.75 eV.

11.11 52.918 pm.

11.13 $\gamma = 1 + 2.66 \times 10^{-5}/n^2$, so $n = 1, 2,$ and 3 gives $1 + 2.66 \times 10^{-5}$, $1 + 6.66 \times 10^{-6}$, and $1 + 2.96 \times 10^{-6}$.

11.15 8.53 kW/m^2.

11.17 $h\nu$ gives the energy difference for $Z = 1$.

11.19 8.0×10^8 photons/m$^2 \cdot$ s.

12.1 a) $p^2/2M = 7.78 \times 10^{-8}$ eV.   b) From $(2mKE_{th})^{1/2} = (m + M)v$ and $KE_{th} = \frac{1}{2}(m + M)v^2 + E_{ex}$.   c) i) 12.10 eV, ii) 24.17 eV.

12.3 12.75 eV, 3.08×10^{15} Hz, and 97.2 nm.

12.5 95.00 MeV/c^2 or $185.9m$, -2.530 keV, 0.256 pm, and 653 pm.

12.7 a) 30.6 eV.   b) 91.8 eV.   c) 91.8 eV, 2.22×10^{16} Hz, and 13.5 nm.

12.9 All r are multiplied by κ, and all E are divided by κ^2.

12.11 The other atoms have different reduced masses.

12.13 a) He, the $n = 4$ state.   b) He has a larger reduced mass than H.

12.15 $0.00027c = 81$ km/s away.

12.17 $v = Zc/137$ and $\gamma = 1 + 2.66 \times 10^{-5} Z^2$.

13.1 No. In his model, $n = 1$ is the minimum, and by his first postulate the angular momentum equals $nh/2\pi$.

13.3 1.0546×10^{-34} J $\cdot$ s and 6.582×10^{-16} eV $\cdot$ s.

13.5 Differentiate $\Psi * \Psi$ and set equal to zero to find maximum and minimum values.

13.7 $x^2 + y^2$.

13.9 Because (probability/volume) times volume equals probability, and if Ψ is normalized, $\Psi * \Psi \, dv$ gives the probability.

13.11 The SWE still holds because every term is simply multiplied by a constant.

13.13 $P = \int_v \Psi * \Psi \, dv / \int_{\text{all space}} \Psi * \Psi \, dv$.

13.15 $A = (96\pi a_0^5)^{-1/2}$.   **13.17** 1/8.

13.19 $\Psi = \psi e^{-i(n+3/2)\omega_c t}$.

13.21 Then $\Psi * \Psi = \psi * \psi$, which is independent of t.

13.23 Substitute V and ψ into the pertinent SWE, Eq. (13.5). $E = \hbar^2 a^2 / 2m$.

13.25 At $x = 0$. This is the least probable allowed classical position.

13.27 a) $A = (2/\pi)^{1/2}$. b) 0.091.

13.29 $25\pi^2\hbar^2/2mL^2$.

13.31 It, and the probability density, approach infinity as x approaches minus infinity.

14.1 The derivatives become infinite at the boundary.

14.3 No. **14.5** $\frac{1}{2}$.

14.7 a) 0.020. b) 8×10^{-5} (or 1.05×10^{-4} by integration).

14.9 a) $\psi = 2.39 \times 10^7 \text{ m}^{-1/2} \sin (9.0 \times 10^{14} \text{ m}^{-1})x$ and $E = 2.7 \times 10^{-12} \text{ J} = 16.7 \text{ MeV}$. b) $\psi = 2.39 \times 10^7 \text{ m}^{-1/2} \sin (1.80 \times 10^{15} \text{ m}^{-1})x$ and $E = 1.07 \times 10^{-11} \text{ J} = 67 \text{ MeV}$.

14.11 Similarities: Quantized energy levels, same quantum numbers for ground and excited levels, and so on. b) Differences: Different V's, different values and spacings of energy levels, and so on.

14.13 a) 15.7 μm. b) Increases it, as λ is proportional to L^2.

14.15 a) 3. b) 0. c) (b). **14.17** $(h/L)L = h$.

14.19 a) In Eq. (13.5), $0 + 0 = 0$. b) $0 = A0 + B$ gives $B = 0$. $0 = AL + 0$ gives $A = 0$. Therefore ψ, ψ^*, and $P = 0$.

14.21 $V = 0.490 h^2 / 8mL^2$.

14.23 For the ground state, substitute into Eq. (14.3). For the first excited state, $k_2 > 3\pi/2L$ gives $E > V$.

14.25 Because $\sin 0 = 0$, at $x = 0$ $B \sin k_n x$ will not allow ψ to be continuous and $A \cos k_n x$ will not allow $d\psi/dx$ to be continuous.

14.27 a) 0.55 nm in regions I and III and 0.27 nm in region II. b) $2L$ is about 1.5 wavelengths, which would give a minimum in transmission.

14.29 In Eq. (14.4), as $V \rightarrow 0$, $T \rightarrow 1$.

14.31 a) No region III, so $F = 0$, and no reflected

particles in region II, so $D = 0$. Continuity of ψ and $d\psi/dx$ at $x = 0$ gives $A + B = C$ and $k_1 A - k_1 B = k_{II} C$. b) Eliminate C from the two equations of part (a) and solve for B/A.

15.1 There are no particles incident from region II.

15.3 Then the reflected particle waves interfere destructively.

15.5 a) $n(0.194 \text{ nm})$. b) $[n - (1/2)](0.194 \text{ nm})$.

15.7 The workers expend energy and leave an actual hole, while the electron doesn't.

15.9 Otherwise the ψ and the probability density $\rightarrow \infty$ as $x \rightarrow \infty$.

15.11 a) 8.4×10^{-6}. b) 2.6×10^{-28}.

15.13 Let $\sinh KL = KL$, with the expression for K in Eq. (15.2) to cancel the $(V - E)$ terms. Then let $V = E$ and use $k_1 = \sqrt{2mE}/\hbar$.

15.15 a) They can tunnel through the insulating layer. b) Because T rapidly approaches zero as L increases.

15.17 The transmission coefficient is large only for small L.

15.19 a) $m\omega_c/2\hbar$. b) $(2m\omega_c/h)^{1/4}$.

15.21 Yes. For the particle in a box or the hydrogen atom, the quantum number for the nth excited level is $n + 1$.

15.23 a) 740 N/m. b) $(n + \frac{1}{2})(83 \text{ meV}) = 41 \text{ meV}, 124 \text{ meV}$, and so on.

15.25 $[(2n + 1)h\nu_c/m]^{1/2}$.

15.27 a) $m\omega_c/2\hbar$. b) $(n + \frac{1}{2})\hbar\omega_c$ but $n = [1 + (1 + d)^{1/2}]/2$, where $d = 8mb/\hbar^2$.

16.1 V depends only on r.

16.3 No change in Eqs. (16.3) or (16.4) or their solutions. Equation (16.5) would then have a different V than that of the H atom, which would change its solutions.

16.5 a) No. b) No. They are for the Coulomb potential energy.

16.7 $\int_0^\infty R_{21}^2(r) \, r^2 \, dr = [(Z/a_0)^5/24]\int_0^\infty e^{-Zr/a_0}r^4 \, dr = 1$.

16.9 Substitute into Eq. (16.5) with $Z = 3$ and $l = 2$ as in Example 16.2 and its following paragraph.

16.11 a) Yes. b) No (see Example 16.2).

16.13 $(1/2\pi)\int_0^{2\pi} e^0\, d\phi = 1$.

16.15 $0/(m_l' - m_l) = 0$, if $m_l' \neq m_l$.

16.17 $(1/2)\int_0^\pi \sin\theta\, d\theta = 1$.

16.19 V must be a function of r only.

16.21 Substitute into Eq. (16.4), with $l = 2$ and $m_l = \pm 2$.

16.23 $n = 1$, 2, or 3 levels.

16.25 a) l can equal 0, 1, 2, 3, or 4. $m_l = 0$ only for $l = 0$, $m_l = -1$, 0, or $+1$ for $l = 1$, $m_l = -2$, $-1, 0, +1$, or $+2$ for $l = 2$, and so on. b) 25.

17.1 It has $-i\phi$ instead of $+i\phi$ in the exponent.

17.3 $(4/81)(2/\pi)^{1/2}(r^2/a_0^{7/2})e^{-2r/3a_0}\sin^2\theta\, e^{\pm i2\phi}$. 52.9 pm (see Example 16.2.).

17.5 $n \geq 5$.

17.7 a) 0. b) $-2, -1, 0, +1, +2$.

17.9 a) and b) $\psi_{320}^*\psi_{320} \to 0$ as $r \to \infty$, and it equals zero at angles of 54.736° and 125.264° with the z axis.

17.11 $\psi_{32\pm1}^*\psi_{32\pm1} = 0$ at $\theta = 0$, 180°, and 90°, while having maxima at 45° and 135°.

17.13 4.8×10^{-4}.   **17.15** 3×10^{-14}.

17.17 a) 52.9 pm. b) 17.6 pm. c) 105.7 pm.

17.19 1.47 and 1.21.

17.21 a) $l_{max} = n - 1$, so check $R_{10}(r)$, $R_{21}(r)$, and $R_{32}(r)$ with $Z = 1$. b) Proceed as in Example 16.3, with $l_{max} = n - 1$. c) Proceed as in Example 17.5, with $l_{max} = n - 1$.

18.1 a) 0. b) $\sqrt{20}\,\hbar$.

18.3 0, 9.49 $\hbar$, and 99.5 $\hbar$ compared to $1\hbar$, 10 $\hbar$, and 100 $\hbar$. The Bohr value is approached as n increases.

18.5 $\sqrt{1 + 2/l} - 1$, which $\to 0$ as $l \to \infty$.

18.7 $L_z = 0$ when $m_l = 0$, and m_l varies through zero from $-l$ to $+l$ for all l.

18.9 2.

18.11 s: $L_z = 0$ and $L = 0$. d: $L_z = -2\hbar, -1\hbar, 0, +1\hbar, +2\hbar$ and $L = \sqrt{6}\,\hbar$. f: $L_z = -3\,\hbar, -2\hbar, -1\hbar, 0, +1\hbar, +2\hbar, +3\,\hbar$ and $L = 2\sqrt{3}\,\hbar$.

18.13 They are spherically symmetric states, which give the electron just as much probability

of moving in any one rotational direction as in the opposite direction. This behavior gives a net probability of zero for any resultant angular momentum.

18.15 $-(e/2m)\mathbf{L} \times \mathbf{B}$.   **18.17** $2l + 1$.

18.19 $\theta = 30°$, 90°, and 150° for $\mathbf{L}$; 150°, 90°, and 30° for μ_l.

18.21 0.8598 T.

18.23 To a p state with $m_l = -1$, 0, and $+1$.

18.25 λ_0 and $1/(1/\lambda_0 \pm \mu_B B/hc)$.

18.27 7.092639×10^{14} Hz, 7.092779×10^{14} Hz, and 7.092919×10^{14} Hz. 422.6811 nm, 422.6728 nm, and 422.6645 nm.

18.29 The Zeeman splitting of the wavelengths and frequencies of light from the stars can be measured and equations such as Eq. (18.9) used to solve for B.

18.31 Yes, to a considerable extent, because most atoms do not exhibit the normal Zeeman effect.

18.33 Like Fig. 18.8 with two extra levels for each state. The selection rules will still allow only three different energy changes.

19.1 s is a dimensionless quantum number, whereas S is the quantized magnitude of the spin angular momentum determined by that quantum number.

19.3 a) $S = 0$, $m_s = 0$ only, and $S_z = 0$ only. b) $S = \sqrt{2}\hbar$, $m_s = -1$, 0, $+1$, and $S_z = -\hbar$, 0, $+\hbar$.

19.5 The z component. Its absolute value is 0.116% larger.

19.7 6.95×10^{-5} eV.

19.9 The ground state is a 1s state, and s states have no orbital angular momentum and thus no spin–orbit interaction.

19.11 a) $(2,0,0,-\frac{1}{2})$, $(2,0,0,+\frac{1}{2})$, $(2,1,-1,-\frac{1}{2})$, $(2,1,-1,+\frac{1}{2})$, $(2,1,0,-\frac{1}{2})$, $(2,1,0,+\frac{1}{2})$, $(2,1,+1,-\frac{1}{2})$, and $(2,1,+1,+\frac{1}{2})$. b) $(2,0,\frac{1}{2},-\frac{1}{2})$, $(2,0,\frac{1}{2},+\frac{1}{2})$, $(2,1,\frac{1}{2},-\frac{1}{2})$, $(2,1,\frac{1}{2},+\frac{1}{2})$, $(2,1,\frac{3}{2},-\frac{3}{2})$, $(2,1,\frac{3}{2},-\frac{1}{2})$, $(2,1,\frac{3}{2},+\frac{1}{2})$, and $(2,1,\frac{3}{2},+\frac{3}{2})$.

19.13 f.

19.15 125.2644° or 54.7356° and 140.7685°, 104.9632°, 75.0368°, or 39.2315°.

19.17 For s states and for states where $S = 0$.

19.19 No. The components of μ_l and μ_s aren't independent but are coupled together.

19.21 a) $1\ ^2S_{\frac{1}{2}}$. b) $3\ ^2D_{\frac{5}{2}}$ and $3\ ^2D_{\frac{3}{2}}$. c) $4\ ^2P_{\frac{3}{2}}$ and $4\ ^2P_{\frac{1}{2}}$. d) No such state, $l = 0$ or 1 for $n = 2$. e) $6\ ^2F_{\frac{7}{2}}$ and $6\ ^2F_{\frac{5}{2}}$.

19.23 $140.7685°$, $104.9632°$, $75.0368°$, and $39.2315°$. $J = (\sqrt{15}/2)\hbar$. $J_z = -\frac{3}{2}\hbar, -\frac{1}{2}\hbar, +\frac{1}{2}\hbar$, and $+\frac{3}{2}\hbar$.

19.25 For $l = 2$, $n \geq 3$. Also, for $l = 2$, $j = \frac{5}{2}$ or $\frac{3}{2}$, not $\frac{1}{2}$.

20.1 1s. **20.3** Yes.

20.5 There can be two electrons in each state, one with $m_s = -\frac{1}{2}$ and one with $m_s = +\frac{1}{2}$.

20.7 The energy splitting is about a thousandth of the energies plotted.

20.9 a) The outer electron has a greater probability of spending time within the 1s shell when it is in a 2s state and therefore the two 1s electrons shield it less from the three protons in the nucleus. b) -3.4 eV.

20.11 4.18 eV. **20.13** 4.50.

20.15 The O shell ($n = 5$).

20.17 a) 1. b) 2. c) 3. d) 4. e) 5.

20.19 Move one or more electrons to higher levels.

20.21 $1s^{92}$.

20.23 Photons emitted when electrons drop down into holes in the L shell from the M, N, O, and so on, shells.

20.25 Yes. When the electron dropped into the hole in the K shell, it left a hole in the L shell to be filled by an electron from a higher shell.

20.27 See the answer to Question 20.25.

20.29 Take the square root of Eq. (20.4). Slope = $(2.47 \times 10^{15}\ \text{Hz})^{1/2}$, vertical intercept = $-(2.47 \times 10^{15}\ \text{Hz})^{1/2}$, and horizontal intercept = -1.

20.31 22.7 pm and 9% larger.

20.33 Ag. **20.35** Ge.

20.37 Yes, but only if the photon were absorbed by a K-shell electron at the instant there was an unfilled L-shell hole.

20.39 a) A large n means a very large r_n, so the probability is overwhelming that the outer electron will spend its time completely outside the $Z - 1$ other electrons that surround the Z protons. Therefore $Z_{\text{eff}} = 1$, giving an ionization energy of 13.6 eV/n^2 for *all* Rydberg atoms. b) Different reduced masses. c) The binding energy of that last electron is so small. d) 0.162 meV. e) About $2r_{732} = 0.06$ mm.

21.1 0.9995.

21.3 It increases (the cause of thermal expansion).

21.5 $m_r = 1.6295 \times 10^{-27}$ kg, $\omega_c = 5.627 \times 10^{14}$ rad/s, $E_{v0} = 0.185$ eV, and $E_{v1} = 0.556$ eV.

21.7 0.135 eV. **21.9** $\omega = \sqrt{l(l+1)}\,\hbar/I$.

21.11 a) 2.97 meV. b) 2.77 meV and 2.87 meV.

21.13 Rotational. Vibrational.

21.15 $l(l+1)h^2\omega_c^2/(8\pi^2 k' r_0^2)$.

21.17 Its m_1, m_2, and therefore m_r are easily the smallest, while its k' is the same order of magnitude as for other molecules.

21.19 a) Those with identical atoms. b) E_e.

21.21 Because of selection rules and because the tendency is for $\Delta E_e \gg \Delta E_v \gg \Delta E_r$.

21.23 If the isotopes are different.

21.25 a) They result from the electron interactions, and isotopes of the same atom have the same number of electrons in virtually the same states.   b) No.

21.27 For a given energy, heavier mass atoms move more slowly and therefore require less centripetal force.

21.29 Use Eqs. (21.4) and (21.5), along with $(1 + x)^n \approx 1 + nx$.

21.31 6.536×10^{13} Hz $\pm (l' + 1)(1.16 \times 10^{11}$ Hz), with $l' = 0, 1, 2, 3, 4, \ldots$

22.1 No, because the third law of thermodynamics says absolute zero cannot be reached in a finite number of operations.

22.3 5.

22.5 The individual particles are all hydrogen atoms. The individual atoms are well separated and are thus distinguishable from one another in principle.

22.7 $[(\Delta E)^2 / kT^2] \, e^{\Delta E/kT} / (e^{\Delta E/kT} + 1)^2$. At $T = 0.417 \Delta E / k$. Proportional to $1/T^2$.

22.9 $N_{-1} = Ne^x / (e^x + 1 + e^{-x})$, $N_0 = N / (e^x + 1 + e^{-x})$, and $N_{+1} = Ne^{-x} / (e^x + 1 + e^{-x})$, where $x = \mu_B B / kT$.

22.11 a) $N\mu_B \tanh(\mu_B B / kT)$.   b) Starts from the origin and asymptotically approaches $N\mu_B$, almost reaching it by $B/T = 4$.   c) $\tanh x \approx x$ for small x gives $N\mu_B^2 B / kT$.   d) Because $\tanh x \to 1$ for large x.

22.13 Then $p/p_0 = N/N_0$. $G = G_0$ and $\Delta E = mgy$ in Eq. (22.3) give the desired result.

22.15 a) 220 K.   b) 3.5×10^3 K.

22.17 $\int_0^\infty v n(v) \, dv / N =$ $(m/kT)^{3/2} \sqrt{2/\pi} \int_0^\infty v^3 e^{-mv^2/2kT} \, dv =$ $(m/kT)^{3/2} \sqrt{2/\pi} / [2(m/2kT)^2]$.

22.19 200 m/s, 225 m/s, and 250 m/s.

22.21 1376 K.

22.23 6030 and 36 in the first two excited states for every million in the ground state.

22.25 Δn for the nth excited state is n, so if the ratio is r in Eq. (22.4) for $\Delta n = 1$, it will be r^n for $\Delta n = n$ (assuming the harmonic oscillator model can be used for all states up to n).

22.27 a) Use Eq. (22.10) with $l_2 = l$ and $l_1 = 0$.   b) Differentiate that expression w.r.t. l and set equal to zero.   c) There will be more

molecules in this level than any other, giving the largest absorption to transitions from it.

22.29 3.9 K.

22.31 A population inversion is needed so that it is more probable that a photon will stimulate an emission than be absorbed.

22.33 No. For the atom to increase its energy by itself would violate conservation of mass–energy.

22.35 When $G_2 = G_1$.

22.37 a) About 70,000 yr.   b) 16.7 hr. In 16.7 hr, 1000 would have arrived and 1 served, leaving 999.

22.39 To reflect some photons back to continue stimulating the emission.

22.41 a) The incoming photons must have the same frequency and wavelength as the emitted photons to cause stimulated emission.   b) The two photons out are in phase, so destructive interference does not occur.   c) The reflections back and forth from the ends keep the light from spreading out.

22.43 a) Phosphorescence also involves metastable states de-exciting to produce photons. Fluorescence also involves de-exciting in steps.   b) No, almost all of the emission is spontaneous, not stimulated, and doesn't involve laser action.

23.1 It's a gas of charged particles.

23.3 Not if the crystal is rotated about a cube axis. However, in rotation about the diagonal of an fcc structure, there is a face-centered point every 60°.

23.5 One.

23.7 Because the basis may not have as much symmetry as the lattice.

23.9 a) 0.39 nm.   b) 124 nm.   c) 9.0 pm.

23.11 0.25 nm.

23.13 It looks like eight small cubes making up a larger cube. The small cubes do not have the same basis associated with each lattice point (half have Na atoms, half Cl atoms).

23.15 Same lattice (fcc). Different bases and crystal structures.

23.17 a) 52%.   b) 68%.   c) 74%.

23.19 a) 0.245 nm.   b) 0.400 nm.

23.21 $109.47°$.

23.23 The metallic bond is less localized and therefore more forgiving of the many local defects found in ordinary materials.

23.25 The strongly bonded planes remain as planes as they slip across one another, easily breaking the weak van der Waals bonds.

23.27 No, it's covalently bonded.

23.29 $(2A/B)^{1/6}$.

23.31 a) Whole crystal: ionic, metallic, hydrogen, and van der Waals (but it drops off rapidly). Nearest neighbors: covalent.   b) Only nearest neighbors have any appreciable overlap.

24.1 Equation (6.5) times V.

24.3 The Einstein model, which assumed a constant frequency.

24.5 1.8%.   **24.7** $E = E_F + kT \ln(1/P - 1)$.

24.9 The triple integral of $\psi^*\psi$ over x, y, and z, each from 0 to L, gives 1.

24.11 a) 5.6×10^7. b) 2.0×10^6 m/s.

24.13 4.70×10^{28} m^{-3}.   **24.15** 2.22×10^{22}/eV.

24.17 $\int_0^\infty E\, n(E)\, dE/N = 0.60E_{F0}$.

24.19 2.4×10^{24} eV $= 3.8 \times 10^5$ J.

24.21 7.0×10^{22} eV.

24.23 $3E_{F0}/10$.   **24.25** $0.629961E_{F0}$.

24.27 Between two points on the Fermi surface, the initial with a larger p_x and the final with a smaller p_x.

24.29 a) 1.55×10^{11} $(\Omega \cdot m)^{-1}$. b) 9.4 eV.
c) 1.82×10^6 m/s, and 76 μm.

24.31 σ and τ are the most temperature dependent for most metals, and V is considerably less dependent.

24.33 The electrons are not really free with $V = 0$ in the material.

24.35 No.

25.1 The 1s, 2s, and 2p levels are barely affected, but the 3s and 3p levels are spread into bands.

25.3 The energy gap decreases to zero.

25.5 1.9×10^{-6} to 1.9×10^{-10} to 4.6×10^{-13} to 8.8×10^{-27} to 1.7×10^{-48}. Same for the holes.

25.7 The increase in the number of collisions per time, due to more phonons, a relatively small effect. The increase in the number of electrons in the conduction band and holes in the valence band, a relatively large effect.

25.9 a) Because **M** opposes **B** not just to slightly reduce **B**, but to make **B** equal to zero.   b) Because the **M** of the disk opposes the **B** of the magnet, which results in like poles of the disk and magnet giving a net repulsion.

25.11 a) 6.60×10^{11} Hz. b) It decreases, approaching zero as T approaches T_c.

25.13 1.591 GHz.   **25.15** $n(17$ nA$)$.

25.17 It greatly increases the number of electrons in the conduction band. The small decrease is due to the extra defects introduced in the doping.

25.19 a) The marbles in A move down and the holes in C move up.   b) Compartment A is analogous to the conduction band, B to the forbidden energy gap, and C to the valence band. The marbles are like electrons, and the absence of a marble is a hole. The holes and marbles move in opposite directions under the influence of the same field.   c) For extrinsic semiconductors, put isolated small compartments in B.   d) You can't put two marbles in the same spot, that is, when C is full, no more marbles can be added.

25.21 1.72 μm. Increases.

25.23 Their net charge is zero.

25.25 5.5 meV.   **25.27** 4.3 nm. $0.19m$.

25.29 A: No elements, undoped. B: Al only. C: More P than Al. D: More Al than P.

25.31 Group II.

26.1 $\mu = (e/m)\tau$.

26.3 a) 4.8×10^2 $(\Omega \cdot m)^{-1}$.
b) 1.28×10^4 $(\Omega \cdot m)^{-1}$.

26.5 Intrinsic material. It will have fewer defects to scatter the charge carriers.

26.7 a) -7.6×10^{-2} m^3/C. b) 5.5×10^4 A/m^2.

26.9 Use $j_x = \sigma E_x$ and $E_H = R_H j_x B$ from Eq. (26.4).

26.11 2.1×10^{21} m^{-3}.

26.13 **B** is parallel to $\mathbf{v}_d$ for longitudinal magneto-resistance. Therefore the $q\mathbf{v}_d \times \mathbf{B}$ force is zero and does not curve the paths of the charge carriers to cause longitudinal magnetoresistance.

26.15 The bands and levels of Figs. 25.12 and 25.13, except a full valence band, an empty conduction band, the Fermi level in the middle of the gap, as many donor states as acceptor states, each donor state with a circled +, and each acceptor state with a circled −. (Every donor electron has dropped down to fill an acceptor level.)

26.17 a) and b) 1.24×10^{-6}.   c) Answer (a) would decrease and (b) would increase.

26.19 The maximum electron energy at $T = 0$, not varying much with T.

26.21 Outside reading.

26.23 The energies of the two sides would approach equality.

26.25 The parallel-plate capacitor has all the sources of the electric field (the plus charges) on one plate and all the sinks (the minus charges) on the other. With no charge between the plates and no spreading of the field lines, the capacitor's field is constant. In the p–n junction, the plus charges are distributed through the n side and the minus charges through the p side, so the electric field increases throughout the n side and then decreases throughout the p side.

26.27 Because that region has been depleted of charge carriers.

26.29 a) $49\,\mu A$.   b) Hole generation current: n to p. Hole recombination current: p to n.

27.1 a) Electrons from the n-type conduction band go to a lower allowed energy by diffusing into the metal. Excess charge goes to the surface of a metal.   b) The previously neutral semiconductor becomes positive when its conduction band electrons diffuse into the metal.   c) The + charge in the semiconductor and the − charge on the metal.   d) An electron traveling in the semiconductor toward the surface must move against the force of the built-in electric field.   e) An external field added to the right decreases both the total electric field and the barrier height, making it easier for electrons to flow to the left. An external field added to the left increases both the total electric field and barrier height, making it very difficult for electrons to flow to the right.   f) Absorbed photons can create electron–hole pairs.

27.3 Equation (27.1) with $x = eV/kT$ gives, for $V \ll kT/e$, $R = V/i \approx kT/i_s e$.

27.5 Because tunneling becomes rapidly less probable as the barrier becomes thicker.

27.7 a) $-i_s = -0.392$ mA.   b) It is the sum of the electron and hole generation currents.

27.9 a) -2.6 mV.   b) -17.3 mV.   c) -58 mV.

27.11 $(kT/i_s e)e^{-eV/kT}$. As $V \to 0$, $e^{-eV/kT} \to 1$.

27.13 The energy released could be going to produce phonons and increase the energy of free charge carriers.

27.15 Its energy gap is 1.45 eV.

27.17 Mainly this rewrite is done by interchanging p with n, positive with negative, and holes with electrons, except that electrons should be removed from the base, giving a current to the base.

27.19 0.29 A.   **27.21** a) No.   b) Yes.   c) Yes.

27.23 The negative charge repels electrons and attracts holes to the channel. Therefore it   a) decreases the current even further as it pinches off the N channel even more, and   b) may increase the current by opening the P channel even more.

27.25 N-channel enhancement type.

27.27 Merely interchange all +'s and −'s.

27.29 Yes.

28.1 12.4 GeV. λ is less than nuclear sizes, so there is little diffraction.

28.3 73. Ta-181.   **28.5** $Z = 19$. $N = 16$ to 31.

28.7 $^{60}_{27}Co$.

28.9 Let positive F be repulsive and negative F be attractive. The exact form of the force is not yet known, but p–n and n–n should be zero for r greater than several fm, while p–p has a positive, $1/r^2$ Coulomb's law force there. The much more negative part of the strong nuclear force should take over until 2 or 3 fm, where the very much stronger repulsive (positive) part of the strong nuclear force comes into play.

28.11 5.18 fm < 40% of 14 fm.

28.13 1.7×10^{-28} m².

28.15 2.224 MeV. 0.5574 pm.

28.17 a) 28.30 MeV. b) 7.074 MeV/nucleon. c) 20.58 MeV. d) 19.81 MeV.

28.19 1766.7 MeV and 7.615 MeV/nucleon.

28.21 The electron binding energies are in the eV to keV range per electron and are therefore small, but measurable, compared to MeV's per nucleon. Greatly compensated for because of the 13.6-eV electron binding energy of each H-1.

28.23 It doesn't in that the whole nuclear mass is less than the sum of the masses of the nucleons by BE/c^2.

28.25 -1.967 MeV. The minus sign means that energy does not have to be added to remove the proton; it can leave with the emission of 1.967 MeV of energy.

28.27 -4.871 MeV. The minus sign means that energy does not have to be added to remove the alpha particle; it can leave with the emission of 4.871 MeV of energy.

29.1 Surface effect (there is no Coulomb effect in H-2).

29.3 The first term recognizes that unstable nuclei result if there are too many more neutrons than protons, an effect less important for large A nuclei. Even–even nuclides are the most stable because of the pairing of nucleons and odd–odd the least stable, explaining the $+$ and $-$ term.

29.5 3.31.

29.7 a) $(42.58 \text{ MHz/T})B$. b) $(29.17 \text{ MHz/T})B$.

29.9 It is a quantum number, and it involves orbital as well as spin angular momentum.

29.11 Because of the Coulomb repulsion of the protons.

29.13 $_2^4$He, $_1^2$H, $_2^3$He, and $_3^7$Li.

29.15 The angular solutions were for any $V(r)$.

29.17 It changes θ to $\pi - \theta$ and ϕ to $\phi + \pi$.

29.19 a) $\frac{7}{2}$, $(\sqrt{63}/2)\hbar$, and $-$. b) $\frac{5}{2}$, $(\sqrt{35}/2)\hbar$, and $+$. c) $\frac{7}{2}$, $(\sqrt{63}/2)\hbar$, and $+$. d) $\frac{1}{2}$, $(\sqrt{3}/2)\hbar$, and $+$ for both.

29.21 a) The value of the integral for one-half of all space will be -1 times that of the other half, so the total integral for all space will be

zero. b) No, the volume of space could give more of a value of one sign than of the other. c) No, part of all space could give a $+$ value and the rest an equal magnitude $-$ value.

29.23 Yes, they assume a completely radial potential energy function.

30.1 Zero. No. **30.3** 113 nCi.

30.5 a) 0.99 Ci. b) 1600 yr. **30.7** 0.10 mg.

30.9 $N_0(1 - e^{-\lambda t})$. **30.11** 29.16%.

30.13 a) $2.6 \times 10^{17} \pm 5 \times 10^8$. b) 187 ± 14.

30.15 0.24 mCi.

30.17 Long. More will be in lines with the slow clerks.

30.19 5.3 dis/s.

30.21 The noble gases will not combine chemically or remain in the body to any extent. Very reactive iodine will form compounds to stay in the body (concentrating in the thyroid gland).

30.23 0.012 Gy = 1.2 rad. 0.012 Sv = 1.2 rem.

30.25 0.45 Gy = 45 rad. 4.5 Sv = 450 rem.

30.27 13.3 nGy = 1.33 μrad. 133 nSv = 13.3 μrem.

30.29 a) 1800. b) 100. c) 0.08.

31.1 a) $_{96}^{247}$Cm. b) $_{92}^{238}$U. c) $_4^8$Be. d) $_1^1$H.

31.3 There is only one KE_α for each Q, but different Q's for the ground state and each excited state of the daughter.

31.5 $M_P < (M_D + M_{He-4})$ or $Q < 0$.

31.7 181.94820 u.

31.9 a) Obtain $KE_D = Q4/A$, then use $KE_D = \frac{1}{2}M_D v^2 = \frac{1}{2}M_D c^2 (v^2/c^2)$. b) 2.67×10^5 m/s, which gives $KE_D = 0.0685$ MeV.

31.11 It decreases rapidly as KE_α decreases. Because the lower the alpha particle energy, the higher and thicker the barrier energy will be with respect to KE_α and the fewer collisions there will be with the barrier per time. This condition makes tunneling in a time interval much less likely, decreasing the probability per time of decay, λ.

31.13 Use a variation of Eq. (31.3), replacing 4 with 14.

31.15 Yes. 1.57 MeV and 0.66 MeV.

31.17 1.391 MeV. **31.19** 93.90966 u.

31.21 $Q = (m_P + m - m_D)c^2 = [(M_P - Zm) + m - (M_D - (Z - 1)m)]c^2 = (M_P - M_D)c^2$.

31.23 $v + {}_0^1n \rightarrow {}_1^1p + {}_{-1}^0e$ (called inverse beta decay).

31.25 a) 1.96 meV. b) 3.25×10^{-27} eV.

31.27 142.7 keV.

31.29 0.600 MeV, 0.414 MeV, 0.256 MeV, and 0.442 MeV. 4.195 MeV, 4.351 MeV, 4.602 MeV, and 4.785 MeV.

31.31 a) Similar to Eq. (31.4) with $Z = (Z + 2) + 2(-1) + 2(0)$. b) $Q = (m_P - m_D - 2m)c^2 = [(M_P - Zm) - (M_D - (Z + 2)m) - 2m]c^2 = (M_P - M_D)c^2$.

32.1 A proton.

32.3 a) The compound nucleus is ${}_2^4He^*$. ${}_2^3He(n,\gamma){}_2^4He$, ${}_2^3He(n,p){}_1^3H$, and so on. b) The compound nucleus is ${}_{72}^{181}Hf^*$. ${}_{72}^{180}Hf(n,\alpha){}_{70}^{177}Yb$, ${}_{72}^{180}Hf(n,d){}_{71}^{179}Lu$, and so on.

32.5 Since they have almost the same mass, the neutron can transfer almost all of its momentum and kinetic energy in a head-on collision with a proton from the H in the paraffin.

32.7 a) ${}_{30}^{63}Zn$. b) 62.933214 u. c) Endoergic. d) 8.435 MeV.

32.9 14 MeV.

32.11 a), b), and c) 0. d) 6.968 MeV. e) 2.224 MeV.

32.13 22.97 MeV.

32.15 a) and b) ${}_{13}^{27}Al$, ${}_{12}^{26}Mg$, ${}_{13}^{26}Al$, ${}_{12}^{25}Mg$, and ${}_{11}^{23}Na$.

32.17 7.115 MeV and 0.061 MeV.

32.19 a) x could tunnel through the barrier. b) No, it will approach zero.

32.21 No. The cross section is not a measure of the actual nuclear cross-sectional area.

32.23 $\Phi_0 = N_0/At$ and $V = Ax$, so $n\sigma\Phi_0 V = n\sigma(N_0/At)Ax = N_r/t$ by Eq. (32.4). Since R is defined as N_r/t, $R = n\sigma\Phi_0 V$. Assumptions: $n\sigma x \ll 1$, and only a small fraction of the original n nuclei/volume are being changed (keeping n almost constant).

32.25 1.88×10^{14}.

32.27 They correspond to compound nucleus excited states.

32.29 $-\frac{1}{2}$. **32.31** 2000 barns.

33.1 An antineutrino is emitted in each beta-minus decay of the fission products.

33.3 Maximum: About 95 and 139. Minimum: About 117.

33.5 No. No. (See the initial paragraphs of Section 33.1.)

33.7 a) To gain more surface area, decreasing BE/A by the surface effect. b) No. c) Strong nuclear. Coulomb. In the dumbbell shape, only those few nucleons in the neck are holding it together while every proton is repelling every other proton. d) That would require the addition of the BE of the nucleus (about 1.8 GeV).

33.9 No. Some neutrons are absorbed by nuclides other than U-235, U-236* doesn't always fission, and some neutrons escape from the mass.

33.11 Radius squared and radius cubed. Therefore the possible fission-to-leakage ratio is proportional to r. As r is increased from a small value, the fraction fissioning increases until, at a certain radius, that fraction becomes large enough to permit a critical reaction. That radius gives a certain volume and a certain mass, called the critical mass.

33.13 a) Because U-235 has a shorter half-life than U-238 and is not in the same radioactive series. b) It acted as the moderator. c) It provided the fissionable nuclei. d) The nuclides to which the fission products decayed. e) n +

U-238 $\rightarrow$ U*-239 $\xrightarrow{\beta^-}$ Np-239 $\xrightarrow{\beta^-}$ Pu-239 $\rightarrow$ U-235 + He-4.

33.15 Low. High. It would remain the same element and not become a defect in the material (a defect might weaken it).

33.17 a) High. Control rods are supposed to capture excess neutrons. b) It can kill the chain reaction by absorbing neutrons. c) No.

33.19 a) 28,000. b) Supercritical. c) The delayed neutrons make control possible, but before any of them have appeared, the reactor is already putting out 28 times its original power.

33.21 a) The I-135 is decaying into Xe-135 (more rapidly than the Xe-135 itself decays), and there are almost no free neutrons around to eliminate the Xe-135 (which has a 9.10 hr half-life). b) It is hard to make the reactor slightly supercritical when so many of the neutrons are being absorbed by so much Xe-135.

33.23 A denser moderator means a shorter distance and time between collisions for the neutrons; there is a lower temperature, so they become thermal faster, giving more reactions and therefore more energy per time, which is more power.

33.25 a) The energies and therefore the temperatures are higher and the water would be steam. Also, water is good at slowing neutrons. b) Na-23 absorbing a neutron to become Na-24, which beta-minus decays to stable Mg-24. c) Fast neutrons are wanted. Moderators slow neutrons down. d) The hot Na is a liquid metal. The neutrons are fast. Nonfissionable U-238 is being converted to fissionable Pu-239, a breeder reaction.

33.27 a) Move all control rods all the way in as fast as possible. b) The heat from the reaction would raise the temperature until liquids boiled and solids melted. The meltdown could continue with the fuel melting its way into the earth. If it melted to groundwater level, the resulting steam explosion could blow radioactive debris over a wide area. c) The ECCS is needed to carry away the heat generated by the continuing decay of the fission fragments.

33.29 a) $Q = 17.59$ MeV. b) r is larger in the PE calculation because H-3 nuclei are larger. c) 3.62 MeV and 14.37 MeV.

33.31 Not in the three preceding Example 33.6, but there is neutrino production in the proton–proton and carbon cycles.

33.33 The $q\mathbf{v} \times \mathbf{B}$ magnetic force provides the centripetal force needed to move the particle in a semicircle.

33.35 Gravitational. No, earth gravity is too weak.

33.37 For example, use the heat generated in stopping the alpha particle and the neutron to produce steam to turn a turbine generator.

33.39 Fusion's problems should be considerably less because not as many fusion products are radioactive.

33.41 10 ns.

34.1 By using phosphors that deexcite by emitting photons in the frequency ranges desired.

34.3 That would shield the radiation from the detector.

34.5 0.67 MeV. Assumption: Every ion pair created is collected.

34.7 Region 1: At the lowest V, no change; every ion pair still recombines. Highest V, $q_{collected}$ increases to match with region 2. Regions 2 and 3: More energy gives proportionally more charge collected. Region 4: Approximately no change in the plateau region. Region 5: No change.

34.9 The positive electrode attracts the electrons, and the negative electrode attracts the ions. The removal happens quickly because the field is strong. The advantage: The tube is rapidly cleared of ion pairs and ready for the next bit of ionizing radiation in a short time, so it can have a high count rate.

34.11 That's where the electric field is the greatest.

34.13 The voltage drops a bit due to the iR drop across R.

34.15 a) 1.7×10^{-14} C. b) 7.1×10^{-15} C.

34.17 Doubly false. All $c > v > c/n$, where $n =$ the index of refraction of the medium. Also, the particles must be charged.

34.19 Leave no tracks.

34.21 The reactions that make the tracks take energy.

34.23 3.6 MeV.

34.25 No. V is the voltage drop. For single ionization, the first increase in KE is $(-e)(-V) = +eV$. The second increase is $(+e)(+V) = +eV$ again, giving a total KE increase of $+2eV$.

34.27 2.0 cm.

34.29 Cyclotron continuously, others pulsed.

34.31 $1.00 : 0.500 : 0.333$ for frequency and KE_{max}. $1.00 : 1.00 : 1.00$ for maximum momentum.

34.33 a) 9.3 MHz. b) 0.11 μs. c) 4.4×10^7 m/s. d) Not quite, to the accuracy required, 10 MeV $\ll$ 938 MeV and $v^2 \ll c^2$.

34.35 No.

34.37 a) Larger numbers of particles can be placed in a bunch, the bunch can be manipulated and compressed, and so on. Particles have more than one chance of colliding. Different types of particles from one accelerator can be made to collide, and so on.   b) More of the kinetic energy brought to the collision is available.

34.39 a) 5.630 GeV.   b) 0.9383 GeV.

34.41 a) $v = 2\pi r/v$, where r is constant and v changes only from $(1 - 27 \times 10^{-11})c$ to $(1 - 1.3 \times 10^{-11})c$.   b) By 100/22.

35.1 1.240 GeV · fm/KE.

35.3 a) Face each other and throw the ball back and forth.   b) Face away from each other and throw the boomerang directly away from the other person so that it comes to the other when moving directly toward the first (quite a trick!).

35.5 Gluons, photons, and gravitons.

35.7 A couple of GeV or so.

35.9 If $\Delta p_x \Delta x < h$, momentum can be momentarily nonconserved.

35.11 One proton every thousand years for an 80-kg person.

35.13 a) Gauge boson and lepton.   b) Hadron.

35.15 There are 10^{78} more baryons than antibaryons.

35.17 $0 = +1 + (-1)$.

35.19 $\pi^+ \to \mu^+ + \nu_\mu$ and $\pi^0 \to \gamma + \gamma$ would be $1 \neq 0 + 0$ and $K_L^0 \to \pi^0 + \pi^0 + \pi^0$ would be $1 \neq 1 + 1 + 1$.

35.21 $D^- \to K^0 + \pi^- + \pi^- + \pi^+$. $D^- \to K^0 + \pi^- + \pi^0$.

35.23 For the K^0 reaction the strangeness would change by -2, but for the $\overline{K}^0$ reaction it doesn't change.

35.25 a) 0, $\frac{1}{2}$, and -1.   b) 0, 0, and 0.   c) $-e$, 0, and -1.   d) 0, $\frac{1}{2}$, and 0.   e) $-e$, 0, and 0.   f) 0, 1, and 0.

35.27 Their two quarks can give $s = \frac{1}{2} - \frac{1}{2}$ or $s = \frac{1}{2} + \frac{1}{2}$.

35.29 Hadron family (meson branch). $+e$, 0, 0, and either 0 or 1.

35.31 They annihilate with their antiquarks.

35.33 uds $\to$ uud $+ \overline{u}$d (s $\to$ d $+$ u $+ \overline{u}$) and uds $\to$ ddu $+ u\overline{u} + d\overline{d}$ (s $\to$ d $+$ u $+ \overline{u} +$ d $+ \overline{d}$).

35.35 Gluons, with their color charge, interact with one another. Neutral photons don't interact with one another.

36.1 The object continues to move at constant velocity while the floor accelerates toward it at 9.8 m/s².

36.3 No. The equivalence principle is for a uniform gravitational field.

36.5 The nonuniformity of the Earth's gravitational field must be ignored.

36.7 For example, let the base of the triangle be $\frac{1}{4}$ of the equator and the other two sides be longitude lines to the N pole.

36.9 Because the clocks in the planes were farther from the center of the Earth than the clock on the surface.

36.11 By the equivalence principle, the energy loss will have to be the same, or $[d(h\nu) = -mg\,dy = -(h\nu/c^2)g\,dy]$, if the acceleration is g in the direction of the photon's motion.

36.13 There would seem to be an outward (centrifugal) force, except at the very center, that increased linearly with distance from the center, as well as a sideways (Coriolis) force on objects moving in a nonvertical direction. These forces aren't real, but are the result of trying to apply inertial reference frame concepts to an accelerating frame. Similarly, gravity isn't a real force, but is the result of trying to apply flat spacetime concepts to a curved spacetime.

36.15 3.53×10^4.

36.17 On almost directly opposite sides of the sun.

36.19 For light, $ds = 0$ so $0 = c^2(dt)^2 - (dx)^2 - (dy)^2 - (dz)^2$.

36.21 $g \to 1$ for $M \to 0$ and/or $r \to \infty$.

36.23 a) Zero.   b) It becomes smaller than 1.

36.25 $dt = dr/gc$. As $r \to r_s$, $g \to 0$.

36.29 0.600.

36.31 Log t_{evap} versus log M will have a slope of 3 and will pass through the point (2×10^{30} kg, 10^{66} yr).

36.33 They push the GTR past its region of validity.

37.1 For the density (= mass/volume) to be constant, the mass must increase as the volume of the universe increases.

37.3 No. Observers on all galaxies see the same effect.

37.5 The relative peculiar velocities of approach of these nearby galaxies are greater than their recession velocities.

37.7 52 km/s, which is about one-sixth and opposite its relative peculiar velocity.

37.9 1.2×10^4 M$c \cdot$ yr.   **37.11** 3.9×10^4 km/s.

37.13 The special theory of relativity speed limit is c for motion through space (peculiar velocities), but the recession velocity is due to the general theory's expansion of space itself and so is not limited by the special theory. The magnitude of the vector sum of the recession and peculiar velocities can exceed c.

37.15 a) $R = R_0 e^{Ht}$.   b) -1. The minus sign means that this universe is accelerating, not decelerating.

37.17 6 H-1 atoms/m^3.   **37.19** Nothing.

37.21 a) $x = y = \frac{1}{2}$ and $z = -\frac{3}{2}$.   b) Substitute in the values.

37.23 8%.

37.25 Because the Coulomb repulsion of the nuclei increases rapidly as the nuclear charge increases.

37.27 a) Use $M = Nm_n$ and $V = \frac{4}{3}\pi R^3$.
b) 9.8 km.   c) $1.66r_s$.

Index